Bullying and Harassment in the Workplace

Developments in Theory, Research, and Practice

Second Edition

Bullying and Harassment in the Workplace

Developments in Theory, Research, and Practice

Second Edition

Edited by
Ståle Einarsen • Helge Hoel
Dieter Zapf • Cary L. Cooper

CRC Press
Taylor & Francis Group
Boca Raton   London   New York

CRC Press is an imprint of the
Taylor & Francis Group, an **informa** business

CRC Press
Taylor & Francis Group
6000 Broken Sound Parkway NW, Suite 300
Boca Raton, FL 33487-2742

© 2011 by Taylor and Francis Group, LLC
CRC Press is an imprint of Taylor & Francis Group, an Informa business

No claim to original U.S. Government works

Printed in the United States of America on acid-free paper
10 9 8 7 6 5 4 3 2

International Standard Book Number: 978-1-4398-0489-6 (Paperback)

Visit the Taylor & Francis Web site at
http://www.taylorandfrancis.com

and the CRC Press Web site at
http://www.crcpress.com

Contents

Section I   The Nature of the Problem

Section II   Empirical Evidence

Section III   Explaining the Problem

Section IV   Managing the Problem

Preface

In 2003 we published the first comprehensive international volume on bullying, emotional abuse, and harassment in the workplace under the title *Bullying and Emotional Abuse in the Workplace: International Perspectives in Research and Practice* (Einarsen, Hoel, Zapf, and Cooper, 2003). The book soon became a highly cited source of knowledge for this new and burgeoning field of research and practice. The present book is a much-needed revision and update of our original 2003 book. The present edition has, in addition to a new title in line with its current content, a range of new chapters together with a collection of much-revised and expanded chapters from the original book. Although now highly revised and updated, Chapters 1, 2, 3, 5, 7, 8, 9, 10, 19, and 21 were also part of the original 2003 book. The remaining 11 chapters are new to the present edition. For those wanting a comprehensive overview of the field, we can still advise you to read the 14 chapters of the original book that do not form a part of the present edition.

Research into bullying and harassment at work, primarily of a nonphysical nature, as distinct from sexual and racial harassment, started in the early 1990s, expanded in the late 1990s, and really took off after the publication of our edited book, and particularly so in Europe, Australia, and North America. Later on interest also spread to South America, with Asia slowly catching up. Already in 1999 the field was nominated by some of the editors of this book as the research topic of the 1990s (Hoel, Rayner, and Cooper, 1999), a conclusion that may seem somewhat hasty in light of recent developments in this field. A meta-analysis of methodological moderators of estimates of prevalence rates of bullying (Nielsen et al., in press) showed that of the 86 independent studies on prevalence of workplace bullying identified, 82% were published after the turn of the century, with 15% published in the 1990s and less than 3% published in the 1980s. In addition to this explosion in the sheer quantity of studies published during the last 10 years, a qualitative development has taken place regarding breadth in the perspectives taken, the theoretical depth of the analysis provided, and the methodological rigor of the empirical contributions. The present edition of the book is a consequence and a reflection of this recent development in the field.

The main aim of this book is to present the reader with a comprehensive review of the literature, the empirical findings, the theoretical developments, and the experience and advice of leading international academics and practitioners in the field of bullying and harassment at work. In this volume, the reader will find chapters examining the concept of bullying and harassment at work itself, as well as its measurement. The reader will also find chapters that document the existence and consequences of the problem. The book also explores a variety of explanatory models as well as presents the available empirical evidence that may help us understand where, when, and why bullying develops. Lastly, the book contains a wide range of contributions on the possible remedies that may exist to prevent and minimize the problem, to manage it when it occurs, and to heal the wounds and scars it may have left on those exposed.

The book has been structured in four sections: (I) the nature of the problem, (II) empirical evidence, (III) explaining the problem, and (IV) managing the problem. In Section I, the concept of bullying and harassment at work itself is introduced and discussed. First, the editors review and discuss the original European tradition on the issue of bullying and harassment at work. Chapter 1 starts by looking at the development and increasing prominence of the concept in Europe. Then discussed are the various key defining characteristics of bullying, such as frequency, duration, power balance, status of bullies and victims, objective versus subjective bullying, and interpersonal versus organisational bullying. This chapter also discusses those conceptual models of bullying that have so far dominated European research on bullying.

In Chapter 2, Keashly and Jagatic review recent research findings, as well as key conceptual and methodological challenges, from the North American literature on hostile behaviours at work that are highly relevant to the issues of workplace bullying. An important aim of this chapter is to facilitate awareness of and communication between the various literatures on hostile workplace interactions.

The next section of the book, titled "Empirical Evidence," contains four chapters that review and summarise empirical evidence on bullying at work. Chapter 3, by Zapf, Escartín, Einarsen, Hoel, and Vartia, summarises descriptive empirical findings, looking at such issues as the frequency and duration of bullying, the gender and status of bullies and victims, how bullying is distributed across different sectors, and the use of different kinds of bullying strategies. Chapter 4, by Hogh, Mikkelsen, and Hansen, reviews the most recent literature on the potential negative effects of bullying on the health and well-being of the individual victim, as well as suggests theoretical explanations for the relationships that have been found to exist between exposure to bullying and symptoms of ill health. In Chapter 5, Hoel, Sheehan, Cooper, and Einarsen review the emerging evidence of the relationship between bullying at work and negative organisational outcomes, such as absenteeism, high turnover, and low productivity. In the last chapter of this section of the book, Nielsen, Notelaers, and Einarsen provide a comprehensive review of the quantitative measurement of bullying and how to estimate prevalence rates. This chapter provides an overview of the many options that exist when measuring exposure to workplace bullying in survey research as well as discusses the methodological challenges associated with them. It also provides guidelines for interpreting and understanding the results of various studies.

So far, much work in the field of bullying at work has been of an empirical nature. A major aim of this book has also been to develop our theoretical understanding of the problem. This is also particularly the aim of Section III, "Explaining the Problem." The first chapter in this section, by Zapf and Einarsen, discusses the role of individual antecedents of bullying at work. The issue of individual causes of bullying at work has been a hot issue for debate in both the popular press and the scientific community. While some argue that individual antecedents, such as the personalities of bullies and victims, may indeed be considered as possible causes of bullying, others have disregarded totally the role of individual characteristics in this respect. In Chapter 7, Zapf and Einarsen provide an in-depth discussion on the role of individual factors among both victims and bullies. Chapter 8 presents a social interactionist approach to bullying behaviours, looking at social and situational antecedents of aggressive behaviours at work. Here, Neuman and Baron address a key question: why do societal norms against aggression fail to apply, or apply only weakly, where workplace bullying is concerned? The authors draw from a substantial literature devoted to interpersonal aggression and examine what they believe to be important social antecedents of bullying. In Chapter 9, Salin and Hoel examine potential organisational antecedents of bullying, focusing on quality of work environment, leadership, organisational culture, and the impact of change.

Then, Schneider, Pryor, and Fitzgerald, in Chapter 10, provide a broad overview regarding a particular kind of harassment at work, namely, sexual harassment. Again, a special focus is on possible explanations for the problem. Because sexual harassment has, at least until recently, been a more extensively researched issue than workplace bullying, the field also has much to offer the more general field of bullying and harassment at work. Another related field is that of discrimination. In Chapter 11, Lewis, Giga, and Hoel discuss how and why bullying and discrimination can be so readily colocated. The authors claim that if one considers organisations as a microcosm of society, then many economic, social, political, historical, and global issues impacting intergroup relationships and experiences of discrimination outside work are just as likely to have an impact on experiences at work, and in particular in cases of workplace bullying.

The next chapter, by David Beale, discusses how workplace bullying may be understood from an industrial relations perspective. In this chapter, Beale argues for a more contextualised and interdisciplinary approach to the study of workplace bullying, in his case drawing from the very developed field of industrial relations and in particular three earlier contributions, that of the hitherto unknown Norwegian pioneer Jon Sjøtveit, the contribution in the original 2003 version of this book by Ironside and Seifert, as well as the mobilisation theory of Kelly. Lastly, in this section of the book, Matthiesen, Bjørkelo, and Burke look at the interrelationships between workplace bullying and whistleblowing in organisations. The authors look at bullying as one of the major downsides and negative consequences after whistleblowing at work. That is, severe bullying may develop as a retaliation against an employee who has reported on some kind of observed wrongdoing in the organisation, be it to senior management or to an external body.

A major aim of this book is to provide some examples of how bullying and harassment at work may be prevented, managed, and healed. Therefore, Section IV, "Managing the Problem," contains eight chapters written by leading practitioners and researchers in the field. First, in Chapter 14, Rayner and Lewis discuss the role of antibullying policies in an organisation's obligation to manage the problem. In this chapter, they discuss how such policies may be developed, implemented, and monitored, as well as look at the pitfalls and limitation of such policies. Chapter 15 outlines the process of investigating complaints of bullying. Here, Hoel and Einarsen describe the basic principles of how to conduct a fair and proper internal investigation in response to complaints about bullying and harassment at work. The chapter also discusses some of the obstacles often encountered in providing a fair hearing in such cases. Chapter 16, by Vartia and Leka, examines the different types of interventions and strategies used in the prevention and management of bullying in organisations as well as discusses their effectiveness. The chapter also looks at the key principles of planning and implementation of interventions for the prevention and management of bullying.

In Chapters 17 and 18, the role of counselling and treatment of targets of bullying is addressed. First, Tehrani looks at various counselling strategies that may be adapted when working with individual targets and perpetrators, looking at the very nature of counselling and what a targeted employee might expect when attending an assessment and counselling session. The argument is made for an integrated approach to counselling in cases of bullying, including the need to counsel and support the organisation in dealing with bullying in a positive and creative way. Chapter 18 takes this concept a step further to describe how one may treat the health-related aftereffects of bullying that haunt many targets of severe bullying. In this chapter, Schwickerath and Zapf describe the procedures and principles of such treatment, as well as the results of the treatment based on an inpatient clinic in Germany specialising in the treatment of victims of bullying.

In Chapter 19, Keashly and Nowell discuss the relationships between the concepts *bullying* and *conflict* and consider the value of a conflict perspective in the study and amelioration of workplace bullying. In Chapter 20, Namie, Namie, and Lutgen-Sandvik share their theoretical basis and personal experiences when adopting an activist and public communication approach to challenge workplace bullying on a national level. The last contribution of the book identifies and discusses some of the central themes concerning bullying and the law. In this chapter, Yamada also investigates some national examples of how the law is used in order to prevent and respond to this problem.

As editors, we are delighted about the enthusiasm we encountered when approaching authors invited to participate in this revision, update, and extension of our original 2003 book. We believe that as part of a collective of leading academics and practitioners in this field, our contributors together provide the reader with the best and most comprehensive information currently available to understand and counteract workplace bullying. Therefore, we hope that this book will be a useful tool for students and academics as well as practitioners in this intriguing and difficult problem area.

References

Einarsen, S., Hoel, H., Zapf, D., and Cooper C. L. (eds.) (2003) *Bullying and emotional abuse in the workplace: International perspectives in research and practice*. London: Taylor & Francis.

Hoel, H., Rayner, C., and Cooper, C. L. (1999) Workplace bullying. In C. L. Cooper and I. T. Robertson (eds.), *International review of industrial and organizational psychology*, vol. 14 (pp. 195–230). Chichester, UK: Wiley.

Nielsen, M. B., Matthiesen, S. B., and Einarsen S. (in press) The impact of methodological moderators on prevalence rates of workplace bullying: A meta-analysis. *Journal of Occupational and Organizational Psychology*.

Editors

Ståle Einarsen is Professor in Work and Organizational Psychology at the University of Bergen, Norway, where he also acts as Head of the Bergen Bullying Research Group. Einarsen is one of the Scandinavian pioneers in research on workplace bullying and has conducted a wide range of studies related to its very nature, its prevalence, its antecedents, and its consequences, as well as its measurement. He has published journal articles, book chapters, and books on issues related to workplace bullying, leadership, psychosocial factors at work, creativity and innovation, and whistleblowing. With Helge Hoel, he has developed methodology and training for investigating bullying complaints and has acted as adviser to the Norwegian government regarding the prevention and management of bullying in Norwegian working life.

Helge Hoel (PhD) is a Senior Lecturer in Work Psychology at Manchester Business School, University of Manchester, United Kingdom. He has carried out several large-scale research projects on workplace bullying, including the first nationwide survey (with Cary L. Cooper), and contributed to a number of books, academic journal articles, and reports of violence, bullying, and harassment. This includes work commissioned by the International Labour Organisation (ILO) and by the European Foundation for the Improvement of Living and Working Conditions. His current research also involves cross-disciplinary work, exploring bullying in the context of industrial relations, law, sexuality, and masculinity. He has developed and delivered courses in the investigation of bullying complaints (with Ståle Einarsen) and acted as adviser to the Norwegian government.

Dieter Zapf is Professor for Work and Organizational Psychology at the Department of Psychology, J. W. Goethe-University Frankfurt, Germany, and Visiting Professor at Manchester Business School, University of Manchester, United Kingdom. He studied psychology and theology and received his diploma in psychology and his PhD from Free University of Berlin, and his habilitation from Giessen University. He has carried out research projects on workplace bullying for more than 15 years, published a number of journal articles and book chapters, and organised symposia on bullying at various international conferences. Other research interests include stress at work, job analysis, human errors and emotion work in service jobs, and emotions in leadership.

Cary L. Cooper is Distinguished Professor of Organizational Psychology and Health at Lancaster University Management School in England. He has written or edited over 100 books and over 400 scholarly articles in the field

of occupational stress, women at work, health and well-being, and bullying in the workplace. He is the Founding President of the British Academy
of Management, has served on the Board of Governors of the Academy
of Management, and is currently President of the British Association for
Counselling and Psychotherapy as well as Chair of the Academy of Social
Sciences (an umbrella body of 37 learned societies in the social sciences).
He was awarded the honor of Commander of the British Empire (CBE) by
the Queen in 2001 for his contribution to occupational and organizational
health.

Contributors

Robert A. Baron
Oklahoma State University
Stillwater, Oklahoma

David Beale
University of Manchester
Manchester, United Kingdom

Brita Bjørkelo
University of Bergen
Bergen, Norway

Ronald J. Burke
York University
Toronto, Ontario, Canada

Cary L. Cooper
University of Lancaster
Lancaster, United Kingdom

Ståle Einarsen
University of Bergen
Bergen, Norway

Jordi Escartín
University of Barcelona
Barcelona, Spain

Louise F. Fitzgerald
University of Illinois at Urbana-
    Champaign
Urbana-Champaign, Illinois

Sabir Giga
University of Bradford
Bradford, United Kingdom

Åse Marie Hansen
National Research Centre for the
    Working Environment
Copenhagen, Denmark

Helge Hoel
University of Manchester
Manchester, United Kingdom

Annie Hogh
University of Copenhagen
Copenhagen, Denmark

Karen Jagatic
Private Consultant
Hoboken, New Jersey

Loraleigh Keashly
Wayne State University
Detroit, Michigan

Stavroula Leka
University of Nottingham
Nottingham, United Kingdom

Duncan Lewis
University of Glamorgan
Pontypridd, Wales,
United Kingdom

Pamela Lutgen-Sandvik
The University of New Mexico
Albuquerque, New Mexico

Stig Berge Matthiesen
University of Bergen
Bergen, Norway

Eva Gemzøe Mikkelsen
CRECEA
Aarhus, Denmark

Gary Namie
Workplace Bullying Institute
Bellingham, Washington

Ruth Namie
Workplace Bullying Institute
Bellingham, Washington

Joel H. Neuman
State University of New York at
    New Paltz
New Paltz, New York

Morten Birkeland Nielsen
University of Bergen
Bergen, Norway

Guy Notelaers
University of Bergen
Bergen, Norway

Branda L. Nowell
North Carolina State University
Raleigh, North Carolina

John B. Pryor
Illinois State University
Normal, Illinois

Charlotte Rayner
University of Portsmouth
Portsmouth, United Kingdom

Denise Salin
Hanken School of Economics
Helsinki, Finland

Kimberly T. Schneider
Illinois State University
Normal, Illinois

Josef Schwickerath
AHG Klinik Berus
Überherrn-Berus, Germany

Michael J. Sheehan
University of Glamorgan
Pontypridd, Wales,
United Kingdom

Noreen Tehrani
Noreen Tehrani Associates Ltd.
Twickenham, Middlesex,
United Kingdom

Maarit Vartia
Finnish Institute of Occupational
    Health
Helsinki, Finland

David C. Yamada
Suffolk University Law School
Boston, Massachusetts

Dieter Zapf
Johann Wolfgang Goethe
    University
Frankfurt, Germany

Section I

The Nature of the Problem

1

The Concept of Bullying and Harassment at Work: The European Tradition

Ståle Einarsen, Helge Hoel, Dieter Zapf, and Cary L. Cooper

CONTENTS

Introduction

During the 1990s, the concept of bullying or mobbing at work found resonance within large sections of the European working population as well as in the academic community. A wide range of popular as well as academic books and articles were published in many European languages (e.g., Ege, 1996; Einarsen et al., 1994b; Field, 1996; Leymann, 1993; Niedl, 1995;

Rayner et al., 2002), and public interest spread from country to country. From being a taboo in both organisational research and organisational life, the issue of bullying and harassment at work became what was called the "research topic of the 1990s" (Hoel et al., 1999). Yet, such a conclusion seems somewhat hasty in hindsight, as the sheer number of publications on workplace bullying since 2000 by far outnumbers those published in the 1990s. In a meta-analysis on the effects of methodological moderators on the observed prevalence of workplace bullying, Nielsen et al. (in press) noted that of the 91 studies identified, the majority (81.3%) of the studies included were published in the period 2000–2008, 16% in the 1990s, and only 2 (2.7%) in the 1980s. The identified studies include samples from over 20 countries across all continents, yet with more than 60% originating in Europe.

The issue of bullying in the workplace, also called "mobbing" in many countries, is a complex one. It may come in many shapes and shades, with multiple causes on many levels, and with diverging views on its very nature (see also Agervold, 2007). Yet, at a basic level it is about the systematic mistreatment of a subordinate, a colleague, or a superior, which, if continued and long-lasting, may cause severe social, psychological, and psychosomatic problems in the target. Exposure to such treatment has been claimed to be a more crippling and devastating problem for employees than all other kinds of work-related stress put together, and it is seen by many researchers and targets alike as an extreme type of social stress at work or even as a traumatic event (Zapf et al., 1996; Høgh et al., this volume).

Whereas this phenomenon is usually referred to as workplace bullying in English-speaking countries and as harassment *(harcèlement morale)* in the French-speaking world, it has mainly been termed *mobbing* in most other European countries, although other nation-specific terms continue to live on side by side, as, for example, *pesten* in Dutch and *acoso* or *maltrato psicológico* in Spanish (Di Martino et al., 2003). The term *mobbing* was coined from the English word *mob* and was originally used to describe animal aggression and herd behaviour (Munthe, 1989). Heinemann (1972) originally adopted the term from the Swedish translation of Konrad Lorenz's 1968 book *On Aggression* to describe victimisation of individual children by a group of peers in a school setting (see also Munthe, 1989; Olweus, 1987, 1991, 2003). Later on, Leymann (1986, 1990a) borrowed the term from the research on bullying in the schoolyard (see Olweus, 1987, 1991, 1994) to describe the systematic mistreatment of organisation members, which, if continued, could cause severe social, psychological, and psychosomatic problems in the target. From Scandinavia, the concept spread to other European countries during the late 1990s.

As a new field of study and public interest, the issue of labelling has been much discussed (see also Keashly and Jagatic, this volume). Yet, even with different words in use in different languages, the European tradition has been characterised by a high degree of unity regarding concepts and features. Hence, Europe has avoided the trap, often observed in social sciences, where new issues are coming into focus and a plethora of competing terms and concepts are introduced. In this respect, the European scene has been quite different from what has happened in the United States, where a range of overlapping constructs falling under a rubric of "hostile relations at work" has been introduced (see Keashly and Jagatic, this volume).

In practice, only minor differences exist between the concepts of bullying, harassment, and mobbing (Zapf and Einarsen, 2005). The term *bully* may more easily lend itself to descriptions of the perpetrator who behaves aggressively in many situations and possibly towards more than one target, whereas the concept of mobbing is more attuned to the experiences of targets who are systematically exposed to harassment by one or more perpetrators and who may, over time, become severely victimised by this treatment. Hence, the concepts seem to focus on the two different but interrelated sides of the same phenomenon, the perpetrators and the targets. According to Leymann (1996), the choice of using *mobbing* in preference to *bullying* was a conscious decision on his part, reflecting the fact that the phenomenon among adults often refers to subtle, less direct forms of aggression as opposed to the more physical forms of aggression that may be associated with bullying. Yet, even among those who use the term *bullying*, empirical evidence suggests that the behaviours involved are often of a verbal, passive, and indirect nature (Einarsen, 1999; Keashly and Harvey, 2005). The common stereotype of a bully as a dominant, rude, and aggressive figure is probably not typical for many bullying cases, at least as seen in most European countries. Hence, in the present chapter the terms *harassment*, *bullying*, and *mobbing* will be used interchangeably to refer to both these phenomena, namely, the systematic exhibition of aggressive behaviour at work directed towards a subordinate, a coworker, or even a superior, as well as the perception of being systematically exposed to such mistreatment while at work.

The purpose of this chapter is, then, to present and discuss the European perspective on bullying, harassment, and mobbing at work. We will start with some historical notes and will then discuss such various key characteristics of bullying as the frequency, duration, power balance, and content of bullying behaviours; objective versus subjective bullying; intentionality of bullying; interpersonal versus organisational bullying; and bullying as a process. A formal definition of the concept will then be proposed, and we will discuss and present various conceptual models of such bullying at work.

The Development of a New Concept: Some Historical Notes

The interest in the issue of workplace bullying originated in Scandinavia in the 1980s, partly inspired by ongoing research and focus on bullying among schoolchildren (see also Olweus 1987, 2003). The late Heinz Leymann worked

as a family therapist in the 1970s. Having had experience with family conflicts, he decided to investigate direct and indirect forms of conflicts in the
workplace (Leymann, 1995). Through empirical work in various organisations he encountered the phenomenon of mobbing, and in 1986 he wrote the
first Swedish book on the subject, entitled *Mobbing: Psychological Violence at
Work*. Leymann soon became convinced that this problem had less to do with
those involved than with extraneous matters, for it was rather deeply rooted
in the organisational factors and qualities of the psychosocial work environment, including leadership practises. Inspired by Leymann and much public
interest and debate, large-scale research projects were initiated in Norway
(Einarsen and Raknes, 1991; Einarsen et al., 1994b; Kile, 1990; Matthiesen
et al., 1989), Sweden (Leymann, 1990b, 1996), and Finland (Björkqvist, 1992;
Björkqvist et al., 1994; Vartia, 1991, 1996), documenting the existence of this
phenomenon and the severe negative effects such treatment and these experiences had on targets as well as observers (for a review, see Høgh et al.,
this volume). The seemingly new phenomenon of bullying, or mobbing, as it
was referred to, also attracted growing interest from the public, from those
responsible for health and safety in the workplace, and from union representatives with regard to legal changes securing workers the right to a work
environment free of harassment (see also Hoel and Einarsen, 2009).

　　Yet, the very phenomenon of bullying at work had been thoroughly
described already in 1976 by the American psychiatrist Carroll M. Brodsky
in an intriguing book entitled *The Harassed Worker*. Brodsky was inspired
by hundreds of years of literature on the cruelty and brutality human beings
sometimes show towards both enemies and friends, sometimes even for
no apparent reason. In this qualitative study, Brodsky described a range of
cases where employees at all organisational levels claimed to have been systematically mistreated and abused by their superiors or coworkers while at
work, with devastating effects on their productivity, health, and well-being.
The mistreatment described by the informants of the study was mainly of
a psychological and nonsexual nature, characterised by rather subtle and
discrete actions, yet causing severe and traumatic effects in the targets by
being repeatedly and persistently aimed at employees who felt unable to
retaliate. Brodsky described five main types of harassment: sexual harassment, scapegoating, name-calling, physical abuse, and work pressure. Yet,
Brodsky's pioneering work did not receive much attention at the time; it was
only rediscovered many years later, long after the work of Heinz Leymann
in Sweden.

　　However, until the early 1990s the interest in this subject was largely
limited to the Nordic countries, with only a few publications available in
English (e.g., Leymann, 1990b). Yet, seemingly in parallel with the development in the Nordic countries, UK journalist Andrea Adams in collaboration with the psychologist Neil Crawford put the issue of bullying at work
firmly on the UK agenda through radio appearances followed by a popular
book (Adams, 1992). Many radio listeners apparently saw this broadcast as

an opportunity to make sense of their own experiences, and many were willing to come forward and bring their experiences into the public domain, as had happened in Scandinavia some years earlier (Hoel and Einarsen, 2009). This development was followed by several large-scale national surveys that clearly documented the presence of the problem in the United Kingdom (Hoel and Cooper, 2001; UNISON, 1997).

Heinz Leymann also published a book aimed at a wider readership in German in 1993. Through the media, within a year his book disseminated awareness of the concept of mobbing. Likewise, the concern about the issue of adult bullying soon spread from Scandinavia to such other countries as Austria (Kirchler and Lang, 1998; Niedl, 1995), the Netherlands (Hubert and van Veldhoven, 2001), and Italy (Ege, 1996) and as far afield as Australia (McCarthy et al., 1996). Later on, interest was also sparked in such countries as Spain (Moreno-Jiménez and Rodriguez-Muños, 2006), Lithuania (Malinauskienë et al., 2005), and Turkey (Demirel and Yoldas, 2008; Yücetürk and Öke, 2005).

Again, seemingly in parallel, interest was boosted in France in the late 1990s thanks to the bestselling book by Marie France Hirigoyen (1998) entitled *Le harcèlement moral: La violence perverse au quotidien*, with interest rapidly spreading throughout the French-speaking world, including Quebec in Canada (Soares, 1999). Although workplace bullying has slowly and much later become a concept used in the United States, important empirical, theoretical, and legal contributions in line with this European tradition have come from such U.S. researchers as Keashly (1998), Lutgen-Sandvik (2007), and Yamada (2004), among others (see also Keashly and Jagatic, this volume). Thus, the very same phenomenon is described, and again, the message is the same: (1) many employees suffer from severe mistreatment at work by superiors or coworkers in the form of systematic exposure to sometimes flagrant as well as subtle forms of aggression, mainly characterised by persistency and long-term duration; (2) the effects on the targets are devastating and traumatic, with negative effects also found on the health, motivation, and well-being of those who witness it, with potential major cost implications for employers; and (3) managers and employers, and sometimes even public sector or government bodies, are often unwilling to accept the very existence of the problem, much less prevent it and manage fairly those cases that come to the fore. Luckily, this last situation has seen dramatic changes lately as bullying and harassment have increasingly become issues acknowledged by researchers, practitioners, employers, and politicians.

A Case of Moral Panic?

The introduction of and public interest in the concept of bullying or mobbing at work emerged in a similar pattern across most countries (Rayner et al., 2002). Such "waves of interest" (Einarsen, 1998) are typically instigated by press reports on high-profile single court cases or cases where one or more employees publicly claim to have been subjected to extreme mistreatment at work. Such public attention is then followed by a study of the magnitude of the problem or the consequences of such experiences. Such, often preliminary, research findings generate further publicity and public attention, especially from union representatives and a large number of human resources professionals. The actions and determination of articulate and high-profile victims fuel the public debate. By exercising continuous pressure on the media in the broadest sense, and through numerous innovative initiatives utilising conference appearances, publications, and the World Wide Web, these activists (e.g., Field, 1996) contribute to informing and educating the public, thereby effectively preventing the issue from disappearing from public view (see also Namie et al., this volume).

In the first decade of the new millennium, antibullying policies and procedures have been produced and implemented in large numbers of organisations throughout many countries in Europe in collaboration with, or under pressure from, local trade unions as well as such international bodies as the European Union and the European Social Partners and their European Social Dialogue (2007). In some cases, for example, Norway and the United Kingdom, large-scale national projects instigated by government departments have been undertaken in order to put the issue of bullying at work on the agenda of the employers.

This swift development and these "waves" of strong public interest, moving from media attention by means of research reports and focus by trade unions to an issue on the political agenda, may call into question whether the rise of the concept of bullying at work and the accompanying high figures highlighting perceived exposure to bullying were fuelled by a wave of "moral panic." The concept of moral panic refers to "events the majority of members of society feel threatened by as a result of wrongdoers who threaten the moral order and that something must be done about such behaviour" (Lewis, 2001, p. 17). The fear or concern created by this public panic is often significantly disproportionate to the actual threat of the particular event or phenomenon. The media focus and their use of inflated headlines and dramatic pictures and stories, political figures (including union representatives) calling for penalties and adopting a vindictive and self-righteous stance, and action groups (e.g., former victims) campaigning for action to be taken against offenders all create a demonising process where "folk devils" and a "disaster mentality" prevail (Lewis, 2001).

In the case of bullying at work, the folk devils were evident in the discourse on "psychopaths at work" in the United Kingdom (e.g., Field, 1996) and in the discourse on the "neurotic victim" in Scandinavian countries (see Einarsen, 2000a; Leymann, 1996). Disaster mentality has been seen both in media headlines and in claims by union representatives that bullying is the most profound work environment problem of our time (Einarsen and Raknes, 1991).

Although the attention given to the phenomenon of bullying at work may at times have elements of such moral panic, Lewis (2001) showed that perceptions of being exposed to bullying at work or perceptions of others being bullied at work do not appear to be a response to a general panic or fear produced by the popular media and the public debate. A study among 415 Welsh teachers in further and higher education showed that being a perceived target of bullying was not related to any particular knowledge of the public debate on bullying at work. Rather, it was based on the teachers' own personal experiences or on information provided by their colleagues. Furthermore, the results of the study by Nielsen and colleagues (2009) suggest that bullying in Norway was reduced by 50% between the early 1990s and 2005.

Hence, bullying at work seems to be a phenomenon that is, in fact, prevalent in many organisations. Although hardly a new phenomenon, many people seem to find this concept useful to describe and label situations at work where someone's behaviour is perceived to be systematically aimed at frustrating or tormenting an employee who is unable to defend him- or herself or escape from this situation (Einarsen et al., 1994b; Olweus, 1991).

From Organisational Psychology to an Interdisciplinary Field

In its early phases, the issue of bullying at work was investigated mainly from a psychological perspective, and particularly so by work and organisational psychologists. Hence, the focus of research in the 1990s was very much about "who does what to whom; when, where, why; and with what kinds of consequences for the organisation and for those targeted." As interest spread, the focus of research has spread and diverged into such domains as sociology, communication theory, law, industrial relations, and medicine, to name but a few. This development has seen the field blossom and contribute to an extensive and varied body of literature, as evident in the present volume.

The Concept of Bullying at Work

Bullying at work is about repeated actions and practises that are directed against one or more workers; that are unwanted by the victim; that may be carried out deliberately or unconsciously, but clearly cause humiliation, offence, and distress; and that may interfere with work performance and/or cause an unpleasant working environment (Einarsen and Raknes, 1997). Hence, the concept of bullying at work relates to persistent exposure to negative and aggressive behaviours of a primarily psychological nature (Leymann, 1996; Olweus, 1987, 1991). It describes situations where hostile behaviours that are directed systematically at one or more colleagues or subordinates lead to a stigmatisation and victimisation of the recipient(s) (Björkqvist et al., 1994; Leymann, 1996).

Target Orientation

From the beginning, the Scandinavian public debate had a target or victim's perspective on bullying (Einarsen et al., 1994b; Leymann, 1993, 1996). People were more alarmed by the serious damage to health reported by the victims than by the often unethical behaviour of the perpetrators. Consequently, bullying or mobbing was seen from a stress perspective (Einarsen and Raknes, 1991; Leymann, 1996; Zapf et al., 1996). Mobbing or bullying was understood as a subset of social stressors at work. The basic characteristic of social stressors is that they are related to the way employees interact socially within an organisation. Applying concepts used in stress research, bullying can manifest itself in the form of daily hassles (Kanner et al., 1981). However, there are also many cases where single episodes of bullying are experienced as critical life events (Dohrenwend and Dohrenwend, 1974): for example, when victims are threatened physically or threatened with dismissal, or when their career prospects are destroyed.

If "normal" social stressors occur in a department, it can be assumed that almost everybody will be negatively affected after some time. In a study by Frese and Zapf (1987), members of the same work group reported more similar levels of social stressors compared with members of different groups. Bullying, however, is targeted at particular individuals. These individuals will show severe health consequences after some time, whereas the perpetrators, observers, or neutral bystanders may not be affected at all. Being singled out and stigmatised has been considered a key characteristic of bullying (Leymann, 1993, 1996; Zapf, 1999a). Furthermore, bullying not only acts as a stressor in its own right but also leads to the loss of personal resources for the targets, including the loss of social support and the ability to control their own situations (Zapf and Einarsen, 2005). Studies have shown, for instance, that such ordinary coping strategies as actively tackling the problem do not work for targets of bullying (ibid.). Taken together, these factors may explain the major detrimental effects bullying may have on the target, even if the behaviours themselves are seen as rather subtle and indirect by nonaffected observers.

This victim's view is in contrast to approaches that focus on the aggressive behaviour of perpetrators. An aggressive perpetrator can frequently harass other persons; however, if this behaviour is distributed across several persons, as in the case of an abusive supervisor targeting subordinates, clearly the case is different from that in which a victim or even a few targeted victims are the focus of aggressive acts by one or more other persons. Although the concept of bullying is used to describe both the aggressive behaviour of certain perpetrators as well as the victimisation process of particular targets, the latter has been the main focus of the European perspective. Yet, in some countries, e.g. the United Kingdom, bullying has been proven to be so closely related to the behaviours of managers and leaders that the term has a connotation more or less synonymous with destructive or highly aggressive leadership (see also Hoel et al., 2009).

The Frequency of Negative Behaviours

Definitions of bullying at work further emphasise two main features: repeated and enduring aggressive behaviours that are intended to be hostile and/or perceived as hostile by the recipient (Leymann, 1990b; Olweus, 1987, 1991; Zapf et al., 1996). In other words, bullying is normally not about single and isolated events but, rather, about behaviours that are repeatedly and persistently directed towards one or more employees. Leymann (1990b, 1996) suggested that to be termed mobbing or bullying, such events should occur at least once a week, which characterises bullying as a severe form of social stress. In many cases this criterion is difficult to apply because not all bullying behaviours are strictly episodic in nature. For example, a rumour can circulate that may be harmful or even threaten to destroy the victim's career or reputation. However, the rumour does not have to be repeated every week. In cases brought to our attention, victims had to work in basement rooms without windows and telephones. Here, bullying consists of a permanent state rather than a series of episodic events. Hence, the main criterion is that the behaviours or their consequences are repeated on a regular as opposed to an occasional basis. In a study by Notelaers and colleagues (2006), based on a study among 6,175 Belgian workers, who had responded to an 18-item version of the Negative Acts Questionnaire employing a statistical technique called latent class cluster analysis, six main clusters of respondents were identified regarding their exposure to bullying behaviours. While only 35% did not experience any kind of bullying behaviours during the preceding six months, some respondents (28%) experienced negative work-related behaviours now and then without seeing that as bullying. Yet another group of 16% experienced some personally related bullying behaviours now and then during the preceding six months. Hence, some 80% reported only marginal exposure to systematic bullying behaviours. However, one group of respondents, labelled "latent victims" and comprising 9% of the total sample, reported exposure to a range of bullying behaviours, although each type of behaviour occurred only now and then. Another group (8%), labelled "work-related bullying," reported high exposure to a few kinds of behaviours; their work situation was manipulated in a negative way combined with social exclusion from the work group. Lastly, 3% of the respondents reported a high exposure to many different kinds of bullying behaviours and with severe symptoms of reduced health and well-being; they were, therefore, labelled the "victims." Following this study, a majority of workers may be exposed to some varying levels of social stress, whilst yet others are more severely affected by systematic and intense bullying.

The Duration of Bullying

A number of studies have shown the prolonged nature of the bullying experience, with a majority of targets reporting an exposure time greater than 12 months (e.g., Leymann, 1992; UNISON, 1997; Zapf, 1999a; Zapf et al., this volume). A mean duration time of 18 months is reported in the case of Einarsen and Skogstad's Norwegian study (1996) and on average not less than 3.4 years in an Irish nationwide study (O'Moore, 2000).

The problem then arises of how to define operationally the duration of bullying behaviours. Leymann (1990b, 1996) suggested exposure for more than six months as an operational definition of bullying at work. Others have used repeated exposure to negative behaviours within a six-month period as the proposed timeframe (Einarsen and Skogstad, 1996). Leymann's strict criterion has been argued to be somewhat arbitrary, as bullying seems to exist on a continuum from occasional exposure to negative behaviours to severe victimisation resulting from frequent and long-lasting exposure to negative behaviours at work (Matthiesen et al., 1989). Yet, the criterion of about six months has been used in many studies in order to differentiate between exposure to social stress at work and victimisation from bullying (e.g., Einarsen and Skogstad, 1996; Mikkelsen and Einarsen, 2001; Niedl, 1995; Vartia, 1996; Zapf et al., 1996). The reason Leymann (1993, 1996) chose this criterion was to argue that mobbing leads to severe psychiatric and psychosomatic impairment, stress effects that would not be expected to occur as a result of such normal occupational stressors as time pressure, role conflicts, or everyday social stressors. Hence, Leymann chose the period of six months because it is frequently used in the assessment of various psychiatric disorders. However, in practice, victims feel bullied after a much shorter time. Also, from a theoretical perspective it is reasonable that exposure to systematic negative treatment may be observed within shorter timeframes. Furthermore, empirically, many studies do not define a specific timeframe (see Zapf et al., this volume, Table 3.3). In particular, when dealing with victims of bullying in organisations, a criterion of six months might not be very helpful.

Nevertheless, the mean duration of bullying in the various empirical studies is relatively high (see Zapf et al., this volume, Table 3.1), ranging from 12 to 62 months. Even without a concrete timeframe, there is consensus among researchers that bullying is a matter of months and years rather than days and weeks.

The duration of the bullying seems to be closely related to the frequency of bullying, with those bullied regularly reporting a longer duration of their experience than those bullied less frequently (Einarsen and Skogstad, 1996). This seems to be in line with a model of bullying highlighting the importance of conflict escalation, with the conflict becoming more intense and more personalised over time (Zapf and Gross, 2001). Based on these findings, we suggest defining behaviour as bullying if somebody is exposed to systematic and prolonged negative behaviour, and to speak of *severe bullying* if the duration is at least six months. Whether or not it is advisable to use a minimum duration criterion (e.g., six months) depends on the practical context. Independently of which definition of bullying is used, one important implication of such a definition is that based on the observation of a *single negative act*, it is impossible to decide whether or not this is bullying. Rather, the *bullying process* has to be taken into account.

The Nature of Behaviours Involved

The negative and unwanted nature of the behaviour involved is essential to the concept of bullying. Victims are exposed to persistent insults or offensive remarks, persistent criticism, and personal, or even, in a few cases, physical abuse (Einarsen, 2000b). Others experience social exclusion and isolation; that is, they are given the "silent treatment" or are "sent to Coventry" (Williams, 1997). These behaviours are "used with the aim or at least the effect of persistently humiliating, intimidating, frightening or punishing the victim" (Einarsen, 2000b, p. 8).

In recent years, many researchers developed instruments to measure bullying (e.g., Einarsen et al., 2009; Escartín et al., in press a; Leymann, 1990a; Zapf et al., 1996). The following types of bullying were repeatedly distinguished by these and other researchers:

1. Work-related bullying versus person-related bullying. The former includes such behaviours as giving unreasonable deadlines or unmanageable workloads, excessive monitoring of work, or assigning meaningless tasks or even no tasks. Empirical studies show that it is often difficult to decide if somebody is bullied based on work-related bullying alone. The reason is that there are also employees who report suffering from an unmanageable workload or excessive monitoring who would not consider themselves victims of bullying. The most extreme form of work-related bullying is probably "not to assign any tasks at all to a person." To the knowledge of the authors, this happens more often than one would believe.

   Person-related bullying consists of such behaviours as making insulting remarks, engaging in excessive teasing, spreading gossip or rumours, hurling persistent criticism, playing practical jokes, and engaging in intimidation. These behaviours are, by and large, independent of the work organisation.

2. In line with research into school bullying (Olweus, 1991; see also Buss, 1961; Baron and Neuman, 1996), person-related bullying

behaviours have repeatedly been defined on a scale ranging from passive and indirect to active and direct. Social isolation (e.g., not communicating with somebody or excluding someone from social events) and gossiping and spreading rumours (e.g., Einarsen et al., 2009; Escartín et al., in press a; Yildirim and Yildirim, 2007; Zapf et al., 1996) are on the passive and indirect end of this dimension. In the middle are such behaviours as belittling, making insulting remarks, making jokes, or engaging in other forms of humiliation (e.g., Einarsen and Raknes, 1997; Escartín et al., in press a; Moreno Jiménez et al., 2007; Zapf et al., 1996). At the active and direct end of the dimension are verbal threats and verbal aggression (e.g., Zapf et al., 1996).

3. With regard to Buss's differentiation between psychological and physical aggression (1961), aggressive acts related to person-related bullying are clearly psychological in nature. Many researchers (e.g., Einarsen 1999; Leymann, 1996; Niedl, 1995; Vartia, 1991; Zapf et al., 1996) also included physical abuse in their categorisation of bullying. However, they all agree that the behaviours involved in workplace bullying are mainly of a psychological rather than a physical nature. In a study among male Norwegian shipyard workers, where 88% had experienced some form of bullying behaviours during the preceding six months, only 2.4% reported having been subjected to physical abuse or threats of such abuse (Einarsen and Raknes, 1997). In Zapf's study (1999a), only about 10% of the bullying victims reported physical violence or the threat of physical violence. There may, however, be cultural differences. In a study comparing Southern European and Latin American cultures, employees from Latin America more often considered the physical component to be central to workplace bullying than did the Southern European employees (Escartín et al., in press b).

Factor-analysing bullying items often leads to a "one-factor solution" (e.g., Niedl, 1995). That is, in large samples people usually differ on a scale from "not being bullied" to "being bullied." If people are bullied, then they are usually exposed to all kinds of bullying behaviours. Cases where victims are exposed to just one kind of bullying behaviour are relatively few. Nevertheless, when *victim samples* are analysed (e.g., Zapf et al., 1996) or when more sophisticated methods such as latent cluster analysis are used (e.g., Einarsen et al., 2009), it is possible to identify groups of targets who are exposed to specific configurations of bullying behaviours.

Many of the acts just described may be relatively common in the workplace (Leymann, 1996). But when frequently and persistently directed towards the same individual, they may be considered an extreme social source of stress (Zapf et al., 1996) and become capable of causing severe harm and damage (Mikkelsen and Einarsen, 2002). In particular, there is some evidence that humiliating behaviour as a medium-active form of person-related bullying is perceived as most severe (Escartín et al., 2009) and has more negative effects on health than passive and indirect person-related bullying (social isolation) or work-related bullying (Zapf et al., 1996). The persistency of these behaviours also seems to drain the coping resources of the victim (Leymann, 1990b). The stigmatising effects of these activities, and their escalating frequency and intensity, make the victims constantly less able to cope with their daily tasks and collaborative requirements of the job, thus leading them to become ever more vulnerable and "a deserving target" (Einarsen, 2000b, p. 8). Hence, the frequency and duration of unwanted behaviours seem to be as important as the actual nature of the behaviours involved.

The Imbalance of Power between the Parties

A central feature of many definitions of bullying is the imbalance of power between the parties (Einarsen et al., 1994b; Leymann, 1996; Niedl, 1995; Zapf et al., 1996). Typically, a victim is constantly teased, badgered, and insulted and perceives that he or she has little recourse to retaliate (Einarsen, 1999). In many cases, it is a supervisor or manager who systematically, and over time, subjects subordinates to highly aggressive or demeaning behaviour (Rayner et al., 2002). In other cases, a group of colleagues bully a single individual, who for obvious reasons finds it difficult to defend him- or herself against this overwhelming group of opponents.

The imbalance of power often mirrors the formal power structure of the organisational context in which the bullying scenario unfolds. This would be the case when someone is on the receiving end of negative acts from a person in a superior position in the organisational hierarchy. Alternatively, the source of power may be informal, based on knowledge and experience as well as access to support from influential persons (Hoel and Cooper, 2000). The imbalance of power may also be reflected in the target's dependence on the perpetrator(s), which may be of a social, physical, economic, or even psychological nature (Niedl, 1995). An employee will in most cases be more dependent on his or her supervisor than vice versa. A single individual will be more dependent on the work group than the other way around. Thus, at times the perceptions of targets may be more dependent on the actual instigator of a negative act than the act itself (Einarsen, 2000a). Einarsen (1999) argues that knowledge of someone's "weak point" may become a source of power in a conflict situation. Bullies typically exploit the perceived inadequacies of the victim's personality or work performance, which in itself indicates the victim's powerlessness.

However, one may argue that in a conflict situation, some individuals may initially feel that they are as strong as their opponent, but they gradually come to realise that their first impression was wrong or that their own or their opponent's moves have placed them in a weaker position. In other words, and at least from the point of view of targets, a power deficit has emerged. Equal balance of power in a harsh conflict may, therefore, be considered hypothetical, since the balance of power in such situations is unlikely to remain stable for any length of time. Hence, bullying may result from the exploitation of power by an individual or by a group, as well as from taking advantage of a power deficit on the part of the target.

Subjective versus Objective Bullying

The distinction between subjective and objective experiences of bullying was first made by Brodsky (1976) and has been an important part of the discussion about the definition of bullying at work. Although most studies theoretically seem to regard bullying as an objective and observable phenomenon, which is not entirely in the eye of the beholder, with only a few exceptions the empirical data have so far been gathered by the use of self-report from victims (Einarsen et al., 1994a; Vartia, 1996). So far, little is known about the *interrater reliability* with regard to bullying, that is, about the agreement of the victim with some external observers of the bullying. There are two things that should be considered in the debate on subjective versus objective bullying: (1) the dependency on subjective appraisal, and (2) the observability of bullying.

1. *Subjective appraisal.* According to Niedl (1995), the "definitional core of bullying at work rests on the subjective perception made by the victim that these repeated acts are hostile, humiliating and intimidating and that they are directed at himself/herself" (p. 49). Yet, situations where one person offends, provokes, or otherwise angers another person often involve substantial discrepancies between the subjective perceptions and interpretations of the conflicting participants. Incidences that may be considered mildly offensive by one individual might be seen as serious enough to warrant a formal complaint by others (Terspptra and Baker, 1991). This variability in perception reflects the discussion on the role of subjective appraisal in psychological stress theories (Lazarus and Folkman, 1984). Frese and Zapf (1988) defined a *subjective stressor* as an event that is highly influenced by an individual's cognitive and emotional appraisal. As we argued before, based on a single act it is impossible to decide whether or not something is bullying because bullying is a longer-term process and the process character of bullying cannot be derived from a single act. In addition, whether or not people perceive *a particular act* as negative often depends strongly on subjective appraisals. Frese and Zapf (1988) suggested speaking of an objective stressor if it can be observed independently of an individual's cognitive and emotional appraisals. In this sense, there are only a few behaviours, such as offending or threatening somebody, that can

be judged as negative without knowing the subjective appraisals of the target person because such behaviours would be experienced as negative by almost everyone. In contrast, it is more difficult to judge such behaviours as not greeting somebody, making a joke, or criticising someone as negative independent of all the contextual information that would lead a victim to judge a particular behaviour as harassment. Yet, in U.S. court cases regarding sexual harassment, a "reasonable woman or man" standard is often used to warrant if something is to be seen as harassment or not (see also Hoel and Einarsen, this volume, regarding investigations into actual complaints of bullying).

2. *Observability.* The second issue is observability. We propose that bullying, in contrast to single acts, is to a large extent an objective stressor. This is so because we assume that the *bullying process*—namely, repeated and systematic negative acts aimed at a target person—would in all likelihood be appraised as such by any observer if the observer had the chance always to be present and to have access to all information. That is, whereas single acts may be ambiguous, we propose that the bullying process is not. There is some evidence from schoolyard bullying. In studies among Scandinavian schoolchildren, teachers tended to agree to a large extent with children's subjective perceptions of victimisation (Olweus, 1987). The problem is, however, that the process of bullying at work is often not observable. Individuals other than the target person receive only limited information about the bullying process. They may occasionally be witnesses of single acts. However, single acts, as described earlier, are ambiguous and prone to subjective appraisal processes.

For empirical and practical purposes, Björkqvist and others (1994) argue strongly against an approach where peer nominations are used as an objective measure of bullying. The economic importance of and dependence on a job would effectively prevent people from being honest in their assessment, in particular with regard to superiors in formal positions of power. It has also been argued that it is often difficult for the observer to stay neutral in cases of bullying (Einarsen, 1996; Neuberger 1999), a fact that is likely to influence the objectivity of such ratings. In the course of time, social perceptions of the victim seem to change, turning the situation gradually into one where even third parties perceive it as no more than fair treatment of a neurotic and difficult person (Einarsen et al., 1994b; Leymann, 1992). As the behaviours involved in bullying are often of a subtle and discrete nature (e.g., not greeting, leaving the table when the victim arrives, not passing on information, or gossiping), and sometimes even exhibited in private (Einarsen, 1999; Neuberger, 1999), they are not necessarily observable to others. Bullying is, therefore, often a subjective process of social reconstruction and thus difficult to prove. Uninformed bystanders could interpret the respective behaviours completely differently (e.g., as forgetful, impolite, careless, or casually humorous). The significance of a particular behaviour may be known only by the perpetrator and the recipient. The fact that the parties have a past and a future together will have a bearing on the perceptions and interpretations of the exhibited behaviours as well as on the process development (Hoel et al., 1999). Imbalances of power are also more evident from the perspective of those experiencing a lack of power. For instance, Hofstede (1980) measures the cultural dimension of power distance as the difference in power between two parties as perceived by the one in the least powerful position.

On the whole, we tend to agree with Lengnick-Hall (1995), who argues, in the case of sexual harassment, that an objective conceptualisation is, of course, necessary in connection with legal issues and cases of internal investigations and hearings. Subjective conceptualisations may, however, be a better prediction of victims' responses and reactions, such organisational outcomes as turnover and absenteeism, and organisational responses. However, when subjective conceptualisations exist in a workplace, procedures for complaint handling, including procedures for the investigation of the complaint, must be evoked securing a fair hearing for both parties with conclusions drawn on an objective basis. Following this, suitable organisational interventions and problem solving must be invoked by the employer if it is concluded that bullying has taken place (see also Hoel and Einarsen, this volume; European Social Dialogue, 2007).

Intentionality of Bullying

Considerable disagreement exists with regard to the issue of intent in the definition of bullying, as will be apparent also in this volume (Hoel et al., 1999). Some contributors, notably those whose approach is anchored in aggression theory, consider the perpetrator's intent to cause harm a key feature of bullying (Björkqvist et al., 1994; Olweus, 2003). A study of lay, or nonacademic, definitions of bullying also pinpointed intent as an important feature of the definition (Saunders et al., 2007). In other words, where there is no intention to cause harm, there is no bullying. Including intent in the definition of bullying will make it distinguishable from such other forms of unpleasant behaviours as incivility at work. It will also distinguish it from episodes of thoughtlessness or from the misperception of innocent or even fairly legitimate behaviours. Theoretically, aggression research has built on the notion that one must distinguish between accidental and intended harm (see also Keashly and Jagatic, this volume).

Yet, intent is generally not considered an essential element in most of the European research on workplace bullying. The problem is that it is normally impossible to verify the presence of intent in cases of bullying

(Hoel et al., 1999). The only one who can actually verify the presence of intent is the alleged perpetrator, creating a situation where the perpetrator has, in fact, veto power over the decision of whether or not something is to be regarded as bullying. For these reasons, intent is also excluded from most definitions of sexual harassment (e.g., Fitzgerald and Shullman, 1993).

To consider intent as a constituent part of the bullying definition, it would be necessary to clarify which "intent" is being referred to, since there are several possibilities: (1) Intent may be referred to a single act. That is, every single act occurring in bullying may be considered as intended or not intended. (2) Intent may be referred to the bullying process: somebody is intentionally exposed to repeated and systematic acts. (3) Intent may be referred to victimise the target, that is, bringing the target into an inferior and defenceless position. (4) Intent may also be referred to the extent of harm done to the victim. Sometimes, as in the case of repeatedly making a fool of someone, the single acts might be intended, but not the victimisation or the harm that is done. In other cases, a group might want to expel a person from the team or the organisation by exhibiting continuous negative behaviours, but the group members may not intend the psychological harm they are causing.

There may also be cases where negative behaviours continuously affect a particular person without any intention to inflict harm on this person. One example is the case of an arrogant and self-aggrandising person who might continuously exhibit behaviours to protect and enhance his own self-esteem. Unwittingly, such behaviours humiliate or ridicule the other person. The arrogant individual constantly criticises the other person to demonstrate his own superiority rather than to bully someone intentionally. In another case, the bullying might be a by-product of "micro-political behaviour" (see Zapf and Einarsen, this volume). The intention here is to protect and increase one's own power; bullying is then more a matter of collateral damage rather than intended hurt. Although the negative effects on the victim might be accepted or even justified in the aftermath, in such cases harming the victim is not the primary intention.

The point is that the lack of intent to harm someone does not change the situation for the target. In the context of bullying and harassment, research that takes the perspective of the target, actual repeated exposure to actual exhibited behaviours that would be seen as inappropriate, unwanted, harmful, or unpleasant by any reasonable man or woman exposed to them must be deemed inappropriate independently of any kind of intent by the perpetrator. In his classic work *The Psychology of Aggression*, Buss (1961) himself argues against intent, even in one-off incidents of aggression, because of the difficulty of including this term in the analysis of any behavioural event. Buss (p. 2) even claims that including intent in the definition of aggression is both "awkward and unnecessary," since among other things, "intent is a private event that may or may not be capable of verbalization, may or may not be accurately reflected in a verbal statement."

Hence, intent is a problem from both a theoretical and an applied perspective. In addition, the concept also poses problems from a methodological perspective. How can intent be measured? A study by Schat and colleagues (2006, p. 49) defined aggression as behaviour intended to cause harm, but this was not included in the measurement, where participants or subjects were asked only if different kinds of aggressive behaviours were enacted at the workplace and who the actor was. The intent of the actor remained unknown.

Whereas intent may not be a necessary feature of bullying definitions, however, there is little doubt that perception of intent may determine whether an individual decides to label his or her experience as bullying or not (see also Keashly, 2001).

Interpersonal versus Organisational Bullying

Following on from the preceding presentation, bullying is an interpersonal phenomenon that evolves from a dynamic interaction between at least two parties. Bullying is exhibited by one or more persons, directed towards another individual, and perceived and reacted to by this individual. However, from the work of Liefooghe and Mackenzie Davey (2001), one may argue that the concept may also refer to incidences of what we may call "organisational bullying" or "structural mobbing" (Neuberger, 1999), a depersonalised form of bullying (Hoel and Beale, 2006). Organisational bullying refers to situations in which organisational practises and procedures, perceived to be oppressive, demeaning, and humiliating, are employed so frequently and persistently that many employees feel victimised by them. Again, we are talking about persistent negative events and behaviours that wear down, frustrate, frighten, or intimidate employees. However, in these situations managers individually or collectively enforce organisational structures and procedures that may torment, abuse, or even exploit employees. Hence, bullying in these cases does not strictly refer to interpersonal interactions; rather, it involves indirect interactions between the individual and management per se in terms of various institutional arrangements. Thus, in a study in a UK telecommunications company using focus groups, employees did use the term *bullying* to account for grievances and discontent with the organisation and its procedures (Liefooghe and Mackenzie Davey, 2001). Examples of such procedures were the excessive use of statistics (e.g., performance targets), rules regarding call-handling times, penalties for not hitting targets (e.g., withdrawing possibilities for overtime), and use of sickness policy. These employees acknowledged the existence of bullying as an interpersonal phenomenon, but they also used the construct as an emotive and highly charged term, which helped them to highlight their discontent with what they perceived to be increasingly difficult work situations.

However, one may of course question the fruitfulness of such a use of the bullying concept, especially since most authors see bullying as an interpersonal phenomenon. Furthermore, we have elsewhere argued that the term *bullying* is easily misused (Einarsen, 1998; Hoel and Beale, 2006; Zapf, 1999a). Therefore, although applying the rhetoric of bullying as a political act to draw attention to oppressive work practises (Liefooghe, 2003) may be understandable, any gains might be shortsighted, contributing to dilute the power of the *bullying* term altogether (Hoel and Beale, 2006).

Bullying as a Process

Empirical studies indicate that bullying is not an either-or phenomenon; rather, it is a gradually evolving process (Björkqvist, 1992; Einarsen, 2000b; Leymann, 1990b; Zapf and Gross, 2001). During the early phases of the bullying process, victims are typically subjected to aggressive behaviour that is difficult to pin down because of its indirect and discrete nature. Later on, more direct aggressive acts appear (Björqkvist, 1992). The victims are clearly isolated and avoided, humiliated in public by excessive criticism or by being made a laughing stock. In the end, both physical and psychological means of violence may be used.

In line with Leymann (1990b), Einarsen (1999) identified four stages of process development and referred to them as *aggressive behaviours, bullying, stigmatisation*, and *severe trauma*. In many cases, the negative behaviours in the first phase may be characterised as indirect aggression. They may be "subtle, devious and immensely difficult to confront" (Adams, 1992, p. 17) and sometimes difficult to recognise for the persons being targeted (Leymann, 1996). Where bullying evolves out of a dispute or a conflict, it may even at times be difficult to tell who may turn out to be the victim (Leymann, 1990b). The initial phase, which in some cases can be very brief, tends to be followed by a stage of more direct negative behaviours, often leaving the target humiliated, ridiculed, and increasingly isolated (Leymann, 1990b, 1996). As a result, the targets become stigmatised and find it more and more difficult to defend themselves (Einarsen, 1999). At this point in the process the victims may suffer from a wide range of stress symptoms. According to Field, himself a victim of bullying:

> The person becomes withdrawn, reluctant to communicate for fear of further criticism. This results in accusations of 'withdrawal', 'sullenness', 'not co-operating or communicating', 'lack of team spirit', etc. Dependence on alcohol, or other substances can then lead to impoverished performance, poor concentration and failing memory, which brings accusations of 'poor performance'. (Field, 1996, p. 128)

It has also been noted that the erratic and obsessive behaviour of many victims in this phase may frequently cut them off from support within their own working environment, exacerbating their isolation and the victimisation process (Leymann, 1986). The situation is frequently marked by helplessness, and, for many, lengthy sickness absences may be necessary to cope with the situation (Einarsen et al., 1994b; Zapf and Gross, 2001). When the case reaches this stage, victims are also often left with no role in the workplace, or they are provided with little or no meaningful work.

Leymann (1990b) refers to this last stage as "expulsion," where victims are forced out of the workplace either directly, by means of dismissal or redundancy, or indirectly, when the victims consider their work situation so impossible that they decide to leave "voluntarily" (constructive dismissal). In a study among German victims of severe bullying, Zapf and Gross (2001) revealed that victims of such bullying strongly advise other victims to leave the organisation and seek support elsewhere.

Despite the severity of the situation, neither management nor work colleagues are likely to interfere or take action in support of the victim. On the contrary, if victims complain, they frequently experience disbelief and questioning of their own role. Such prejudices against victims often extend beyond management to include work colleagues, trade union representatives, and the medical profession (Leymann, 1996; O'Moore et al., 1998).

Such rather depressing outcomes have been observed in many countries (Björqkvist, 1992; Einarsen et al., 1994b; Leymann, 1990b, 1996; Zapf and Gross, 2001). The prejudices against the victim produced by the bullying process seem to cause the organisation to treat the victim as the source of the problem (Einarsen, 1999). When the case comes to their attention, senior management, union representatives, or personnel administrations tend to accept the prejudices espoused by the offenders, thus blaming victims for their own misfortune. Third parties and managers may see the situation as no more than fair treatment of a difficult and neurotic person (Leymann, 1990b).

A Definition of Bullying at Work

Building on the foregoing argument, we suggest the following definition of bullying (cf. Einarsen and Skogstad, 1996; Leymann, 1996; Olweus, 1987, 1991, 1994; Zapf, 1999b):

> Bullying at work means harassing, offending, or socially excluding someone or negatively affecting someone's work. In order for the label *bullying* (or *mobbing*) to be applied to a particular activity, interaction, or process, the bullying behavior has to occur repeatedly and regularly (e.g., weekly) and over a period of time (e.g., about six months). Bullying is an escalating process in the course of which the person confronted ends up in an inferior position and becomes the target of systematic negative social acts. A conflict cannot be called bullying if the incident is an isolated event or if two parties of approximately equal strength are in conflict.

Conceptual Models of Bullying at Work

The Leymann Model

Heinz Leymann (1990b, 1993, 1996), who has been influential in many European countries, argued strongly against individual factors as antecedents of bullying, especially when related to issues of victim personality. Instead, he advocated a situational outlook, where organisational factors relating to leadership, work design, and the morale of management and workforce are seen as the main factors. He asserted that four factors are prominent in eliciting bullying behaviours at work (Leymann, 1993): (1) deficiencies in work design, (2) deficiencies in leadership behaviour, (3) the victim's socially exposed position, and (4) low departmental morale. Leymann (1996) also acknowledged that poor conflict management might be a source of bullying, but in combination with inadequate organisation of work. However, he again strongly advocated that conflict management is an organisational issue and not an individual one (Leymann, 1996). Conflicts escalate into bullying only when the managers or supervisors either neglect or deny the issue, or if they themselves are involved in the group dynamics, thereby fuelling conflict. Since bullying takes place within a situation regulated by formal behavioural rules and responsibilities, it is always and by definition the responsibility of the organisation and its management.

Some research has been conducted on the work environment hypothesis of Leymann (e.g., Agervold, 2009; Hauge et al., 2007). Some 30 Irish victims of bullying described their workplace as a highly stressful and competitive environment, plagued with interpersonal conflicts and a lack of a friendly and supportive atmosphere, undergoing organisational changes, and managed by means of an authoritarian leadership style (O'Moore et al., 1998). In a Norwegian survey of 2,200 members of seven different trade unions, both victims and observers of bullying at work reported a lack of constructive leadership, lack of possibilities to monitor and control their own work tasks, and, in particular, a high level of role conflict (Einarsen et al., 1994a). In a Finnish survey, victims and observers of bullying described their work unit as having the following features: poor information flow, an authoritative way of settling differences of opinion, a lack of discussions about goals and tasks, and a lack of opportunity to influence matters affecting themselves (Vartia, 1996). A few studies have also shown a link between organisational changes and bullying at work (e.g., Skogstad et al., 2007).

It should be noted, however, that these results are based on cross-sectional studies that do not allow us to interpret relations as cause and effect. Although we believe that in many cases organisational deficiencies contribute substantially to the development of bullying, it is equally plausible that severe social conflicts at work may be the cause, rather than the effect, of organisational problems (Zapf, 1999b). Conflicts may, for example, negatively affect the information flow and thus impair leader–member relationships. Moreover, relations between bullying and low levels of control can also be explained by the fact that restricting someone's opportunities to affect decision making has been described as a bullying strategy (Leymann, 1990a). Yet, a study by Agervold (2009) supported the hypothesis that departments with high levels of bullying also have a poorer psychosocial work environment. Agervold studied the relationships between organisational factors and bullying on a departmental level, ruling out the effect of dissatisfied victims and supporting the hypothesis that the pressure of work, performance demands, autocratic management, and role conflict and lack of role clarity, as well as a poor social climate in a working group, can contribute to the emergence of higher incidences of bullying. The same conclusion was reached by Hauge and colleagues (2007) in a representative sample when comparing targets and observers of bullying with noninvolved workers. Furthermore, the latter study showed that the lack of leadership in the form of laissez-faire leadership, as predicted by Leymann, moderated the relationship between role conflicts and exposure to bullying.

Yet, Einarsen and colleagues (1994a) found that work environment factors explained only 10% of the variance in the prevalence of workplace bullying within seven different organisational settings, and in no subsetting was this greater than 24%. Thus, there is certainly room for alternative explanations. Leymann himself never presented any empirical evidence for his strong focus on organisational factors and his disregard for the role of personality. Hence, bullying is the product neither of chance nor of destiny (Einarsen et al., 1994a). Instead, bullying should be understood primarily as a dyadic interplay between people, where neither situational nor personal factors are entirely sufficient to explain why it develops. Although one may agree that the organisation and its management are responsible for intervening in cases of interpersonal conflict and bullying, this may still be caused by a wide range of factors, both on an individual level and on dyadic, group, organisational, and societal levels (Hoel and Cooper, 2001; Zapf, 1999b). Zapf provided some preliminary evidence for these various potential causes of bullying by identifying subgroups of bullying victims for which such causal factors as the organisation or characteristics of victims themselves were likely to dominate.

Assuming that the concept of bullying at work refers to a range of situations and contexts where repeated aggressive behaviour may occur and where the targets are unable to defend themselves, Einarsen (1999) introduced the concepts of *dispute-related* and *predatory* bullying to broaden the perspectives and to account for the two main classes of situations where bullying may seem to originate.

Predatory Bullying

In cases of predatory bullying, the victim has personally done nothing provocative that may reasonably justify the behaviour of the bully. In such cases, the victim is accidentally in a situation where a predator is demonstrating power or is exploiting the weakness of someone who has become a victim by accident. The concept of petty tyranny proposed by Ashforth (1994) seems to refer to such kinds of bullying. Petty tyranny refers to leaders who lord their power over others through arbitrariness and self-aggrandisement, the belittling of subordinates, lack of consideration, and the use of an authoritarian style of conflict management. In some organisations, bullying is more or less institutionalised as part of their leadership and managerial practise, sometimes in the guise of "firm and fair" management (Brodsky, 1976). However, firm and fair may easily turn into harsh and unfair management, which may, again, turn into bullying and the victimisation of subordinates.

A person may also be singled out and bullied because he or she belongs to a certain outsider group, for instance by being the first woman in a local fire brigade. If perceived as a representative of a group or a category of people who are not approved by the dominant organisational culture, such employees may indeed be bullied without doing anything other than merely showing up at work (Archer, 1999). A study in UK fire brigades, for example, revealed an environment where bullying of females, nonwhites, and nonconforming whites prevailed as a mechanism to ensure the preservation and dominance of the white male culture (Archer, 1999). As such, the individual victim of bullying was in fact a coincidental target.

An employee may even be bullied by being an easy target of frustration and stress caused by other factors. In situations where stress and frustration are caused by a source that is difficult to define, inaccessible, or too powerful or respected to be attacked, the group may turn its hostility towards a suitable scapegoat (Thylefors, 1987). Björkqvist (1992) argues that such displaced aggressiveness may act as a collective defence mechanism in groups where much unstructured aggression and hostility prevail.

Allport's study (1954) details the process involved in acting out prejudices and seems to describe very well how such bullying evolves. In his first phase, *antilocation*, prejudicial talk starts but is restricted to small circles of the "in group" and takes place "behind the back of the victim." This stage is followed by a second phase in which one moves beyond talking and starts to avoid the victim. In the third phase, the victim is openly harassed and discriminated against by being alienated and excluded or subjected to offensive remarks and jokes. In the fourth phase, physical attacks occur, which may lead to the final stage, *extermination*. Although victims of bullying are not literally killed, some do commit suicide (Leymann, 1990b); others are permanently expelled from working life (Leymann, 1996) or at least driven out of their organisation (Zapf and Gross, 2001). Hence, these examples of predatory bullying suggest the following subcategories: exposure to a destructive and aggressive leadership style, being singled out as a scapegoat, and the acting out of prejudice.

Dispute-Related Bullying

Dispute-related bullying, on the other hand, occurs as a result of highly escalated interpersonal conflicts (Einarsen, 1999; Zapf and Gross, 2001). Although interpersonal struggles and conflicts are a natural part of all human interaction and must not be considered bullying, there may be a thin line between the disagreements between two parties in an interpersonal conflict and the aggressive behaviour used in bullying (see also Zapf and Gross, 2001). The difference between interpersonal conflicts and bullying is to be found not in what is done or how it is done, but in the frequency and duration of what is done (Leymann, 1996), as well as in the ability of both parties to defend themselves in the situation (Zapf, 1999a). In some instances, the social climate at work turns sour and creates differences that may escalate into harsh personalised conflicts and even "office wars," where the total annihilation of the opponent is seen as the ultimate goal to be gained by the parties (van de Vliert, 1998).

In highly escalated conflicts, both parties may deny their opponent any human value, thus clearing the way for highly aggressive behaviours. The party placed at a disadvantage in this struggle may, of course, become the victim of bullying (Zapf, 1999b). It may also be the case that one of the parties exploits his or her own power or a potential power imbalance, leading to a situation where the other is unable to mount a defence or retaliate against increasingly aggressive behaviours. The defenceless position will then lead to a victimisation of one of the parties.

Interpersonal conflicts where the identity of the protagonists is at stake— for instance, when one party attacks the self-esteem or self-image of the other—are often characterised by intense emotional involvement (Glasl, 1994). The latter includes feelings of being insulted, fear, suspicion, resentment, contempt, anger, and so forth (van de Vliert, 1998). In such cases, people may subject each other to bullying behaviour or resent the behaviour of their opponent to a degree where they feel harassed and victimised even though there are few observable signs of bullying behaviour by the alleged offender. It may also be true that claiming to be a victim of bullying may be a very effective strategy in interpersonal conflicts, in some cases even used by both parties. The conflict escalation model of Glasl (1982, 1994) has been proposed as a model suitable to explain how conflicts may escalate into bullying (Einarsen et al., 1994b; Neuberger, 1999; Zapf and Gross, 2001). The model differentiates between three phases and nine stages (Figure 1.1). According to this model, conflicts are inevitable in organisations, and under certain circumstances they are even fruitful, contributing to innovation, performance, and learning (de Dreu, 1997). However, if allowed to

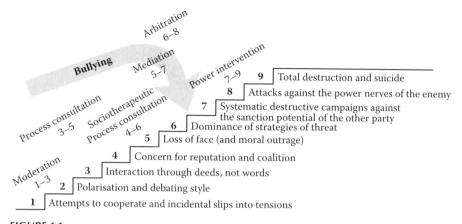

FIGURE 1.1
The conflict escalation model of Glasl (1994). (After Zapf, D., and Gross, C., *European Journal of Work and Organizational Psychology*, 10, 501, 2001. With permission.)

escalate, conflicts may turn into so-called office wars and become extremely harmful and destructive on an individual as well as organisational level (see Figure 1.1).

In the first stages of a conflict, the parties are still interested in a reasonable resolution of conflict about tasks or issues. Although they may experience and acknowledge interpersonal tension, they mainly focus on cooperation to solve the problems in a controlled and rational manner. However, this joint effort becomes increasingly more difficult as the interpersonal tensions escalate (Zapf and Gross, 2001).

The second phase of conflict is characterised by a situation where the original issue of the conflict has more or less vanished, while the interpersonal tension between the parties and their increasingly difficult relationship becomes the heart of the problem. Now the issue of the conflict has more to do with who is the problem than what is the problem. The parties cease to communicate and start to seek allies and support from others. They become increasingly more concerned about their own reputation and about losing face; and they experience moral outrage against their opponents, perceiving them as immoral, as having a personality deficit, or as being plain stupid. At this point, disrespect, lack of trust, and finally overt hostility evolve. Ultimately, the interaction is dominated by threat as well as openly hostile and aggressive behaviour. In the following phase, the confrontations become increasingly more destructive until the total annihilation of the opponent is the sole aim of the parties. Both the parties in this struggle are willing to risk their own welfare, even their own existence, in order to annihilate the opponent.

Zapf and Gross (2001) argue that bullying may be seen as a kind of conflict at the boundary between phases 2 and 3. In their interview study of 19 German victims of bullying, 14 victims reported a continuously escalating situation, in which the situation became worse over the course of time. Almost 50% of the victims described a sequence of escalation resembling Glasl's model.

Although Glasl argues that the latter stages of the model may not be reached in organisations, we will argue that they are, in fact, reached in the more extreme cases of bullying. Some victims commit suicide (Leymann, 1990b), and many others consider it (Einarsen et al., 1994b). Some victims go to court even though they may be unable to afford a solicitor and are likely to lose. Others refuse a reasonable settlement out of court because they want to take their employer to court at any cost (cf. the case described in Diergarten, 1994). Some work groups even take pleasure in the suffering of the victim.

A Theoretical Framework

In the following section, we will argue that such a complex social phenomenon as bullying is characterised by multicausality, involving a range of factors found at many explanatory levels, depending on whether we focus on the behaviour of the actor or on the perceptions, reactions, and responses of the target (see also Einarsen, 1999, 2000b; Hoel and Cooper, 2001; Zapf, 1999b). On an individual level, the personalities of both the perpetrator and the victim may be involved as causes of both bullying behaviour and perceptions of being bullied. Individual factors may also contribute to the victim's potential lack of coping strategies as well as other emotional and behavioural reactions to the perceived treatment. On a dyadic level, the focus is on the relationship and the interaction between the alleged perpetrator and the alleged victim. Since a power differential between the parties is central to the definition of bullying, a dyadic perspective is vital to the understanding of the concept of bullying at work. According to Brodsky (1976), many cases of bullying involve an artless teaser who meets a humourless target. Also, to focus on a potential clash or mismatch in terms of personalities and power may be as relevant as to focus on the pathological and deviant personality of the perpetrator or the victim. On a dyadic level, we may also focus on the dynamics of conflict escalation and the dynamic transaction between the perpetrator and the victim in the course of the conflict (Glasl, 1994; Zapf and Gross, 2001). In most cases, and especially those involving disputes, the victim is not an entirely passive recipient of negative acts and behaviours (Hoel and Cooper, 2001). The victim's responses are likely to impinge upon the further responses of the perpetrator. As shown by Zapf and Gross (2001), those victims who successfully coped with bullying fought back with similar means less often and avoided further escalation of the conflict. Less successful victims in terms of coping often contributed to the escalation of the bullying by their aggressive counterattacks and "fights for justice" (p. 497).

On a social-group level, bullying may be explained in terms of scapegoating processes in groups and organisations. Such witch-hunting processes arise when groups displace their frustration and aggression onto a suitable and less powerful group member. Being seen as an outsider or as part of a minority may be one criterion for this choice (Schuster, 1996). Another may be outdated behaviours that do not keep pace with the developments of the group. Also, being too honest or unwilling to compromise may also contribute to being put into the role of a scapegoat (Thylefors, 1987). On the organisational level, many factors may contribute to explain cases of bullying at work (Hoel and Cooper, 2001). Archer (1999) has shown how bullying may become an integrated part of an organisational culture, whilst Zapf and others (1996) have shown that requirements for a high degree of cooperation combined with restricted control over one's own time may contribute to someone's becoming a victim of bullying. This situation may lead to many minor interpersonal conflicts and may simultaneously undermine the possibilities for conflict resolution. Similarly, Vartia (1996) has shown that the work environment in organisations with bullying is characterised by a general atmosphere experienced by employees as strained and competitive, where everyone pursues his or her own interests.

In Figure 1.2 we present a theoretical framework that identifies the main classes of variables to be included in future research efforts and future theoretical developments in this field. The model may also be utilised in order to guide and structure future organisational action programmes. This model pinpoints another level of explanation, the societal level, consisting of national

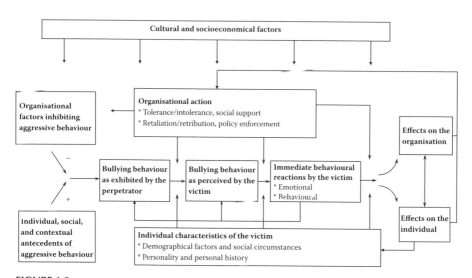

FIGURE 1.2
A theoretical framework for the study and management of bullying at work.

culture and historical, legal, and socioeconomic factors (see also Hoel and Cooper, 2001). Although it has not been much studied yet, the occurrence of bullying must always be seen against such a background (see also Beale, this volume; Ironside and Seifert, 2003; McCarthy, 2003). The high pace of change, intensifying workloads, increasing working hours, and uncertainty with regard to future employment that characterise contemporary working life in many countries influence the level of stress of both perpetrator and victim. Hence, both the level of aggression and one's coping resources may be influenced by such factors. In addition, the tolerance of organisations and their management of cases of bullying must also, to some extent, be seen in light of prevailing societal factors.

Following on from the debate on objective and subjective bullying, this model distinguishes between the nature and causes of bullying behaviour as exhibited by the alleged offender and the nature and causes of the perceptions of the target of these behaviours. Furthermore, it distinguishes between the perceived exposure to bullying behaviours and the reactions to these kinds of behaviour.

Looking at the behaviour of the perpetrator first, Brodsky (1976) claimed that although bullies may suffer from personality disorders, they will act as bullies only when the organisational culture permits or even rewards this kind of misbehaviour. Although there may be situational and contextual as well as personal factors that may cause a manager or an employee to act aggressively towards subordinates or colleagues, such behaviour will not be exhibited systematically if there are factors in the organisation that hinder or inhibit it. On the basis of survey data on the experiences and attitudes of British union members, Rayner (1998) concluded that bullying prevails because of an organisational tolerance of such behaviour. Ninety-five percent of the respondents in her study claimed that bullying was caused by the facts that "bullies can get away with it" and "victims are too scared to report it." Hence, bullying behaviour may be a result of the combination of a propensity to bully because of either personal or situational factors and the lack of organisational inhibitors of bullying behaviour (see also Pryor et al., 1993).

Furthermore, the model shows that such organisational prohibitive factors, as well as an effective support system for victims, are key factors that may moderate the perceptions and reactions of the victim. The presented model argues that attention to organisational response patterns and other contextual issues within the organisation are highly important when understanding the many different aspects of bullying at work. The latter part of the model has clearly an individual, subjective, and most of all, reactive focus (Einarsen, 2000b). Although bullying at work may to some degree be a subjectively experienced situation in which the meaning assigned to an incident will differ depending on the persons and the circumstances involved, this part of the model highlights the necessity for any strategy against bullying to take the perceptions and reactions of the victims seriously and functions as a real description of how they experience their work environment. Second, this part of the model argues for inclusion of a rehabilitation programme in an effective organisational strategy against bullying.

This theoretical framework also gives some credit to the dynamic process involved in the interaction among perpetrator, victim, and organisation (see earlier). Even Leymann (1986, 1992) argued that the stress reaction of the victim to the perceived bullying, and the consequential effects on the victim, may backfire and justify the treatment of the victim. The process of stigmatisation may also alter the perception of the victim, which may, again, change how an organisation tolerates, reacts to, and manages a particular case of bullying. Hence, the behaviour of the perpetrator and the personal characteristics of the victim as well as the organisation's responses to bullying may be altered in the course of the process. We believe that knowledge of the escalation and the dynamics of interaction involved in the victimisation process is essential to the understanding of this phenomenon.

Conclusion

In spite of the cultural diversity in Europe and the evolving multidisciplinary approach to the study of bullying and harassment in organisations, the European tradition in this field of research has largely succeeded in developing and maintaining a shared conceptual understanding of the underlying phenomenon. Some authors have argued that there is, in fact, a difference between the UK concept of bullying and the Scandinavian and German concept of mobbing at work (Leymann, 1996). According to Leymann (1996), the choice of using the term *mobbing at work* in preference to *bullying* was made consciously. It reflected the fact that the phenomenon in question very often refers to subtle, less direct aggression as opposed to the more physical aggression commonly identified as bullying, but with the same debilitating and stigmatising effects. The term *bullying* is also particularly concerned with aggression from someone in a managerial or supervisory position (Zapf, 1999b).

Although the concept of bullying as used in English-speaking countries and mobbing as used in many other European countries may have semantic nuances and different connotations depending on where they are used, to all intents and purposes they refer to the same phenomenon. Any differences in the use of the terms may be concerned as much with cultural differences related to the phenomenon in the different countries as with real differences in the concepts themselves (Di Martino et al., 2003). Whilst *bullying* may better fit predatory situations, *mobbing* may be more attuned to dispute-related scenarios. Then again, Scandinavian and German cases of bullying seem to be mainly of a dispute-related kind (Einarsen et al., 1994b; Leymann, 1996; Zapf and Gross, 2001). In any case, the nature of the bullying situation seems to be the same—the persistent and systematic victimisation of a colleague or a subordinate through repeated use of various kinds of aggressive behaviours over a long period of time and in a situation where victims have difficulty in defending themselves.

Because much of the data on the process of bullying still stems from the victims of long-lasting bullying, the processes described in this chapter may seem more deterministic than they are in real life. Most conflicts in organisations do not escalate into bullying. Not all cases of bullying will prevail for years, leading to severe trauma in the victim. However, so far we still know little of the processes and factors involved when cases of bullying and potential bullying take alternative routes. Research efforts must therefore still be directed at such issues.

References

Adams, A. (1992) *Bullying at work: How to confront and overcome it*. London: Virago Press.
Agervold, M. (2007) Bullying at work: A discussion of definitions and prevalence, based on an empirical study. *Scandinavian Journal of Psychology, 48*, 161–172.
—— (2009) The significance of organizational factors for the incidence of bullying. *Scandinavian Journal of Psychology, 50*, 267–276.
Allport, G. (1954) *The nature of prejudice*. Reading, MA: Addison Wesley.
Archer, D. (1999) Exploring "bullying" culture in the para-military organisation. *International Journal of Manpower, 20*, 94–105.
Ashforth, B. E. (1994) Petty tyranny in organizations. *Human Relations, 47*, 755–778.
Baron, R. A., and Neuman, J. H. (1996) Workplace violence and workplace aggression: Evidence on their relative frequency and potential causes. *Aggressive Behavior, 22*, 161–173.
Björkqvist, K. (1992) Trakassering förekommer bland anställda vid AA [Harassment exists among employees at Abo Academy]. *Meddelande från Åbo Akademi, 9*, 14–17.
Björkqvist, K., Österman, K., and Hjelt-Bäck, M. (1994) Aggression among university employees. *Aggressive Behavior, 20*, 173–184.
Brodsky, C. M. (1976) *The harassed worker*. Lexington, MA: D. C. Heath.
Buss, A. H. (1961) *The psychology of aggression*. New York: Wiley.
de Dreu, C. K. W. (1997) Productive conflict: The importance of conflict management and conflict issue. In C. K. W. de Dreu and E. van de Vliert (eds.), *Using conflict in organizations* (pp. 9–22). London: Sage.
Demirel, Y., and Yoldas, M. A. (2008) The comparison of behavior at health institutions of Turkey and Kazakhstan. *International Journal of Human Sciences, 5* (2), 2–3.

Di Martino, V., Hoel, H., and Cooper, C. L. (2003) *Preventing violence and harassment in the workplace*. Luxembourg: European Foundation for the Improvement of Living and Working Conditions.

Diergarten, E. (1994) *Mobbing: Wenn der Alltag zum Alptraum wird* [Mobbing: When everyday life becomes a nightmare]. Cologne, Germany: Bund Verlag.

Dohrenwend, B. S., and Dohrenwend, B. P. (eds.) (1974) *Stressful life events: Their nature and effects*. New York: Wiley.

Ege, H. (1996) *Mobbing: Che cosé il terrore psicologico sull posto di lavoro* [Mobbing: What is psychological terror on the job?]. Bologna, Italy: Pitagora Editrice.

Einarsen, S. (1996) Bullying and harassment at work: Epidemiological and psychosocial aspects. PhD thesis, Department of Psychosocial Science, University of Bergen, Norway.

—— (1998) Dealing with bullying at work: The Norwegian lesson. In C. Rayner, M. Sheehan, and M. Barker (eds.), *Bullying at work 1998 research update conference: Proceedings* (pp. 28–33). Stafford, UK: Staffordshire University.

—— (1999) The nature and causes of bullying. *International Journal of Manpower, 20,* 16–27.

—— (2000a) Bullying and harassment at work: Unveiling an organizational taboo. In M. Sheehan, S. Ramsey, and J. Partick (eds.), *Transcending boundaries: Integrating people, processes and systems*. Brisbane, Australia: School of Management, Griffith University.

—— (2000b) Harassment and bullying at work: A review of the Scandinavian approach. *Aggression and Violent Behavior, 4,* 371–401.

Einarsen, S., Hoel, H., and Notelaers, G. (2009) Measuring exposure to bullying and harassment at work: Validity, factor structure and psychometric properties of the Negative Acts Questionnaire–Revised. *Work & Stress, 23* (1), 24–44.

Einarsen, S., and Raknes, B. I. (1991) *Mobbing i arbeidslivet* [Bullying at work]. Bergen, Norway: Research Centre for Occupational Health and Safety, University of Bergen.

—— (1997) Harassment at work and the victimization of men. *Violence and Victims, 12,* 247–263.

Einarsen, S., Raknes, B. I., and Matthiesen, S. B. (1994a) Bullying and harassment at work and their relationships to work environment quality: An exploratory study. *European Work and Organizational Psychologist, 4,* 381–401.

Einarsen, S., Raknes, B. I., Matthiesen, S. B., and Hellesøy, O. H. (1994b) *Mobbing og harde personkonflikter: Helsefarlig samspill pa arbeidsplassen* [Bullying and severe interpersonal conflicts: Unhealthy interaction at work]. Soreidgrend, Norway: Sigma Forlag.

Einarsen, S., and Skogstad, A. (1996) Prevalence and risk groups of bullying and harassment at work. *European Journal of Work and Organizational Psychology, 5,* 185–202.

Escartín, J., Rodríguez-Carballeira, A., Gómez-Benito, J., and Zapf, D. (in press a) Development and validation of the Workplace Bullying Scale "EAPA-T." *International Journal of Clinical and Health Psychology.*

Escartín, J., Rodríguez-Carballeira, A., Zapf, D., Porrúa, C., and Martín-Peña, J. (2009) Perceived severity of various bullying behaviours at work and the relevance of exposure to bullying. *Work & Stress, 23,* 191–205.

Escartín, J., Zapf, D., Arrieta C., and Rodríguez-Carballeira, A. (in press b) Workers' perception of workplace bullying: A cross-cultural study. *European Journal of Work and Organizational Psychology.*

European Social Dialogue (2007) Framework Agreement on harassment and violence at work. http://www.etuc.org/a/3524.

Field, T. (1996) *Bullying in sight*. Wantage, UK: Success Unlimited.

Fitzgerald, L. F., and Shullman, S. (1993) Sexual harassment: A research analysis and agenda for the 1990s. *Journal of Vocational Behaviour*, 42, 5–27.

Frese, M., and Zapf, D. (1987) Eine Skala zur Erfassung von Sozialen Stressoren am Arbeitsplatz [A scale for the assessment of social stressors at work]. *Zeitschrift für Arbeitswissenschaft*, 41, 134–141.

—— (1988) Methodological issues in the study of work stress: Objective vs. subjective measurement of work stress and the question of longitudinal studies. In C. L. Cooper and R. Payne (eds.), *Causes, coping, and consequences of stress at work* (pp. 375–411). Chichester, UK: Wiley.

Glasl, F. (1982) The process of conflict escalation and roles of third parties. In G. B. J. Bomers and R. Peterson (eds.), *Conflict management and industrial relations* (pp. 119–140). Boston: Kluwer-Nijhoff.

—— (1994) *Konfliktmanagement. Ein Handbuch für Führungskräfte und Berater* [Conflict management: A handbook for managers and consultants], 4th ed. Bern, Switzerland: Haupt.

Hauge, L. J., Skogstad, A., and Einarsen, S. (2007) Relationships between stressful work environments and bullying: Results from a large representative study. *Work and Stress*, 21, 220–242.

Heinemann, P. (1972) *Mobbning—gruppvåld bland barn och vuxna* [Mobbing—group violence by children and adults]. Stockholm: Natur and Kultur.

Hirigoyen, M. F. (1998) *Le harcélement moral: La violence perverse au quotidien* [Bullying: The daily perverse violence]. Paris: La Découvert et Syros.

Hoel, H., and Beale, D. (2006) Workplace bullying, psychological perspectives and industrial relations: Towards a contextualised and interdisciplinary approach. *British Journal of Industrial Relations*, 44, 239–262.

Hoel, H., and Cooper, C. L. (1999) The role of "intent" in perceptions of workplace bullying. Presented at the 9th European Congress on Work and Organizational Psychology: Innovations for Work, Organization and Well-being, May 12–15, Espoo-Helsinki, Finland.

—— (2000) Destructive conflict and bullying at work. Unpublished report, Institute of Science and Technology, University of Manchester.

—— (2001) Origins of bullying: Theoretical frameworks for explaining bullying. In N. Tehrani (ed.), *Building a culture of respect: Managing bullying at work* (pp. 3–19). London: Taylor & Francis.

Hoel, H., and Einarsen, S. (2010) Shortcomings of antibullying regulations: The case of Sweden. *European Journal of Work and Organisational Psychology*, 19 (1), 30–50.

Hoel, H., Glasø. L., Hetland, J., Cooper, C. L., and Einarsen, S. (2009) Leadership styles as predictors of self-reported and observed bullying. *British Journal of Management*. doi: 10.1111/j.1467-8551.2009.00664.x.

Hoel, H., Rayner, C., and Cooper, C. L. (1999) Workplace bullying. In C. L. Cooper and I. T. Robertson (eds.), *International review of industrial and organizational psychology*, vol. 14 (pp. 195–230). Chichester, UK: Wiley.

Hofstede, G. (1980) *Culture's consequences*. New York: Sage.

Hubert, A. B., and van Veldhoven, M. (2001) Risk sectors for undesired behaviour and mobbing. *European Journal of Work and Organizational Psychology*, 10, 415–424.

Ironside, M., and Seifert, R. (2003) Tackling bullying in the workplace: The collective dimension. In S. Einarsen, H. Hoel, D. Zapf, and C. L. Cooper (eds.), *Bullying and emotional abuse in the workplace: International perspectives in research and practice* (pp. 383–398). London: Taylor & Francis.

Kanner, A. D., Coyne, J. C., Schaefer, C., and Lazarus, R. S. (1981) Comparison of two modes of stress measurement: Daily hassles and uplifts versus major life events. *Journal of Behavioral Medicine, 4,* 1–39.

Kaucsek, G., and Simon, P. (1995) Psychoterror and risk-management in Hungary. Paper presented as a poster at the 7th European Congress of Work and Organizational Psychology, April 19–22, Györ, Hungary.

Keashly, L. (1998) Emotional abuse in the workplace: Conceptual and empirical issues. *Journal of Emotional Abuse, 1,* 85–117.

—— (2001) Interpersonal and systemic aspects of emotional abuse at work: The target's perspective. *Violence and Victims, 16,* 233–268.

Keashly, L., and Harvey, S. (2005) Emotional abuse in the workplace. In S. Fox and P. E. Spector (eds.), *Counterproductive work behavior: Investigations of actors and targets* (pp. 201–235). Washington, DC: American Psychological Association.

Kile, S. M. (1990) *Helsefarleg leiarskap: Ein eksplorerande studie* [Health endangering leadership: An exploratory study]. Bergen, Norway: Department of Psychosocial Science, University of Bergen.

Kirchler, E., and Lang, M. (1998) Mobbingerfahrungen: Subjektive Beschreibung und Bewertung der Arbeitssituation [Bullying experiences: Subjective description and evaluation of the work situation]. *Zeitschrift für Personalforschung, 12,* 352–262.

Lazarus, R. S., and Folkman, S. (1984) *Stress, appraisal and coping.* New York: Springer.

Lengnick-Hall, M. L. (1995) Sexual harassment research: A methodological critique. *Personnel Psychology, 48,* 841–864.

Lewis, D. (2000) Bullying at work: A case of moral panic? In M. Sheehan, S. Ramsey, and J. Partick (eds.), *Transcending boundaries: Integrating people, processes and systems.* Brisbane, Australia: School of Management, Griffith University.

—— (2001) Perceptions of bullying in organizations. *International Journal of Management and Decision Making, 2* (1), 48–63.

Leymann, H. (1986) *Vuxenmobbning—psykiskt våld i arbetslivet* [Bullying—psychological violence in working life]. Lund, Sweden: Studentlitterature.

—— (1990a) *Handbok för användning av LIPT-formuläret för kartläggning av risker för psykiskt vald* [Manual of the LIPT questionnaire for assessing the risk of psychological violence at work]. Stockholm: Violen.

—— (1990b) Mobbing and psychological terror at workplaces. *Violence and Victims, 5,* 119–126.

—— (1992) *Från mobbning til utslagning i arbetslivet* [From bullying to exclusion from working life]. Stockholm: Publica.

—— (1993) *Mobbing—Psychoterror am Arbeitsplatz und wie man sich dagegen wehren kann* [Mobbing—psychoterror in the workplace and how one can defend oneself]. Reinbeck bei Hamburg, Germany: Rowohlt.

—— (1995) Einführung: Mobbing. Das Konzept und seine Resonanz in Deutschland [Introduction. Mobbing: The concept and its resonance in Germany]. In H. Leymann (ed.), *Der neue Mobbingbericht. Erfahrungen und Initiativen, Auswege und Hilfsangebote* (pp. 13–26). Reinbeck bei Hamburg, Germany: Rowohlt.

—— (1996) The content and development of mobbing at work. *European Journal of Work and Organizational Psychology, 5,* 165–184.

Liefooghe, A. P. D. (2003) Employee accounts of bullying at work. *International Journal of Management and Decision Making, 4,* 24–34.

Liefooghe, A. P. D., and Mackenzie Davey, K. (2001) Accounts of workplace bullying: The role of the organization. *European Journal of Work and Organizational Psychology, 10,* 375–392.

Lorenz, K. (1968) *Aggression: Dess bakgrund och natur* [On aggression]. Stockholm: Natur & Kultur.

Lutgen-Sandvik, P. (2007). But words will never hurt me: Abuse and bullying at work, a comparison between two worker samples. *Ohio Communication Journal, 45,* 27–52.

Malinauskienė, V., Obelenis, V., and Đopagienė, D. (2005) Psychological terror at work and cardiovascular diseases among teachers. *Acta Medica Lituanica, 12* (2), 20–25.

Matthiesen, S. B., Raknes, B. I., and Røkkum, O. (1989) Mobbing på arbeidsplassen [Bullying in the workplace]. *Tidsskrift for Norsk Psykologforening, 26,* 761–774.

McCarthy, P. (2003) Bullying at work: A postmodern experience. In S. Einarsen, H. Hoel, D. Zapf, and C. L. Cooper (eds.), *Bullying and emotional abuse in the workplace: International perspectives in research and practice* (pp. 231–244). London: Taylor & Francis.

McCarthy, P., Sheehan, M. J., and Wilkie, W. (eds.) (1996) *Bullying: From backyard to boardroom.* Alexandria, Australia: Millenium Books.

Mikkelsen, G. E., and Einarsen, S. (2001) Bullying in Danish work-life: Prevalence and health correlates. *European Journal of Work and Organizational Psychology, 10,* 393–413.

—— (2002) Basic assumptions and symptoms of post-traumatic stress among victims of bullying at work. *European Journal of Work and Organizational Psychology, 11,* 87–111.

Moreno-Jiménez, B., and Rodríguez-Muñoz, A. (2006) Número monográfico sobre acoso psicológico en el trabajo: Una perspectiva general [Monograph on bullying at work: An overview]. *Revista de Psicología del Trabajo y de las Organizaciones, 22,* 245–432.

Moreno-Jiménez, B., Rodríguez-Muñoz, A., Martínez, M., and Gálvez, M. (2007) Assessing workplace bullying: Spanish validation of a reduced version of the negative acts questionnaire. *Spanish Journal of Psychology, 10* (2), 449–457.

Munthe, E. (1989) Bullying in Scandinavia. In E. Roland and E. Munthe (eds.), *Bullying: An international perspective* (pp. 66–78). London: David Fulton.

Neuberger, O. (1999) *Mobbing: Übel mitspielen in Organisationen* [Mobbing: Playing bad games in organizations], 3rd ed. Munich: Hampp.

Niedl, K. (1995) *Mobbing/Bullying am Arbeitsplatz. Eine empirische Analyse zum Phänomen sowie zu personalwirtschaftlich relevanten Effekten von systematischen Feindseligkeiten* [Mobbing/bullying at work: An empirical analysis of the phenomenon and of the effects of systematic harassment on human resource management]. Munich: Hampp.

Nielsen, M. B., Matthiesen, S. B., and Einarsen, S. (in press) The impact of methodological moderators on prevalence rates of workplace bullying: A meta-analysis. *Journal of Occupational and Organizational Psychology.*

Nielsen, M. B., Skogstad, A., Matthiesen, S. B., Glasø, L., Schanke, M. A., Notelaers, G., and Einarsen S. (2009) The prevalence of bullying in Norway: Comparisons across time and estimation methods. *European Journal of Work and Organizational Psychology, 18* (1), 81–101.

Notelaers, G., Einarsen, S., De Witte, H., and Vermunt, J. (2006) Measuring exposure to bullying at work: The validity and advantages of the latent class cluster approach. *Work & Stress, 20* (4), 289–302.

Olweus, D. (1987) Bullying/victim problems among school children in Scandinavia. In J. P. Myklebust and R. Ommundsen (eds.), *Psykologprofesjonen mot ar 2000* (pp. 395–413). Oslo: Universitetsforlaget.

—— (1991) Bullying/victim problems among school children. In I. Rubin and D. Pepler (eds.), *The development and treatment of childhood aggression* (pp. 411–447). Hillsdale, NJ: Erlbaum.

—— (1994) Annotation: Bullying at school—basic facts and effects of a school based intervention program. *Journal of Child Psychology and Psychiatry, 35,* 1171–1190.

—— (2003) Bully/victim problems in school: Basic facts and an effective intervention programme. In S. Einarsen, H. Hoel, D. Zapf, and C. L. Cooper (eds.), *Bullying and emotional abuse in the workplace: International perspectives in research and practice* (pp. 62–77). London: Taylor & Francis.

O'Moore, M. (2000) *National survey on bullying in the workplace.* Dublin: Anti-Bullying Research Centre, Trinity College.

O'Moore, M., Seigne, E., McGuire, L., and Smith, M. (1998) Victims of bullying at work in Ireland. In C. Rayner, M. Sheehan, and M. Barker (eds.), *Bullying at work 1998 research update conference: Proceedings* (pp. 70–77). Stafford, UK: Staffordshire University.

Pikas, A. (1989) The common concern method for the treatment of mobbing. In E. Roland and E. Munthe (eds.), *Bullying: An international perspective* (pp. 91–104). London: David Fulton.

Pryor, J. B., LaVite, C., and Stoller, L. (1993) A social psychological analysis of sexual harassment: The person/situation interaction. *Journal of Vocational Behavior, 42* (Special issue), 68–83.

Rayner, C. (1997) The incidence of workplace bullying. *Journal of Community and Applied Social Psychology, 7,* 199–208.

—— (1998) Workplace bullying: Do something! *Journal of Occupational Health and Safety–Australia and New Zealand, 14,* 581–585.

Rayner, C., Hoel, H., and Cooper, C. L. (2002) *Workplace bullying: What we know, who is to blame, and what can we do?* London: Taylor & Francis.

Rayner, C., Sheehan, M., and Barker, M. (1998) *Bullying at work 1998 research update conference: Proceedings.* Stafford, UK: Staffordshire University.

Roland, E., and Munthe, E. (eds.) (1989) *Bullying: An international perspective.* London: David Fulton.

Saunders, P., Huynh, A., and Goodman-Delahunty, J. (2007) Defining workplace bullying behaviour: Professional lay definitions of workplace bullying. *International Journal of Law and Psychiatry, 30,* 340–354.

Schat, A. C. H., Frone, M. R., and Kelloway, E. K. (2006) Prevalence of workplace aggression in the U.S. workforce: Findings from a national study. In E. K. Kelloway, J. Barling, and J. J. Hurrell Jr. (eds.), *Handbook of workplace violence* (pp. 47–90). Thousand Oaks, CA: Sage.

Schuster, B. (1996) Rejection, exclusion, and harassment at work and in schools. *European Psychologist, 1,* 293–317.

Skogstad, A., Einarsen, S., Torsheim, T., Aasland, M. S., and Hetland, H. (2007) The destructiveness of laissez-faire leadership behavior. *Journal of Occupational Health Psychology, 12,* 80–92.

Soares, A. (1999) La violence (in)visible au travail: Le cas du harcèlement psychologique au Québec [The [in]visible violence at work: The case of psychological harassment in Quebec]. In P. Molinar and V. Weber-Herve (eds.), *Violence et Travail* (pp. 191–201). Paris: CNAM.

Terpstra, D. E., and Baker, D D. (1991) Sexual harassment at work: The psychosocial issues. In M. J. Davidson and J. Earnshaw (eds.), *Vulnerable workers: Psychosocial and legal issues* (pp. 179–201). Chichester, UK: Wiley.

Thylefors, I. (1987) *Syndabockar: Om utstötning och mobbning i arbetslivet* [Scapegoats: On expulsion and bullying in working life]. Stockholm: Natur och Kultur.

UNISON (1997) *UNISON members' experience of bullying at work.* London: UNISON.

van de Vliert, E. (1998) Conflict and conflict management. In P. J. D. Drenth, H. Thierry, and C. J. J. Wolff (eds.), *Handbook of work and organizational psychology,* vol. 3: *Personnel psychology,* 2nd ed. (pp. 351–376). Hove, UK: Psychology Press.

Vartia, M. (1991) Bullying at workplaces. In S. Lehtinen, J. Rantanen, P. Juuti. A. Koskela, K. Lindström, P. Rehnström, and J. Saari (eds.), *Towards the 21st century: Proceedings from the International Symposium on Future Trends in the Changing Working Life* (pp. 131–135). Helsinki: Institute of Occupational Health.

—— (1996) The sources of bullying—psychological work environment and organizational climate. *European Journal of Work and Organizational Psychology, 5,* 203–214.

Williams, K. (1997) Social ostracism. In R. M. Kowalski (ed.), *Aversive interpersonal behaviors* (pp. 133–170). New York: Plenum.

Yamada, D. (2004) Crafting a legislative response to workplace bullying. *Employee Rights and Employment Policy Journal, 8,* 475–521.

Yildirim, A., and Yildirim, D. (2007) Mobbing in the workplace by peers and managers: Mobbing experienced by nurses working in healthcare facilities in Turkey and its effect on nurses. *Journal of Clinical Nursing, 16,* 1444–1453.

Yücetürk, E. E, and Öke, M. K. (2005) Mobbing and bullying: legal aspects related to workplace bullying in Turkey. *South-East Europe Review, 2/2005,* 61–70.

Zapf, D. (1999a) Mobbing in Organisationen. Ein Überblick zum Stand der Forschung [Mobbing in organisations. A state of the art review]. *Zeitschrift für Arbeits- und Organisationspsychologie, 43,* 1–25.

—— (1999b) Organizational, work group related and personal causes of mobbing/ bullying at work. *International Journal of Manpower, 20,* 70–85.

Zapf, D., and Einarsen, S. (eds.) (2001) Bullying in the workplace: Recent trends in research and practice. *European Journal of Work and Organizational Psychology, 10* (Special issue), 369–525.

—— (2005) Mobbing at work: Escalated conflicts in organisations. In S. Fox and P. E. Spector (eds.), *Counterproductive work behaviour: Investigations of actors and targets* (pp. 271–295). Washington, DC: American Psychological Association.

Zapf, D., and Gross, C. (2001) Conflict escalation and coping with workplace bullying: A replication and extension. *European Journal of Work and Organizational Psychology, 10,* 497–522.

Zapf, D., and Leymann, H. (eds.) (1996) Mobbing and victimization at work. *European Journal of Work and Organizational Psychology*, 5 (Special issue), 161–307.

Zapf, D., Knorz, C., and Kulla, M. (1996) On the relationship between mobbing factors, and job content, the social work environment and health outcomes. *European Journal of Work and Organizational Psychology*, 5, 215–237.

2

North American Perspectives on Hostile Behaviors and Bullying at Work

Loraleigh Keashly and Karen Jagatic

CONTENTS

Introduction

When we first began looking at persistently hostile and abusive behaviors at work in the early 1990s (what we called "emotional abuse"), the North American literature on aggressive behaviors at work had focused primarily on acts of physical aggression or violence (e.g., Fitzgerald, 1993), and there was little attention outside of the sexual harassment literature given to nonphysical forms of hostility, and even less exploring the enduring hostility that characterized abusive relationships at work. Indeed, in our own early research (Keashly et al., 1994, 1997), we looked for guidance to the domestic violence construct of emotional abuse and the workplace bullying and mobbing constructs that were being discussed in the developing European literature (e.g., Einarsen et al., 1994; Leymann, 1996; Vartia, 1996). Fortunately, North American attention to nonphysical forms of aggression and, specifically, persistent forms of aggression like bullying has increased dramatically, particularly in the past 10 years, as noted in terms of published research, dissertations, and conference presentations. Thus, while workplace aggression and workplace abuse literatures once operated somewhat in parallel, they are now more intimately connected and increasingly cohesive. This research interest has coincided with increased public interest as a result of high-profile legal cases (e.g., *Raess v. Doescher-Swaitek*, 2008) and popular books with such attention-grabbing titles as Robert Sutton's *The No Asshole Rule* (2007). This increased attention has been characterized by a number of features, including a profusion of constructs (Aquino and Lamertz, 2004; Griffin and Lopez, 2005; Ferris et al., 2007), which we will discuss shortly; explicit empirical attention to the relational nature of aggression that clearly has implications for understanding and examining workplace bullying (Aquino and Lamertz, 2004; Harlos and Pinder, 1999; Meares et al., 2004); recognition of the systemic or communal nature of persistent hostility generally and bullying in particular, implicating organizational culture and norms in the process (Keashly, 2001; Lutgen-Sandvik and Namie, 2009); and drawing from the revenge and retaliation literature (e.g., Bies and Tripp, 2005; Lutgen-Sandvik, 2006; Mitchell and Ambrose, 2007), increased attention to target responses that move beyond simply victimization. Thus, the North American literature on workplace aggression is varied, rich, and complex, but also somewhat fragmented.

To facilitate our own exploration through this rich landscape and to provide the findings most relevant to an understanding of workplace bullying, we focused specifically on research in which the behaviors and the relationships under investigation

1. Include nonphysical forms of hostility and aggression (i.e., this review is not about physical violence at work but about aggression in various forms)

2. Do not include the sexual harassment literature, because that is comprehensively discussed elsewhere (e.g., Cortina and Berdahl, 2008; Schneider et al., this volume)

3. Focus on actions that occur between such organizational insiders as coworkers, supervisors, and subordinates

4. Cause harm, broadly defined (i.e., physical, psychological, individual, and organizational)

Within these parameters, we are aware that our survey is not exhaustive of all relevant North American literature; rather, it is exemplary of it. We have been aided considerably in this synthesis by the presence of several recent reviews, meta-analyses, and edited volumes on aggression-related constructs (Aquino and Thau, 2009; Hershcovis and Barling, 2008; Kelloway et al., 2006; Lapierre et al., 2005; Lutgen-Sandvik et al., 2009; Martinko et al., 2006; Tepper, 2007), as well as recent nationwide surveys (Lutgen-Sandvik and Namie, 2009; Rospenda et al., 2009; Schat et al., 2006). For that which we do not discuss, we refer readers to these excellent resources.

We have chosen to organize this chapter around the core features of workplace bullying (Keashly, 1998), specifically time, intention, power differences, source, and norm violation. The direct applicability of North American research findings on hostile workplace behaviors to the discussion of workplace bullying will be dependent upon the consideration and assessment of these elements in this research. To facilitate this presentation, we first need to address the issue of the construct space as reflected in this literature. It is to this issue we now turn our attention.

The Labeling and Definitional Dilemma: Construct Profusion and Confusion

The variety of constructs falling under the rubric of "hostile workplace behaviors" has expanded significantly in the past decade as a result of a welcome increase in research attention (Keashly, 1998). As exciting as this increasing interest has been, the proliferation of constructs complicates identifying and building on relevant literature, which is necessary in order to develop a fuller picture of what is known and not known about hostile behaviors at work. Fortunately, we are pleased to see that increasingly, scholars are recognizing this issue and are developing frameworks for understanding the interconnections and conceptual overlaps among these various constructs (Aquino and Thau, 2009; Barling et al., 2009; Griffin and Lopez, 2005; Keashly and Jagatic, 2003).

In Table 2.1, we have provided a sampling of terms and definitions from North American research, including *harassment, workplace deviance, workplace incivility, workplace bullying*, and *employee mistreatment*. Whereas workplace deviance encompasses both hostile behaviors and other forms of deviant workplace behaviors (e.g., theft and sabotage), the other terms and definitions focus on types and combinations of hostile behaviors that occur at work to varying degrees. Beyond the variability in breadth and type of behaviors, these definitions differ most fundamentally in the acknowledgement of contextual features related to the "experience" of these behaviors (Hershcovis and Barling, 2008; Lutgen-Sandvik, 2006). More specifically, these definitions vary in the degree to which elements of time, intention, power differences, source, and norm violation are incorporated as central features of

TABLE 2.1

Select Definitions of Hostile Workplace Behaviors

Harassment (Brodsky, 1976)	"repeated and persistent attempts by one person to torment, wear down, frustrate, or get a reaction from another. It is treatment which persistently provokes, pressures, frightens, intimidates, or otherwise discomforts another person" (p. 2).
Workplace deviance (Robinson and Bennett, 1995)	"voluntary behavior that violates significant organizational norms and, in so doing, threatens the well-being of the organization or its members, or both" (p. 555).
Workplace aggression (Baron and Neuman, 1996)	"efforts by individuals to harm others with whom they work, or have worked, or the organizations in which they are currently, or were previously, employed. This harm-doing is intentional and includes psychological as well as physical injury" (p. 38).
Generalized workplace abuse (Richman et al., 1999)	"violations of workers' physical, psychological and/or professional integrities . . . nonsexual yet psychologically demeaning or discriminatory relationships" (p. 392).
Workplace incivility (Andersson and Pearson, 1999)	"low-intensity deviant behavior with ambiguous intent to harm the target, in violation of workplace norms for mutual respect. Uncivil behaviors are characteristically rude and discourteous, displaying a lack of regard for others" (p. 457).
Employee mistreatment (Meares et al., 2004)	"the interactional, distributive (lack of access to resources), procedural, or systemic abuse of employees that takes place at both interpersonal and institutional levels" (p. 6).
Abusive supervision (Tepper, 2000)	"subordinates' perceptions of the extent to which supervisors engage in the sustained display of hostile verbal and nonverbal behaviors, excluding physical contact" (p. 178).
Workplace bullying (Namie and Namie, 2000)	"is the deliberate, hurtful and repeated mistreatment of a Target (the recipient) by a bully (the perpetrator) that is driven by the bully's desire to control the Target . . . encompasses all types of mistreatment at work . . . as long as the actions have the effect, intended or not, of hurting the Target, if felt by the Target" (p. 17).
Leader bullying (Ferris et al., 2007)	"strategically selected tactics of influence by leaders designed to convey a particular image and place targets in a submissive, powerless position whereby they are more easily influenced and controlled, in order to achieve personal and/or organizational objectives" (p. 197).
Workplace bullying (Lutgen-Sandvik et al., 2007)	"a type of interpersonal aggression at work that goes beyond simple incivility and is marked by the characteristic features of frequency, intensity, duration and power disparity" (p. 837).
Emotional abuse at work (Keashly, 2001)	"interactions between organizational members that are characterized by repeated hostile verbal and nonverbal, often nonphysical behaviors directed at a person(s) such that the target's sense of him/herself as a competent worker and person is negatively affected" (p. 212).

TABLE 2.1 (continued)

Select Definitions of Hostile Workplace Behaviors

Social undermining (Duffy et al., 2002)	"behavior intended to hinder, over time, the ability to establish and maintain positive interpersonal relationships, work-related success, and favorable reputation" (p. 331).
Dysfunctional behavior (Griffin and Lopez, 2005)	"motivated behavior by an employee or group of employees that is intended to have negative consequences for another individual and/or group and/or organization itself" (p. 1000).
Workplace victimization (Aquino and Lamertz, 2004)	"an employee's perception of having been the target, either momentarily or over time, of emotionally, psychologically, or physically injurious actions by another organizational member with whom the target has an ongoing relationship" (p. 1023).
Workplace harassment (Bowling and Beehr, 2006)	"interpersonal behavior aimed at intentionally harming another employee in the workplace" (p. 998).
Disruptive practitioner behavior (Cawley, n.d.)	"a 'chronic' pattern of contentious, threatening, intractable, litigious behavior that deviates significantly from the cultural norm of the peer group, creating an atmosphere that interferes with the efficient function of the healthcare staff and the institution" (p. 5).
Emotional tyranny (Waldron, 2009)	"use of emotion by powerful organization members in a manner that is perceived to be destructive, controlling, unjust, and even cruel" (p. 9).

the construct they reflect (Keashly, 1998; Keashly and Harvey, 2006). These features are key elements of the definition of workplace bullying used in this volume.

Chapter Overview

We will begin with a brief discussion of the types of behaviors that are considered as hostile actions within the North American literature. We will then focus on the notion of hostile relationships as revealed by attention to single versus patterned and persistent hostile behaviors. This distinction has important implications for the degree to which research on understanding hostile incidents can be extended to understanding such hostile relationships as workplace bullying. We will then discuss research evidence regarding the other contextual elements of intent, power, source, and norm violation. Actor intent has been critical in many definitions yet is rarely assessed empirically (Aquino and Thau, 2009; Hershcovis and Barling, 2007; Tepper, 2007). Any discussion of relationship requires an exploration of the role of power in the experience. We will focus on moving from the simplistic assessment of relative power to include the dynamics of power. This discussion leads naturally into a fuller consideration of source, which we are expanding to include the nature of the actor and the target vis-à-vis one another (Aquino and Thau, 2009; Hershcovis and Barling, 2007). The element of norm violation will be examined by reviewing evidence on organizational culture and hostile work climate that suggests these behaviors and relationships may actually be consistent with norms rather than deviate from them (e.g., Ferris et al., 2007; Neuman and Baron, this volume; Opotow, 2006). Throughout our discussion, we will identify methodological and measurement challenges in the research on hostile workplace interactions. We will wrap up this chapter with a summary of the unique perspectives and ideas that the North American literature has to offer in the continuing effort to understand and ultimately address workplace bullying.

The Behavioral Domain: Actual Conduct

For quite some time, research attention focused almost exclusively on workplace violence (e.g., homicide, assault; Baron and Neuman, 1996; Kelloway et al., 2006). This attention has been fueled by several high-profile shootings in the workplace ("US: Eight dead in shooting rampage in the workplace"; IANSA, 2006) and schoolyard (e.g., Columbine High School, April 1999) and university (e.g., Virginia Tech, April 2007; Northern Illinois University, February 2008) killings that have occurred in the United States in recent years. It has been with a sense of shock that people have come to realize that workplaces, schools, and other public spaces are not the safe places they were assumed to be. Journalistic analyses of these incidents found evidence in some cases that the actors had been marginalized, harassed, or bullied by other workers and/or managers over an extended period of time prior to their deadly acts. For instance, in early April 1999, Pierre LeBrun shot and killed four coworkers and then shot himself at a public transportation facility in Ottawa, Ontario, Canada (McLaughlin, 2000). The coroner's inquest revealed that LeBrun had been mocked and teased over the years for stuttering and that the company had done little in response to his complaints. This identification of hostile mistreatment by coworkers as contributing to violent episodes has resulted in increased media attention to and public awareness of bullying as a workplace phenomenon.

There has been a growing recognition that physical violence was merely the tip of the iceberg concerning hostile behavior at work (Neuman and Baron, 1997). This expansion of the range and type of hostile behaviors studied was spurred by Baron and Neuman's application (1996) of Buss's conceptualization of human aggression to hostile workplace behaviors. Buss (1961) argued that aggressive behavior could be conceptualized along three dimensions: physical–verbal, active–passive, and direct–indirect. When fully crossed, there are eight categories of behavior. Several studies investigating this expanded domain (e.g., Greenberg and Barling, 1999; Neuman and Baron, 1997; Schat et al., 2006) have documented what has been long known by workers: most hostile behavior in the workplace is verbal, indirect, and passive. These types of behaviors have been variously labeled as "psychological aggression" (Barling, 1996), "emotional abuse" (Keashly, 1998), "generalized workplace abuse" (Richman et al., 1999), "workplace victimization" (Aquino and Lamertz, 2004), and "social undermining" (Duffy et al., 2002). As a result of the accumulating evidence that workplace hostility has many behavioral faces, most research on hostile behavior at work in the United States and Canada now examines a variety of behaviors reflecting these dimensions concurrently (e.g., Glomb, 2001; Keashly and Neuman, 2004; Rospenda et al., 2008; Schat et al., 2006). This broadened focus enhances the applicability of North American research on hostile workplace behaviors to workplace bullying.

Incidence of Hostile Workplace Behaviors

While there are myriad constructs and definitions regarding persistent forms of workplace hostility, they tend to be measured using similar items (Aquino and Thau, 2009; Barling et al., 2009). Thus, there is sufficient overlap in the behaviors studied to hazard sharing some incidence data regarding hostile behavior in American workplaces. The Northwest National Life Insurance survey (1993) is frequently cited as an illustration of the pervasiveness of aggression in American workplaces. Based on a sample of 600 full-time workers, the researchers concluded that one in four workers reported being harassed, threatened, or physically attacked on the job in the previous 12 months. More recently, a representative national survey of over 2,500 U.S. wage and salaried workers found that 6% experienced workplace violence and 41.4% experienced psychological aggression at work during the previous 12 months (Schat et al., 2006). Of particular relevance to workplace bullying, and in keeping with the European tradition of at least weekly exposure, 13% of the sample reported experiencing psychological aggression on a weekly basis. Two other nationwide surveys have focused specifically on assessing the incidence of generalized workplace harassment, a construct related to bullying (Rospenda et al., 2008; N = 2,151) and of workplace bullying specifically (Lutgen-Sandvik and Namie, 2009; N = 6,263). In their telephone survey of U.S. workers regarding harassment and discrimination experiences at work, Rospenda et al. (2008) report that 63% of respondents indicated at least one or more experiences of behaviors categorized as generalized workplace harassment in the previous 12 months, a notably higher rate than that cited by Schat et al. This discrepancy is likely a reflection of the number and type of behavioral items utilized in their measures: Schat et al.'s measure involved five items, three of which involved threats of physical contact or harm, which tend to be notably less frequent in workplaces, whereas the measure used by Rospenda et al. contained 10 items. Unlike the measure used by Schat et al. (2006), the scale anchors that Rospenda et al. used do not permit identification of exposure to persistent hostility (i.e., at least weekly).

The method used in these two surveys to assess exposure to hostile behaviors utilizes a list of behaviors given to respondents, who are then required to indicate the frequency with which each behavior had been directed at them (i.e., the behavioral exposure method). Another assessment method involves providing respondents with a definition of the construct and asking whether it describes their experience at work (i.e., self-labeling). Lutgen-Sandvik and Namie (2009) utilized the self-labeling method in their nationwide survey and found that 12.6% of their sample considered themselves as having been bullied in the past 12 months. Thus, by two different methods, similar rates of exposure to persistent aggression are found. To give a sense of the magnitude of these findings, assuming an American working population of approximately 140 million, these figures translate to 17–18 million American workers who have been bullied in the past year alone. This number expands even more when we consider those who witness bullying of others. Lutgen-Sandvik and Namie (2009) found that another 12.3% of their sample reported having seen others being bullied at work during their working career.

From an organizational perspective, Grubb et al. (2004) report that based on a nationally representative sample of U.S. organizations (N = 494), 24.5% reported some degree of bullying in the past year, with 7.1% indicating it occurred sometimes or often. These estimates are based on incidents about which a key informant, such as a human resources professional or company owner, would know. Given the evidence that employees do not tend to report such incidents formally (e.g., Keashly and Neuman, 2004; Keashly et al., 1994; Meares et al., 2004), this rate is likely an underestimate.

These data are gathered from individuals who have experienced these bullying behaviors in the workplace or who deal with those who have experienced aggression. Of particular interest in this field of research is incidence from the perspective of those engaging in hostility (i.e., enacted aggression). Greenberg and Barling (1999) found that over 75% of male workers sampled reported engaging in some form of psychologically aggressive behaviors against their fellow employees. Similar rates were found in Glomb's 2002 study of specific angry incidents in which participants reported on their own behavior. Using the self-labeling method, Lutgen-Sandvik and Namie (2009) report that only 0.4% of respondents self-identified as having bullied others. It is clear from these different operationalizations of incidence that hostile workplace behaviors of a variety of forms, as well as behaviors identified as bullying, are a notable part of many people's working experiences in North

America either as a direct target or as someone in the immediate organizational environment.

Discrete Event or Pattern: Hostile Relationships

The critical dimension in extending North American research findings to workplace bullying involves the issue of hostility as a discrete event versus a pattern of events. It is with this dimension that we found the greatest difference among concepts utilized in the literature. For example, unlike the workplace abuse literature, workplace aggression theorizing has tended to focus implicitly, if not explicitly, on understanding single incidents or aggregate levels of aggression (Glomb, 2002; Neuman and Baron, 1997; O'Leary-Kelly et al., 1996). Fortunately, recent research in this literature has begun to recognize and focus on persistent hostility with a particular actor or actors (e.g., Hershcovis and Barling, 2007; Tepper, 2007).

Repetition and Duration

Hostile workplace behaviors have been operationalized in the workplace aggression and workplace abuse literatures by their frequency over some specified time period. Thus, even though these literatures differ conceptually in how central the facet of time is to their particular constructs, they all measure *repetitive* hostile behaviors and link increased frequency to negative impact.

Although the facet of *duration* is included in definitions of workplace abuse and harassment (e.g., Duffy et al., 2002; Keashly, 2001; Tepper, 2007), it has essentially been ignored from a measurement perspective in both literatures. Duration appears primarily as a timeframe over which respondents are asked to assess the frequency of their experience (e.g., in the past six or twelve months, two years, or five years). With few exceptions (e.g., Neuman and Keashly, 2004; Schat et al., 2006), the anchors on the frequency scales are not often tied to such specific time referents as monthly, weekly, or daily occurrence. Such specificity would be necessary for determining duration of exposure.

The importance of duration of exposure to hostile behavior is highlighted by several recent studies. In a Web-based survey of people who self-identified as having been bullied at work, Namie (2000) reported the average exposure to bullying behaviors as 16.5 months. In recent work with a university sample, Keashly and Neuman (2008), in a study of faculty and staff, found that over a third of those who said they had been bullied indicated the bullying had been going on for over three years. This is a much longer time period than the one-year timeframe most of the workplace aggression and abuse research typically uses. In terms of the significance of duration of exposure, Kathy Rospenda, Judy Richman, and their colleagues have utilized a longitudinal methodology in their multiyear study of university employees (e.g., Rospenda et al., 2000, 2006), in which they assessed exposure to generalized workplace abuse and sexual harassment over a two-year period. Respondents were surveyed at two points, one year apart. Participants who reported exposure to generalized workplace abuse at both time periods (i.e., two years' duration) were found to be more likely to report one or more indicators of problem drinking at time 2, compared to participants who experienced abuse only at one time or not at all. Their more recent work utilizes a nationwide representative sample (Rospenda et al., 2009) and provides further evidence of the connection between duration of exposure (chronicity) and outcomes. Finally, in their study of women exposed to multiple forms of harassment, Schneider et al. (2000a) found that the longer the period over which the women experienced the behaviors, the more upset they became. Based on this evidence, it is clear that duration of exposure is a key and influential aspect of the experience of hostile behavior.

Patterning and Escalation

The frequency indicator as currently utilized also does not capture the element of a *pattern* of hostility. Doing so would require a recognition that behaviors may co-occur and may be sequenced, as opposed to occurring independently or in isolation (Berdahl and Moore, 2006; Glomb and Miner, 2002; Keashly and Rogers, 2001). For example, Barling (1996) suggests that just as with family violence, psychological aggression may well precede, and therefore be predictive of, physical violence in the workplace. Research assessing a variety of forms of harassment concurrently (e.g., Lim and Cortina, 2005; Richman et al., 1999; Rospenda et al., 2000) recognizes this idea of variety and the potential for a pattern. For example, Schneider et al. (2000a) found that employees who experienced one form of harassment were also likely to experience other forms of harassment. Lim and Cortina (2005) found that incivility tended to co-occur with gender harassment and sexual harassment. In fact, abusive *relationships* may only become clear when whole patterns of behavior are examined (Bassman, 1992). In the related literature on perceptions of justice, Lind (1997) argues that people's judgments of fair or unfair treatment are based on the patterning of everyday social interaction. As an illustration of the "whole being more than a sum of its parts" notion of hostile relationships, Keashly (2001) found that while the respondents could articulate specific and discrete examples of behavior, they struggled with communicating that their experience was more than this specific focus suggested. It was necessary for participants to describe the entire set of behaviors and their interrelationships.

A relatively simple indicator of patterning arising from frequency data involves counting the number of different events that have occurred to each respondent. A few authors have used this indicator to illustrate patterning (Glomb, 2002; Keashly and Jagatic, 2000; Keashly et al., 1994), although they have interpreted its meaning differently. For example, while we have defined it as a pattern or a variety of behaviors, Glomb (2002) views this indicator as a measure of severity and ties it to escalatory sequencing by examining both the number and the forms of behavior involved. She then classifies study respondents by whether they report psychologically aggressive behaviors, physically aggressive behaviors, or a combination of both, varying in severity. This measure of patterning is limited, however, because it does not allow the differentiation between situations of multiple behaviors occurring within a single incident versus multiple incidents. The distinction between single-incident and enduring hostile interactions is important because these are qualitatively different phenomena that may have different antecedents and outcomes (Glomb, 2002; Keashly, 1998).

Glomb's (2002) work forms the bridge to the fourth facet of aggression over time—*escalation*. While escalation is a key aspect of the conflict literature (see Keashly and Nowell, this volume), it has only relatively recently made its way into the research literature on workplace aggression (Baron and Neuman, 1996; Bassman and London, 1993; Glomb, 2001). Discussions of escalation carry implicit assumptions of dynamic interaction between an actor or actors and a target, mutuality of these actions, and increasing severity of behavior. Andersson and Pearson (1999) describe an incivility spiral in which parties start out with a retaliatory exchange of uncivil behaviors ("tit for tat") until one party perceives that the other's behavior directly threatens his or her identity (i.e., the tipping point). At this point, the parties engage in increasingly more coercive and severe behaviors with presumably greater risk of injury. Glomb and Miner (2002) make distinctions between independent aggressive acts and escalatory or sequenced aggression, with the latter focused on the movement from low-level, relatively minor behaviors to such more severe and damaging behaviors as physical violence. This innovation in the workplace aggression and abuse literatures links to some extent with European research on the evolving process of workplace bullying. This evolving process involves an escalation of the bully's behaviors from indirect and subtle behaviors, to more direct psychologically aggressive acts, and to ultimately severe psychological and physical violence (e.g., Einarsen, 1999).

While the discussion of dynamic escalatory processes in the workplace aggression literature focuses attention on the missing element of the relationship between the parties, it appears to imply the ability of both parties to *mutually* engage in negative behaviors. Glomb (2002) argues that actors become targets and targets become actors such that there are no pure actors or pure targets; rather, there are actor-targets. This interpretation is also reflected in research examining the elements of victims' personalities (e.g., negative affectivity and conflict management style) that may relate to their experiences of being treated aggressively, thus highlighting the victim's role as a precipitator of and active contributor to events (e.g., Aquino and Thau,

2009; Tepper, 2007). (We will discuss this developing focus on the actor-target dynamics a bit later.) This assumption of mutuality in the escalatory process is consistent with the literature on escalated conflict where at some point, regardless of who initiated the actions, the actions become increasingly reciprocated and occur at more severe levels (e.g., Rubin et al., 1994). However, this interpretation stands in sharp contrast to the construct of workplace bullying, which is often characterized as involving a clearly identified actor (bully) and an identifiable target (victim). Indeed, it could be argued that what is being described in the escalation of aggression is the escalation of a conflict, rather than a situation of the escalatory processes involved in workplace bullying. In this latter situation, the actor is primarily the provocateur, whereas the target is attempting to defend against or deter further behavior (coping and management behaviors), rather than to initiate or retaliate (see Einarsen, 1999). This focus on mutuality may well be an artifact of having focused on anger-related or reactive aggression that is reciprocal in nature. This mutuality is in contrast to more proactive and purposeful instigating of aggression, which tends to be more unidirectional and which is more consistent with descriptions of predatory workplace bullying (Opotow, 2006).

Intent (Harm by Design)

A cornerstone of the definition of workplace aggression is that the behavior or behaviors must be intended to cause harm, thus distinguishing them from behaviors that may cause harm but were not intended to do so, that is, that were accidental (Neuman and Baron, 1997). In the current definition of workplace bullying, intent to harm is revealed in the reference to deliberate or premeditated action. However, other authors suggest that intent is not a defining element. To the extent that there are s[...] to these behaviors or relationships, as with ra[...] may be conforming to broader norms withou[...] 1998, 2001; Richman et al., 2001; Wright and Sn[...] gests that some elements of abusive supervision[...] desire to harm but are more a result of indiff[...] Keashly (1998), in a preliminary study of gra[...] hostility, found that students tended to repo[...] neglect rather than actively hostile behaviors. [...] and others (2007) suggest that politically skill[...] use bullying not to harm but to influence low-r[...] form and perform. This is not to suggest the be[...] cause harm; rather, the intent to harm is not clear or is nonexistent. Indeed, Andersson and Pearson (1999) distinguish workplace incivility from workplace aggression by arguing they are referring to behavior for which intent is ambiguous. From the perspective of the targets, Keashly (2001) found that intent did not figure prominently in their experience of feeling abused.

While this conceptual debate continues (for more detailed discussion, see Aquino and Thau, 2009; Hershcovis and Barling, 2007; Pearson et al., 2005), the issue is moot at the measurement level. Rarely do any of the workplace aggression and abuse studies measure either actual or perceived intent. It appears as though intent is considered implicit in the particular behaviors studied. However, it could be argued that many of the behaviors encompassed under workplace aggression and abuse are ambiguous for intent (Aquino and Thau, 2009). Rather, as was noted earlier regarding severity, it is the context in which the behaviors occur (e.g., patterning, duration, relationship to and history with actor, and organizational norms) that affects attribution of intent (Hershcovis and Barling, 2008; Keashly, 2001; Keashly et al., 1994). This attribution, in turn, affects both researchers' and workers' assessments as to whether the behavior is experienced as hostile. The concept of threat appraisal of stressors, from the organizational stress literature (e.g., Barling, 1996; Lazarus and Folkman, 1984), is extremely useful in helping to discern which factors relate to perception of intent to harm and how this attribution subsequently relates to the subjective experience of the hostile event and the outcomes of that experience. Regarding the latter connection, Keashly and Rogers (2001) found that those incidents in which the actor was perceived as intending harm were evaluated as more threatening, and therefore more hostile, than those where no intention was perceived. This appraisal was subsequently related to greater perceived stress.

Another important aspect to consider when discussing intent is the distinction between intent and motive (Fox and Spector, 2005). While many behaviors may be intended to harm someone, the motive for that harm may be different. For example, Neuman and Baron (1997) highlight the distinction made by Buss (1961), namely, the distinction between reactive aggression (annoyance motivated, anger related) and instrumental aggression (proactive or incentive motivated). In the former, the goal is to harm the person. In the latter, harming the person is a means of obtaining something of value, such as promotions, or valued self-presentational goals, such as heightened self-image. The focus in popular writing on the psychopathology and personality of bullies (e.g., that they are power addicted or controlling) suggests that such hostile treatment is viewed as instrumental in nature and, in some cases, fun for the actor (see, e.g., Hornstein, 1995; Namie and Namie, 2009; Sutton, 2007). Further, references to these hostile behaviors as exercises of power (e.g., Namie and Namie, 2009; Cortina et al., 2001; Lewis and Zare, 1999), and as efforts to control and create target dependency (e.g., Bassman, 1992; Hornstein, 1995) are consistent with an instrumental aggression perspective. It may be that workplace bullying is more illustrative of instrumental and proactive aggression and would be suggestive of different antecedents than might be true for reactive aggression. There is evidence that different motives for aggression may affect targets' experiences.

Crossley (2009) empirically examined how an actor's perceived motive (malice or greed) affected the targets' emotional and behavioral reactions to social undermining. Perceived malice (intentional desire to harm another out of ill will or hatred) was associated with anger and subsequent revenge or avoidance actions, whereas perceived greed (instrumental desire for gain at expense of another) resulted in sympathy and attempts at reconciliation. Just as targets consider underlying motive, researchers need to consider this as well. Understanding the motives of actors has implications for how to intervene and address such behaviors.

Power

In the North American literature, power has not explicitly been a key aspect of definitions of hostile workplace behavior, with the exception of those concerned with supervisor behavior (e.g., Ashforth, 1997; Hornstein, 1995; Tepper, 2000). However, power is embedded in the manner in which authors discuss actor motive as well as the vulnerability or protective factors for targets. As suggested in the earlier discussion of motive and instrumental aggression, one conceptualization of power is the unwanted exertion of influence over another (i.e., coercive action; Tedeschi and Felson, 1994). Thus, hostile behaviors are used to demonstrate the actor's ability to control others (e.g., Ferris et al., 2007). Supportive of the notion of power as control, almost 60% of the targets responding to a Web-based survey of workplace bullying indicated that they were being bullied because they refused to be subservient (Namie, 2000). This conceptualization of power rarely appears in definitions of the constructs related to hostile workplace behaviors, but it does have a place in the theoretical discussions of why some people behave in these ways (for a fuller discussion of antecedents, see Neuman and Baron, this volume; Martinko et al., 2006).

Another manifestation of power is much more concrete and is focused on identifying the relative power differential between the actor and the target within the organizational and social contexts as a potential vulnerability or as a protective factor. In the workplace aggression and abuse literatures, power has typically been operationalized as organizational position and occasionally as gender and race/ethnicity (e.g., Aquino, 2000; Cortina et al., 2001; Keashly, 1998; Schneider et al., 2000b). The proposition is that those in low-power positions (subordinates, entry-level employees, women) are more vulnerable to being the target of hostile behaviors than those in higher power positions (supervisors, bosses, men). Conversely, those in high-power positions are hypothesized as more likely to be the instigators of hostile workplace behaviors. Thus, by virtue of position and the access to resources and influence that the position entails, the potential exists for the abuse of power (Bassman, 1992; Cortina et al., 2001). Support for this proposition is mixed, suggesting that operationalizing power in terms of organizational position and gender may be too limiting. Regarding gender, Cortina (2008) and Rospenda and colleagues (2000) report that women were more likely to be targets of workplace hostility. Conversely, other research (Keashly et al., 1994, 1997; Lutgen-Sandvik and Namie, 2009; Richman et al., 1999) finds that men and women report similar rates of exposure to these behaviors.

Regarding organizational position, some studies found that bosses are identified as actors more often than not (e.g., Keashly et al., 1994; Lutgen-Sandvik and Namie, 2009). However, several studies have found that coworkers are the most frequent source of hostile workplace behaviors (e.g., Cortina et al., 2001; Keashly and Rogers, 2001; Keashly and Neuman, 2004; Schat et al., 2006). Aquino (2000), in a study of coworker victimization, examined power in terms of relative and absolute position in the organization. Higher status workers were just as likely to be victimized by their coworkers as lower status workers. Some studies have even identified subordinates as active perpetrators (Inness et al., 2008; Keashly et al., 1996; Namie, 2000).

The existence of lateral (coworker) and upward (subordinate) aggression challenges the conceptualization and operationalization of power as primarily a structural quality. In their treatise on bases of social power, French and Raven (1959) identified at least five bases of social power: legitimate, reward, coercion, referent, and expert. The first three coincide easily with hierarchical sources of power; that is, bosses are considered more powerful by virtue of their (legitimate) organizational position and their control of resources that are meaningful to employees. Expert power refers to the influence that arises from someone having specialized knowledge or experience in a particular area, which is something coworkers and subordinates certainly can manifest. Referent power is influence grounded in one's likeability (we allow friends and loved ones to influence us) and connections to and positions within various social networks in the organization. Thus, lower status individuals may enjoy positions of influence by virtue of their personality and their connections. Drawing on this work on social power and social networks, Lamertz and Aquino (2004) developed and found support for a social stratification model of workplace victimization wherein "different social status/power conditions lead to differences in the way social actors experience protection from and vulnerability to the capricious, abusive or potentially harmful behaviors emanating from others in the social system" (p. 798).

Lamertz and Aquino's work represents a more nuanced articulation of power within the context of hostile workplace relationships. There is a recognition of power as a process of dependency, creating a dominant–subordinate structure in the relationship (Tepper et al., 2009). Bassman (1992) suggests "one common thread in all abusive relationships is the element of dependency. The abuser controls some important resources in the victim's life, and the victim is therefore dependent on the abuser" (p. 2). Tepper (2000) echoes this theme in his work on abusive supervisors when he theorizes that targets remain in these abusive relationships because of economic dependence, learned helplessness, and fear of the unknown. Reduced job mobility has been linked to intensification of the experience of abusive supervision (Tepper, 2000). In essence, worker dependency on others creates a situation for power (and the misuse of that power in the form of hostile treatment) to become an issue (Keashly, 2001).

The notion of the central role of power as dependency in the experience of workplace abuse is further illustrated in analogies used to portray the experience. In Keashly's interview study with targets (2001), respondents used analogies to feeling treated as a child or feeling diminished or discounted to describe their experiences in hostile interactions. Hornstein (1995) speaks of how "brutal bosses" infantilize their subordinates by controlling their behavior off the job. These are characteristic feelings and features of dominant–subordinate relationships.

While we have some inkling from the research examining organizational position and gender as target and instigator characteristics that power is an important feature of hostile workplace experiences, much of the work has been conceptual in nature. Though the nature and types of effects on targets reveal a diminishment and disempowerment pattern (Keashly and Jagatic, 2003), there is little systematic empirical documentation of the process by which target dependencies or vulnerabilities are formed and utilized by the actors. Some preliminary suggestions of how this process may work come from interviews with targets. Keashly (2001) found that behaviors from more powerful others affected targets by reducing their ability to respond. For example, if the actor had important information the target required to complete a task, the target felt less willing to confront the actor for fear that critical information would be withheld. Utilizing Emerson's power-dependence theory (1972), Tepper et al. (2009) examined the circumstances under which a more dependent and hence less powerful subordinate will retaliate against a more powerful, abusive supervisor. Specifically, they argued that a strong intention to quit makes subordinates less dependent on their supervisors. As a result, they will experience less constraint on their ability to act in their own interests and will be more likely to retaliate against an abusive supervisor and the organization by engaging in deviant behavior. The researchers' findings supported this hypothesis in two different samples. While preliminary, evidence of the process of dependency and its relationship to effects is critical for making the distinction between aggression or conflict between equals and workplace bullying, as noted in the European literature (Einarsen, 1999). Further, empirical documentation of the process of abuse of power and disempowerment will be critical in identifying prevention and intervention efforts directed at reducing the presence of these types of behaviors and relationships.

The nuances and dynamics of power point directly to the relationship between the actor and the target. What is becoming clear is that who the actor and target are as individuals, with respect to each other and in the larger organizational context, will have a profound impact on their behaviors and experiences. Aquino and Lamertz (2004) propose a relational model of workplace victimization, arguing that the nature of the dyadic relations creates a context that increases the likelihood of workplace victimization occurring over the course of ongoing workplace interactions. Of particular relevance to workplace bullying is their identification of the types of dyads that are likely to manifest institutionalized (persistent, enduring, and patterned) aggression as opposed to episodic (occasional) aggression. Specifically, they suggest that the pairing of a dominating perpetrator with a submissive victim or a reactive perpetrator with a provocative victim will result in behavior that might be considered workplace bullying. The former is consistent with Einarsen's notions of predatory bullying (1999), and the latter with dispute-related bullying. While these dyadic combinations may lay the groundwork for bullying, the (im)balance of social power in the dyad and the presence and involvement of others who have relationships to both parties can influence whether and how bullying actually occurs (Lamertz and Aquino, 2004). What impresses us about this approach is that it draws attention to how bullying is molded by both parties and those around them (Hodson et al., 2006; Venkataramani and Dalal, 2007). It also provides a perspective that envisions the target as not simply a helpless "victim" but, rather, as an actor capable of influencing and challenging the treatment he or she receives (Lutgen-Sandvik, 2006). Such a model provides guidance for future empirical research by focusing on how bullying actually evolves, which is something that has been relatively unexamined in both the North American and European literatures. To the extent there is empirical support, this model could be useful in assessing a potential situation and possible places for taking action to prevent or address bullying.

Violation of Norms

The theme that abusive behaviors or patterns of behavior violate some set of understood norms is pervasive in much of the workplace aggression literature. For example, Andersson and Pearson (1999) define uncivil behaviors as those that are in violation of workplace norms for mutual respect. Workplace deviance, which includes interpersonal aggression, refers to behaviors that violate norms (Robinson and Bennett, 2000). Hornstein's belief (1995) is that "brutal" bosses violate norms of decency and civility. However, other authors explicitly note that some of these behaviors may not be deviant from organizational norms at all (Keashly, 2001; Tepper, 2000). As noted earlier in the discussion of intent, to the extent that these behaviors are systemic in nature, they are by definition conforming rather than deviating from organizational and social norms. Neuman and Baron (this volume) and Opotow (2006) note that norms that typically counteract aggressive behavior either do not apply or apply weakly in the case of workplace bullying. In their examination of global organizations, Harvey and his colleagues (Harvey et al., 2007; Heames and Harvey, 2006) postulate that given the diverse nature of the workforce, miscommunication arises as to what behaviors are acceptable in the workplace because norms vary among cultures. As a result, a broader sense of understanding what is acceptable behavior (i.e., norms) is not established, providing an opportunity for bullying to occur unchecked. The organizational culture evidence (e.g., Andersson and Pearson, 1999; Aquino and Lamertz, 2004; Keashly and Jagatic, 2000) and the work on organizational conditions as antecedents to aggression appear to be supportive of this claim.

Organizational Culture

The overall workplace bullying literature continues to support the notion that an organization's culture and related climate play a large and important role in the manifestation of hostile behaviors at work (e.g., Aquino and Lamertz, 2004; Harvey et al., 2007; Martinko et al., 2006; O'Leary-Kelly et al., 1996). However, there are still relatively few empirical studies in which organizational culture (or climate) is measured and the relationship tested among its dimensions and workplace hostility. In representative surveys of Michigan workers, Keashly and her colleagues (Burnazi et al., 2005; Keashly and Jagatic, 2000) documented links between indicators of organizational culture and climate and exposure to aggression. Specifically, they found that higher rates of hostile behaviors were reported in organizations in which respondents perceived that employee involvement was not facilitated, morale was low, teamwork was not encouraged, and supervision was problematic (Keashly and Jagatic, 2000). Burnazi and others (2005) found that organizational climate, as manifested in such higher involvement work practices as employee involvement and empowerment, was negatively related to experienced workplace aggression.

Taking a somewhat different tack, Schneider and others (2000a) conceived of a hostile workplace climate as one characterized by multiple forms of harassment as assessed from both target and bystander perspectives (see Robinson and O'Leary-Kelly, 1998, for a similar approach). They found that female employees exposed to multiple types of harassment reported more negative outcomes than those reporting fewer types of harassment. Further, Cortina and others (2001) found that an organizational climate characterized by fair interpersonal treatment was negatively related to personal experiences of rude, uncivil behavior.

While the foregoing limited research suggests that a relationship between overall organizational culture and hostile workplace behaviors may indeed exist, the direction of the relationship is unclear. Does something about the organization's culture lead to greater incidence of hostility, or does a prevalence of abuse poison an organization's culture? The current data on the role of organizational culture and bullying behavior are correlational (Keashly and Harvey, 2005) and indicate a relationship, but longitudinal research is necessary to explore whether the relationship is bidirectional, as suggested by some researchers (Glomb, 2001; Keashly and Harvey, 2005). The current North American literature offers several theoretical explanations for organizational culture as both a cause and an effect of hostile workplace behavior and relationships.

One argument in favor of organizational culture as supportive of hostile work behavior comes from Brodsky (1976), who suggests that hostile behaviors may be a result of the belief in industrial society that "workers are most productive when subjected to the goad or fear of harassment" (p. 145). From this perspective, harassment is viewed by management as functional, and perhaps necessary, to achieve productivity and acceptable performance from employees. Aquino and Lamertz (2004) suggest that norms that develop in an organization in which harassment is deemed necessary for worker motivation legitimize and provide justification for harmful behaviors. Consistent with this perspective, Ferris et al. (2007) have argued that "strategic" leader bullying can have positive consequences, including temporary increases in productivity of both targets and witnesses, voluntary attrition of underperforming employees, and increased power for the bully-leader. It is not clear whether the employees who leave organizations as they flee leader bullying are truly underperforming, and whether the temporary increases in production are of a desirable quality or merely an attempt to ward off negative attention from a bully. As suggested by the revenge and retaliation literature (e.g., Bies and Tripp, 2005; Lawrence and Robinson, 2007), employees may not react favorably to such efforts. Further, high turnover rates resulting from employees leaving the organization may mitigate any positive effects achieved in coercion of targets. Bullying may also result in difficulty in attracting and retaining new employees as organizations develop reputations as supportive of a bullying culture (Heames and Harvey, 2006).

The second influence of organizational culture on hostile work behavior involves what Brodsky refers to as a "sense of permission to harass" (1976, p. 84). Brodsky suggests that harassment at work cannot occur without the direct or indirect agreement of management. This argument thus includes the involvement of individuals within the broader organization, apart from the actor and target involved in the abusive situations (Aquino and Douglas, 2003; Glomb and Liao, 2003). Specifically, the responses of these others to bullying behaviors, or their lack of a response, communicate the organization's (in)tolerance for such treatment (Ferris, 2004; Harvey et al., 2007; Heames and Harvey, 2006). In effect, an organization can enable bullying (Ferris, 2004; Keashly and Harvey, 2005). Andersson and Pearson's notion of incivility spirals or cascades (1999) provides a possible explanation for how "permission to harass" may develop. They argue that a "climate of informality" pervades some organizations and eventually encourages employees to behave toward one another in a disrespectful manner. In such climates, incivility spirals may form in which one employee treats another employee disrespectfully in response to the disrespectful behavior to which he or she had been subjected. Unless there is intervention or action taken to interrupt these spirals and thus communicate the organization's lack of tolerance for these behaviors, secondary incivility spirals may develop as other employees who observe the initial disrespectful behavior begin treating others similarly. Ultimately, an organization becomes uncivil at a point in which a number of these incivility spirals occur simultaneously and perpetuate each other, causing employees to believe that the organization itself disrespects its employees. Recent empirical research provides support for this "aggression is contagious" notion (Bowling and Beehr, 2006; Glomb and Liao, 2003).

What is clear from this discussion of the relationship between organizational culture and hostile workplace behavior is that these behaviors may not be deviant from workplace norms. Rather, they may be consistent with them. This conclusion challenges definitions of workplace hostility that label it as deviant behavior to provide evidence of the existence of the norms that are being contravened (Wright and Smye, 1996). It also highlights that addressing workplace bullying and other hostility may require fundamental organizational culture change (Keashly and Neuman, 2004).

By defining aggression and, by extension, bullying as relational and potentially systemic in nature, we are advocating for an interactionist perspective. That is, aggression generally, but particularly persistent aggression such as bullying, arises from a complex interaction of actor, target, and situational and structural forces. Workplace bullying is a hostile relationship that *occurs* and *is maintained* through a variety of mechanisms that we can identify as sources of the condition. While a full discussion of research on these sources is beyond the scope of this chapter, we have discussed the connection of status and power to aggression as well as the potential role of norms and organizational culture. For information on the nature of antecedents associated with aggression, we refer readers to chapters in this volume and reviews and meta-analyses in the literature on aggression-related constructs (Aquino and Thau, 2009; Hershcovis and Barling, 2008; Kelloway et al., 2006; Keashly and Harvey, 2006; Lapierre et al., 2005; Lutgen-Sandvik et al., 2009; Martinko et al., 2006; Tepper, 2007).

Measurement and Methods of Studying Hostile Relationships: A Caution

The direct relevance and usefulness of the North American literature on workplace aggression to an understanding of the nature, dynamics, and effects of workplace bullying hinge upon the methods and measurements utilized. In the North American literature, the reliance on aggregate measures of hostility does not allow for the differentiation between hostile behaviors on the part of different actors or multiple behaviors on the part of a single actor. Indeed, much of workplace aggression and abuse research utilizing behavioral checklists does not specifically focus on aggressive behaviors from particular actors (e.g., Brotheridge and Lee, 2006; Cortina et al., 2001; Keashly et al., 1997). This makes it difficult to operationalize facets of pattern and escalation (core aspects of workplace bullying), which require the establishment and assessment of a specific relationship between actors and targets. In addition, it is not unreasonable to suppose that a variety of behaviors coming from one actor may be experienced differently than behaviors coming from a variety of actors. Indeed, simply measuring frequency without relating it to a specific actor or actors is measuring something much different, namely, hostile workplace climate (Schneider et al., 2000a) or ambient hostility (Robinson and O'Leary-Kelly, 1998). Fortunately, as we noted earlier, recent workplace aggression and bullying studies have embraced the relational nature of aggression, and aggressive behaviors are examined within the context of the source of the aggression vis-à-vis the target (e.g., Grandey et al., 2007; Greenberg and Barling, 1999; Inness et al., 2008; Keashly and Neuman, 2004).

To achieve a better understanding of these more specific hostile relationships and, indeed, the experience of hostile behaviors generally, more qualitative methods, such as interviews focused on the details of specific incidents and relationships, are more appropriate than the broader survey methodology (Glomb, 2002; Keashly, 1998; Lutgen-Sandvik, 2006). For example, Keashly (2001) used semistructured interviews to probe targets about their experiences with abusive colleagues. She focused on identifying those features of the hostile interaction that lead targets to label their experiences as abusive. Among other findings, this study revealed that prior history with the actor (i.e., an enduring relationship) was key in targets' labeling of their experience. Interestingly, respondents considered the actor's history with other targets in the workplace in their assessment of their experience as well. This finding is consistent with Name's survey of targets (2000), which found that 77% of actors were identified as having also harassed others. Further, Schneider and others (2000a) found that having been a bystander to others' harassment experiences was related to a degree of upset at one's own experience of harassment. This aspect of actor history with other targets highlights the importance of exploring more fully the persons involved in the behaviors. As we noted earlier, while more work of this nature is being focused on targets, little attention has been paid to the actor's perspective and interpretations here. This lack of attention appears to be an issue in the European literature as well (Einarsen et al., 2003).

With regard to specific incident descriptions, Keashly and Rogers (2001) had respondents complete checklists on each of five working days. Respondents were then interviewed in detail about one or two particular incidents. Interviewees were asked about the status and gender of the actor, the history with the actor, and the importance and salience of the situation, as well as how stressful the situation was. This information permitted the researchers to explore the role that features of the hostile relationship play in respondents' appraisals of these incidents as threatening and stressful. Using a similar method, Glomb (2002) had participants discuss actual "angry incidents" in which they were *involved* as either the target or the actor. They were questioned about personal characteristics, job attitudes and experience, and interpersonal conflict, anger, and aggression. Individuals were then asked to select a particular incident that affected them the most, the "specific aggressive incident," and to speak in detail about the people involved in their actions and the outcomes of these actions. The information garnered suggests a progression of hostility such that severe behaviors are preceded by "less severe" behaviors. These results would not have been revealed by sole reliance on the more aggregate measure of hostile workplace behaviors.

It is also important to address the notion of severity or intensity of the hostile behaviors. A consensus seems to have developed that these various forms of hostile behaviors arrange themselves along a continuum of increasing severity. For example, behaviors captured under the rubrics of "emotional abuse," "psychological aggression," and "incivility" are often characterized as low intensity or low level (see Andersson and Pearson, 1999; Glomb, 2002) relative to such forms of physical violence as assault, rape, and homicide, which are considered extreme forms. Recently, several authors have proposed that aggression is a continuum anchored on one end by incivility and on the other end by physical violence, with harassment and bullying falling between these two extremes (Namie, 2003; Lutgen-Sandvik, 2004; Tracy et al., 2005). The notion of a continuum is premised on the assumption that these forms reflect increasing intensity or severity and that if low-intensity behaviors are left unaddressed, they could lead to increasingly severe behaviors. While we do not necessarily disagree with this depiction, this assignment of varying degrees of severity is rarely based on empirical evidence (for exceptions, see Glomb and Miner, 2002; Meglich-Sespico, 2006; Rogers, 1998).

Glomb and Miner (2002) included ratings of severity of each of 18 behaviors when assessing patterning of hostile behaviors within computer models. The ranking of these behaviors by severity ratings was consistent with what has been assumed in the literature. More specifically, talking behind someone's back was rated as less severe, whereas angry gestures, insulting or criticizing, and yelling were rated as moderately severe. Making threats and physical assaults were rated as most severe. In another study, Rogers (1998) had a group of people with work experience rate 49 physically, emotionally, and sexually abusive behaviors on a variety of dimensions, including severity. She found that while all emotionally abusive and sexually harassing behaviors tended to be rated as less severe than the physical behaviors of rape, homicide, and assault; some emotionally abusive items had similar ratings to some physical items (e.g., being publicly belittled, verbally insulted versus bumped into, grabbed, and subjected to threatening gestures). Thus, severity did vary across types, though it also varied within each type of hostile behavior. Further, these types of behaviors overlapped in terms of severity ratings. Within the realm of psychologically aggressive behaviors, Meglich-Sespico (2006) found personal-related behaviors (e.g., gossip and rumors) were rated as more severe and harassing than task-related behaviors (e.g., interfering with an assignment and withholding information). Indeed, while there may be a normative aspect to severity (what we as a group, occupation, organization, or society see as appropriate), the target's perception of severity is influenced by such contextual factors as actor power (e.g., Lamertz and Aquino, 2004; Keashly and Neuman, 2002; Meglich-Sespico, 2006) and the target's access to personal and organizational resources for coping and responding (e.g., Schat and Kelloway, 2003). Taken together, these studies reveal a much more complex and nuanced understanding of severity than has been assumed thus far in discussions of workplace aggression and abuse. Given that workplace bullying is cast as a severe form of workplace hostility, it is important that research empirically assess severity and its link to actors' and targets' interpretations and experiences of these hostile relationships.

Conclusion

In this chapter, we have sought to identify and share relevant concepts and research from the North American literatures on workplace aggression and abuse. As an overall assessment, there are several findings that research on this side of the pond has to offer discussions on workplace bullying. First, consideration of a variety of hostile behaviors and a range of effects has now become the mainstay in the literature, providing more points of behavioral comparison with the other global literatures and creating opportunities to explore the inherently multivariate nature of this phenomenon. Second, the recognition of the relational and communal nature of workplace aggression has led to exploration of the dynamic process of bullying development and the contextual factors that support and thus can mitigate such hostile and damaging behaviors. Third, the utilization of such diverse measures and methodologies as surveys, interviews, longitudinal studies, case studies, and computer modeling bodes well for more accurate and precise data, as well as the development and testing of more complex models of these hostile behaviors.

In addition to the existing research findings, we need more research from the perspectives of the expanding net of stakeholders. While our understanding of the immediate target's experiences and responses is developing nicely, we need to know more about the actors not so much in terms of "who they are" but, rather, how they understand and interpret their own behaviors (e.g., motive). We also need to focus on those who witness or observe and others who are connected directly and indirectly to those involved (e.g., family, friends, upper management, health professionals, and the broader community). Although the North American literature on workplace aggression is beginning to examine these stakeholders more closely, there is still much work to be done about the roles they play and how they play them, as well as how these different players interconnect. Lamertz and Aquino's model (2004) is a good beginning and needs to be carried further. There also needs to be development and systematic evaluation of the myriad actions proposed to ameliorate bullying, and these efforts need to be grounded in our understanding of the sources, contexts, and mechanisms of bullying.

In conclusion, the growing evidence is that hostile, bullying behaviors are more frequent and, if persistent, can have profound negative implications for individual, organizational, familial, and community well-being. This may be one of those rare situations in which the research can help drive public opinion, rather than the other way around. The efforts of several activists such as Gary and Ruth Namie (see Namie et al., this volume), through the popular literature, the media, unions, and community trainings, are raising public consciousness of these behaviors and communicating the message of their potential for great damage. We hope our review of the North American literature will contribute to the broader global conversation and efforts needed to understand and address this devastating phenomenon.

References

Andersson, L. M., and Pearson, C. M. (1999) Tit for tat? The spiraling effect of incivility in the workplace. *Academy of Management Review, 24,* 452–471.

Aquino, K. (2000) Structural and individual determinants of workplace victimization: The effects of hierarchical status and conflict management style. *Journal of Management, 26,* 171–193.

Aquino, K., and Douglas, S. (2003) Identity threat and antisocial behavior in organizations: The moderating effects of individual differences, aggressive modeling, and hierarchical status. *Organizational Behavior and Human Decision Processes, 90,* 195–208.

Aquino, K., and Lamertz, K. (2004) A relational model of workplace victimization: Social roles and patterns of victimization in dyadic relationships. *Journal of Applied Psychology, 89* (6), 1023–1034.

Aquino, K., and Thau, S. (2009) Workplace victimization: Aggression from the target's perspective. *Annual Review of Psychology, 60,* 717–741.

Ashforth, B. (1997) Petty tyranny in organizations: A preliminary examination of antecedents and consequences. *Canadian Journal of Administrative Sciences, 14,* 126–140.

Barling, J. (1996) The prediction, psychological experience, and consequences of workplace violence. In G. VandenBos and E. Q. Bulatao (eds.), *Violence on the job: Identifying risks and developing solutions* (pp. 29–50). Washington, DC: American Psychological Association.

Barling, J., Dupre, K. E., and Kelloway, E. K. (2009) Predicting workplace aggression and violence. *Annual Review of Psychology*, 60, 671–692.

Baron, R. A., and Neuman, J. H. (1996) Workplace violence and workplace aggression: Evidence on their relative frequency and potential causes. *Aggressive Behavior*, 22, 161–173.

Bassman, E. (1992) *Abuse in the workplace*. Westport, CT: Quorum Books.

Bassman, E., and London, M. (1993) Abusive managerial behavior. *Leadership and Organization Development Journal*, 14 (2), 18–24.

Berdahl, J. L., and Moore, C. (2006) Workplace harassment: Double jeopardy for minority women. *Journal of Applied Psychology*, 91 (2), 426–436.

Bies, R. J., and Tripp, T. M. (2005) The study of revenge in the workplace: Conceptual, ideological, and empirical issues. In S. Fox and P. E. Spector (eds.), *Counterproductive work behavior: Investigations of actors and targets* (pp. 65–81). Washington, DC: American Psychological Association.

Bowling, N. A., and Beehr, T. A. (2006) Workplace harassment from the victim's perspective: A theoretical model and meta-analysis. *Journal of Applied Psychology*, 91 (5), 997–1012.

Brodsky, C. M. (1976) *The harassed worker*. Lexington, MA: D. C. Heath.

Brotheridge, C. M., and Lee, R. T. (2006) Examining the relationship between the perceived work environment and workplace bullying. *Canadian Journal of Community Mental Health*, 25 (2), 31–44.

Burnazi, L., Keashly, L., and Neuman, J. H. (2005) Aggression revisited: Prevalence, antecedents, and outcomes. Paper presented as part of the symposium on Workplace Bullying and Mistreatment, Academy of Management, Oahu, HI.

Buss, A. H. (1961) *The psychology of aggression*. New York: Wiley.

Cawley, P. J. (n.d.) "Disruptive clinicians": Dealing with low performers. Presentation for the Center for Clinical Effectiveness and Patient Safety, Medical University of South Carolina, Charleston.

Coombs, W. T., and Holladay, S. J. (2004) Understanding the aggressive workplace: Development of the Workplace Aggression Tolerance Questionnaire. *Communication Studies*, 55 (3), 481–497.

Cortina, L. M. (2008) Unseen injustice: Incivility and discrimination in organizations. *Academy of Management Review*, 33 (1), 55–75.

Cortina, L. M., and Berdahl, J. L. (2008) Sexual harassment in organizations: A decade of research in review. In C. Cooper and J. Barling (eds.), *Handbook of organizational behavior*. Thousand Oaks, CA: Sage.

Cortina, L. M., Magley, V. J., Williams, J. H., and Langhout, R. D. (2001) Incivility in the workplace: Incidence and impact. *Journal of Occupational Health Psychology*, 6, 64–80.

Crossley, C. D. (2009) Emotional and behavioral reactions to social undermining: A closer look at perceived offender motivates. *Organizational Behavior and Human Decision Processes*, 108, 14–24.

Duffy, M. K., Ganster, D. C., and Pagon, M. (2002) Social undermining in the workplace. *Academy of Management Journal*, 45, 331–351.

Einarsen, S. (1999) The nature and causes of bullying at work. *International Journal of Manpower, 20* (1/2), 16–27.

Einarsen, S., Hoel, H., Zapf, D., and Cooper, C. L. (eds.) (2003) *Bullying and emotional abuse in the workplace: International perspectives in research and practice.* London: Taylor & Francis.

Einarsen, S., Raknes, B. I., and Matthiesen, S. B. (1994) Bullying and harassment at work and their relationships to work environment quality: An exploratory study. *European Work & Organizational Psychology, 4,* 381–401.

Emerson, R. M. (1972) Exchange theory. Part 1: A psychological basis for social exchange. In J. Berger, M. Zelditch, and B. Anderson (eds.), *Sociological theories in progress* (Vol. 2, pp. 38–57). Boston: Houghton Mifflin.

Ferris, G. R., Zinko, R., Brouer, R. L., Buckley, M. R., and Harvey, M. G. (2007) Strategic bullying as a supplementary, balanced perspective on destructive leadership. *The Leadership Quarterly, 18,* 195–206.

Ferris, P. (2004) A preliminary typology of organizational response to allegations of workplace bullying: See no evil, hear no evil, speak no evil. *British Journal of Guidance and Counselling, 32* (3), 389–395.

Fitzgerald, L. F. (1993) Sexual harassment: Violence against women in the workplace. *American Psychologist, 48,* 1070–1076.

Fox, S., and Spector, P. E. (2005) Introduction. In S. Fox and P.E. Spector (eds.), *Counterproductive work behavior: Investigations of actors and targets* (pp. 3–10). Washington, DC: American Psychological Association.

French, R. J. P., and Raven, B. (1959) The bases of social power. In D. Cartwright (ed.), *Studies in social power* (pp. 150–167). Ann Arbor: University of Michigan.

Glomb, T. M. (2002) Workplace aggression: Informing conceptual models with data from specific encounters. *Journal of Occupational Health Psychology, 7* (1), 20–36.

Glomb, T. M., and Liao, H. (2003) Interpersonal aggression in work groups: Social influence, reciprocal, and individual effects. *Academy of Management Journal, 46* (4), 486–496.

Glomb, T. M., and Miner, A. G. (2002) Exploring patterns of aggressive behaviors in organizations: Assessing model-data fit. In J. M. Brett and F. Drasgow (eds.), *The psychology of work: Theoretically based empirical research* (pp. 235–252). Mahwah, NJ: Lawrence Erlbaum.

Grandey, A. A., Kern, J. H., and Frone, M. R. (2007) Verbal abuse from outsiders versus insiders: Comparing frequency, impact on emotional exhaustion, and the role of emotional labor. *Journal of Occupational Health Psychology, 12* (1), 63–79.

Greenberg, L., and Barling, J. (1999) Predicting employee aggression against coworkers, subordinates, and supervisors: The roles of person behaviors and perceived workplace factors. *Journal of Organizational Behavior, 20,* 897–913.

Griffin, R. W., and Lopez, Y. P. (2005) "Bad behavior" in organizations: A review and typology for future research. *Journal of Management, 31,* 988–1005.

Grubb, P. L., Roberts, R. K., Grosch, J. W., and Brightwell, W. S. (2004) Workplace bullying: What organizations are saying. *Employee Rights and Employment Policy Journal, 8,* 407–417.

Harlos, K. P., and Pinder, C. (1999) Patterns of organizational injustice: A taxonomy of what employees regard as unjust. In J. Wagner (ed.), *Advances in qualitative organization research* (pp. 97–125). Stamford, CT: JAI Press.

Harvey, M., Treadway, D. C., and Heames, J. T. (2007) The occurrence of bullying in global organizations: A model and issues associated with social/emotional contagion. *Journal of Applied Social Psychology, 31* (11), 2576–2599.

Heames, J., and Harvey, M. (2006) Workplace bullying: A cross-level assessment. *Management Decision, 44* (9), 1214–1230.

Hershcovis, M. S., and Barling, J. (2007) Towards a relational model of workplace aggression. In J. Langan-Fox, C. L. Cooper, and R. J. Klimoski (eds.), *Research companion to the dysfunctional workplace: Management challenges and symptoms* (pp. 268–284). Northampton, MA: Edward Elgar.

——— (2008) Comparing the outcomes of sexual harassment and workplace aggression: A meta-analysis to guide future research. Paper presented at the Seventh Annual Work, Stress and Health Conference, Washington, DC.

Hodson, R., Roscigno, V. J., and Lopez, S. H. (2006) Chaos and the abuse of power: Workplace bullying in organizational and interactional context. *Work and Occupations, 33* (4), 382–416.

Hornstein, H. (1995) *Brutal bosses and their prey.* New York: Riverhead Books.

Inness, M., LeBlanc, M. M., and Barling, J. (2008) Psychosocial predictors of supervisor-, peer-, subordinate-, and service-provider-targeted aggression. *Journal of Applied Psychology, 93* (6), 1401–1411.

International Action Network on Small Arms (IANSA, 2006) US: Eight dead in shooting rampage at workplace. http://www.iansa.org/regions/namerica/workplace-shooting-rampage.htm (accessed August 5, 2009).

Jagatic, K., and Keashly, L. (1998) The nature and effects of negative incidents by faculty members toward graduate students. Unpublished manuscript, Wayne State University, Detroit.

Keashly, L. (1998) Emotional abuse in the workplace: Conceptual and empirical issues. *Journal of Emotional Abuse, 1,* 85–117.

——— (2001) Interpersonal and systemic aspects of emotional abuse at work: The target's perspective. *Violence and Victims, 16* (3), 233–268.

Keashly, L., and Harvey, S. (2005) Emotional abuse in the workplace. In S. Fox and P. Spector (eds.), *Counterproductive work behaviors* (pp. 201–236). Washington, DC: American Psychological Association.

——— (2006) Workplace emotional abuse. In E. K. Kelloway, J. Barling, and J. J. Hurrell Jr. (eds.), *Handbook of workplace violence* (pp. 95–120). Thousand Oaks, CA: Sage.

Keashly, L., Harvey, S., and Hunter, S. (1997) Abusive interaction and role state stressors: Relative impact on student residence assistant stress and work attitudes. *Work & Stress, 11,* 175–185.

Keashly, L., and Jagatic, K. (2000) The nature, extent, and impact of emotional abuse in the workplace: Results of a statewide survey. Paper presented at the Academy of Management Conference, Toronto, Canada.

——— (2003) By any other name: American perspectives on workplace bullying. In S. Einarsen, H. Hoel, D. Zapf and C. L. Cooper (eds.), *Bullying and emotional abuse in the workplace: International perspectives in research and practice* (pp. 31–61). London: Taylor & Francis.

Keashly, L., and Neuman, J. H. (2002) Exploring persistent patterns of workplace aggression. Paper presented as part of symposium Workplace Abuse, Aggression, Bullying, and Incivility: Conceptual Integration and Empirical Insights, Annual meeting of the Academy of Management, Denver.

———— (2004) Bullying in the workplace: Its impact and management. *Employee Rights and Employment Policy Journal, 8* (2), 335–373.

———— (2008) Final report: Workplace Behavior (Bullying) Project Survey. Unpublished report, Minnesota State University, Mankato.

———— (2009) Building constructive communication climate: The U.S. Department of Veterans Affairs Workplace Stress and Aggression Project. In P. Lutgen-Sandvik and B. D. Sypher (eds.), *The destructive side of organizational communication: Processes, consequences and constructive ways of organizing* (pp. 339–362). New York: Routledge/LEA.

Keashly, L., and Rogers, K. A. (2001) Aggressive behaviors at work: The role of context in appraisals of threat. Unpublished manuscript, Wayne State University, Detroit.

Keashly, L., Trott, V., and MacLean, L. M. (1994) Abusive behavior in the workplace: A preliminary investigation. *Violence and Victims, 9*, 125–141.

Keashly, L., Welstead, S., and Delaney, C. (1996) Perceptions of abusive behaviors in the workplace: Role of history, emotional impact and intent. Unpublished manuscript, University of Guelph, Ontario, Canada.

Kelloway, E. K., Barling, J., and Hurrell, J. J. Jr. (2006) *Handbook of workplace violence.* Thousand Oaks, CA: Sage.

Lamertz, K., and Aquino, K. (2004) Social power, social status and perceptual similarity of workplace victimization: A social network analysis of stratification. *Human Relations, 57* (7), 795–822.

Lapierre, L. M., Spector, P. E., and Leck, J. D. (2005) Sexual versus nonsexual workplace aggression and victims' overall job satisfaction: A meta-analysis. *Journal of Occupational Health Psychology, 10*, 155–169.

Lawrence, T. B., and Robinson, S. L. (2007) Ain't misbehavin': Workplace deviance as organizational resistance. *Journal of Management, 33* (3), 378–394.

Lazarus, R. S., and Folkman, S. (1984) *Stress, appraisal, and coping.* New York: Springer.

Lewis, G., and Zare, N. (1999) *Workplace hostility: Myth and reality.* Muncie, IN: Accelerated Development.

Leymann, H. (1996) The content and development of mobbing at work. *European Journal of Work and Organizational Psychology, 5*, 165–184.

Lim, S., and Cortina, L. M. (2005) Interpersonal mistreatment in the workplace: The interface and impact of general incivility and sexual harassment. *Journal of Applied Psychology, 90* (3), 483–496.

Lind, T. (1997) Litigation and claiming in organizations: Antisocial behavior or quest for justice? In R. A. Giacalone and J. Greenberg (eds.), *Antisocial behavior in organizations* (pp. 150–171). Thousand Oaks, CA: Sage.

Lutgen-Sandvik, P. (2003) The communicative cycle of employee emotional abuse: Generation and regeneration of workplace mistreatment. *Management Communication Quarterly, 16*, 471–501.

———— (2006) Take this job and . . .: Quitting and other forms of resistance to workplace bullying. *Communication Monographs, 4*, 406–433.

Lutgen-Sandvik, P., and Namie, G. (2009) Workplace bullying from start to finish: Bullies' position and supporters, organizational responses and abuse cessation. Paper presented at the Western States Communication Association, Phoenix, AZ.

Lutgen-Sandvik, P., Namie, G., and Namie, R. (2009) Workplace bullying: Causes, consequences and corrections. In P. Lutgen-Sandvik and B. D. Sypher (eds.), *The destructive side of organizational communication: Processes, consequences and constructive ways of organizing* (pp. 27–52). New York: Routledge/LEA.

Lutgen-Sandvik, P., Tracy, S. J., and Alberts, J. K. (2007) Burned by bullying in the American workplace: Prevalence, perception, degree, and impact. *Journal of Management Studies, 44* (6), 837–862.

Martinko, M. J., Douglas, S. C., and Harvey, P. (2006) Understanding and managing workplace aggression. *Organizational Dynamics, 35* (2), 117–130.

McLaughlin, J. (2000) Anger within. *OHS Canada, 16* (8), 30–36.

Meares, M. M., Oetzel, J. G., Torres, A., Derkacs, D., and Ginossar, T. (2004) Employee mistreatment and muted voices in the culturally diverse workplace. *Journal of Applied Communication Research, 31* (1), 4–27.

Meglich-Sespico, P. A. (2006) Perceived severity of interpersonal workplace harassment behaviors. Unpublished dissertation, Kent State University, Kent, Ohio.

Mitchell, M. S., and Ambrose, M. L. (2007) Abusive supervision and workplace deviance and the moderating effects of negative reciprocity beliefs. *Journal of Applied Psychology, 92* (4), 1159–1168.

Namie, G. (2000) U.S. Hostile Workplace Survey 2000. Paper presented at the New England Conference on Workplace Bullying, Suffolk University Law School, Boston.

——— (2003) Workplace bullying: Escalated incivility. *Ivey Business Journal*, November/December, 1–6.

Namie, G., and Namie, R. (2000; 2009) *The bully at work: What you can do to stop the hurt and reclaim your dignity on the job.* Naperville, IL: Sourcebooks.

Neuman, J. H., and Baron, R. A. (1997) Aggression in the workplace. In R. A. Giacalone and J. Greenberg (eds.), *Antisocial behavior in organizations* (pp. 37–67). Thousand Oaks, CA: Sage.

Neuman, J. H., and Keashly, L. (2006) Exploring the patterns, prevalence, persistence, and perceptions of workplace aggression within and across superior-subordinate, peer, and customer/client relationships. Unpublished manuscript. State University of New York, New Paltz.

Northwest National Life Insurance Company (1993) *Fear and violence in the workplace.* Minneapolis: Northwest National Life Insurance Company.

O'Leary-Kelly, A. M., Griffin, R. W., and Glew, D. J. (1996) Organization-motivated aggression: A research framework. *Academy of Management Review, 21*, 225–253.

Opotow, S. (2006) Aggression and violence. In M. Deutsch and P. T. Coleman (eds.), *The handbook of conflict resolution,* 2nd ed. (pp. 509–532). San Francisco: Jossey-Bass.

Pearson, C. M., Andersson, L. M., and Porath, C. L. (2005) Workplace incivility. In S. Fox and P. E. Spector (eds.), *Counterproductive work behavior: Investigations of actors and targets* (pp. 177–200). Washington, DC: American Psychological Association.

Richman, J., Rospenda, K. M., Flaherty, J. A., and Freels, S. (2001) Workplace harassment, active coping, and alcohol-related outcomes. *Journal of Substance Abuse, 13* (3), 347–366.

Richman, J., Rospenda, K. M., Nawyn, S. J., Flaherty, J. A., Fendrich, M., Drum, M. L., and Johnson, T. P. (1999) Sexual harassment and generalized workplace abuse among university employees: Prevalence and mental health correlates. *American Journal of Public Health, 89* (3), 358–363.

Robinson, S. L., and Bennett, R. J. (1995) A typology of deviant workplace behaviors: A multidimensional scaling study. *Academy of Management Journal, 38*, 555–572.

——— (2000) Development of a measure of workplace deviance. *Journal of Applied Psychology, 85* (3), 349–360.

Robinson, S. L., and O'Leary-Kelly, A. (1998) Monkey see, monkey do: The influence of work groups on the antisocial behavior of employees. *Academy of Management Journal, 41*, 658–672.

Rogers, K. A. (1998) Toward an integrative understanding of workplace mistreatment. Unpublished dissertation, University of Guelph, Ontario, Canada.

Rospenda, K., Richman, J. A., and Shannon, C. A. (2006) Patterns of workplace harassment, gender, and use of services: An update. *Journal of Occupational Health Psychology, 11* (4), 379–393.

——— (2009) Prevalence and mental health correlates of harassment and discrimination in the workplace: Results from a national study. *Journal of Interpersonal Violence, 24*, 819–843.

Rospenda, K., Richman, J. A., Wislar, J. S., and Flaherty, J. A. (2000) Chronicity of sexual harassment and generalized workplace abuse: Effects on drinking outcomes. *Addiction, 95* (12), 1805–1820.

Rubin, J. Z., Pruitt, D. G., and Kim, S. H. (1994) *Social conflict: Escalation, stalemate, and settlement*, 2nd ed. New York: McGraw-Hill.

Schat, A. C. H., Frone, M. R., and Kelloway, E. K. (2006) Prevalence of workplace aggression in the US workforce: Findings from a national study. In E. K. Kelloway, J. Barling, and J. J. Hurrell Jr. (eds.), *Handbook of workplace violence* (pp. 47–90). Thousand Oaks, CA: Sage.

Schat, A. C. H., and Kelloway, E. K. (2003) Reducing the adverse consequences of workplace aggression and violence: The buffering effects of organizational support. *Journal of Occupational Health Psychology, 8* (2), 110–122.

Schneider, K. T., Hitlan, R. T., Delgado, M., Anaya, D., and Estrada, A. X. (2000a) Hostile climates: The impact of multiple types of harassment on targets. Paper presented at the Society for Industrial and Organizational Psychology, April, New Orleans.

Schneider, K. T., Hitlan, R. T., and Radhakrishnan, P. (2000b) An examination of the nature and correlates of ethnic harassment experiences in multiple contexts. *Journal of Applied Psychology, 85* (1), 3–12.

Sutton, R. (2007) *The no asshole rule: Building a civilized workplace and surviving one that isn't*. New York: Business Plus.

Swaitek, J. (2008) Jury award upheld against doctor in bullying case. *Indianapolis Star*, April 9.

Tedeschi, J. T., and Felson, R. B. (1994) *Violence, aggression, and coercive actions.* Washington, DC: American Psychological Association.

Tepper, B. J. (2000) Consequences of abusive supervision. *Academy of Management Journal, 43* (92), 178–190.

——— (2007) Abusive supervision in the workplace: Review, synthesis and research agenda. *Journal of Management, 33* (3), 261–289.

Tepper, B. J., Carr, J. C., Breaux, D. M., Geider, S., Hu, C., and Hua, W. (2009) Abusive supervision, intentions to quit, and employees' workplace deviance. Unpublished manuscript, Georgia State University, Atlanta.

Tracy, S., Lutgen-Sandvik, P., and Alberts, J. K. (2005) Escalated incivility: Analyzing workplace bullying as a communication phenomenon. Unpublished manuscript, Arizona State University, Tempe.

Vartia, M. (1996) The sources of bullying—psychological work environment and organizational climate. *European Journal of Work and Organizational Psychology, 5*, 203–214.

Venkataramani, V., and Dalal, R. S. (2007) Who helps and harms whom? Relational antecedents of interpersonal helping and harming in organizations. *Journal of Applied Psychology*, 92 (4), 952–966.

Waldron, V. (2009) Emotional tyranny at work: Suppressing the moral emotions. In P. Lutgen-Sandvik and B. D. Sypher (eds.), *The destructive side of organizational communication: Processes, consequences and constructive ways of organizing* (pp. 7–26). New York: Routledge/LEA.

Wright, L., and Smye, M. (1996) *Corporate abuse: How "lean and mean" robs people and profits*. New York: Macmillan.

Section II

Empirical Evidence

3

Empirical Findings on Prevalence and Risk Groups of Bullying in the Workplace

Dieter Zapf, Jordi Escartín, Ståle Einarsen, Helge Hoel, and Maarit Vartia

CONTENTS

Introduction

This chapter aims at summarising some descriptive empirical findings of bullying in the workplace. We will start with the frequency and the duration of bullying. This is followed by an examination of the number, gender, and status of bullies and victims; the distribution of bullying across industries and occupations; and the use of various categories of bullying. The empirical basis of this chapter is restricted to studies carried out in Europe (see Table 3.3, Appendix, for an overview of the included studies).

The phenomenon of bullying, which includes being exposed to persistent insults or offensive remarks, persistent criticism, and personal or even physical abuse, has been labelled "mobbing at work" in some Scandinavian and German countries (Leymann, 1996) and "bullying at work" in many English-speaking countries (Liefooghe and Olafsson, 1999). Typically, a victim is constantly teased, badgered, and insulted and perceives that he or she has little recourse to retaliate in kind. Bullying may take the form of open verbal or physical attacks on the victim, but it may also take the form of more subtle acts, such as excluding or isolating the victim from his or her peer group (Einarsen et al., 1994; Leymann, 1996; Zapf et al., 1996). The following definition of bullying or mobbing seems to be widely agreed upon:

> Bullying at work means harassing, offending, socially excluding some-one or negatively affecting someone's work tasks. In order for the label bullying (or mobbing) to be applied to a particular activity, interaction or process it has to occur repeatedly and regularly (e.g. weekly) and over a period of time (e.g. about six months). Bullying is an escalating process in the course of which the person confronted ends up in an inferior position and becomes the target of systematic negative social acts. A conflict cannot be called bullying if the incident is an isolated event or if two parties of approximately equal "strength" are in conflict. (Einarsen et al., this volume, p. 22)

It should be noted that the increased attention that bullying has received in research and practice during recent years has not led to an agreement on how to define and operationalise the phenomenon. Rather, there are authors who use more or less strict definitions with regard to the timeframe (e.g., less than six months or at least six months) and the frequency of the bullying behaviour (e.g., less often than once a week) (cf. Einarsen, 2000; Einarsen et al., this volume; Hoel et al., 1999; Keashly and Jagatic, this volume; Zapf and Einarsen, 2005).

The Frequency of Bullying

For practical reasons, in particular it is important to know how frequently bullying actually occurs in organisations because efforts to develop measures against it would depend on this information. However, it is not easy to provide reliable numbers. The problem is that the frequency of bullying depends very much on how it is measured (cf. Hoel et al., 1999; Nielsen, Notelaers, and Einarsen, this volume). Furthermore, the measurement method employed is influenced by the general understanding of what constitutes bullying.

One of the major approaches in measuring bullying is using a questionnaire consisting of a list of bullying behaviours. Nielsen et al. (in press) called this the "behavioural experience method." Another approach is to use a precise definition—for example, the definition presented earlier—and then ask the respondents to label themselves as bullied or otherwise not bear this definition in mind at all. This method has frequently been referred to as the "self-labelling method" (Nielsen et al., in press). In the meta-analysis of

Nielsen et al., the behavioural experience method led to a prevalence rate of 14.7% bullying, whilst the self-labelling method led to a prevalence rate of 10.6% when a definition of bullying was used, compared with 19.8% if no definition was given (see also Nielsen et al., this volume). In the latter case, researchers have asked directly, "Have you been bullied during the last six months?" (e.g., Rayner, 1997). This, typically, leads to a comparatively high amount of bullying because people will also tend to say that they have been bullied when only occasional, minor negative acts have occurred.

Some researchers who administered questionnaires using the behavioural method have used a fixed cut-off point (e.g., Björkqvist et al., 1994). Respondents scoring higher than the cut-off point were considered to be victims of bullying. Usually, these studies report a prevalence rate as high as 10%–17% bullying (cf. Table 3.3, Appendix). Other researchers using the behavioural experience method applied a strategy developed by Leymann (1996), which we will call the "Leymann criterion": Here, the Leymann Inventory of Psychological Terrorization (LIPT), a questionnaire of 45 items (Leymann, 1990, 1996), or a similar questionnaire such as the Negative Acts Questionnaire (NAQ) (Einarsen et al., 2009; Einarsen and Raknes, 1997) is administered. To be considered a bullying victim, the respondent must mark at least one item or one general item on the frequency of bullying actions as "at least once a week," and the duration of bullying should be at least six months. Weighted for sample size, the mean prevalence rate for studies using this strategy (see Table 3.3, Appendix) is 7.7% (total N = 8,270, k = 20 samples). Those studies using the weekly criterion but asking for bullying "within the last six months" had a weighted prevalence rate of 11.9% (total N = 34,018, k = 17 samples). In the meta-analysis of Nielsen et al. (in press), the self-labelling-method led to a lower prevalence rate than the behavioural experience method. Combining the self-labelling method with the behavioural experience criterion of bullying "at least once a week" led to a weighted prevalence rate of 3.7% bullying in the respective studies included in the present review (total N = 44,878, k = 15 samples) (see Table 3.3, Appendix).

Taking the Leymann criterion and self-labelling in addition to weekly bullying as indicators of severe bullying, it can be concluded that a figure of between 3% and 4% serious bullying has emerged in Europe in the sense of the foregoing definition. For somewhat less severe cases (including bullying experienced less often than weekly and a duration of less than six months), the meta-analytical results of Nielsen et al. (2009), as well as our own results based on the studies in Table 3.3, suggest a figure of between 10% and 15% bullying. Moreover, the meta-analytical results of Nielsen et al. on self-labelled bullying *without a definition* suggests that in many organisations, up to 20% of the employees are occasionally exposed to negative social acts such as being yelled at, teased, or humiliated. Although such acts do not fall within the stringent definition of bullying, they do imply that these employees are exposed to severe social stressors at work (for a discussion from a methodological perspective, see Nielsen et al., this volume).

A rate of 4% serious bullying in a 1,000-employee organisation means that around 40 people are involved at any point in time. Given that not only these 40 people but also usually a larger group of bullies as well as bystanders are, in one way or another, negatively affected by the bullying situation, we consider this a serious number.

The Duration of Bullying

In many European countries, meanwhile, even minor conflicts and arguments in everyday life may be described as mobbing or bullying. Therefore, the duration of bullying is an important criterion to differentiate between bullying and everyday conflicts in organisations. Studies reporting on the duration of bullying are summarised in Table 3.1. The studies show that bullying is a long-lasting conflict. Looking at some large representative samples in Sweden (Leymann, 1996), Norway (Einarsen and Skogstad, 1996), and Germany (Meschkutat et al., 2002), the average duration of bullying was 15, 18, and 16 months, respectively. Among bullied Finnish prison officers, 66% of the females and 53% of the males had been bullied for more than two years (Vartia and Hyyti, 2002). In the study by Hoel and Cooper (2000), 39% of the victims had been bullied for more than two years. Among victims in a Finnish municipal institution, 29% had been bullied for two to five years and as many as 30% for over five years (Vartia, 2001). In studies of victims only, the average duration was much higher, with a mean of more than three years (e.g., Leymann and Gustafsson, 1996; Zapf, 1999a). This difference is probably attributable to method: If one tries to identify bullying victims and contact them by means of help lines, self-help groups, and the like, one will receive a self-selected sample of more severe bullying cases (see also Nielsen and Einarsen, 2008). As shown in Table 3.1, studies that were carried out more recently confirm previous findings. The numbers on the duration underscore that bullying is not a short episode but a long-lasting process that "wears down" its victims, mostly lasting much longer than one year.

Gender Effects of Bullying

A frequently asked question among the public is whether there are gender effects of bullying. Although data exist on the gender of bullies and victims, there has still not been much theorising or in-depth research on this issue (Vartia and Hyyti, 2002).

TABLE 3.1

Average Duration of Workplace Bullying in Months

Study	No.	Duration in Months
Finland (Salin, 2001)	34	32
Finland (Vartia and Hyyti, 2002)	896	24
Germany (DAG-Study, Zapf, 1999a)	56	47
Germany (Gießen Study, Zapf, 1999a)	50	40
Germany (Halama and Möckel, 1995)	183	40
Germany (Konstanz Study, Zapf, 1999a)	87	46
Germany (Stuttgart Study, Zapf et al., 1996)	188	29
Germany (communal administration, zur Mühlen et al., 2001)	55	34
Germany (army administration, zur Mühlen et al., 2001)	55	24
Germany (representative study, Meschkutat et al., 2002)	356	16
Ireland (O'Moore, 2000)	248	41
Norway (Einarsen and Skogstad, 1996)	268	18
Spain (González and Graña, 2009)	2861	12
Spain (Segurado et al., 2008)	235	30
Switzerland (Holzen-Beusch et al., 1998)	28	36
Switzerland (Kudielka and Kern, 2004)	28	62
Sweden (Leymann, 1996)	85	15
Turkey (Ozturk et al., 2008)	162	36

Gender of the Victims

One can argue that there exists some relation between female socialisation and the victim role because women are said to be taught to be less
self-assertive and less aggressive, and they tend to be more obliging than
men (Björkqvist, 1994). Consequently, women are even less able than men to
defend themselves when bullying emerges. Moreover, for various reasons,
women hold less powerful positions in organisations. For example, they are
less often in managerial or supervisor positions (Davidson and Cooper, 1992).
We carried out an analysis based on 53 samples of bullying victims, most of
them listed in Table 3.3, which reported gender distribution among victims.
Weighted percentages with regard to sample size (total sample size N = 8,169
victims) showed that 62.5% of victims were women and 37.5% men. In the
large majority of samples, one-third of the victims were men and two-thirds
women. However, several Scandinavian samples (Einarsen and Skogstad,
1996; Leymann, 1996; Leymann and Tallgren, 1993) and the UK studies by
Hoel and Cooper (2000) and Rayner (1997) showed a more balanced picture.

An analysis of the aforementioned studies, where the gender distribution of victims and the gender distribution of the total samples were available, led to the following results: Of the more than 5,000 victims (N = 5,679)

from 30 separate samples, a total of 60.9% were women and 39.1% were men. These victims emerged from a total sample of 54,775 employees with a gender distribution of 60.7% women and 39.3% men. These numbers show that the men–women ratio of the victims corresponds to the respective ratio in the overall sample, suggesting that the overrepresentation of women among victims is due to the overrepresentation of women in the respective populations. Of course, one could argue that bullying in some sectors of industries and occupations is higher because of their overrepresentation of women. However, if women's attitudes and behaviours played a role, there should still have been an effect if the baseline (and thus the industry or occupation) is controlled for. All in all, there seems to be little evidence that women are more at risk because of specific female socialisation.

Nevertheless, in some samples there exists a higher risk for women to be victimised. In the case of the police sample in Nuutinen et al.'s (1999) study, the explanation of women's higher risk of victimisation may lie in their visibility in a male-dominated organisation (see also Archer, 1999). Minority groups who differ from the main groups in salient characteristics carry a higher risk of being socially excluded from the group (Schuster, 1996; see also Zapf and Einarsen, this volume). Women may be seen as intruders in the male-dominated cultures of researchers, business professionals, or the police force (Archer, 1999). Yet, in a study among a large representative sample of assistant nurses where men are only a small minority of less than 3%, male nurses had nearly three times the risk of being a target of bullying compared to female nurses (Eriksen and Einarsen, 2004).

Gender of the Bullies

Less information is available on the gender of the bullies. In the Zapf (1999a) studies (N = 209), altogether 26% of the bullies were men only, 11% were women only, and in 63% of all cases men and women were among the bullies. Einarsen and Skogstad (1996) reported that 49% of the victims were bullied by men, 30% by women, and in 21% of all cases the bullies were both men and women. The respective numbers in the study by Mackensen von Astfeld (2000) were 32% men, 27% women, and 37% both men and women. All in all, men seem to be clearly overrepresented among the bullies in most studies (see also Meschkutat et al., 2002; Rayner, 1997; an exception is UNISON, 1997). This result corresponds to similar findings in research on bullying at school (Olweus, 1994). Bullying, at least in part, includes forms of direct aggression such as shouting or humiliating somebody. There is substantial empirical evidence that this kind of aggression is much more typical for men than for women, who tend to make more use of indirect aggression such as social exclusion or spreading rumours (Björkqvist, 1994). Moreover, managers and supervisors appear to play a dominant role in bullying (see later); that men are overrepresented in such positions may explain why men are more often among the bullies than women.

Finally, Leymann (1993a, 1993b) reported that women are more often bullied by other women and that men are more often bullied by other men, which he explained in terms of labour market segregation. Similar results were reported by Einarsen and Skogstad (1996), Hoel et al. (2001), Mackensen von Astfeld (2000), Meschkutat et al. (2002), Niedl (1995), Rayner (1997), and Zapf (1999a). Whereas women are sometimes exclusively bullied by men, it appears to seldom be the case that men are exclusively bullied by women. This finding may be explained by the different power positions of men and women in organisations.

The Number of Bullies

Although bullying can be a conflict between two people, some victims report that the entire organisation is bullying them. Data on the number of bullies in various studies are summarised in Table 3.2. In most studies, in 20%–40% of all cases there was only one person bullying the victim, whereas in 15%–25% there were more than four bullies involved. In the German victims studies (Zapf, 1999a), bullying by only one person was much rarer. In fact, in these

TABLE 3.2

The Number of Bullies (%)

Bullies	N	1 Bully	2–4 Bullies	More than 4 Bullies
Austria (hospital, Niedl, 1995)	82	20	52	28
Austria (research institute, Niedl, 1995)	11	55	27	18
Germany (DAG Study, Zapf, 1999a)	55	9	35	56
Germany (Gießen Study, Zapf, 1999a)[a]	50	10	50	40
Germany (Konstanz Study, Zapf, 1999a)	78	9	32	59
Germany (Mackensen von Astfeld, 2000)	115	38	46	16
Hungary (Army Study, Kauscek and Simon, 1995)	18	23	62	14
Ireland (O'Moore, 2000)	248	62	38	0
Ireland (O'Moore et al., 1998)	30	63	33	3
Italy (Ege, 1998)	301	20	46	34
Norway (Einarsen and Skogstad, 1996)[b]	392	42	43	15
Spain (González and Graña, 2009)	234	51	27	22
Sweden (Leymann and Tallgren, 1993)	24	43	50	7
Sweden (Leymann, 1993b)	85	34	43	23

Notes: N = sample size.
[a] The middle category of this study was "2–5 Bullies."
[b] The third category of this study was "4 and More Bullies."

studies, in more than 50% of all cases more than four bullies were involved. These differences may be explained as follows: As described earlier, samples consisting of bullying victims usually consist of more serious bullying cases, which, for example, show a longer mean duration of the bullying conflict.

There is some evidence that bullying becomes more and more serious the longer it lasts. Studies by Einarsen and Skogstad (1996) and Zapf and Gross (2001) showed that bullying incidents and negative acts occurred more often the longer the bullying behaviour lasted. In the study by Zapf (1999a), the duration of bullying correlated positively with the number of bullies. The average duration of bullying for those who were bullied by only one person was 28 months, and for those who were bullied by two to four and more than four people, it was 36 months and 55 months, respectively. These data suggest that it is getting more and more difficult to remain a neutral bystander the longer bullying goes on. Therefore more and more people may become involved as bullies in the course of time. This analysis may explain the higher mean number of bullies in the pure victim samples, which show a higher mean duration of bullying.

Some studies, especially the British ones (Hoel and Cooper, 2000; Rayner, 1997), report that many victims share their experience with other colleagues. For example, in the study by Hoel and Cooper, as much as 55% of the bullying victims reported that they shared their experience with other work colleagues, and 15% reported that everyone in the work group was bullied. Similar results were reported in the UNISON studies (1997, 2000). In such other countries as Austria (Niedl, 1995) or Germany (Zapf, 1999a, 1999b), sharing is reported only occasionally. Sharing may be a country-specific phenomenon; however, it may have to do with the definition of bullying. The more stringent the definition of bullying, the more likely it is that it involves only one victim. While a perpetrator may occasionally bully everyone in the work group for months and years, it seems much more unlikely that he or she can bully to such an intensity that everyone in the work group is exposed to bullying at least on a weekly basis.

The Organisational Status of Bullies and Victims

The relationship between bullying and organisational status has so far received limited attention (Hoel et al., 2001). Organisational status in this respect refers to the formal position within the organisational hierarchy.

The Status of the Victim

Relatively little has been reported about the status of the victim. Einarsen and Raknes (1997), in a study of male employees at a Norwegian engineering plant, found no difference between the experience of negative behaviours for workers, on the one hand, and supervisors and managers, on the other. Similar results were found by Hoel et al. (2001), who present the most elaborate study on this issue. They found similar numbers of bullying for workers, supervisors, and middle and senior management. A representative sample of Finnish employees showed that white-collar employees in higher ranks experienced bullying somewhat more often than lower-ranked white-collar employees or workers (Piirainen et al., 2000). Salin (2001), however, found less bullying at the higher levels of the organisation. Skogstad et al. (2007) showed that although managers in a representative sample of the Norwegian workforce reported the same level of exposure to bullying behaviours, they labelled their experiences less as bullying compared to nonmanagers. Hoel et al. (2001) report some interesting interaction effects with gender: Whereas male workers and supervisors were bullied more than women, this was the other way round at the management level. The largest differences occurred for the senior management level, where 16% of the female senior managers reported having been bullied. This finding may be attributable to the visibility of women at this male-dominated hierarchical level and may reflect widespread prejudice against women in leadership positions (see Davidson and Cooper, 1992).

All in all, the findings of Hoel et al. (2001) question a common assumption in various European countries that the weak and defenceless, in terms of organisational status, become the primary victims of bullying. Rather, there seem to be similar risks at all organisational status levels. Supervisors and senior managers may also experience a power imbalance relative to their colleagues and superiors.

The Status of the Bully

By contrast, the issue of perpetrator status has received considerable attention. Interestingly, the findings vary across countries. Leymann (1993b) introduced the term *mobbing* as the definition of a lasting conflict among colleagues. Yet even in his study, there were only marginally more colleagues among the bullies than there were supervisors. However, the Scandinavian studies as a whole identified people in superior positions as offenders in approximately equal numbers to peers, with only a small number bullied by a subordinate (Einarsen and Skogstad, 1996; Leymann, 1992, 1993b). Hence, supervisors are seen as the perpetrators relatively more often than subordinates, although an equal number of cases are reported where the perpetrator is a supervisor as where the perpetrator is a colleague. In contrast, British studies have consistently identified people in superior positions as perpetrators in an overwhelming majority of cases (Cowie et al., 2002; Hoel et al., 2001; Rayner, 1997). Analysing the available samples listed in Table 3.3, Appendix (total N = 6,783 victims, k = 40 samples), the percentages weighted by sample size were as follows: 65.4% were bullied by supervisors, 39.4% were bullied by colleagues, and 9.7% were bullied by subordinates. Thus, the overall picture across countries is that supervisors are more often among the bullies than are colleagues. However, the numbers for colleagues and subordinates involved clearly speak against the view that bullying is primarily a top-down process.

Occasionally, gender differences are reported. For example, Vartia and Hyyti (2002) found that women were more often bullied by coworkers, whereas men were more often bullied by immediate supervisors or managers.

Einarsen (2000), referring to Hofstede (1993), argued that some cultural differences between the Nordic and the central European countries may explain some of the different findings with regard to the organisational status of the bullies. Hofstede's studies suggest that low power differentials and feminine values prevail in the Scandinavian countries. The abuse of formal power is much more sanctioned in such countries. Power differences between immediate supervisors and their colleagues are small, hence producing more similar numbers of perpetrators for supervisors and colleagues. In a Danish study by Ortega et al. (2008), peer bullying was found to be the most typical kind of bullying, with colleagues being the main perpetrators in more than 70% of the cases.

Generally, superiors are seldom bullied by subordinates. In particular, there are only a small number of cases reported where superiors were *exclusively* bullied by their subordinates. Usually, subordinates bully a superior in conjunction with other supervisors or managers. The reason for this is, of course, that it is not easy to overcome the formal power of a superior by using informal power. Although doing so is possible if the superior is socially isolated (which points at tensions or conflicts within management), it is almost impossible to bully a superior if the superior is backed up by superiors at the same level and/or by senior management. One can certainly say that only superiors who have lost the support of their colleagues and of senior management carry the risk of becoming the victims of bullying by subordinates.

With regard to negative supervisory behaviour, there is still a lack of knowledge on its prevalence and risk factors (see Aasland et. al., in press, for an exemption). As stated by Ashforth (1994) and Tepper (2000, 2007), leadership studies have focused almost exclusively on the positive aspects of leadership, that is, on factors that contribute to job satisfaction and performance—obviously, on the implicit assumption that negative leadership behaviour can be equated with a lack of positive behaviour such as task orientation or consideration. Although some of the common leadership questionnaires, such as the Leader Behaviour Description Questionnaire, or LBDQ (Fleishman, 1953), contain some items similar to those that appear in workplace bullying questionnaires, negative leadership behaviour has not really been investigated in this tradition. Humiliating, yelling at, or threatening somebody is not simply the absence of consideration or employee orientation. Bullying by superiors is, therefore, a leadership research issue in its own right (see Aasland et al., in press; Einarsen et al., 2007; Tepper, 2000, 2007). In a study employing a large-scale sample of UK workers, Hoel and colleagues (in press) showed that authoritarian, laissez-faire, and inconsistent leadership were all associated with experiences of bullying as reported by both victims and observers. Yet, while observers regarded authoritarian leadership to be most strongly associated with bullying, victims' reports of bullying were most closely related to inconsistent leadership in the form of inconsistent punishment. In a representative study of Norwegian workers, reports of bullying, as made by both victims and observers, were strongly correlated with reports of tyrannical leadership from one's immediate supervisor (Hauge et al., 2007), while Skogstad and colleagues (2007), employing the same data source, showed that laissez-faire leadership was associated with reports of bullying through its effect on role stressors and interpersonal conflicts in the work environment. Hence, there appears to be strong support for a view that leadership styles are related to experiences of bullying among observers as well as targets.

The Frequency of Bullying in Various Sectors

In this section, some findings on the frequency of bullying in various sectors are summarised. Leymann (1993a, 1993b) reported an overrepresentation of bullying in the educational (approximately 2:1) and administrative (1.5:1) sectors and an underrepresentation in the trade and retail, production, and health sectors. The prevalence of bullying in Swedish public administration was 1% higher than the average of 3.5% (Leymann, 1993a). However, in other studies Leymann also found a high level of occurrence in the health sector. In the study by Leymann and Gustafsson (1996), public administration and the social and health sectors, as well as religious organisations, showed greater prevalence, whereas trade and industry were underrepresented. Meschkutat et al. (2002), Niedl (1995), Piirainen et al. (2000), and Vartia (1993, 1996) also reported high levels of bullying in the health and social sector. Similarly, in the studies by Einarsen and Skogstad (1996), the highest rate of frequent bullying (weekly or more often) was found among clerical workers (3.9%) and within trade and commerce (3.5%). For occasional bullying, the results were different. Here, in contrast to Leymann's Swedish study, there were significantly fewer respondents from public-sector organisations who reported bullying than from private enterprises. The highest prevalence rate was found amongst industrial workers, where 17.4% reported having been occasionally bullied during the last six months. Bullying was also frequent among those who did graphical work and among hotel and restaurant workers. The lowest rate of bullying was found amongst psychologists and university employees.

In Germany, analyses based on almost 400 victims of serious bullying (Zapf, 1999a) showed that employees of the health and social sector had a sevenfold risk of being bullied. Other risk factors were being in public administration (3.5:1) and the educational sector (3:1). Moreover, there was an increased risk of being bullied in the banking and insurance sectors. In contrast, the risk was relatively low in the traffic, trade, and farming industries; in the hotel and restaurant sector; and in the building industry. Hubert and van Veldhoven (2001) found increased risks of aggressive and unpleasant behaviour in service organisations, in industry, and in education. Salin (2001) reported more frequent bullying in the public sector than in the private sector; and Piirainen et al. (2000), in the municipal sector rather than the private sector or the civil service.

Taking the studies together, a higher risk of being bullied is reported for the social and health, public administration, and education sectors, which all belong to the public sector. Various reasons may explain the differences between sectors. First, one may assume that bullying is less frequent in small family enterprises in the hotel and restaurant business, for example, or in the building sector. Here, personal relationships can be expected to develop between employees and between employers. If severe conflicts arise, one party may leave the "family." Moreover, in these and other areas, short-term job contracts prevail; conflicts lasting several years are almost impossible because employees find it relatively easy to leave their jobs.

On the other hand, in many European countries, for example, Germany, Norway, and Sweden, working in public administration means having a secure, lifelong job that usually compensates for a somewhat lower-than-average salary. In this case, it is much more complicated to give up one's job when bullying occurs because doing so would involve giving up the high job security that is among the most important aspects of state jobs. Frequently, the specific knowledge gained in such jobs cannot easily be applied in the private sector. Moreover, moving to another job within the public sector is often difficult because one still finds oneself within the same organisation. An example would be the case of a bullied police officer. In a single organisation such as the police force, rumours spread fast and the officer's potential new superior might receive biased information and, to be on the safe side, reject the bullied officer's application (cf. Leymann, 1993b).

Yet another aspect may be inherent in the very nature of the job itself. Some jobs in the service sector, and in particular in the social and health sectors, require a high level of personal involvement, that is, a form of emotional labour (Hochschild, 1983; Zapf, 2002), which means sensing and expressing emotions and building up personal relationships. In such other jobs as manufacturing work, a much more instrumental job attitude may suffice. The higher the level of personal involvement, the more personal information is available, and the more possibilities for being attacked exist. Moreover, it is much more difficult to evaluate or appraise these jobs objectively. Again, this lack of objectivity offers a lot of opportunity for attacking someone. If a production worker is accused of doing a bad job, he or she can easily mount a self-defence by referring to the job description, whereas a teacher or a nurse may have much greater difficulties proving that he or she is doing a good job.

All in all, in looking at the distribution of bullying across sectors, bullying seems to be a greater problem among white-collar workers, service employees, and employees in supervisory positions than it is among blue-collar workers. Still, a representative study of the Norwegian workforce from 2005 showed that bullying prevailed in all kinds of organisations, with no particular sector being "bully proof" (Einarsen et al., 2007).

Categories of Bullying

The final question addressed in this chapter is whether bullying is a homogeneous construct or can instead be identified by typical types or categories. Homogeneity of bullying would imply that all bullying actions show similar frequencies, have similar causes and consequences, and occur under the same circumstances (Zapf et al., 1996). Leymann (1996) differentiated between five classes of bullying behaviour, which he referred to as the manipulation of (1) the victim's reputation, (2) the victim's possibilities of communicating with coworkers, (3) the victim's social relationships, (4) the quality of the victim's occupational and life situation, and (5) the victim's health. In an empirical study, Leymann (1992) found factors that he labelled as negative communication: humiliating behaviour, isolating behaviour, frequent changes of tasks to punish someone, and violence or threat of violence. Using factor analyses, Zapf et al. (1996) found seven factors in two samples: *Organisational measures* consist of behaviours initiated by the supervisor or aspects directly related to the victim's tasks. *Social isolation* is related to informal social relationships at work. The third factor is related to individual attributes of the victim and the victim's private life. *Physical violence* includes two items of sexual harassment as well as general physical violence or threat of violence. *Attacking the victim's attitudes* is related to political, national, and religious attitudes. The factor *verbal aggression* consists of items related to verbal attacks. Finally, there is a factor consisting of two items related to spreading rumours (for comparable results, see Niedl, 1995; Vartia, 1991, 1993; zur Mühlen et al., 2001). More recent studies have come to similar results employing different questionnaires (cf. Nielsen et al., this volume, Table 6.1).

Factor analysis of the Negative Acts Questionnaire, or NAQ (Einarsen and Raknes, 1997), identified five factors, four of which appear to overlap with attacking the private person, social isolation, work-related measures, and physical violence. Based on a revised version of the NAQ applied to a random sample of 5,288 UK employees, Einarsen et al. (2009) found three major factors: personal bullying, work-related bullying, and intimidation.

Taking the existing studies together, most researchers have suggested to differentiate between *work-related* bullying and *person-related* bullying. For work-related bullying, researchers used a general factor in most of the cases, as in the case of Einarsen et al. (2009); and in some cases, authors suggested various categories such as control and manipulation of information and control and abuse of working conditions (Escartín et al., in press a). With regard to person-related bullying, various categories were suggested, including verbal aggression (e.g., Zapf et al., 1996); isolation or social exclusion (e.g., Einarsen and Raknes, 1997; Escartin et al., in press a; Yildirim and Yildirim, 2007; Zapf et al., 1996); emotional abuse (Keashly, 1998); and humiliation, which may involve attacks towards self-esteem (Ozturk et al., 2008), attacking the victim's private life (Jiménez et al., 2007; Zapf et al., 1996), and personal derogation (Einarsen and Raknes, 1997). Moreover, most bullying categories can be considered active forms of aggression (most kinds of work-related bullying, verbal aggression, and emotional abuse). Occasionally, there are such passive forms of aggression as not passing on information. Moreover, both direct forms of aggression (e.g., verbal aggression and most forms of emotional abuse) and indirect forms of aggression (e.g., spreading rumours: Zapf et al., 1996; most forms of work-related bullying) occur. Finally, physical and psychological bullying can be distinguished, as can social exclusion and ostracism. In the shipyard study by Einarsen and Raknes (1997), physical violence was reported by 2.4%; whilst in the various studies reported by Zapf (1999a), physical aggression occurred between 3.6% and 9.1% of the bullying cases. Thus the results underline that in the first instance, bullying is primarily a form of psychological rather than physical aggression, although some cultural differences exist (Escartín et al., in press b).

Correlational analyses of overall samples (e.g., Niedl, 1995) show that the bullying categories are very highly correlated. This means that if people are bullied, they tend to experience a large number of bullying behaviours from different behavioural categories. With regard to gender-specific bullying categories, Leymann and Tallgren (1993) report that women used slander and making someone look a fool, whereas men preferred social isolation. Mackensen von Astfeld (2000) found that women used significantly more strategies affecting communication, social relationships, and social reputation, whereas men preferred strategies affecting the victim's work. In a sense, these results correspond to findings regarding schoolyard bullying. Here, Björkqvist et al. (1992) found that boys used physical aggression more often, whereas girls preferred such more indirect strategies as rumours and social exclusion. In Vartia's study (1993), women were more often the victims of strategies of such indirect aggression as spreading rumours and social isolation, whereas men were more often the victims of threats and criticism. However, to challenge possible stereotyping, Hoel and Cooper's nationwide British study (2000) reported that spreading negative rumours and gossiping were particularly widespread in the police service, a highly male-dominated organisation.

Work-related strategies including such acts as being given tasks with impossible targets or deadlines, having one's opinions and views ignored, and being given work clearly below one's level of competence seem to be experienced more often amongst persons in superior positions (Hoel et al., 2001; Salin, 2001). In the studies reported by Zapf (1999a) and Zapf et al. (1996), coworkers used social isolation and attacking the private sphere more often than did the supervisors or managers. Bullying was most frequent when coworkers as well as supervisors were among the bullies. When only supervisors engaged in bullying, strategies such as social isolation, attacking the private sphere, and spreading rumours occurred less often.

One explanation for the foregoing findings may be that some categories, such as social isolation and spreading rumours, work only if many people are involved. It is far more difficult for a single supervisor to isolate somebody. For other bullying categories, such as attacking the private sphere, personal and private information about the victim is necessary, which may be less often at hand for superiors.

Conclusion

This chapter has summarised empirical findings of bullying studies in European countries over the past 20 years. Although different definitions and measures were used in these studies, and although there may be some cultural differences, a converging picture emerges showing that between 3% and 4% of employees may experience serious bullying, and somewhere between 9% and 15% may experience occasional bullying. Between 10% and 20% (or even higher) of employees may occasionally be confronted with negative social behaviour at work that does not correspond to strict definitions of bullying but is nevertheless still very stressful for the persons concerned. In most countries, there seems to be a tendency for bullying to occur more often in the public sector, although bullying seems to exist in all sectors of working life. Bullying occurs on all organisational levels and finds its targets among both young and old as well as among women and men. Yet, men seem to be more often among the perpetrators. Perpetrators, for their part, may be supervisors or colleagues. Most studies report an average duration of bullying well beyond one year. Bullying can be a conflict between just two people; however, very often, there is more than one perpetrator. More and more people seem to become involved the longer bullying lasts. Finally, there is some empirical evidence showing that there exist a variety of bullying behaviours. At least some of the variations found in separate studies may be attributable to cultural differences. It is also important to note that overall findings may mask underlying trends with regard to prevalence as well as the nature of experience, for example, with respect to gender and occupational status. Summarising the existing results on workplace bullying shows that great progress has been made during the last 20 years, which, overall, has led to converging results in the various European countries.

Appendix

TABLE 3.3

Studies on the Frequency of Workplace Bullying

Country	Authors	Sample	No.	Definition*	Prevalence
Austria	Niedl (1995)	Hospital employees	368	1b + 3a	26.6% in sample; 7.8% of the population
		Research institute employees	63	1b + 3a	17.5% in sample; 4.4% of the population
Belgium	Notelaers and De Witte (2003)	Association of local government, consulting office, nonprofit organisation, print office, chemical production	873	8	16%
Belgium	Notelaers et al. (2006)	18 organisations	6175	1a + 3a 7	20.6% 3.1%
Croatia	Russo et al. (2008)	Schoolteachers	764	1b + 3b	22.4%
Denmark	Hogh and Dofradottir (2001)	Randomised sample	1857	5	2%
	Mikkelsen and Einarsen (2001)	Course participants at the Royal Danish School of Educational Studies	99	1b + 3a + 4	4: 2%; 1b 3a: 14% (7.8% for a more stringent criterion)
		Hospital employees	236	1b + 3a + 4	4: 3% now and then; 1b 3a: 16% (2%)
		Manufacturing company	224	1b + 3a + 4	4: 4.1% now and then; 1b 3a: 8% (2.7%)
		Department store	215	1a + 3a + 4	4: 0.9%; 1b 3a: 25% (6.5%)
	Mikkelsen and Einarsen (2002)	Danish manufacturing company	224	1a + 3a + 6a 1a + 3a + 6b	8% 2.7%
	Agervold and Mikkelsen (2004)	Danish manufacturing company	186	1a + 3a + 6a 3a + 4 3b + 4	13% 1.6% 10.3%

Country	Study	Population	N		Prevalence
	Agervold (2007)	Small rural authorities	3024	3a + 4	1%
		State institutions		3b + 4	2.7%
		Day-care institutions		1a + 3a + 6a	4.7%
		Psychiatric wards in hospitals		1a + 3a + 6b	1.2%
	Hansen et al. (2008)	General working population	3363	1a + 3a + 4	1.5%
				1a + 3b + 4	8.5%
	Ortega et al. (2008)	Danish elderly care sector	6301	1b + 4	11.9%
				1a + 4	1.6%
	Ortega et al. (2009)	General working population	3429	1b + 4	8.3%
				1b + 3a + 4	1.6%
Finland	Björkqvist et al. (1994)	University employees	338	1a + 2	16.9%
	Vartia (1996)	Municipal employees	949	4	10.1%
	Kivimäki et al. (2000)	Hospital staff	5655	4	5.3%
	Piirainen et al. (2000)	Representative of employed	1991	4	4.3%
	Salin (2001)	Random sample of business professionals holding a university degree	385	1b + 4	1.6%; 8.8% occas.; 1b + 3a 24.1%
	Vartia and Hyyti (2002)	Prison officers	896	1a + 4	20%; 11.8% bullied several times a month
	Kivimäki et al. (2004)	Hospital employees	4791	4	4.8%
	Varhama and Björkqvist (2004a)	Municipal Finnish employees	1961	1b + 4	16%
	Varhama and Björkqvist (2004b)	General working population	330	1b + 4	14%
	Vartia and Giorgiani (2008)	Immigrants	208	3b + 4	18%
		Finnish employees	600	3b + 4	10%
France	Niedhammer et al. (2007)	General working population	7694	1a + 3a	11.86%
				1a + 3a + 4	9.74%

(continued)

TABLE 3.3 (continued)

Studies on the Frequency of Workplace Bullying

Country	Authors	Sample	No.	Definition*	Prevalence
Germany	Minkel (1996)	Employees of a rehabilitation clinic	46	1b + 3a	8.7%
	Mackensen von Astfeld (2000)	Administration	1989	1b + 3a	2.9%
	Meschkutat et al. (2002)	Representative sample of general working population	2765	4	2.7% (currently) 5.5% (whole year)
	zur Mühlen et al. (2001)	Communal administration	552	1b + 3a	10%
		Administration within federal armed forces	511	1b + 3a	10.8%
Greece	Apospori and Papalexandris (2008)	General working population in Athens area	3301	1b + 3b	30%
Hungary	Kaucsek and Simon (1995)	Army	323	1b + 3a	5.6%
		Bank employees	41	1b + 3a	4.9%
		Bank inspectors	43	1b + 3a	2.5%
Ireland	O'Moore (2000)	Random national sample	1009	4	16.9% occasionally 6.2% frequently
	O'Connell and Williams (2002)	General working population	5252	1a + 3b + 4	7%
	O'Moore et al. (2003)	General working population	1057	1b + 4	6.2%
	O'Connell et al. (2007)	General working population	3579	1a + 4 + 6a	7.9%
Italy	Campanini et al. (2008)	General working population in Lombardy	9229	6b	7%
Lithuania	Malinauskiene et al. (2005)	Schoolteachers from Kaunas City	475	1a + 3a + 4 1a + 3b + 4	2.6% 23%
The Netherlands	Hubert et al. (2001)	Mixed production office business	427	4	4.4%
		Financial institutions; stacked sample	3011	3a + 4	1%

		Sample	N	Code	Percentage
	Hubert and van Veldhoven (2001)	Sample including a variety of branches	66764	2 + 5	2.2% mean of 4 items referring to aggressive and unpleasant situations often or always
Norway	Matthiesen et al. (1989)	Nurses and assistant nurses	99	1a + 4	3.9%
		Teachers	84	1a + 4	10.3%
	Einarsen and Skogstad (1996)	14 different samples; total	7787	1a + 4	Weekly 1.2% (yes, now and then: 3.4%)
		Health and welfare managers	344		8.6% occasional bullying
		Psychologists' union	1402		0.3% (12.0%)
		Employers' federation	181		0.6% (2.3%)
		University	1470		0.6% (2.3%)
		Electricians' union	480		0.7% (2.8%)
		Health-care workers	2145		0.8% (3.1%)
		Industrial workers	485		1.1% (2.2%)
		Graphical workers' union	159		1.3% (6.5%)
		Teachers' union	554		1.9% (8.9%)
		Trade and commerce	383		2.4% (2.0%)
		Union of hotel and restaurant workers	172		2.9% (4.3%)
		Clerical workers and officials	265		2.9% (4.1%)
	Einarsen et al. (1998)	Representative sample from a county	745	1a + 4	3%. 8.4% with previous experience
	Eriksen and Einarsen (2004)	Nurses	6485	3a + 4	4.5%
	Hauge et al. (2007)	General working population	2539	1a + 3a	1.9%
	Matthiesen and Einarsen (2007)	Six Norwegian labour unions	4742	1a + 4	8.3%
	Glasø et al. (2009)	General working population	2539	1a +3a + 4	4.1%
	Magerøy et al. (2009)	Royal Norwegian Navy	1604	1a + 4	2.5%

(continued)

TABLE 3.3 (continued)

Studies on the Frequency of Workplace Bullying

Country	Authors	Sample	No.	Definition*	Prevalence
	Nielsen et al. (2009)	General working population	2539	1a + 4	4.6%
				1b + 4	5.2%
				1a + 3a + 4	0.6%
				1a + 3a + 6a	14.3%
				1a + 3a + 6b	5.2%
				7	6.8%
Poland	Varhama and Bjorkqvist (2004b)	General working population	66	1b + 4	23%
	Merecz et al. (2006)	Nursing staff	413	1b	69.6%
	Warszewska-Makuch (2008)	Polish teachers	1080	1a + 3a	9.3%
Spain	Fidalgo and Piñuel (2004)	General working population	1303	1a + 3a + 8	16%
	Moreno-Jiménez et al. 2005)	Transport and communication–sector employees	103	1a + 3a	26%
	Gil-Monte et al. (2006)	Employees working with disabled people from Valencia	696	3b	19%
				3a	12%
	Justicia et al. (2006)	University staff	548	3b	9%
	Piñuel (2006)	General working population	4250	1a + 6a	9.2%
	Justicia et al. (2007)	University employees	325	4	24.1%
				1a + 3a	11%
	Meseguer et al. (2007)	Fruits and vegetables producers' sector employees	396	1a + 3a	28%
	Escartín et al. (2008)	General working population	300	4	10%
	Fornés et al. (2008)	Professional school nurses	464	1b	17.2%
	Segurado et al. (2008)	Local police	235	1c	57%
	González and Graña (2009)	General working population	2861	1a + 3b	8.2%
				1a + 3a	5.8%

Country	Author(s)	Population	N	Operationalisation	Prevalence
Sweden	Leymann (1992)	Handicapped employees; nonprofit organisation	179	1b + 3a	8.4%; 21.6% handicapped; 4.4% not handicapped
	Leymann and Tallgren (1993)	Steelworks employees	171	1b + 3a	3.5% (probably lower because of dropouts)
	Leymann (1993a)	Sawing factory	120	1b + 3a	1.7%
	Leymann et al. in Leymann (1993b)	Nursery schools	37	1b + 3a	16.2%
	Leymann (1993a, 1993b)	Representative of employed except self-employed	2438	1b + 3a	3.5%
	Lindroth and Leymann (1993)	Nursery school teachers	230	1b + 3a	6%
	Hansen et al. (2006)	Pharmaceutical	91	1a + 4	2%
		Telecommunication	101	1a + 4	5%
		High school	172	1a + 4	7%
		Wood industry	34	1a + 4	6%
		Social insurance	39	1a + 4	3%
Turkey	Cemaloglu (2007)	Schoolteachers	337	1a + 3b	6.4%
	Soylu et al. (2008)	General working population	152	1a + 3a	48%
	Ozturk et al. (2008)	Academic nurses	162	1c + 3b	20.4%
	Yildirim et al. (2007)	University nursing school academics	210	1b	17%
	Yildirim and Yildirim (2007)	Nurses from the European side of Istanbul Province	505	1c + 3b	86.5%
United Kingdom	Bilgel et al. (2006)	Public-sector organisations	877	1b + 3b	55%
	Rayner (1997)	Part-time students	581	1c + 4	53%
	UNISON (1997)	Public-sector union members	736	1 + 4	14%; 1c+4: 50%
	Quine (1999)	National Health Service	1100	3b	38% persistently bullied within last 12 months
	Cowie et al. (2000)	International organisation	386	4	15.4%

(*continued*)

TABLE 3.3 (continued)

Studies on the Frequency of Workplace Bullying

Country	Authors	Sample	No.	Definition*	Prevalence
	Hoel et al. (2001)	Representative sample	5288	1a + 3a + 4	1.4%; 3b: 10.6%
	Baruch (2005)	Multinational corporation	649	8	22.8%
	Tehrani (2004)	Care professionals	162	1c	40%
	Simpson and Cohen (2004)	University teachers	378	8	25%
	Coyne et al. (2004)	Firefighters	288	1a + 4	16%
	Jennifer et al. (2003)	Three large European organisations (Portugal, Spain, United Kingdom)	677	4	21.1%
	Paice et al. (2004)	21 hospitals from London north of the Thames	2730	4	18%
	Quine (2002)	Junior doctors of the British Medical Association	594	4 1b + 3b	37% 84%
	Lewis and Gun (2007)	13 public organisations in South Wales (UK)	247	1a + 3a	20%
	Thomas (2005)	Employees educational sector	42	8	45%
	Coyne et al. (2003)	Public-sector organisation	288	4 1a + 3a + 4	39.6% 3.9%

Legend:

1. denotes duration of acts: 1a within the last six months; 1b over six months; 1c ever in the career
2. denotes type of acts included in judgements (it is phrased as, "intention to harass")
3. denotes frequency of acts: 3a at least weekly; 3b less frequently than weekly
4. denotes victims labelled themselves as bullied based on a definition
5. denotes approximate criterion
6. denotes number of negative acts per week: 6a, one negative act; 6b, two negative acts or more
7. Latent-class cluster analysis (LCC)
8. denotes victims labelled themselves (without a definition)

References

Aasland, M. S., Skogstad, A., Notelaers, G., Nielsen, M. B., and Einarsen, S. (in press) The prevalence of destructive leadership behaviour. *British Journal of Management*.

Agervold, M. (2007) Bullying at work: A discussion of definitions and prevalence, based on an empirical study. *Scandinavian Journal of Psychology*, 48, 161–172.

Agervold, M., and Mikkelsen, G. E. (2004) Relationships between bullying, psycho-social work environment and individual stress reactions. *Work & Stress, 18* (4), 336–351.

Apospori, E., and Papalexandris, N. (2008) Workplace bullying and organizational culture: A multi-level approach. *Sixth International Conference on Workplace Bullying* (pp. 52–54), June 4–6, Montreal, Canada.

Archer, D. (1999) Exploring "bullying" culture in the para-military organisation. *International Journal of Manpower, 20*, 94–105.

Ashforth, B. E. (1994) Petty tyranny in organizations. *Human Relations, 47*, 755–778.

Balducci, C., Alfano, V., and Fraccaroli, F. (2009) Relationships between mobbing at work and MMPI-2 personality profile, posttraumatic stress symptoms, and suicidal ideation and behavior. *Violence and Victims, 24* (1), 52–67.

Baruch, Y. (2005) Bullying on the net: Adverse behavior on e-mail and its impact. *Information & Management, 42*, 361–371.

Bilgel, N., Aytac, S., and Bayram, N. (2006) Bullying in Turkish white-collar workers. *Occupational Medicine, 56*, 226–231.

Björkqvist, K. (1994) Sex differences in aggression. *Sex Roles, 30*, 177–188.

Björkqvist, K., Lagerspetz, K. M. J., and Kaukiainen, A. (1992) Do girls manipulate and boys fight? Developmental trends in regard to direct and indirect aggression. *Aggressive Behavior, 18*, 117–127.

Björkqvist, K., Österman, K., and Hjelt-Bäck, M. (1994) Aggression among university employees. *Aggressive Behavior, 20*, 173–184.

Campanini, P. M., Punzi, S., Costa, G., and Conway, P. M. (2008) Workplace bullying in a large sample of Italian workers. *Sixth International Conference on Workplace Bullying*, June 4–6, Montreal, Canada.

Cemaloglu, N. (2007) The exposure of primary school teachers to bullying: An analysis of various variables. *Social Behavior and Personality: An International Journal, 35*, 789–802.

Cowie, H., Jennifer, D., Neto, C., Angula, J. C., Pereira, B., del Barrio, C., and Ananiadou, K. (2000) Comparing the nature of workplace bullying in two European countries: Portugal and the UK. In M. Sheehan, S. Ramsey, and J. Patrick (eds.), *Transcending the boundaries: Integrating people, processes and systems. Proceedings of the 2000 Conference* (pp. 128–133). Brisbane, Australia: Griffith University.

Coyne, I., Craig, J., and Smith-Lee, P. (2004) Workplace bullying in a group context. *British Journal of Guidance & Counselling, 32* (3), 301–317.

Coyne, I., Smith-Lee, P., Seigne, E., and Randall, P. (2003) Self and peer nominations of bullying: An analysis of incident rates, individual differences, and perceptions of the working environment. *European Journal of Work and Organizational Psychology, 12* (3), 209–228.

Cubela, V., and Kvartuc, T. (2007) Effects of mobbing on justice beliefs and adjustment. *European Psychologist*, 12 (4), 261–271.

Davidson, M. J., and Cooper, C. L. (1992) *Shattering the glass ceiling*. London: Paul Chapman.

Dick, U., and Dulz, K. (1994) *Zwischenbericht Mobbing-Telefon für den Zeitraum 23.8.93–22.2.1994* [Intermediate report of the mobbing telephone]. Hamburg: AOK.

Ege, H. (1998) *I numeri del mobbing: La prima ricera italiana* [The frequency of bullying: The first Italian study]. Bologna: Pitagora Editrice.

Einarsen, S. (2000) Harassment and bullying at work: A review of the Scandinavian approach. *Aggression and Violent Behavior*, 4, 371–401.

Einarsen, S., Aasland, M. S., and Skogstad, A. (2007) Destructive leadership: A definition and a conceptual model. *Leadership Quarterly*, 18, 207–216.

Einarsen, S., Hoel, H., and Notelaers, G. (2009). Measuring exposure to bullying and harassment at work: Validity, factor structure and psychometric properties of the Negative Acts Questionnaire–Revised. *Work & Stress*, 23 (1), 24–44.

Einarsen, S., Matthiesen, S. B., and Skogstad, A. (1998) Bullying, burnout and wellbeing among assistant nurses. *Journal of Occupational Health and Safety*, 14, 563–568.

Einarsen, S., and Raknes, B. I. (1991) *Mobbing i arbeidslivet* [Bullying in working life]. Bergen, Norway: University of Bergen.

—— (1997) Harassment at work and the victimization of men. *Violence and Victims*, 12, 247–263.

Einarsen, S., Raknes, B. I., Matthiesen, S. B., and Hellesøy, O. H. (1994) *Mobbing og harde personkonflikter: Helsefarlig samspill pa arbeidsplassen* [Bullying and severe interpersonal conflicts: Unhealthy interaction at work]. Soreidgrend, Norway: Sigma Forlag.

Einarsen, S., and Skogstad, A. (1996) Prevalence and risk groups of bullying and harassment at work. *European Journal of Work and Organizational Psychology*, 5, 185–202.

Einarsen, S., Tangedal, M., Skogstad, A., Matthiesen, S. B., Aasland, M. S., Nielsen, M. B., Bjørkelo, B., Glaso, L., and Hauge, L. J. (2007) *Et brutalt arbeidsmiljø? En undersøkelse av mobbing, konflikter og destruktiv ledelse i norsk arbeidsliv* [A brutal work life? An investigation of bullying, conflicts and destructive leadership in Norwegian working life]. Bergen, Norway: University of Bergen.

Eriksen, W., and Einarsen, S. (2004) Gender minority as a risk factor of exposure to bullying at work: The case of male assistant nurses. *European Journal of Work and Organizational Psychology*, 13 (4), 473–492.

Escartín, J., Rodríguez-Carballeira, A., Gómez-Benito, J., and Zapf, D. (in press a) Development and validation of the Workplace Bullying Scale "EAPA-T". *International Journal of Clinical and Health Pychology*.

Escartín, J., Rodríguez-Carballeira, A., Porrúa, C., and Martín-Peña, J. (2008) Estudio y análisis sobre cómo perciben el mobbing los trabajadores [Study and analysis of the way mobbing is perceived by workers]. *Revista de Psicología Social*, 23 (2), 203–211.

Escartín, J., Zapf, D., Arrieta C., and Rodríguez-Carballeira, A. (in press b) Workers' perception of workplace bullying: A cross-cultural study. *European Journal of Work and Organizational Psychology*.

Fidalgo, A., and Piñuel, I. (2004) La escala Cisneros como herramienta de valoración del mobbing [Cisneros scale as a tool to assess psychological harassment or mobbing at work]. *Psicothema, 16* (4), 615–624.

Fleishman, E. A. (1953) The description of supervisory behavior. *Journal of Applied Psychology, 37*, 1–6.

Fornés, J., Martínez-Abascal, M., and De la Banda, G. (2008) Análisis factorial del cuestionario de hostigamiento psicológico en el trabajo en profesionales de enfermería [Factor analysis of the questionnaire of psychological harassment at work in clinic employees]. *International Journal of Clinical and Health Psychology, 8* (1), 267–283.

Gil-Monte, P. R., Carretero, N., and Luciano, J. V. (2006) Prevalencia del mobbing en trabajadores de centros de asistencia a personas con discapacidad [Prevalence of mobbing in centers assisting people with disabilities]. *Revista de Psicología del Trabajo y de las Organizaciones, 22* (3), 275–292.

Glaso, L., Nielsen, M. B., and Einarsen, S. (2009) Interpersonal problems among perpetrators and targets of workplace bullying. *Journal of Applied Social Psychology, 39* (6), 1316–1333.

González, D., and Graña, J. L. (2009) El acoso psicológico en el lugar de trabajo: Prevalencia y análisis descriptivo en una muestra multiocupacional [Bullying in the workplace: Prevalence and descriptive analysis of a sample with multiple occupations]. *Psicothema, 21* (2), 288–293.

Halama, P., and Möckel, U. (1995) "Mobbing": Acht Beiträge zum Thema Psychoterror am Arbeitsplatz ["Mobbing": Eight contributions to the issue of psychological terror at work]. In Evangelischer Pressedienst (ed.), *epd-Dokumentation*, vol. 11/95. Frankfurt am Main: Gemeinschaftswerk der Evangelischen Publizistik.

Hansen, A., Hogh, A., Persson, R., and Garde, A. (2008) Associations between bullying, witnessing bullying and sleep problems. *Sixth International Conference on Workplace Bullying* (pp. 133–134), June 4–6, Montreal, Canada.

Hansen, A., Hogh, A., Persson, R., Karlson, B., Garde, A., and Orbaek, P. (2006) Bullying at work, health outcomes, and physiological stress response. *Journal of Psychosomatic Research, 60*, 63–72.

Hauge, L., Skogstad, A., and Einarsen, S. (2007) Relationships between stressful work environments and bullying: Results of a large representative study. *Work & Stress, 21* (3), 220–242.

Hochschild, A. R. (1983) *The managed heart*. Berkeley: University of California Press.

Hoel, H., and Cooper, C. L. (2000) *Destructive conflict and bullying at work*. Manchester, UK: Manchester School of Management, University of Manchester Institute of Science and Technology.

Hoel, H., Cooper, C. L., and Faragher, B. (2001) The experience of bullying in Great Britain: The impact of organisational status. *European Journal of Work and Organizational Psychology, 10*, 443–465.

Hoel, H., Glaso, L., Einarsen, S., Hetland, J., and Cooper, C. L. (in press). Leadership style as a predictor of bullying in the workplace. *British Journal of Management*.

Hoel, H., Rayner, C., and Cooper, C. L. (1999) Workplace bullying. In C. L. Cooper and I. T. Robertson (eds.), *International review of industrial and organizational psychology*, vol. 14 (pp. 195–230). Chichester, UK: Wiley.

Hofstede, G. (1993) Cultural constraints in management theories. *The Executive, 7*, 84–91.

Hogh, A., and Dofradottir, A. (2001) Coping with bullying in the workplace. *European Journal of Work and Organizational Psychology, 10*, 485–495.

Holzen Beusch, E. V., Zapf, D., and Schallberger, U. (1998) Warum mobbingopfer ihre Arbeitsstelle nicht wechseln [Why the victims of bullying do not change their job]. Konstanz, Germany: Department of Psychology, University of Konstanz.

Hubert, A. B., Furda, J., and Steensma, H. (2001) Mobbing, systematisch pestgedrag in organisaties [Mobbing: Systematic harassment in organisations]. *Gedrag & Organisatie, 14*, 378–396.

Hubert, A. B., and van Veldhoven, M. (2001) Risk sectors for undesired behaviour and mobbing. *European Journal of Work and Organizational Psychology, 10*, 415–424.

Jennifer, D., Cowie, H., and Ananiadou, K. (2003) Perceptions and experience of workplace bullying in five different working populations. *Aggressive Behavior, 29*, 489–496.

Justicia, F. J., Benítez Muñoz, J. L., and Fernández de Haro, E. (2006) Caracterización del acoso psicológico en el contexto universitario [Characterisation of bullying in the university context]. *Revista de Psicología del Trabajo y de las Organizaciones, 22* (3), 293–308.

Justicia, F. J., Benítez Muñoz, J. L., Fernández de Haro, E., and Berbén, A. G. (2007) El fenómeno del acoso laboral entre los trabajadores de la universidad [The phenomenon of mobbing among university employees]. *Psicologia em Estudo, 12* (3), 457–463.

Kaucsek, G., and Simon, P. (1995) Psychoterror and risk-management in Hungary. Paper presented as poster at the Seventh European Congress of Work and Organizational Psychology, April 19–22, Györ, Hungary.

Keashly, L. (1998) Emotional abuse in the workplace: Conceptual and empirical issues. *Journal of Emotional Abuse, 1*, 85–117

Kivimäki, M., Elovainio, M., and Vahtera, J. (2000) Workplace bullying and sickness absence in hospital staff. *Occupational and Environmental Medicine, 57*, 656–660.

Kivimäki, M., Leino-Arjas, P., Virtanen, M., Elovainio, M., Keltikangas-Jarvinen, L., Puttonen, S., Vartia, M., Brunner, E., and Vahtera, J. (2004) Work stress and incidence of newly diagnosed fibromyalgia: Prospective cohort study. *Journal of Psychosomatic Research, 57*, 417–422.

Kudielka, B., and Kern, S. (2004) Cortisol day profiles in victims of mobbing (Bullying at the work place): Preliminary results of a first psychobiological field study. *Journal of Psychosomatic Research, 56*, 149–150.

Lewis, D., and Gunn, R. (2007) Workplace bullying in the public sector: Understanding the racial dimension. *Public Administration, 85* (3), 641–665.

Leymann, H. (1990) *Handbok för användning av LIPT-formuläret för kartläg-gning av risker för psykiskt vald* [Manual of the LIPT questionnaire for assessing the risk of psychological violence at work]. Stockholm: Violen.

—— (1992) *Fran mobbning till utslagning i arbetslivet* [From bullying to exclusion from working life]. Stockholm: Publica.

—— (1993a) Ätiologie und Häufigkeit von Mobbing am Arbeitsplatz - eine Übersicht über die bisherige Forschung [Etiology and frequency of bullying in the workplace—an overview of current research]. *Zeitschrift für Personalforschung, 7*, 271–283.

—— (1993b) *Mobbing—Psychoterror am Arbeitsplatz und wie man sich dagegen wehren kann* [Mobbing—psychoterror in the workplace and how one can defend oneself]. Reinbeck, Germany: Rowohlt.

—— (1996) The content and development of mobbing at work. *European Journal of Work and Organizational Psychology, 5*, 165–184.

Leymann, H., and Gustafsson, A. (1996) Mobbing and the development of post-traumatic stress disorders. *European Journal of Work and Organizational Psychology*, 5, 251–276.

Leymann, H., and Tallgren, U. (1990) Investigation into the frequency of adult mobbing in a Swedish steel company using the LIPT questionnaire. Unpublished manuscript.

—— (1993) Psychoterror am Arbeitsplatz [Psychological terror in the workplace]. *Sichere Arbeit*, 6, 22–28.

Liefooghe, A. P. D., and Olaffson, R. (1999) "Scientists" and "amateurs": Mapping the bullying domain. *International Journal of Manpower*, 20, 16–27.

Lindroth, S., and Leymann, H. (1993) *Vuxenmobbning mot en minoritetsgrupp av män inom barnomsorgen: Om mäns jämställdhet i ett kvinnodominerat yrke* [Bullying of a male minority group within child-care: On men's equality in a female-dominated occupation]. Stockholm: Arbetarskyddstyrelsen.

Mackensen von Astfeld, S. (2000) *Das Sick-Building-Syndrom unter besonderer Berücksichtigung des Einflusses von Mobbing* [The sick building syndrome with special consideration of the effects of mobbing]. Hamburg: Verlag Dr Kovac.

Magerøy, N., Lau, B., Riise, R. T., and Moen, B. (2009) Association of psychosocial factors and bullying at individual and department levels among naval military personnel. *Journal of Psychosomatic Research*, 66, 343–351.

Malinauskienë, V., Obelenis, V., and Đopagienë, D. (2005) Psychological terror at work and cardiovascular diseases among teachers. *Acta Medica Lituanica*, 12 (2), 20–25.

Mathisen, G. E., Einarsen, S., and Mykletun, R. (2008) The occurrences and correlates of bullying and harassment in the restaurant sector. *Scandinavian Journal of Psychology*, 49, 59–68.

Matthiesen, S. B., and Einarsen, S. (2001) MMPI-2-configurations among victims of bullying at work. *European Journal of Work and Organizational Psychology*, 10, 467–484.

—— (2007) Perpetrators and targets of bullying at work: Role stress and individual differences. *Violence and Victims*, 22 (6), 735–753.

Matthiesen, S. B., Raknes, B. I., and Rökkum, O. (1989) Mobbing på arbeid-splassen [Bullying in the workplace]. *Tidsskrift for Norsk Psykologforening*, 26, 761–774.

Merecz, D., Rymaszewska, J., Moscicka, A., Kiejna, A., and Jarosz-Nowak, J. (2006) Violence at the workplace: A questionnaire survey of nurses. *European Psychiatry*, 21, 442–450.

Meschkutat, B., Stackelbeck, M., and Langenhoff, G. (2002) *Der Mobbing-Report: Repräsentativstudie für die Bundesrepublik Deutschland* [The mobbing report: Representative study for the Federal Republic of Germany]. Bremerhaven: Wirtschaftsverlag.

Meseguer, M., Soler, M., Sáez, M., and García, M. (2007) Incidencia, componentes y origen del mobbing en el trabajo en el sector hortofrutícola [Incidence, components and source of bullying at work in an agrofruit sector]. *Anales de Psicología*, 23 (1), 92–100.

Mikkelsen, G. E., and Einarsen, S. (2001) Bullying in Danish work-life: Prevalence and health correlates. *European Journal of Work and Organizational Psychology*, 10, 393–413.

—— (2002) Relationships between exposure to bullying at work and psychological and psychosomatic health complaints: The role of state negative affectivity and generalized self-efficacy. *Scandinavian Journal of Psychology*, 43, 397–405.

Minkel, U. (1996) Sozialer Stress am Arbeitsplatz und seine Wirkung auf Fehlzeiten [Social stress at work and its consequences for sickness absence]. Unpublished diploma thesis, Social Science Faculty, University of Konstanz, Germany.

Moreno-Jiménez, B., Rodríguez-Muñoz, A., Garrosa, E., Morante, M., and Rodríguez, R. (2005) Diferencias de género en el acoso psicológico en el trabajo: Un estudio en población española [Gender differences in bullying at work: A study in the Spanish population]. *Psicología em Estudo, 10* (1), 3–10.

Moreno-Jiménez, B., Rodríguez-Muñoz, A., Martínez, M., and Gálvez, M. (2007) Assessing workplace bullying: Spanish validation of a reduced version of the negative acts questionnaire. *Spanish Journal of Psychology, 10* (2), 449–457.

Niedhammer, I., David, S., and Degioanni, S. (2007) Economic activities and occupations at high risk for workplace bullying: Results from a large-scale cross-sectional survey in the general working population in France. *International Archives of Occupational and Environmental Health, 80,* 346–353.

Niedl, K. (1995) *Mobbing/Bullying am Arbeitsplatz. Eine empirische Analyse zum Phänomen sowie zu personalwirtschaftlich relevanten Effekten von systematischen Feindseligkeiten* [Mobbing/bullying at work: An empirical analysis of the phenomenon and of the effects of systematic harassment on human resource management]. Munich: Hampp.

—— (1996) Mobbing and well-being: Economic and personnel development implications. *European Journal of Work and Organizational Psychology, 5,* 239–249.

Nielsen, M. B., and Einarsen, S. (2008) Sampling in research on interpersonal aggression. *Aggressive Behaviour, 34,* 265–272.

Nielsen, M. B., Matthiesen, S. B., and Einarsen, S. (2008) Sense of coherence as a protective mechanism among targets of workplace bullying. *Journal of Occupational Health Psychology, 13* (2), 128–136.

—— (in press) The impact of methodological moderators on prevalence rates of workplace bullying: A meta-analysis. *Journal of Occupational and Organizational Psychology.*

Nielsen, M. B., Skogstad, A., Matthiesen, S. B., Glaso, L., Aasland, M. S., Notelaers, G., and Einarsen, S. (2009) Prevalence of workplace bullying in Norway: Comparisons across time and estimation methods. *European Journal of Work and Organizational Psychology, 18* (1), 81–101.

Notelaers, G., and De Witte, H. (2003) Pesten op het werk: Omvang en welke gedragingen? [Bullying at work: Frequency and negative acts involved]. *Over.Werk, Tijdschrift van het Steunpunt Werkgelegenheid Arbeid en Vorming, 4,* 165–169.

Notelaers, G., Einarsen, S., De Witte, H., and Vermunt, J. (2006) Measuring exposure to bullying at work: The validity and advantages of the latent class cluster approach. *Work & Stress, 20* (4), 289–302.

Nuutinen, I., Kauppinen, K., and Kandolin, I. (1999) *Tasa-arvo poliisitoimessa* [Equality in the police force]. Helsinki: Työterveyslaitos, Sisäasiainministeriö.

O'Connell, P. J., Calvert, E., and Watson, D. (2007) *Bullying in the workplace: Survey Reports, 2007.* Dublin: Economic and Social Research Institute.

O'Connell, P. J., and Williams, J. (2002) *The incidence and correlates of workplace bullying in Ireland.* Dublin: Economic and Social Research Institute.

Olafsson, R., and Johannsdottir, H. (2004) Coping with bullying in the workplace: The effect of gender, age and type of bullying. *British Journal of Guidance and Counselling, 32* (3), 319–333.

Olweus, D. (1994) Annotation: Bullying at school: Basic facts and effects of a school based intervention program. *Journal of Child Psychology and Psychiatry, 35*, 1171–1190.

O'Moore, M. (2000) *Summary report on the national survey on workplace bullying.* Dublin: Trinity College.

O'Moore, M., Lynch, J., and Nic Daeid, N. (2003) The rates and relative risks of workplace bullying in Ireland, a country of high economic growth. *International Journal of Management and Decision Making, 4* (1), 82–95.

O'Moore, M., Seigne, E., McGuire, L., and Smith, M. (1998) Victims of bullying at work in Ireland. *Journal of Occupational Health and Safety–Australia and New Zealand, 14*, 569–574.

Ortega, A., Hogh, A., and Borg, V. (2008) Bullying, absence and presenteeism in Danish elderly care sector: A one-year follow-up study. *Sixth International Conference on Workplace Bullying* (pp. 84–86), June 4–6, Montreal, Canada.

Ortega, A., Hogh, A., Pejtersen, J., and Olsen, O. (2009) Prevalence of workplace bullying and risk groups: A representative population study. *International Archives of Occupational and Environmental Health, 82*, 417–426.

Ozturk, H., Sokmen, S., Yilmaz, F., and Cilingir, D. (2008) Measuring mobbing experiences of academic nurses: Development of a mobbing scale. *Journal of the American Academy of Nurse Practitioners, 20*, 435–442.

Paice, E., Aitken, M., Houghton, A., and Firth-Cozens, J. (2004) Bullying among doctors in training: Cross sectional questionnaire survey. *British Medical Journal, 329*, 658–659.

Piirainen, H., Elo, A.-L., Hirvonen, M., Kauppinen, K., Ketola, R., Laitinen, H., Lindström, K., Reijula, K., Riala, R., Viluksela, M., and Virtanen, S. (2000) *Työ ja terveys—haastattelututkimus* [Work and health—an interview study]. Helsinki: Työterveyslaitos.

Piñuel, I. (2006) *Mobbing, acoso psicológico en el trabajo* [Mobbing—psychological harassment at work]. Madrid: Instituto Regional de Seguridad y Salud en el Trabajo, Conserjería de Empleo y Mujer.

Quine, L. (1999) Workplace bullying in NHS community trust: Staff questionnaire survey. *British Medical Journal, 3*, 228–232.

——— (2002). Workplace bullying in junior doctors: Questionnaire survey. *British Medical Journal, 324*, 878–879.

Rayner, C. (1997) The incidence of workplace bullying. *Journal of Community and Applied Social Psychology, 7*, 199–208.

Russo, A., Milic, R., Knezevic, B., Mulic, R., and Mustajbegovic, J. (2008) Harassment in workplace among school teachers: Development of a survey. *Croatian Medical Journal, 49*, 545–552.

Salin, D. (2001) Prevalence and forms of bullying among business professionals: A comparison of two different strategies for measuring bullying. *European Journal of Work and Organizational Psychology, 10*, 425–441.

Schuster, B. (1996) Rejection, exclusion, and harassment at work and in schools. *European Psychologist, 1*, 293–317.

Schwickerath, J., Riedel, H., and Kneip, V. (2006) Le harcelement moral sur le lieu de travail: Fondements et therapie cognitivo-comportementale des maladies psychosomatiques liees au harcelement moral dans le milieu hospitalier [Bullying in the workplace: Principles and cognitive-behavioral therapy of psychosomatic disorders in relation to bullying in an inpatient setting]. *Journal de Therapie Comportementale et Cognitive, 16* (3), 108–112.

Segurado, A., Agullo, E., Rodríguez, J., Agulló, M., Boada, J., and Medina, R. (2008) Las relaciones interpersonales como fuente de riesgo de acoso laboral en la Policía Local [Interpersonal relations as a source of risk of mobbing in the local police]. *Psicothema, 20* (4), 739–744.

Simpson, R., and Cohen, C. (2004) Dangerous work: The gendered nature of bullying in the context of higher education. *Gender, Work and Organization, 11* (2), 163–186.

Skogstad, A., Einarsen, S., Torsheim, T., Aasland, M. S., and Hetland, H. (2007) The destructiveness of laissez-fair leadership. *Journal of Occupational Health Psychology, 12,* 80–92.

Skogstad, A., Glaso, L., and Hetland, J. (2008) Er ledere i kraft av sin stilling beskyttet mot mobbing? [Are leaders protected against bullying?]. *Søkelys på arbeidslivet, 25* (1), 119–142.

Soylu, S., Peltek, P., and Aksoy, B. (2008) The consequences of bullying at work on organization-based self-esteem, negative affectivity, and intentions to leave: A study in Turkey. *Sixth International Conference on Workplace Bullying* (pp. 6–8), June 4–6, Montreal, Canada.

Tehrani, N. (2004) Bullying: A source of chronic post traumatic stress? *British Journal of Guidance & Counselling, 32* (3), 357–366.

Tepper, B. J. (2000) Consequences of abusive supervision. *Academy of Management Journal, 43* (2), 178–190.

—— (2007) Abusive supervision in work organization: Review, synthesis, and research agenda. *Journal of Management, 33* (3), 261–281.

Thomas, M. (2005) Bullying among support staff in a higher education institution. *Health Education, 105* (4), 273–288.

UNISON (1997) *UNISON members' experience of bullying at work.* London: UNISON.

—— (2000) *Police staff bullying report* (No. 1777). London: UNISON.

Varhama, L. M., and Bjorkqvist, K. (2004a) Conflicts, burnout, and bullying in a Finnish and a Polish company: A cross-national comparison. *Perceptual and Motor Skills, 98,* 1234–1240.

—— (2004b) Conflicts, workplace bullying and burnout problems among municipal employees. *Psychological Reports, 94,* 1116–1124.

Vartia, M. (1991) Bullying at workplaces. In S. Lehtinen, J. Rantanen, P. Juuti, A. Koskela, K. Lindström, P. Rehnström, and J. Saari (eds.), *Towards the 21st century: Proceedings from the International Symposium on Future Trends in the Changing Working Life* (pp. 131–135). Helsinki: Institute of Occupational Health.

—— (1993) Psychological harassment (bullying, mobbing) at work. In K. Kauppinen-Toropainen (ed.), *OECD Panel group on women, work, and health* (pp. 149–152). Helsinki: Ministry of Social Affairs and Health.

—— (1996) The sources of bullying: Psychological work environment and organizational climate. *European Journal of Work and Organizational Psychology, 5,* 203–214.

—— (2001) Consequences of workplace bullying with respect to well-being of its targets and the observers of bullying. *Scandinavian Journal of Work Environment and Health, 27,* 63–69.

Vartia, M., and Giorgiani, T. (2008) Bullying of immigrant workers. *Sixth International Conference on Workplace Bullying* (pp. 149–150), June 4–6, Montreal, Canada.

Vartia, M., and Hyyti, J. (1999) *Väkivalta vankeinhoitotyössä* [Violence in prison work; English summary]. Helsinki: Oikeusministeriön vankeinhoito-osaston julkaisuja 1.

—— (2002) Gender differences in workplace bullying among prison officers. *European Journal of Work and Organizational Psychology, 11,* 1–14.

Warszewska-Makuch, M. (2008) Workplace bullying, the big five personality dimensions, and job insecurity findings from a Polish teachers' sample. *Sixth International Conference on Workplace Bullying* (pp. 72–73), June 4–6, Montreal, Canada.

Yildirim, A., and Yildirim, D. (2007) Mobbing in the workplace by peers and managers: Mobbing experienced by nurses working in healthcare facilities in Turkey and its effect on nurses. *Journal of Clinical Nursing, 16,* 1444–1453.

Yildirim, D., Yildirim, A., and Timucin, A. (2007) Mobbing behaviours encountered by nurse teaching staff. *Nursing Ethics, 14* (4), 447–463.

Zapf, D. (1999a). Mobbing in Organisationen. Ein Überblick zum Stand der Forschung [Mobbing in organisations: A state of the art review]. *Zeitschrift für Arbeits- and Organisationspsychologie, 43,* 1–25.

—— (1999b). Organizational, work group related and personal causes of mobbing/bullying at work. *International Journal of Manpower, 20,* 70–85.

—— (2002) Emotion work and psychological strain: A review of the literature and some conceptual considerations. *Human Resource Management Review, 12,* 237–268.

Zapf, D., and Einarsen, S. (2005) Mobbing at work: Escalated conflicts in organizations. In S. Fox and P. E. Spector (eds.), *Counterproductive work behaviour* (pp. 237–270). Washington, DC: American Psychological Association.

Zapf, D., and Gross, C. (2001) Conflict escalation and coping with workplace bullying: A replication and extension. *European Journal of Work and Organizational Psychology, 10,* 497–522.

Zapf, D., Knorz, C., and Kulla, M. (1996) On the relationship between mobbing factors, and job content, the social work environment and health outcomes. *European Journal of Work and Organizational Psychology, 5,* 215–237.

Zapf, D., Renner, B., Bühler, K., and Weinl, E. (1996) *Ein halbes Jahr Mobbingtelefon Stuttgart: Daten und Fakten* [Half a year mobbing telephone Stuttgart: Data and facts]. Konstanz, Germany: Social Science Faculty, University of Konstanz.

zur Mühlen, L., Normann, G., and Greif, S. (2001) Stress and bullying in two organisations. Unpublished manuscript, Faculty of Psychology, University of Osnabrück

4

Individual Consequences of Workplace Bullying/Mobbing

Annie Hogh, Eva Gemzøe Mikkelsen, and Åse Marie Hansen

CONTENTS

Introduction

Over the past three decades, research into the consequences of bullying at work has shown that exposure to systematic negative behavior at work such as bullying or mobbing may have devastating effects on the health and well-being of the exposed individuals. The early research consisted mostly of cross-sectional studies, a few case-control studies, and clinical interviews with bullied people. The studies found correlations between exposure to bullying and chronic fatigue; psychosomatic, psychological, and physical symptoms; general stress; insomnia; and mental stress reactions (for reviews, see, e.g., Dofradottir and Høgh, 2002; Einarsen and Mikkelsen,

2003; Moayed et al., 2006). Common symptoms such as musculoskeletal complaints, anxiety, irritability, and depression were reported by targets in different European countries (Einarsen et al., 1996; Niedl, 1996; O'Moore et al., 1998; Zapf et al., 1996b). Some victims displayed a pattern of symptoms indicative of post-traumatic stress disorder (PTSD) (Björkqvist et al., 1994; Einarsen et al., 1999; Leymann and Gustafsson, 1996; Mikkelsen and Einarsen, 2002). Self-hatred and suicidal thoughts have also been reported (Einarsen et al., 1994; Thylefors, 1987). Qualitative studies (Kile, 1990; Mikkelsen and Iversen, 2002; O'Moore et al., 1998; Price Spratlen, 1995; Thylefors, 1987) have demonstrated consequences such as reduced self-confidence, low self-worth, shyness, and an increased sense of vulnerability as well as feelings of guilt and self-contempt. Indeed, many targets reported that their physical health and mental well-being had been permanently damaged (Mikkelsen, 2001b). Not surprisingly, exposure to bullying seemed to be associated with increased sickness absence (Kivimäki et al., 2000), risk of unemployment (Einarsen et al., 1999; Mikkelsen and Iversen, 2002), and early retirement (Matthiesen et al., 1989).

Bullying at work not only affects the targets. Nonbullied witnesses or bystanders also reported significantly more general stress and mental stress reactions than employees from workplaces without bullying (Vartia, 2001), and some might leave their jobs as a result of witnessing bullying (Rayner, 1999).

All together, these early studies suggest that bullying at work is a potent stressor that negatively affects the health and well-being of the targets and, to some degree, the witnesses as well. Thus, based on the foregoing, the aim of the present chapter is to review and discuss the more recent research findings on individual consequences of bullying or mobbing at work and to relate these findings to different theories as well as to the methods used to investigate bullying at work.

Stress Theory

The early research demonstrated that exposure to bullying may have highly detrimental effects on targets' psychological and physiological health. Yet, the studies also show that some targets exhibit only moderate levels of stress reactions whereas other victims develop such severe symptoms of stress as, for instance, depression or PTSD.

According to transactional stress models, the nature and severity of emotional reactions following exposure to bullying are functions of a dynamic interplay between event characteristics and individual appraisal and coping processes (Folkman and Lazarus, 1991; Lazarus, 1999; Zapf and Einarsen, 2003). By definition, bullying at work is defined as a *prolonged exposure* to *interpersonal acts* of a *negative nature*, with which the target is *unable to cope*. Together, these four factors are likely to make up a highly stressful situation characterized by lack of control. Attributions of control and predictability are salient features of the individual's appraisal processes (Joseph, 1999; Lazarus, 1999). People have a strong need to feel that they can predict the future and control the events that happen to them (Janoff-Bulman and Thomas, 1989; Joseph, 1999). When facing uncontrollable and unpredictable events such as bullying (Zapf and Einarsen, 2003), the target will be highly motivated to try to understand what is happening (Weiner, 1985) with the purpose of reestablishing perceptions of the world as predictable and controllable (Janoff-Bulman, 1989). Although influenced by personality factors attributions of control are very stressor specific (Miller and Porter, 1983; Peterson et al., 1981), exposure to long-term bullying by colleagues and/or superiors typically connotes being the subject of frequent personal attacks and social isolation that increases the target's perception of being a helpless victim of intentional psychological abuse. Moreover, long-term, high-intensity bullying signals an overwhelming danger of exclusion from the company. As such, the threat potentials in bullying are abundant, yet individual differences may still influence the extent to which targets develop severe stress symptoms.

The New Evidence

Cross-Sectional Studies

A number of studies have been published recently that explore exposure to bullying and potential psychological and physiological consequences. Bullying in the more recent studies has been measured by self labeling (asking about exposure to bullying, usually by one question that is preceded by a definition) or by self-reported exposure to different negative acts.

Psychological Stress Reactions, Absence and Substance Abuse

An interview study with 20 targets found that they usually developed psychological and psychosomatic symptoms just a few months after the bullying started (Hallberg and Strandmark, 2006). At first, the symptoms would only appear when the targets were at work. However, over time the symptoms became more chronic. The targets described a number of psychological symptoms (e.g., inability to concentrate, mood swings, anxiety, sleep problems, fear, and depressive symptoms) as well as psychosomatic symptoms (e.g., headaches, respiratory and cardiac complaints, hypertension, and hypersensitivity to sounds). If the targets had a chronic disease, the symptoms often became worse.

A few studies show that targets may cope with the bullying by using sickness absence or taking drugs. Some may even contemplate suicide. With respect to sickness absenteeism, a large Swedish study of postal workers found that bullied women had a 1.6 times higher risk of taking sickness absence than nonbullied women (Voss et al., 2004). Similar results were found among women in the Swedish public sector (Vingård et al., 2005). Among Danish health care workers in care of the elderly, targets of frequent bullying had a four times' higher risk of taking many sick days (> 15 days) (Høgh et al., 2007). The risk was 2.3 times higher among the occasionally bullied. In addition, compared to nonbullied employees, more targets went to work even when they were sick (i.e., presenteeism).

Among Tyrolean targets of bullying, 20% took drugs repeatedly because of job problems as compared to 4.1% of the nonbullied employees. The feeling of being a victim of bullying at work appeared to be particularly destructive for the individuals' well-being. The prevalence of drug consumption increased from 5.6% on average to 85% when the employees were discontented with their jobs as well as exposed to bullying at work (Traweger et al., 2004).

In his early work, Leymann stressed the risk of targets committing suicide (Leymann, 1988; Leymann, 1992). Recent research confirms that being bullied at work may increase the risk of suicide even among men and women with no prior psychiatric disorder. Among 102 interviewed individuals, 48% had no suicide risk, 31.4% had a low risk, 16.7% had a medium risk, and 3.9% had a high risk; 65% also reported physical complaints. The targets with suicide risk reported greater psychopathology as well (Pompili et al., 2008).

Sleep Problems

Sleep problems interfere significantly with daytime functioning and overall well-being and may lead to serious clinical consequences. Sleep deprivation may be regarded as a neurobiological and physiological stressor. Sleep has important homeostatic functions, and sleep deprivation is a stressor that may have consequences for the brain, for instance, memory and cognitive functions as well as regulation of neuroendocrine systems (McEwen, 2006). Targets of bullying are more likely than nontargets to have sleep difficulties (Rafnsdóttir and Tómasson, 2004), a lower sleep quality (Notelaers et al., 2006), and more often use of sleep-inducing drugs and sedatives (Vartia, 2001) compared to nonbullied respondents. A recent study in 60 Danish workplaces found that targets report more sleep problems, poorer general health, and more somatic symptoms after controlling for gender and age (Hansen et al., 2009). The precise pathological processes behind the target's poorer health are not fully understood, but it is commonly conceived that stress reactions including sleep problems play a major role. Because sleep is a major path for restitution and is vital for health and well-being, it seems plausible that sleep problems might provide a link between work-related bullying and poor health.

Negative Acts and Health Problems

Some studies have used different negative acts questionnaires to assess exposure to bullying and its potential consequences. They all used different types of inventories. For instance, in a study of the Norwegian restaurant sector, a positive correlation was found between exposure to negative acts (measured by the Negative Acts Questionnaire [NAQ)]: 27 items) and individual burnout (Mathisen et al., 2008). Using a 20-item inventory and including a definition of bullying, a Turkish study of employees in public sectors found a correlation between bullying and anxiety, depression, and job-induced stress. Respondents with low social support reported the poorest mental health (Bilgel et al., 2006). Using a 12-item negative acts questionnaire including both work- and person-related acts, a Danish study among 186 blue-collar workers found that targets (i.e., workers with exposure to one negative act at least weekly) reported more psychological and psychosomatic stress symptoms and mental fatigue than did the nonbullied. Moreover, targets also reported more sick leave (Agervold and Mikkelsen, 2004). Still, using a 24-item version of the revised NAQ (Einarsen and Hoel, 2001), a study from a Spanish production sector found a significant association with psychosomatic symptoms among targets of bullying. Various types of personal, work-related, and other forms of bullying accounted for 27% of the global variance of psychosomatic symptoms (Meseguer et al., 2008). In another Spanish study, bullying (as measured by six items of social forms of negative acts), was found to impact the experience of tension (mental strain and anxiety) and burnout (Meliá and Becerril, 2007). Using the Leymann Inventory of Negative Acts (with repeated exposure for almost six months), two studies found depressive symptoms (Niedhaminer et al., 2006; Nolfe et al., 2008) and symptoms of PTSD (Nolfe et al., 2008) among targets.

Finally, a large British study in 70 organizations within the private, public, and voluntary sectors investigated health consequences of bullying as measured by 29 negative acts as well as self-labeled bullying. Results showed a strong correlation between self-reported frequency of bullying and physical and mental well-being. The correlation was slightly larger for men than for women. Also, larger correlations were found between frequency of negative behaviors and health effects than between self-labeled bullying and health effects. The associations between negative behavior and outcomes were slightly larger for women than for men. Some negative acts also seemed to be more detrimental than others. Men reacted more strongly to persistent criticism of work and effort, being ignored, attempts to find faults, and hints that they should quit, whereas women reacted more to hints that they should quit, pressure not to claim something that by right they were entitled to, having allegations made against them, and repeated reminders of errors or mistakes (Hoel et al., 2004).

Summary

The preceding cross-sectional studies confirm results of previous studies in as much as bullying seems to have serious negative effects on the health and well-being of the targets, including effects on the pattern and quality of sleep. Moreover, targets seem to be at a high risk of sickness absence as well as presenteeism. However, the studies are not directly comparable, since they use different methods of data collection (i.e., questionnaires, telephone interviews, and personal interviews), which have different sensitivity to the reporting of sensitive questions (Feveile et al., 2007). Using different inventories, studies have found that certain negative acts are detrimental to the targets' health and well-being. Moreover, with the exception of the studies on sleep problems, most of the studies add little knowledge about the individual consequences of bullying. However, at least three new contributions have been demonstrated: (1) there is more diversity in the way bullying or mobbing is measured, (2) associations are found between both self-labeled bullying and exposure to negative behavior and health effects, and (3) health effects may be different among women and men.

Although cross-sectional studies cannot determine the direction of cause and effect, they demonstrate associations, which may be tested in longitudinal studies.

Longitudinal Studies

Longitudinal studies offer the possibility of measuring exposure and effects at different time points, thus making it easier to conclude on the direction of the relationship between exposure to bullying and health effects (Zapf et al., 1996a). However, as Zapf and colleagues point out, merely applying a longitudinal design does not suffice, since certain methodological issues must be considered. For instance, all variables should be measured at all time points. Third variables should be considered as potential confounders. Moreover, the time lag should be carefully considered, and assumptions of the time course of the variables under study should be made.

A longitudinal Finnish hospital study of primarily female employees (Kivimäki et al., 2003) showed a much higher risk of cardiovascular disease for targets of prolonged bullying compared to nontargets (Odds Ratio [OR] = 2.3). Adjustment for overweight at baseline attenuated the OR to 1.6. The targets had a four times' higher risk of developing depressive symptoms controlled for gender, age, salary, overweight, and illness at baseline. The longer time the bullying had taken place, the higher the risk of depression. The strong association between bullying and subsequent depression suggests that bullying is an etiological factor for mental health problems. Retirement may be the end result of exposure to bullying at work. As such, a retrospective case-control study showed that bullied home-care workers had twice as high a risk of ending up with a disability pension five years later than those who had not been bullied (Dellve et al., 2003).

A five-year follow-up study of a representative sample of the Danish working population showed that exposure to a negative act such as nasty teasing may generate long-term stress reaction even when adjusted for work environment factors, social climate, and psychological health at baseline and for nasty teasing at follow-up. Direct effects were found for women. For both sexes, education, job, and social class had no impact on the relationships between nasty teasing at baseline and health at follow-up (Hogh et al., 2005). Being bullied during education may also have long-term health effects, as demonstrated by a prospective study of health care workers in care of the elderly (Høgh et al., 2007). Respondents who had been bullied during their education had a higher risk of mental health problems, somatic stress reactions, or fatigue during the first year at work after graduation from college when adjusted for health problems at baseline.

Whereas cross-sectional studies show a higher risk of sickness absence among targets of bullying than among nontargets, a longitudinal study of health care workers showed an indirect relationship between bullying and sickness absence among targets through such mental health symptoms as depression and burnout. However, compared to nonbullied health care workers, targets went to work when sick (presenteeism) to a higher degree the more they were bullied (OR = 2.4 among the frequently bullied and 1.5 among the occasionally bullied adjusted for gender, age, occupational group, and poor health at baseline) (Ortega et al., 2009).

In a descriptive study, bullied targets (36 women and 12 men) with no previous significant psychiatric history were followed for a year (Brousse et al., 2008). At the first interview, 81% reported high levels of stress at work; 52% and 83% reported depression and anxiety disorders, respectively. The targets were recommended to create a distance to the bully. At follow-up, only 23 still worked. These respondents had less depressive symptoms than the nonworkers, and only 19% reported a feeling of stress at work. In general, there was a significant change in symptoms of anxiety (60%) from baseline to follow-up, while there was no change in symptoms of depression. Many of the targets reported various persistent somatic symptoms or disorders, 88% had a high neuroticism score, and around half reported feelings of shame and guilt for having been bullied; they also presented loss of self-confidence. A quarter of the targets still had suicidal ideation at follow-up. A key finding of this study is the demonstrated severity of the mental health problems following exposure to bullying, leading to serious psychiatric pathologies in people who previously presented no significant psychiatric history.

Summary

The foregoing longitudinal studies indicate that from a methodological point of view, progress has been made in the investigation of bullying and health. Methodologically, the reviewed studies follow some of the recommendations of Zapf and colleagues (1996a), yet no study follows all of them. The studies show that being exposed to bullying at work may have long-term health effects and that there seems to be some sort of gender difference in the symptoms presented. Accordingly, future studies should aim at stratifying analyses for gender. The studies also indicate that prolonged exposure is more detrimental and that bullying may result in early retirement. Lastly, one important result is that getting away from the bullying while continuing to work is beneficial to the targets' health and well-being.

Physiological Reactions

The fight-or-flight response is the classical way of envisioning the behavioral and physiological response to a threat from a challenging or dangerous situation. The stress response is the activation of the autonomic nervous system and hypothalamo-pituitary-adrenal (HPA) axis. Activation is a normal response and as such not unhealthy. However, inadequate or excessive adrenocortical and autonomic function is deleterious for health and survival. It is when the fight-or-flight response occurs too frequently or is greatly prolonged that we begin to experience the negative effects of stress. This prolonged elevation may be due to anxiety, to constant exposure to adverse environments involving interpersonal conflict, and to changes in lifestyle and health-related behaviors that result from being under chronic stress (McEwen, 2007). Recent research has pointed to a functional link between stress, disturbed sleep, psychiatric disorders, ageing, and neuroendocrine dysfunctions. In particular, elevated plasma cortisol levels have been shown in physiological ageing and patients with psychiatric disorders. Salivary cortisol has increasingly been used to study occupational stress and the responsiveness of the HPA axis in both field studies and experimental studies (Aardal-Eriksson et al., 1999; Evans and Steptoe, 2001; Kirschbaum et al., 1989; van Eck et al., 1996; Zeier, 1994).

Cortisol secretion in saliva exhibits a distinct diurnal variation with the highest concentration in the morning at awakening and lowest in the evening (dynamic HPA axis) (Björntorp and Rosmond, 2000; Kirschbaum and Hellhammer, 1994; Nikolajsen et al., 2003). A few studies have addressed the physiological consequence of workplace bullying with biological measurements among targets who were still working (Hansen et al., 2006; Kudielka and Kern, 2004). Kudielka and Kern presented tentative evidence of an altered circadian cycle of cortisol secretion among targets. Likewise, Hansen and colleagues observed signs of an altered HPA-axis activity among targets manifested as a lower excreted amount of salivary cortisol in the morning. Similar results were reported recently among young adults in as much as salivary cortisol levels and systolic blood pressure were lower in male targets who reported having no feelings of anger about their experience compared to controls and those who did report anger (Hamilton et al., 2008). While these observations are interesting and potentially clarifying as regards how bullying might get "under the skin," it is equally clear that the study designs and methods used have limitations. In order to confirm their validity, the results must be replicated in other populations and with stronger study designs. As such, in the Hansen et al. (2006) study, the definition of bullying did not account for frequency or duration, which are often considered important aspects despite controversies as to how they should be incorporated in a definition (Leymann, 1996; Zapf and Einarsen, 2005). A recent study of a large number of occupationally active persons was designed to counter methodological weaknesses inherent in previous studies involving salivary cortisol (Hansen et al., 2010). Results showed that irrespective of gender, frequently bullied employees had poorer psychological health and a lower level of salivary cortisol compared to a nonbullied reference group. Occasionally bullied employees had only a poorer psychological health compared to a reference group. Similar results were reported among young adults (Hamilton et al., 2008), where the most affected individuals showed long-term effects on salivary cortisol.

Bullying as a Traumatic Event

The studies outlined in this chapter give evidence of bullying at work as an extreme social stressor with traumatic potential. Traumatic stressors are events that are overwhelming to such a degree that the individual feels unable to function without substantial help from others (Lazarus, 1999). According to Janoff-Bulman (1989, 1992), traumatic events threaten to shatter basic cognitive schemas involving fundamental beliefs that the world is benevolent and meaningful and that we, as individuals, are worthy, decent, and capable human beings deserving other people's affection and support. Insofar as we need stability in our conceptual system (Epstein, 1985), sudden and forced changes in core schemas are deeply threatening and may result in an intense psychological crisis (Janoff-Bulman, 1992). The conceptual incongruity between the trauma-related information and prior schemas leads to cognitive disintegration (Epstein, 1985; Janoff-Bulman and Thomas, 1989), which gives rise to stress responses requiring reappraisal and revision of the basic schemas. During recovery, victims must rebuild core schemas that account for the experience of being victimized. Yet, instead of resolving the cognitive-emotional crisis, some victims remain in a chronic state of cognitive confusion and anxiety characteristic of PTSD.

Post-Traumatic Stress Disorder

Post-traumatic stress disorder, or PTSD, is a diagnosis that refers to a constellation of late stress reactions among people who have experienced, witnessed, or been confronted with one or more events that involve actual or threatened death or serious injury or threat to the physical integrity of self or others. The person's response should involve intense fear, helplessness, or horror (McFarlane, 2008). Three types of stress reactions are included in the diagnosis: (1) persistent reexperiencing of the traumatic event in either memories, dreams, or flashbacks or severe distress when reminded of the traumatic event(s); (2) avoidance of reminders of the stressors and a reduction in the ability to feel positive emotions (emotional numbing); and (3) persistent physical arousal that was not present before the stressor occurred. The symptoms should be present for at least one month. Symptoms do not have to be present every day, but they do have to cause disturbance to the person for the entire duration (Ford, 2008). Surveys have shown that targets of bullying have symptoms consistent with PTSD and only fail to fulfill the so-called stressor criterion.

Symptoms of PTSD were investigated among 102 targets of bullying (74% were women), among whom 64% had been bullied for two years or more (Matthiesen and Einarsen, 2004). The targets had higher post-traumatic stress scores as measured on the Revised Impact of Event Scale (IES-R) and the Post-Traumatic Stress Scale (PTSS-10) than three comparison groups (i.e., parents of schoolchildren involved in a fatal bus accident, UN personnel one year after returning from a war zone, and a group of medical students). A majority of the targets exceeded recommended threshold values indicating PTSD, yet the study also showed that symptoms seemed to diminish over time. Similar results were found in a study of 165 care professionals, among whom 65 had been bullied during a two-year period. Levels of symptoms consistent with a diagnosis of PTSD were found among 36 of the bullied care professionals (Tehrani, 2004).

Possible PTSD symptoms attributable to exposure to bullying by superiors, as measured by the NAQ, were investigated among 199 members of two support and bullying victims' organizations. Results showed that 84% had experienced PTSD symptoms as measured on the IES-R scale. This was more than what has been previously found in studies of victims of other traumatic events, among whom 65%–75% had PTSD. The present study showed that managers displaying behaviors compatible with the criteria for tyrannical leadership seemed to constitute a severe stress factor, inasmuch as this leadership style was associated with all three PTSD symptom groups among the targets. This association did not change when controlling for the effects of the bullying behaviors (Nielsen et al., 2005).

A few recent studies have investigated relationships between bullying, basic assumptions, and PTSD. As such, a matched case–control questionnaire study among 41 women and 13 men showed that prolonged exposure to bullying at work (as measured by an inventory of 29 negative acts and self-labeled bullying) adversely affected beliefs in a just world (BJW) and its function in sustaining a positive outlook and mental health (depression, optimism, pessimism, trust, and cynicism). Perceived victimization, rather than the mere frequency of exposure to negative acts, was associated with worse adjustment and a weaker belief in a just world. Targets of self-labeled bullying tended to be more depressed and pessimistic, and they were less trustful in the goodness of people (Adoric and Kvartuc, 2007).

Another matched case–control study of 72 targets and 72 nonbullied controls also explored relationships among bullying, basic assumptions, and PTSD (Glasø et al., 2009). Results showed that the targets could be separated in two clusters with respect to basic assumptions. The first cluster (43.7%) had higher scores on all assumptions and a higher level of PTSD symptoms than the second cluster (56.3%). The control group could also be separated in two clusters, with the second cluster (76.4%) scoring higher on the assumptions "World as meaningful" and "Self as worthy." Both victim clusters had lower scores on the World Assumptions Scale than the largest cluster in the control group. Yet, the smallest control group cluster scored lower on "World as meaningful" and "Self as worthy" than victim cluster 1 and control cluster 1. However, the study failed to support assumed relationships between exposure to bullying and basic assumptions, since only the assumption of "World as benevolent" correlated significantly with exposure to negative acts. Differences in assumptions in the target group irrespective of level of exposure to negative acts could be read as supporting a vulnerability hypothesis (Glasø et al., 2009). For instance, some targets may have had negative assumptions prior to their victimization, as would be the case of individuals high in trait-negative affectivity (trait-NA) or neuroticism (Watson and Clark, 1984). Given that targets of bullying have been found to score higher on trait-NA or neuroticism (Persson et al., 2009), this personality trait may be *the* hidden causal agent behind the supposed relationship between basic assumptions and PTSD (Mikkelsen, 2001a). Moreover, some victims may have had highly rigid or unrealistic assumptions prior to the bullying, as suggested by Thylefors (1987). For these victims, exposure to bullying may bring about an overwhelming feeling of generalized vulnerability (Mikkelsen, 2001a).

Stress and the Coping Process

Although individual coping strategies play a central role in the stress process, research on targets' coping with bullying is far from abundant.

The early coping studies indicate that even though most targets make an effort to stop the bullying by means of various strategies, their efforts are usually to no avail, and hardly any targets manage to stop the bullying without help from others (Hogh and Dofradottir, 2001; Niedl, 1996; Zapf and Gross, 2001). A more recent qualitative study demonstrated similar findings among 30 self-selected targets of bullying (Mikkelsen, 2004). These targets would initially try to solve the problem by means of confronting the bully or reporting to management. When this strategy failed, they often attempted to control or suppress their negative feelings while trying to live up to the bully's demands. Alternatively, they sought support among colleagues and/or superiors. After some time, they once again tried to solve the problem. When this approach failed, they contacted the union. This last step was usually followed by long-term sick leave. Subsequently, the targets either were fired or quit themselves.

In a cross-sectional questionnaire study of 127 students who had been employed during the 12 months preceding the study, Djurkovic and colleagues (2005) explored how targets would use assertiveness, avoidance, or seeking formal help when exposed to five categories of bullying; threats to professional status or to personal standing, isolation, overwork, and destabilization. Results revealed that avoidance was the most common strategy used, suggesting that many targets felt unable to defend themselves. Assertiveness was mostly used when the targets were exposed to work-related behaviors, the consequences of which would lead to a reduction of the targets' ability to perform their jobs. In such a situation, a presumably effective reaction would be to confront the bully. Likewise, targets sought formal help only when exposed to violence, an extreme form of bullying that compels filing a complaint (Djurkovic et al., 2005).

Despite methodological limitations such as the use of cross-sectional designs, convenience sample targets, self-report measure, and possible self-recall bias, the foregoing studies give a clear picture of massive coping failure, which is likely to be traumatizing for at least some individuals inasmuch as it shatters basic assumptions or needs of controllability and predictability, for example.

In addition to the need for perceived controllability and predictability, people also have a need for self-esteem; for that reason, they will tend to construe events in a way that enhances and maintains self-esteem (Brewin, 1988). Attributions of causality are highly relevant in this respect. Prolonged exposure to acts with which the targets cannot cope is likely to instigate the use of stable, internal causal attributions (Peterson et al., 1981). Following the hopelessness theory (Abramson et al., 1988), internal, stable, and global causal attributions for a negative event such as bullying is likely to lead to feelings of hopelessness, reduced self-worth, and depression (Abramson et al., 1978; see also Nielsen et al., 2005). Internal attributions of causality are more likely when the bully is in a superior position, given that within work contexts being the boss implies a right to define what is right and what is wrong (Nielsen et al., 2005). Single targets are also more likely to make internal attributions (Keashly, 1998; Mikkelsen, 2001a; Nielsen et al., 2005).

Many targets are subjected to bullying over a long time period—often several years. This longevity increases the probability of targets blaming themselves for being bullied and for not being able to put an end to the bullying. A young woman put it this way:

> Actually the only one I let down was myself—because I didn't say NO! Because I know that the person I was before would never have put up with it. Because I didn't say anything! I never said: "Don't yell at me!" or "Answer me when I'm talking to you!" (Mikkelsen et al., 2007)

In an early qualitative study, Kile (1990) found self-blame to be a typical reaction to long-term bullying. In fact, attributions of self-blame may be one way of dealing with highly stressful events such as bullying. According to Janoff-Bulman (1992; Janoff-Bulman and Thomas, 1989), there are two kinds of self-blame. Behavioral self-blame is related to internal, unstable, and specific causal attributions and is considered adaptive given that it allows victims to uphold a belief in the world as benevolent, predictable, and controllable. Moreover, by blaming their own actions rather than their character, victims may sustain their self-worth. Conversely, characterological self-blame refers to internal, stable, and global causal attributions and is considered maladaptive, since it implies that the victim is unable to change the situation or avoid similar circumstances in the future. Moreover, blaming stable personal attributes may damage the victims' self-worth. Indeed, characterological self-blame is associated with low self-worth (Janoff-Bulman and Thomas, 1989), increased hopelessness (Peterson et al., 1981), and depression (Glinder and Compas, 1999).

The fact that many targets fail to put an end to the bullying no doubt contributes to the high level of stress symptoms demonstrated in the foregoing studies. However, Djurkovic et al. (2005) also showed that targets do not act uniformly to bullying. As such, this study reveals a more fine-grained picture of the process of coping with bullying. Nonetheless, although stressing the importance of situational factors in determining coping strategies, the study also showed that of the 38 targets who did not use assertiveness or seek help at all, 32 persons used avoidance as their first and only strategy. Hence, this finding may also be read as indicative of personality playing a role in determining targets' coping strategies.

Personality and Individual Differences

Whereas it is still unclear how and to what extent aspects of the target's personality predicts exposure to bullying, it is fairly safe to say that individual dispositions may act as vulnerability factors insofar as such individual variables are likely to affect appraisal and coping processes (Cox and Ferguson, 1991; Lazarus and Folkman, 1984; Spector et al., 2000).

A recent cross-sectional study of 183 targets of bullying (Moreno-Jiménez et al., 2007) showed that social anxiety and assertiveness moderated the relationship between bullying and health effects. Another personality trait that might also be linked to poor social skills is neuroticism or trait-negative affectivity (NA). Individuals high in trait-NA typically tend to have negative views of themselves, other people, and the world in general (Watson and Clark, 1984). Moreover, they are more likely to experience negative emotions such as distress, dissatisfaction, anxiety, and anger, and to be less prone to requesting social support (Hansson et al., 1984; Hobfoll, 1985). A question-naire study of 437 employees in five Swedish companies showed that state NA may partially mediate the association between bullying and health symptoms (Hansen et al., 2006). Conversely, in a questionnaire study of 102 targets of bullying, state NA and state positive affectivity (PA) neither mod-erated nor mediated relationships between bullying and PTSD symptoms (Matthiesen and Einarsen, 2004).

Lastly, in a questionnaire study involving 127 students employed dur-ing the year preceding the study, Djurkovic and colleagues (2005) explored relationships between bullying, neuroticism, and psychological and physi-cal symptoms. Results showed that bullying and neuroticism contributed independently to negative affect. However, NA did not moderate the rela-tionship between bullying and psychological symptoms. In fact, NA did not correlate significantly with bullying at all. As suggested by the authors, these results indicate that personality does not influence targets' emotional reactions to bullying. Given that the effects of bullying on psychological symptoms were larger than the effects of NA, this finding suggests that bul-lying is the primary contributor to negative affect and hence to the physical symptoms.

Sense of coherence (SOC) is another individual disposition that might influence the degree to which targets develop stress reactions following exposure to bullying. Sense of coherence is thought to be a stable global construct comprising three interrelated aspects: sense of comprehensibil-ity, meaningfulness, and manageability (Antonovsky, 1987). Given their tendency to perceive the world as meaningful and manageable, individu-als with a strong SOC will be less likely to feel threatened by bullying, less vulnerable after it has occurred, and more able to cope with potential future assaults and hence exhibit fewer symptoms of stress than individuals with a weak SOC.

Nielsen and colleagues (2008) explored whether SOC acted as a moderator between exposure to bullying and symptoms of PTSD. Their study used a convenience sample of 221 members of two support organizations for targets of bullying. Results revealed a reverse buffering effect inasmuch as higher levels of exposure to bullying were related to significantly elevated symptoms of PTSD among targets with mean and high SOC but not for targets with low SOC. Low levels of bullying seemed to have a stronger effect on targets with low SOC than it did on targets with high SOC. Conversely, increased levels of bullying had a greater effect on targets with mean or high SOC. This find-ing implies that SOC offers most protective benefits when exposed to mild bullying, yet when exposed to severe bullying the protective effects weaken

(Nielsen et al., 2008). The authors suggest two possible explanations: First, results may indicate that for targets with high SOC, exposure to severe, long-term bullying may be particularly detrimental given that it is unexpected and represents an incongruity between their normal positive self-image and the adverse, no-control situation in which they find themselves. Second, the results could be read as supporting the argument that long-term bullying is a traumatic stressor (see Mikkelsen, 2001a), one likely to affect anyone irrespective of the individual's disposition.

Thus, given the foregoing studies, individual dispositions are likely to play a role in explaining reactions following exposure to bullying at work. Nonetheless, several studies also indicate that as the level and frequency of bullying increase, other processes may interfere with these relationships. Hence, when exposed to severe, long-term bullying, targets are left with few, if any, internal and external coping resources (Leymann and Gustafsson, 1996), and they will be painfully aware of this. Besides, it is likely that prolonged victimization affects targets' appraisal of the negative acts in a way that increases their sense of threat potential (Lazarus and Folkman, 1984). The result is a pervasive feeling of hopelessness and lack of control. In such situations, some targets report feeling psychologically dead. We must, however, bear in mind the methodological limitations of the cited studies. For instance, it is not surprising that we find high levels of PTSD reactions among self-reported targets of bullying who have sought assistance in support organizations. One the other hand, recent studies such as those of Hansen and colleagues (2006, 2010) show that exposure to bullying may lead to a lower physiological stress response and to sleep problems even in groups of otherwise healthy working targets.

Conclusion

The research documented in this chapter has demonstrated that exposure to bullying may have highly detrimental effects on the targets' health and well-being. Given the methodological shortcomings of the early research—such as the use of cross-sectional studies and nonrepresentative populations—the recent longitudinal studies as well as the physiological studies yield invaluable contributions to research on bullying. Nonetheless, despite having controlled for individual and situational factors that might play a role in determining the outcome of bullying, the studies fail to identify the dynamic interplay between these and other factors in determining an individual's relative risk of developing health problems. In relation to this, few studies have as yet explored individual consequences of bullying using theoretical models to guide the analyses in a longitudinal design. Thus, there is still need for theory-driven longitudinal studies that focus on exploring relationships among exposure to bullying, the different acts and mechanisms used, and the consequences for the targets' psychological and physical health. Moreover, mediating or moderating effects of personal dispositions, personality patterns, and coping behavior should also be explored. It is also important to take into consideration how organizational actions may impact on these processes and whether preventive measures in organizations actually reduce bullying at work and diminish the effects on targets and bystanders. Further research on rehabilitation of targets is also needed, especially studies that focus on people who are still working.

Research on the individual consequences of bullying is still needed in order to increase our knowledge of the interplay of various factors. Our common efforts are also important with respect to increasing awareness of the phenomenon in companies as well as in society as a whole. This focus is especially important given the recent economic crisis, which may create conditions likely to increase the risk of bullying. In addition, there is possibly an increased risk of individuals using such coping strategies as presenteeism, which might be ineffective or even maladaptive. Finally, there is a heightened risk of organizations using less time and resources on prevention and intervention of bullying as well as on rehabilitation of targets.

References

Aardal-Eriksson, E., Eriksson, T. E., Holm, A.-C., and Lundin, T. (1999) Salivary cortisol and serum prolactin in relation to stress rating scales in a group of rescue workers. *Biological Psychiatry, 46*, 850–855.

Abramson, L. Y., Alloy, L. B., and Metalsky, A. (1988) The cognitive diathesis-stress theories of depression: Toward an adequate evaluation of the theories' validities. In L. B. Alloy (ed.), *The cognitive processes in depression* (pp. 3–30). New York: Guilford.

Abramson, L. Y., Seligman, M. E. P., and Teasdale, J. D. (1978) Learned helplessness in humans: Critique and reformulation. *Journal of Abnormal Psychology, 87*, 49–74.

Adoric, V. C., and Kvartuc, T. (2007) Effects of mobbing on justice beliefs and adjustment. *European Psychologist, 12*, 261–271.

Agervold, M., and Mikkelsen, E. G. (2004) Relationships between bullying, psychosocial work environment and individual stress reactions. *Work and Stress, 18*, 336–351.

Antonovsky, A. (1987) Health promoting factors at work: The sense of coherence. In R. Kalimo, M. Eltatawi, and C. L. Cooper (eds.), *Psychosocial factors at work and their relation to health* (pp. 153–167). Geneva: World Health Organization.

Bilgel, N., Aytac, S., and Bayram, N. (2006) Bullying in Turkish white-collar workers. *Occupational Medicine (London), 56*, 226–231.

Björkqvist, K., Österman, K., and Hjelt-Bäck, M. (1994) Aggression among university employees. *Aggressive Behavior, 20,* 173–184.

Björntorp, P., and Rosmond, R. (2000) Neuroendocrine abnormalities in visceral obesity. *International Journal of Obesity, 24* (Suppl. 2), S80–S85.

Brewin, C. R. (1988) *Cognitive foundations of clinical psychology.* Hove, UK: Lawrence Erlbaum.

Brousse, G., Fontana, L., Ouchchane, L., Boisson, C., Gerbaud, L., Bourguet, D., et al. (2008) Psychopathological features of a patient population of targets of workplace bullying. *Occupational Medicine, 58,* 122–128.

Cox, T., and Ferguson, E. (1991) Individual differences, stress and coping. In C. L. Cooper and R. L. Payne (eds.), *Personality and stress: Individual differences in the stress process* (pp. 7–30). Chichester, UK: John Wiley.

Dellve, L., Lagerström, M., and Hagberg, M. (2003) Work-system risk factors for permanent disability among home-care workers: A case-control study. *International Archives of Occupational and Environmental Health, 76,* 216–224.

Djurkovic, N., McCormack, D., and Casimir, G. (2005) The behavioral reactions of victims to different types of workplace bullying. *International Journal of Organization Theory and Behaviour, 8,* 439–460.

Dofradottir, A., and Høgh, A. (2002) *Mobning på arbejdspladsen: En kritisk gennemgang af dansk og international forskningslitteratur* [Bullying at work: A critical review of Danish and international research] (Rep. No. 10). Copenhagen: Danish Research Centre for the Working Environment.

Einarsen, S., and Hoel, H. (2001) *The Negative Acts Questionnaire: Development, validation and revision of a measure of bullying at work.* Prague: 10th European Congress on Work and Organisational Psychology.

Einarsen, S., Matthiesen, S. B., and Mikkelsen, E. G. (1999). *Tiden leger alle sår? Senvirkninger av mobbing i arbetslivet* [Time heals? Late effects of bullying at work]. Bergen, Norway: Institutt for Samfunnspsykologi, Universitetet i Bergen.

Einarsen, S., and Mikkelsen, E. G. (2003) Individual effects of exposure to bullying at work. In S. Einarsen, H. Hoel, D. Zapf, and C. L. Cooper (eds.), *Bullying and emotional abuse in the workplace. International perspectives in research and practice* (pp. 127–144). London: Taylor & Francis.

Einarsen, S., Raknes, B. I., Matthiesen, S. B., and Hellesøy, O. (1994) *Mobbing og harde personkonflikter: Helsefarlig samspill på arbeidsplassen* [Bullying and serious interpersonal conflicts: Unhealthy interactions at the workplace]. Bergen, Norway: Sigma Forlag A.S.

———— (1996) Helsemessige aspekter ved mobbing i arbeidslivet: Modererende effekter av sosial stotte og personlighet [The health-related aspects of bullying in the workplace: The moderating effects of social support and personality]. *Nordisk Psykologi, 48,* 116–137.

Epstein, S. (1985) The implications for cognitive-experimental self-theory for research in social psychology and personality. *Journal for the Theory of Social Behaviour, 15,* 283–310.

Evans, O., and Steptoe, A. (2001) Social support at work, heart rate, and cortisol: A self-monitoring study. *Journal of Occupational Health Psychology, 6,* 361–370.

Feveile, H., Olsen, O., and Hogh, A. (2007) A randomized trial of mailed questionnaires versus telephone interviews: Response patterns in a survey. *BMC Medical Research Methodology, 7,* 27.

Folkman, S., and Lazarus, R. S. (1991) Coping and emotion. In A. Monat and R. S. Lazarus (eds.), *Stress and coping: An anthology* (pp. 207–227). New York: Columbia University Press.

Ford, J. D. (2008) Diagnosis of traumatic stress disorder (DSM and ICD). In G. Reyes, J. D. Elhai, and J. D. Ford (eds.), *The encyclopedia of psychological trauma* (pp. 200–208). Hoboken, NJ: John Wiley.

Glasø, L., Nielsen, M. B., Einarsen, S, Haugland, K., and Matthiesen, S. B. (2009) Grunnleggende antagelser og symptomer på posttraumatisk stresslidelse blant mobbeofre [Basic assumptions and symptoms of posttraumatic stress disorder among victims of bullying]. *Tidsskrift for Norsk Psykologforening*, 46, 153–160.

Glinder, J. G., and Compas, B. E. (1999) Self-blame attributions in women with newly diagnosed breast cancer: A prospective study of psychological adjustment. *Health Psychology*, 18, 475–481.

Hallberg, L., and Strandmark, M. (2006) Health consequences of workplace bullying: Experiences from the perspective of employees in the public service sector. *International Journal of Qualitative Studies on Health and Well-being*, 1 (2), 109–119.

Hamilton, L. D., Newman, M. L., Delville, C. L., and Delville, Y. (2008) Physiological stress response of young adults exposed to bullying during adolescence. *Physiological Behaviour*, 95, 617–624.

Hansen, Å. M., Hogh, A., and Garde, A. H. (2009) Sleep problems and somatic stress symptoms among bullied persons in 60 Danish workplaces. *International Journal of Behavioral Medicine*, submitted.

Hansen, Å. M., Hogh, A., and Persson, R. (2010) Physiological and psychological consequences of bullying at work. *Journal of Psychosomatic Research*, conditionally accepted.

Hansen, Å. M., Hogh, A., Persson, R., Karlson, B., Garde, A. H., and Ørbæk, P. (2006) Bullying at work, health outcomes, and physiological stress response. *Journal of Psychosomatic Research*, 60, 63–72.

Hansson, R. O., Jones, W. H., and Carpenter, B. N. (1984) Relational competence and social support. In P. H. Shaver (ed.), *Review of personality and social psychology*, vol. 4 (pp. 265–290). Beverly Hills, CA: Sage.

Hobfoll, S. E. (1985) Personal and social resources and ecology of stress resistance. In P. H. Shaver (ed.), *Review of personality and social psychology*, vol. 5 (pp. 265–290). Beverly Hills, CA: Sage.

Hoel, H., Faragher, B., and Cooper, C. L. (2004) Bullying is detrimental to health, but all bullying behaviours are not necessarily equally damaging. *British Journal of Guidance and Counselling*, 32, 367–387.

Hogh, A., and Dofradottir, A. (2001) Coping with bullying in the workplace. *European Journal of Work and Organizational Psychology*, 10, 485–495.

Hogh, A., Henriksson, M. E., and Burr, H. (2005) A 5-year follow-up study of aggression at work and psychological health. *International Journal of Behavioral Medicine*, 12, 256–265.

Høgh, A., Ortega, A., Giver, H., and Borg, V. (2007) *Mobning af personale i ældreplejen* [Bullying of staff in care of the elderly] (Rep. no. 17). Copenhagen: National Research Centre for the Working Environment.

Janoff-Bulman, R. (1989) Assumptive worlds and the stress of traumatic events: Application of the schema construct. *Social Cognition*, 7, 113–136.

—— (1992) *Shattered assumptions: Towards a new psychology of trauma*. New York: Free Press.

Janoff-Bulman, R., and Thomas, C. E. (1989) Toward an understanding of self-defeating responses following victimization. In R. C. Curtis (ed.), *Self-defeating behaviors— experimental research, clinical impressions, and practical implications* (pp. 215–234). New York: Plenum.

Joseph, S. (1999) Attributional processes, coping and post-traumatic stress disorders. In W. Yule (ed.), *Post-traumatic stress disorders: Concepts and therapy* (pp. 52–70). Chichester, UK: John Wiley.

Keashly, L. (1998) Emotional abuse in the workplace: Conceptual and empirical issues. *Journal of Emotional Abuse, 1*, 85–117.

Kile, S. M. (1990) *Helsefarlig lederskab: Ein eksplorerande studie* [Health endangering leadership: An exploratory study]. Bergen, Norway: Department of Psychosocial Science, University of Bergen.

Kirschbaum, C., and Hellhammer, D. (1994) Salivary cortisol in psychoneuroendocrine research: Recent developments and applications. *Psychoneuroendocrinology, 19* (4), 313–333.

Kirschbaum, C., Strasburger, C. J., Jammers, W., and Hellhammer, D. (1989) Cortisol and behavior: 1. Adaptation of a radioimmunoassay kit for reliable and inexpensive salivary cortisol dertermination. *Pharmacology, Biochemistry and Behavior, 34*, 747–751.

Kivimäki, M., Elovainio, M., and Vahtera, J. (2000) Workplace bullying and sickness absence in hospital staff. *Occupational Environmental Medicine, 57*, 656–660.

Kivimäki, M., Virtanen, M., Vartia, M., Elovainio, M., Vahtera, J., and Keltikangas-Järvinen, L. (2003) Workplace bullying and the risk of cardiovascular disease and depression. *Occupational and Environmental Medicine, 60*, 779–783.

Kudielka, B. M., and Kern, S. (2004) Cortisol day profiles in victims of mobbing (bullying at the work place): Preliminary results of a first psychobiological field study. *Journal of Psychosomatic Research, 56*, 149–150.

Lazarus, R. S. (1999) *Stress and emotion—a new synthesis*. London: Free Associations Books.

Lazarus, R. S., and Folkman, S. (1984) *Stress, appraisal and coping*. New York: Springer.

Leymann, H. (1988) *Ingen annan utväg: Om utslagning och självmord som följd av mobbning i arbetslivet* [No other way: About social exclusion and suicide as a result of bullying at work]. Stockholm: Wahlström and Widstrand.

——— (1992) *Från mobbning till utslagning i arbetslivet* [From bullying to exclusion at work]. Stockholm: Publica.

——— (1996) The content and development of mobbing at work. *European Journal of Work and Organizational Psychology, 5*, 165–184.

Leymann, H., and Gustafsson, A. (1996) Mobbing at work and the development of post-traumatic stress disorder. *European Journal of Work and Organizational Psychology, 5*, 251–275.

Mathisen, G. E., Einarsen, S., and Mykletun, R. (2008) The occurrences and correlates of bullying and harassment in the restaurant sector. *Scandinavian Journal of Psychology, 49*, 59–68.

Matthiesen, S. B., and Einarsen, S. (2004) Psychiatric distress and symptoms of PTSD among victims of bullying at work. *British Journal of Guidance and Counselling, 32*, 335–356.

Matthiesen, S. B., Raknes, B. I., and Rokkum, O. (1989) Mobbing på arbeidsplassen [Bullying at the worksite]. *Tidsskrift for Norsk Psykologforening, 26*, 761–784.

McEwen, B. S. (2006) Sleep deprivation as a neurobiologic and physiologic stressor: Allostasis and allostatic load. *Metabolism*, 55, S20–S23.

—— (2007) Physiology and neurobiology of stress and adaptation: Central role of the brain. *Physiological Review*, 87, 873–904.

McFarlane, A. C. (2008) Posttraumatic stress disorder. In G. Reyes, J. D. Elhai, and J. D. Ford (eds.), *The encyclopedia of psychological trauma* (pp. 483–491). Hoboken, NJ: John Wiley.

Meliá, J. L., and Becerril, M. (2007) Psychosocial sources of stress and burnout in the construction sector: A structural equation model. *Psicothema*, 19, 679–686.

Meseguer, de, P. M., Soler Sanchez, M. I., Saez Navarro, M. C., and Garcia, I. M (2008) Workplace mobbing and effects on workers' health. *Spanish Journal of Psychology*, 11, 219–227.

Mikkelsen, E. G. (2001a) Mobning i arbejdslivet: Hvorfor og for hvem er den så belastende [Bullying at work: Why and for whom is it so stressful]. *Nordisk Psykologi*, 53, 109–131.

—— (2001b) Workplace bullying: Its prevalence, aetiology and health correlates. Unpublished PhD thesis, Department of Psychology, University of Århus, Denmark.

—— (2004) Coping with bullying at work: Results from an interview study. In S. Einarsen and M. B. Nielsen (eds.), *Fourth International Conference on Bullying and Harassment in the Workplace* (pp. 90–91). Bergen, Norway: University of Bergen.

Mikkelsen, E. G., and Einarsen, S. (2002) Basic assumptions and symptoms of posttraumatic stress among victims of bullying at work. *European Journal of Work and Organizational Psychology*, 11, 87–111.

Mikkelsen, E. G., and Iversen, G. F. (2002) Bullying at work: Perceived effects on health, well-being and present job situation. In A. P. D. Liefooghe and H. Hoel (eds.), *International Conference on Bullying at Work* (p. 31). London: Birkbeck, University of London.

Mikkelsen, E. G., Kullberg, K., and Eriksen-Jensen, I. L. (2007) *Mobning på arbejdspladsen—årsager og løsninger* [Bullying in the workplace—causes and solutions]. Copenhagen: Forlaget Munksgaard.

Miller, D. T., and Porter, C. A. (1983) Self-blame in victims of violence. *Journal of Social Issues*, 39, 139–152.

Moayed, F. A., Daraiseh, N., Shell, R., and Salem, S. (2006) Workplace bullying: A systematic review of risk factors and outcomes. *Theoretical Issues in Ergonomics Science*, 7, 311–327.

Moreno-Jiménez, B., Rodríguez-Muñoz, A., Moreno, Y., and Garrosa, E. (2007) The moderating role of assertiveness and social anxiety in workplace bullying: Two empirical studies. *Psychology in Spain*, 11, 85–94.

Niedhammer, I., David, S., and Degioanni, S. (2006) Association between workplace bullying and depressive symptoms in the French working population. *Journal of Psychosomatic Research*, 61, 251–259.

Niedl, K. (1996) Mobbing and well-being: Economic and personnel development implications. *European Journal of Work and Organizational Psychology*, 5, 239–249.

Nielsen, M. B., Matthiesen, S. B., and Einarsen, S. (2005) Ledelse og personkonflikter: Symptomer på posttraumatisk stress blandt ofre for mobbing fra ledere [Leadership and person-related conflicts: Symptoms of posttraumatic stress among victims of bullying from superiors]. *Nordisk Psykologi*, 57, 319–415.

———— (2008) Sense of coherence as a protective mechanism among targets of workplace bullying. *Journal of Occupational and Health Psychology, 13,* 128–136.

Nikolajsen, R. P. H., Booksh, K. S., Hansen, Å. M., and Bro, R. (2003) Quantifying catecholamines using multi-way kinetic modelling. *Analytica Chimica Acta, 475,* 137–150.

Nolfe, G., Petrella, C., Blast, F., Zontini, G., and Nolfe, G. (2008) Psychopathological dimensions of harassment in the workplace (mobbing). *International Journal of Mental Health, 36,* 67–85.

Notelaers, G., Einarsen, S., De Witte, H., and Vermunt, J. K. (2006) Measuring exposure to bullying at work: The validity and advantages of the latent class cluster approach. *Work and Stress, 20,* 288–301.

O'Moore, M., Seigne, E., McGuire, L., and Smith, M. (1998) Victims of bullying at work in Ireland. *Journal of Occupational Health and Safety–Australia and New Zealand, 14,* 569–574.

Ortega, A., Hogh, A., and Borg, V. (2009) Workplace bullying, sickness absence and sickness presenteeism: A one-year follow-up study in the elderly-care sector. *Archives of Occupational and Environmental Health,* submitted.

Persson, R., Hogh, A., Hansen, A. M., Nordander, C., Ohlsson, K., Balogh, I., et al. (2009) Personality trait scores among occupationally active bullied persons and witnesses to bullying. *Motivation and Emotion, 33,* 387–399.

Peterson, C., Schwartz, S. M., and Seligman, M. E. P. (1981) Self-blame and depressive symptoms. *Journal of Personality and Social Psychology, 41,* 253–259.

Pompili, M., Lester, D., Innamorati, M., De Pisa, E., Iliceto, P., Puccinno, M., et al. (2008) Suicide risk and exposure to mobbing. *Work, 31,* 237–243.

Price Spratlen, L. (1995) Interpersonal conflict which includes mistreatment in a university workplace. *Violence and Victims, 10,* 285–297.

Rafnsdóttir, G. L., and Tómasson, K. (2004) Bullying, work organization and employee well-being. *Laeknabladid, 90,* 847–851.

Rayner, C. (1999) From research to implementation: Finding leverage for prevention. *International Journal of Manpower, 20,* 28–38.

Spector, P. E., Zapf, D., Chen, P. Y., and Frese, M. (2000) Why negative affectivity should not be controlled in job stress research: Don't throw out the baby with the bath water. *Journal of Organizational Behavior, 21,* 79–95.

Tehrani, N. (2004) Bullying: A source of chronic post traumatic stress? *British Journal of Guidance and Counselling, 32,* 357–366.

Thylefors, I. (1987) *Syndabockar: Om utstödning och mobbning i arbetslivet* [Scapegoats: About social exclusion and bullying at work]. Stockholm: Natur och Kultur.

Traweger, C., Kinzl, J. F., Traweger-Ravanelli, B., and Fiala, M. (2004) Psychosocial factors at the workplace—do they affect substance use? Evidence from the Tyrolean workplace study. *Pharmacoepidemiology and Drug Safety, 13,* 399–403.

van Eck, M., Berkhof, H., Nicolson, N., and Sulon, J. (1996) The effects of perceived stress, traits, mood states, and stressful daily events on salivary cortisol. *Psychosomatic Medicine, 58,* 447–458.

Vartia, M.-L. (2001) Consequences of workplace bullying with respect to the well-being of its targets and the observers of bullying. *Scandinavian Journal of Work Environment and Health, 27,* 63–69.

Vingård, E., Lindberg, P., Josephson, M., Voss, M., Heijbel, B., Alfredsson, L., et al. (2005) Long-term sick-listing among women in the public sector and its associations with age, social situation, lifestyle, and work factors: A three-year follow-up study. *Scandinavian Journal of Public Health, 33,* 370–375.

Voss, M., Floderus, B., and Diderichsen, F. (2004) How do job characteristics, family situation, domestic work, and lifestyle factors relate to sickness absence? A study based on Sweden Post. *Journal of Occupational and Environmental Medicine*, 46, 1134–1143.

Watson, D., and Clark, L. A. (1984) Negative affectivity: The disposition to experience aversive emotional states. *Psychological Bulletin*, 96, 465–490.

Weiner, B. (1985) Spontaneous causal thinking. *Psychological Bulletin*, 97, 74–84.

Zapf, D., Dorman, C., and Frese, M. (1996a) Longitudinal studies in organizational stress research: A review of the literature with reference to methodological issues. *Journal of Occupational Health Psychology*, 1, 145–169.

Zapf, D., and Einarsen, S. (2003) Individual antecedents of bullying: Victims and perspectives. In S. Einarsen, H. Hoel, D. Zapf, and C. L. Cooper (eds.), *Bullying and emotional abuse in the workplace: International perspectives in research and practice* (pp. 165–184). London: Taylor & Francis.

—— (2005) Mobbing at work: Escalated conflicts in organizations. In S. Fox and P. E. Spector (eds.), *Counterproductive work behavior: Investigations of actors and targets* (pp. 237–270). Washington, DC: American Psychological Association.

Zapf, D., and Gross, C. (2001) Conflict escalation and coping with workplace bullying: A replication and extension. *European Journal of Work and Organizational Psychology*, 10, 497–522.

Zapf, D., Knorz, C., and Kulla, M. (1996b) On the relationship between mobbing factors, and job content, social work environment, and health outcomes. *European Journal of Work and Organizational Psychology*, 5, 215–237.

Zeier, H. (1994) Workload and psychophysiological stress reactions in air traffic controllers. *Ergonomics*, 37, 525–539.

5

Organisational Effects of Workplace Bullying

Helge Hoel, Michael J. Sheehan, Cary L. Cooper, and Ståle Einarsen

CONTENTS

Introduction

In February 2007, Gerd Liv Valla, frequently described as Norway's most powerful person, had to step down from her position as leader of the Norwegian Trade Union Confederation (LO) as a result of accusations of bullying. Her meteoric career, which had seen Valla rise to the top of the country's power elite, including a short stint as Secretary of State for Justice, came to an abrupt end as an "official investigation" instigated by Valla and LO itself found her guilty of bullying. In addition to the personal tragedy this represented for Valla herself and the political cost of losing what was considered the Norwegian trade union movement's brightest and most influential strategist for many years, the organisational and economic costs of the case were immense, considering the internal upheaval that followed her resignation, as a result of which the organisation was landed with a bill of close to US$2 million for the investigation alone (Tranøy, 2007).

Whilst much effort has gone into investigating and empirically exploring the individual effects of bullying, comparatively little attention has been paid to assessing the organisational consequences and economic costs of bullying. This inattention is somewhat surprising given the combined effects of the very substantial scale or magnitude of the problem (see Zapf et al., this volume) and the effect this has on the individuals targeted and victimised (Hogh et al., this volume), which would undoubtedly represent a substantial cost to the organisation. In this respect, being exposed to bullying in whatever form is likely to manifest itself behaviourally and attitudinally (Cooper et al., 1996; Hoel et al., 2001), making targets constantly less able to cope with the daily tasks and cooperation required of the job (Einarsen, 2000) and reflected in reduced satisfaction (e.g., Hoel and Cooper, 2000; Price Spratlen, 1995; Quine, 1999) as well as commitment to the organisation (Hoel and Cooper, 2000).

Based on his own and other victims' accounts, Field (1996) suggests that "the person becomes withdrawn, reluctant to communicate for fear of further criticism. This results in accusations of 'withdrawal', 'sullenness', 'not co-operating or communicating', 'lack of team spirit', etc." (p. 128). Although such accounts are often anecdotal in nature, similar effects have been established empirically (e.g., Niedl, 1996). Research has also concluded that bullying can be associated with cognitive effects such as concentration problems, insecurity, and lack of initiative (Leymann, 1992; O'Moore et al., 1998), which may manifest itself organisationally as reduced motivation and creativity, as well as a rise in errors and accidents. Whether deemed deconstructive, negative, or maladaptive, many behavioural and attitudinal responses may have a bearing upon the organisation by affecting levels of absenteeism, turnover, and productivity as well as team and group performance.

Furthermore, any accusation of bullying is likely to subsequently affect the organisation directly. Even where cases are seemingly satisfactorily resolved, there may be a price to pay in terms of organisational upheaval. Moreover, in cases where organisational procedures and responses fail to satisfy the need of the complainant, or where organisations deny targets their right to complain, litigation has increasingly become an option, often with negative implications for public relations.

In line with the previously mentioned Norwegian high-profile bullying case, the organisational cost in monetary terms can be immense. In this respect, Leymann (1990) argues that an average case of bullying may cost an organisation around US$30,000 to US$100,000. Similarly, Knott et al. (2004) reported that official claims lodged because of bullying (24 claims in the period 1993–2002) cost the Australian Correctional Services AUS $736,513. Specifying the costs involved with a typical bullying case, Hoel et al. (2003) arrived at a sum of £28,109 (approximately US$50,000), which included a series of organisational costs in terms of absenteeism, replacement as a result of turnover and formal investigation (but not including loss of productivity), potential costs to observers and coworkers, and harm to public relations (Table 5.1). As this case was at the time still dealt with internally, any costs arising from litigation were also excluded.

The aim of this chapter is to provide the reader with a picture of the totality of organisational costs involved with bullying. For this purpose, we initially review the evidence and examine the theoretical explanation for the

TABLE 5.1

Cost of the Case to the Organisation

Absence	£6.972
Replacement costs	£7,500
Reduced productivity	?
Investigators' time for grievance investigation	£2,110
Local management line management time	£1,847
Head office personnel	£2,600
Corporate officers' time (including staff welfare)	£2,100
Cost of disciplinary process (hearing, solicitor)	£3,780
Witness interview costs	£1,200
Transfers	0
Litigation	?(0)
Effects on those indirectly involved	?
Miscellaneous (effects on public relations, etc.)	?
Total costs (minimum)	£28,109

Note: ? = cost unknown or difficult to assess.

contribution of the various cost elements, such as absenteeism, turnover, and productivity. Based on previous cost models and findings from research on organisational costs related to bullying, harassment, and other associated fields of research (e.g., occupational stress), we will put forward an updated and integrated model and provide examples in monetary terms. Finally, in a discussion section we will reflect on some key issues and possible pitfalls associated with estimations of costs associated with intangible issues such as bullying.

Organisational Effects of Bullying: The Evidence

Absenteeism

From the evidence of the effects of bullying on health, one would intuitively expect that bullying would lead to increased sickness absenteeism. However, in the studies that have explored this relationship, the association between bullying and sickness absenteeism has been found to be relatively weak (Hoel and Cooper, 2000; Price Spratlen, 1995; UNISON, 1997; Vartia, 2001). Yet, Thyholdt et al. (1986) argue that a common finding in health research is that people typically tend to report more specific symptoms as opposed to general negative health effects such as absenteeism. Another factor is the tendency of people to underreport their own absence (Johns, 1994).

Notwithstanding, an Irish telephone survey of a nationally representative sample (N = 3,500) found that 20% of respondents had taken sick leave directly as a result of bullying (O'Connell et al., 2007). A strong relationship between bullying and sickness absenteeism was also reported in a Finnish study of hospital staff (Kivimäki et al., 2000), in which it was possible to match sickness absence records (certified and self-certified) to company records on absence. It emerged that the risk of medically certified sickness absence was 51%, or 1.5 times higher, for those who had been bullied than for those who had not been targeted. When adjusted for a baseline one year prior to the survey, the risk was still 1.2, or 26% more absences. The difference was smaller for self-certified sickness absence but still substantial (16%). In total, bullying accounted for 2% of total sickness absenteeism, at a cost of £125,000 annually (other related costs not incorporated).

Similarly, Hoel and Cooper (2000) found that victims of bullying took on average seven days more sick leave per year than those who neither were bullied nor had witnessed bullying taking place. Based on a bullying rate of 10%, this would account for 18 million lost working days on a UK basis. Furthermore, Quine (2001), in a UK study of bullying amongst nurses, reported that 8% had taken time off because of bullying, whilst Vartia (2001), in a Finnish study of municipal employees, reported that 17% had taken time off for the same reason, 10% several times.

Other studies have reported that approximately one-third of targets reported having taken time off as a result of their experience (McCarthy et al., 1995; UNISON, 1997, 2000). Moreover, amongst targets taking time off work because of bullying, a substantial number reported prolonged sickness absence, with 29% being absent for more than 30 days and 13% for more than 60 days (UNISON, 2000).

When considering the costs of absenteeism in the broadest terms, it should be noted that the unpredictability and unexpectedness of unscheduled absenteeism may represent a particular problem for organisations, interfering with the normal operation of the organisation, and where applicable, the quality of service provision (Seago, 1996).

According to Steer and Rhodes (1978), absenteeism can be seen as a result of a combination of two factors: the possibility or opportunity to be present at work, and the motivation of the individual to go to work. As far as the first factor is concerned, the presence of illness would be a deciding factor for nonattendance. The motivation for attending work appears to hinge upon two factors: the degree of job satisfaction and the pressure to attend. Pressure to attend work, for its part, may relate to social norms or what may be described as normative commitment (Allen and Meyer, 1990), as well as organisational measures in terms of absence control systems and punitive responses to nonlegitimate absence. Following the model of Steer and Rhodes, the relatively strong relationship between bullying and ill health suggests that for many of the victims, sickness absence may be a necessity and possibly a direct outcome of status of health. Health problems inflicted by perceived bullying may undermine individuals' motivation to attend work because of reduced job satisfaction. As the causal relationship between job satisfaction and absence may be reversible and dynamic, growing health problems resulting from bullying may gradually affect job satisfaction and motivation to attend, with absence as a possible result. However, real or perceived pressure to attend may act as a motivator, albeit a negative one, pushing targets to go to work even if, strictly speaking, they would be better off being absent. Using absence as a coping mechanism to recover from a stressful experience may also be considered counterproductive, contributing to ostracism and social isolation, whether real or perceived, with the bullying process escalating with the bullying victim's return to work. Also, in organisational terms, such coping strategies may be considered ineffective and costly, an issue we will return to later. Indirectly, this finding also supports a recent finding by Zapf and Gross (2001), who revealed that victims who successfully coped with their situation used frequent sickness absence significantly less often than victims who were unsuccessful.

Where organisational norms consider absence as deviance, targets may turn up for work to demonstrate their commitment and loyalty and to avoid being characterised as a malingerer, even if in terms of medical recovery they would benefit from staying at home. In some cases, personal guilt in the guise of self-inflicted pressure may be another factor preventing people from taking time off. Even if such mechanisms are negative in their own right, they may actually be beneficial in bullying scenarios as they may prevent absenteeism or sick leave turning into a form of self-exclusion from the work group or the workplace altogether.

Turnover

Among potential organisational outcomes of bullying, turnover has been of particular interest to researchers, with a number of studies reporting a positive relationship between bullying and intention to leave and turnover respectively (e.g., Djurkovic, et al., 2004; Hoel and Cooper, 2000; Keashly and Jagatic, 2000; Quine 1999, 2001). Similar conclusions have frequently been reached in studies of related concepts such as abusive supervision (Tepper, 2000), mistreatment (Boswell and Olson-Buchanan, 2004) and incivility at work (Cortina et al., 2001), and they have been supported by the results of a meta-analysis incorporating 24 samples (N = 13,961) of harassment and bullying (Bowling and Behr, 2006). Some studies have even suggested that as many as one in four people leave their organisations because of bullying (Rayner, 1997; UNISON, 1997). To some extent substantiating such claims, in O'Connell et al.'s (2007) Irish study (2007) reported previously, 60% of respondents considered leaving whilst 15% actually left the organisation.

Whilst some factors appear to mediate the relationship between bullying and intention to leave, others may moderate the effects. Hence, whilst a Norwegian study of restaurant employees (N = 200) (Mathisen et al., 2008) found that bullying may have a negative impact on job satisfaction and commitment, which again increases targets' intention to leave, an Australian study of schoolteachers (Djurkovic et al., 2008) concluded that perceived organisational support moderated the effect of bullying on intention to leave.

However, it is important to note that with a few exceptions, most studies have focused on information about intention to leave rather than about actual turnover or quitting. This point is noteworthy because although intention to leave is considered a key predictor of turnover (Begley, 1998), the shared variation between the two is modest, with Lee et al. (1999) reporting a figure of just 12%. A longitudinal study conducted in Norway showed that although some targets had actually left their organisation two years after they first reported experiencing bullying at work, the majority of targets were still in the same job two years later (Berthelsen et al., 2009). By contrast, Schwickerath and Zapf (this volume) reported that in their German study of workplace bullying victims who were taking part in a therapeutic treatment programme, 36% actually left their job for a position with another organisation, whilst only 18% returned to their former job.

Since their emergence in the 1950s, theories of turnover have discussed the process in terms of "push and pull" factors (March and Simon, 1958). In this respect, push factors refer to a perceived wish or desire to leave, whilst pull factors are associated with ease of movements and perception of the availability of alternative jobs and other labour-market considerations. More recently, Lee et al. (2008) have introduced a different approach, conceptualising turnover in terms of "shock." In this respect, it is suggested that turnover processes, broadly speaking, fall into a few main categories or paths, one of which appears to correspond well with bullying processes. It is noted that such shocks can be perceived as positive or negative, being related to the job or not, as well as being internal or external to the individual. Different from most other paths, job dissatisfaction appears to play a much lesser role in turnover than do other factors. Instead, the turnover process is interpreted in the light of what is referred to as *image violation*, which is seen to offset negative feelings and reactions towards the organisation leading to withdrawal. In this respect, the notion of shock appears to fit well with victims' accounts of bullying, where the initial part of the process is often described in terms of surprise or shock, or even a series of shocks (Leymann, 1990).

Several factors may account for relatively strong associations between bullying and intention to leave: for some, leaving the organisation may be seen as a positive coping strategy as leaving removes them from the source of the problem altogether, which resonates with the advice often given by victims to other victims (Zapf and Gross, 2001). In this respect, such a decision might be a consequence or outcome of receiving therapy (see Schwickerath and Zapf, this volume). Others may resign because of prolonged health problems. In other cases, stigmatized victims may be expelled from the organisation (Leymann, 1996; Zapf and Gross, 2001), for example, where employers use bullying as a strategy to rid the organisation of unproductive or uncooperative employees (Ferris et al., 2007) or where the organisation would otherwise have to pay redundancy compensation (Di Martino et al., 2003; Lee, 2000).

However, other factors such as tight labour markets and lack of mobility (Tepper, 2000) would work against leaving, thus contributing to explaining why the association between bullying and turnover is not even stronger. It also remains true that many victims would be reluctant to leave until they have achieved redress and justice (Kile, 1990). Moreover, where bullying behaviours occur intermittently, victims may also cling to the belief that the problem will eventually disappear. In many cases, this notion may also hold true as more organisations develop and implement policies to handle bullying cases in a constructive way.

Productivity

Although it is assumed that bullying would impact factors such as job satisfaction, commitment, innovation, and creativity (e.g., Bassmann, 1992), empirical evidence of a relationship between bullying and productivity remains sparse. In part, this sparsity reflects the fact that productivity is hard to measure, a reality that has also been pointed out in occupational stress research (Cooper et al., 1996). Furthermore, the few pieces of evidence that have emerged focus on self-reported measures of perceived change in performance because of bullying. Thus, in a study of Norwegian trade union members, a total of 27% of respondents agreed with the statement, "Bullying at my workplace reduces our efficiency" (Einarsen et al., 1994). When applied in a nationwide British study, the same measure yielded a figure of 33% (Hoel and Cooper, 2000). In the latter study, respondents were also asked to assess their performance, independently of their experience of bullying, as a percentage estimate of normal capacity. Applying this figure as a measure of productivity, a moderate negative correlation was found between self-rated performance and bullying, with the "currently bullied" on average reporting a decrease of productivity of approximately 7% compared with those who were neither bullied nor had witnessed bullying taking place (Hoel, Sparks, et al., 2001).

That people may be bullied irrespective of their organisational status or rank, including senior managers (Hoel, Cooper, et al., 2001), indicates the possibility of a negative domino effect, where bullying may be cascaded downwards as the targeted supervisors might offload their own aggression on their subordinates (Tepper et al., 2006). In such situations, a bullying scenario in the boardroom may actually threaten the productivity of the entire organisation.

Where reduced commitment or withdrawal is used as a coping strategy, one would expect a negative impact on performance and productivity. In some cases, targets of bullying will respond by increasing their effort. However, even where targets adopt a strategy of working harder in order to demonstrate their loyalty to the organisation, the net outcome in terms of productivity may be negative, because mental exhaustion may reduce concentration and vigilance, possibly causing irritation and frustration among coworkers. Problems with concentration may also, in their own right, increase the propensity to make mistakes, thus increasing the likelihood of accidents, as well as reduction in output and quality of the product.

In connection with reduced commitment to work or withdrawal, it appears reasonable to invoke the concept of presenteeism. This concept refers to situations where individuals extend their time at work beyond their official working day in order to demonstrate commitment to the organisation. According to Brun and Lamarche (2006), however, presenteeism is likely to have the opposite effect, reducing output and lowering the standard of production. Thus, according to Caverley et al. (2007), sickness presenteeism, where employees are working less effectively because of health-related or medical problems, such as headaches, migraines, and gastrointestinal ailments, represents hidden costs of substantial proportions, actually outstripping the cost of sickness absenteeism. They argue that presenteeism is particularly widespread where people work with or provide services to clients, for example in the health service, where there is unlikely to be any cover, where work tends to accumulate when one is away, or in situations where organisations are undergoing downsizing.

Effects on Observers and Witnesses

A large number of people report having witnessed bullying taking place (Hoel and Cooper, 2000; Rayner, 1997). The fact that "discussing the problem with colleagues" was found to be the response most frequently chosen by targets when faced with bullying (Hoel and Cooper, 2000), however, also suggests that a large number of people will be indirectly affected by bullying. Nevertheless, although it is assumed that witnessing or observing bullying would have a negative effect on third parties, evidence that such is the case is relatively scarce. According to Rayner (1999), in a study of public-sector union members, approximately one in five workers reported having considered leaving the workplace as a result of witnessing bullying taking place. Rayner explained these figures by pointing to the presence of a climate of fear in which employees considered reporting to be unsafe, where bullies had "got away with it" previously despite management knowing of the presence of bullying.

In a British study (Hoel and Cooper, 2000), respondents were divided into four groups: "currently bullied," "previously bullied," "witnessed bullying only," and "neither experienced nor witnessed bullying." Individuals currently exposed to bullying were found to have the worst health, the highest sickness absenteeism and "intention to leave" rates, and the lowest productivity, as well as the lowest organisational satisfaction and commitment. The second-most affected group consisted of those who had been "bullied in the past," followed by "witnessed bullying only." Similarly, in a study by Vartia

(2001), observers of bullying reported higher levels of stress than did those who had not experienced or observed bullying taking place. These findings lend support to the idea that bullying may affect third persons to the detriment of the individual as well as the organisation.

Because of a seemingly strong need for targets to seek support for their cases, it is also difficult to remain uninvolved or neutral in such cases (Einarsen, 1996). With this in mind, it should come as no surprise that there may be a ripple effect involved with bullying (Hoel et al., 1999; Rayner, 1999). With teamwork becoming an increasingly common work strategy, the ripple effect of bullying may turn out to have serious organisational implications (Hoel et al., 1999). Thus, enforced team working may increase the likelihood of bullying in the first place, as well as the chance of third parties being drawn into emerging conflict (Zapf, Knorz, et al., 1996). Also, when the complexity of the operation increases and worker interdependence is central to outcome, employee absence may force colleagues to undertake operations they are less familiar with, with implications for social cohesion and the overall productivity of the team (Johns, 1997).

Other Organisational Effects

Where national customs and practice allow victims to file a complaint or a grievance, then any subsequent investigation and/or hearing will tend to be time-consuming and a drain on organisational resources because of the commitment of all those involved. Without proper policies and procedures in place, cases can remain unresolved for years, with organisational indecision and paralysis contributing to partiality and increasing animosity and internal conflict. However, even in cases where procedures are strictly adhered to and where cases are brought to a conclusion within a reasonable timeframe, the process tends to be destructive for all those concerned (Ishmael, 1999). Depending upon the seriousness of the case, in some instances the alleged offender might be suspended (normally on full pay) for the duration of the internal investigation, a provision that might in some cases be extended to the complainant (see Hoel and Einarsen, this volume). Throughout the investigation process, counselling or other support mechanisms might have to be put in place, with further costs to the organisation (Tehrani, this volume).

A typical outcome of a bullying complaint is the physical, permanent separation of the perpetrator and victim. In this respect, a German questionnaire study (Meschkutat et al., 2002) found that 11.1% of the victims reported that the bullies were transferred within the company, whilst 8.2% of bullies were dismissed. In terms of physical separation, it has been reported that victims are often more likely to be moved, particularly where the perpetrator is a superior (Rayner et al., 2002). Such transfers can be costly because they may encompass replacement costs as well as extra training costs for two or more individuals (Dalton, 1997). And whilst some turnover may be beneficial to the organisation, enforced transfers are, according to Dalton, entirely dysfunctional. As transfers are unlikely to take place without disruption, in this case involving two work groups, they would tend to affect productivity negatively. In some cases, a domino effect may result, as more than one move may be necessary to accommodate the situation with respect to relevant skills and experience.

Replacing an individual may also lead to a chain reaction, with implications for other employees as well (Gordon and Risley, 1999). Further, there may be possible side effects for the organisation.

The use of the legal system to settle work-related disputes and claims varies between countries and depends upon national litigation practices, available legislation, and individual countries' national industrial relations practices. There is evidence to suggest that cases associated with bullying are increasingly taken to court, be it industrial tribunals or labour courts (Yamada, this volume), resulting in increasing costs to the organisation, with compensation approaching the £1 million mark. In a high-profile court ruling, an employee of Deutsche Bank in the United Kingdom was awarded £800,000 in compensation for bullying inflicted by some of her peers and for the lack of intervention by her line manager. In another case, the company, Cantor Fitzgerald International, was ordered to pay the complainants close to £1 million for mistreatment the complainant suffered at the hands of the company president (BBC, 2003). Such cases are also likely to have further implications for the organisations involved in terms of negative publicity and damage to reputation and may lead to potential loss of trade and customers as well reducing the pool of job applicants as potential employees look elsewhere for work (Giga et al., 2008).

In some cases, bullying may also lead to industrial action and unrest as a result of alleged victimisation (Beale, this volume). For example, in one particularly high-profile case in the United Kingdom, a trade union at Ford balloted its members on possible strike action, with allegations that managers tolerated and even encouraged bullying on the shop floor (*Guardian*, 9 October 1999). Similarly, it is also argued that perceptions of unfair treatment and bullying may increase the likelihood of organisation-targeted revenge in terms of theft and sabotage (Neuman and Baron, 2003), with abusive supervision offsetting what is referred to as "dysfunctional resistance" (Tepper, 2007), for example, not following up on supervisors' requests. Hence, bullying may lead to counteraggression from the target (Lee and Brotheridge, 2006) as well as to displaced aggression towards other and even less powerful organisation members or towards customers or clients (see also Neuman and Baron, this volume).

A further factor often overlooked because of its positive connotations is the cost associated with intervention (Giga et al., 2008; World Health Organization, 2004). According to Waters et al. (2004), by reducing levels of bullying and harassment by 1%, the costs to the UK National Health Service would amount to a saving of £9 million annually in costs associated with intervention programmes alone. However, it is important to bear in mind that intervention programmes may have objectives and positive outcomes beyond bullying reduction that may not easily be assessed. For example, in an evaluation of the Norwegian Olweus programme against bullying in schools, a range of positive side effects were documented in addition to the positive effects it had on levels of bullying. This school-based programme has been shown to reduce bullying by 50%, simultaneously reducing such other forms of antisocial behaviours as vandalism, fighting, and truancy and improving a range of positive aspects of the "social" climate in the school (Olweus, 2003).

Assessing the Financial Cost of Bullying

Some attempts have been made to assess the costs of bullying at corporate as well as national levels. For example, Leymann (1990) calculated that every victimised individual would produce a cost to the organisation of approximately $30,000 to $100,000 annually. Kivimäki et al. (2000), in a study of bullying at two Finnish hospitals, estimated that the annual cost of absence from bullying alone accounted for costs of £125,000. Furthermore, Rayner (2000) estimated a cost of approximately £1 million annually in replacement costs for staff who left the organisation as a result of bullying. Similarly, based on a meta-analysis of workplace bullying research and taking into consideration prevalence and severity data, Sheehan et al. (2001), using what they considered to be a conservative prevalence estimate of 3.5%, demonstrated that workplace bullying cost Australian employers between AUS $6 and AUS $13 billion every year (approximately US $5–US $10 billion annually). This figure included hidden and lost opportunity costs as well as costs that had been overlooked in previous studies.

Sheehan et al. (2001) then calculated a unit cost per bullying case. They did so by dividing the annual total, as mentioned, by the relevant number of victims based on Australian working population data at that time. They argued that at the lower range of 3.5% prevalence rate, each case of bullying cost Australian employers at least AUS $16,977. At the higher prevalence rate of 15%, the cost equated to AUS $24,256.

In a report commissioned by the International Labour Organisation (ILO), Hoel et al. (2001) took this argument one step further and estimated the costs of bullying at a national level in Great Britain, concluding that costs related to absence and replacement because of bullying alone may account for close to £2 billion annually. However, they acknowledged that any attempt to assess the costs of a complex problem such as bullying was fraught with difficulties, pointing to factors such as questionable quality of data and unclear connections between cause and effect. An even higher figure of £3.7 billion is suggested by Beswick et al. (2006), a cost calculation made on the assumption that 10%–20% of stress-related absenteeism is associated with bullying.

As noted in the following list, a number of cost factors need to be considered when the financial cost is calculated. In this case, it is worth bearing in mind that, in general, staff account for 50%–80% of organisational costs (Cooper et al., 1996).

- Sickness absenteeism: Employers respond to absenteeism in many ways. The response might be through finding cover for the absentee by means of voluntary or compulsory cover by colleagues, through overtime, or through replacement. However, in most cases the real cost of absenteeism is, by and large, linked to the cost of sick pay, and such systems will vary from country to country (Gordon and Risley, 1999).

- Turnover and replacement costs: These primarily comprise recruitment costs (advertising and selection), as well as the cost of induction, training, and development. In addition to direct costs arising from these activities, administrative costs (e.g., testing, candidates' travel expenses, termination of contracts, issuing of new contracts) will also have to be included in total costs. Replacement costs will tend to rise according to the experience and skills of the appointee. It should be noted that turnover is not necessarily dysfunctional to the effectiveness of an organisation, since a certain amount of turnover allows the organisation the opportunity to bring in new knowledge, skills, and talents (Gordon and Risley, 1999). Turnover costs also include the lower initial efficiency of replacement employees until those employees reach the same level of performance as those individuals who have been replaced (Sheehan et al., 2001). Such an argument may also extend to situations where the person who has experienced being bullied transfers to a new position within the same organisation. It also includes lost productivity among coworkers affected by workplace bullying, the impact of absenteeism on supervisor time in managing the problem of absenteeism and other absenteeism effects, such as a reduction in the quality of the work output.

- Impact on productivity and performance: Any impact of bullying is likely to be considerable, but largely intangible, and will require some element of informed guessing. Nonetheless, it is worth bearing in mind that earlier studies have reported a number of impacts relevant to calculating lost productivity as a result of poorer work performance by those who experience being bullied at work. They include reports of reduced efficiency, lower work output, decreased effort, and decline in work quality (e.g., McCarthy and Barker, 2000; McCarthy et al., 1995). To make this picture complete, the negative impact on productivity may to some extent be cancelled out by the targets' compensatory behaviour as they put in extra effort or do their best to demonstrate their commitment to the organisation (Hoel et al., 1999). However, as such a strategy is unlikely to be effective in the long run, the overall effect on productivity is likely to be negative.

- Grievance, compensation, and litigation: Practices will vary greatly between countries. It is worth noting that for every case that ends in court or an industrial or employment tribunal, there is likely to be a large number of cases resolved at the level of the organisation. Grievance procedures may give rise to administrative costs in connection with the implementation of internal investigation and mediation procedures, where such procedures are in place, and these may in many instances be greater than any compensation package. With grievance procedures, there tends to be a range of hidden direct costs. Such costs include those associated with pursuing formal grievance procedures, other staff time in addressing bullying-related incidents, and employees' compensation costs (Hoel et al., 2003; Sheehan et al., 2001). There also are costs relating to counselling or employee assistance, mediation, or grievance procedures, compensation claims, or other actions (McCarthy and Barker, 2000).

- Loss of public goodwill and reputation: With the high profile of many individual compensation cases, the potential damage to reputation may motivate employers to deal with issues. The detrimental impact of the negative publicity that accompanies workplace bullying cases ought not to be underestimated, particularly in the current difficult economic climate where negative images of the behaviours of some managers in some organisations may result in a negative perception of their organisations and their professions. Such loss of reputation impacts not only staff, customers and suppliers but also shareholders.

Discussion

Our initial examination of the various cost factors involved with bullying tended to look at each factor in isolation, although in reality it is likely to be a dynamic relationship that interweaves those factors together. For example, if work colleagues are not replaced when absent, pressure is likely to mount on their coworkers to cover for those absences whilst simultaneously performing their own work. Such increased workload may contribute to some people possibly reaching their breaking point and, in turn, having to take time off to recuperate. Additional extra duties are likely to increase tension among

coworkers, possibly reducing productivity and increasing sickness absence and turnover rates. By contrast, in cases where victims decide to be present although, strictly speaking, they would benefit physically and mentally from being off work, they may be unproductive at their jobs because of lack of concentration (Leymann, 1992) or for fear of making mistakes or drawing attention to themselves. This may also affect their relationships with coworkers and supervisors, possibly sharpening and broadening, rather than reducing and limiting, the conflict. Such outcomes will, in turn, affect the total productivity of the work unit, increase absenteeism and turnover rates of third parties, and increase the potential scale and scope of any investigation following a complaint in a never-ending spiral of destruction weakening the individual and the organisation.

When discussing the various cost factors involved with bullying, it emerged that the association between bullying and various organisational outcome variables remains, in statistical terms, relatively weak or at best moderate. Despite this limitation, correlations of this order indicate an effect of considerable magnitude (Zapf, Dormann, et al., 1996). It is also argued that people's reactions to their experiences tend to be complex and idiosyncratic, which would then tend to affect overall correlations (Sparks et al., 1997). Use of self-report data as a way of measuring performance may also weaken the true relationship, as individuals may underestimate their own absenteeism and overestimate their own productivity. This relatively weak relationship may also be an artifact of the cross-sectional method normally applied in such studies, where performance is measured entirely at the level of the individual. According to Daniels and Harris (2000), one should bear in mind that a relatively small impact at the individual level may have a substantial aggregated or cumulative effect within the wider work environment.

Most of the data produced as evidence in this chapter are the outcomes of cross-sectional studies and therefore do not allow for inferability or causal relationships. In order to increase our understanding of organisational effects in connection with bullying and the complex interactions involved, studies of a qualitative nature may represent a way forward, not least to explore the significance to the individual of behaviours such as absenteeism and turnover. Diary studies could be one way forward, although ethical consideration may reduce their effectiveness. Studies involving witnesses or observers could also throw important light on the complexity of the interaction between bullying, on the one hand, and factors such as absenteeism, productivity and turnover, on the other.

Moreover, in monetary terms, the calculations and assessment of costs involved with bullying can only be as good as the research on which they are based, and the many uncertainties exposed in previous studies mean that such cost estimations at best represent what has been referred to as well-informed guesses (Giga et al., 2008; Hoel et al., 2001). Despite methodological reservations, and there are many, this chapter has presented evidence to suggest that it makes good business sense for organisations to prevent and stop workplace bullying. Such a view has also influenced the bullying debate in many countries and the extent to which researchers, practitioners, and advocate groups have been able to engage employers or instill management motivation for action. Still, some critics may argue that some of the cost estimates entered into the public debate are inaccurate at best and spurious at worst. In that respect, Beale and Hoel (in press) point out that in terms of days lost because of bullying alone, some of the figures are on par with the total days lost because of strikes in the United Kingdom in the most strike-prone years after the war. Still, it is noteworthy that employers have made little attempt to challenge such cost estimates, suggesting that despite potential inaccuracies, the figures are somewhat realistic in that they refer to a widespread problem with multiple, interrelated organisational consequences that come with a high price tag attached. Moreover, by presenting some tangible estimates of the financial damage caused by workplace bullying to their organisations, the cost estimates offer employers a more realistic picture than those shaped by moral or ethical arguments alone and could thus act as motivation for action by management.

Nevertheless, although organisational cost is an important rationale or reason for counteracting bullying in the workplace, we would like to emphasise that even if there were no financial costs attached, the problem should be challenged on the grounds of equality and fairness. The prevention and management of bullying are, therefore, first and foremost ethical or moral issues, and in many countries they are legal issues as well as bullying increasingly becomes prohibited and regulated by law (Yamada, this volume). We also argue that where organisations succeed in preventing bullying from happening in the first place or successfully intervene when conflict escalates and bullying scenarios begin to unfold, it is more likely that employees' energy, experience, creativity, and innovativeness may be utilised more effectively and productively than in those organisations where the issue is ignored.

References

Allen, N. J., and Meyer, J. P. (1990) The measurement and antecedents of affective, continuance and normative commitment to the organization. *Journal of Occupational and Organizational Psychology*, 63, 18–38.

Bassman, E. (1992) *Abuse in the workplace*. Westport, CT: Quorum Books.

BBC (2003) "City Worker Wins Bullying Damages." http://news.bbc.co.uk/1/hi/england/london/3112899.stm (accessed August 2, 2007).

Beale, D., and Hoel, H. (in press) Workplace bullying and British employers: Exploring questions of cost, policy, context and control. *Work, Employment & Society*.

Begley, T. M. (1998) Coping strategies as predictors of employee distress and turnover after an organizational consolidation: A longitudinal analysis. *Journal of Occupational and Organizational Psychology*, 71, 305–329.

Berthelsen, M., Skogstad, A., Hauge, L. J., Nielsen, M. B., and Einarsen, S. (2009) *Mobbing og utstøtning fra arbeidslivet* [Bullying and exclusion from working life]. Bergen, Norway: University of Bergen.

Beswick, J., Gore, J., and Palferman, D. (2006) *Bullying at work: A review of the literature* (WPS/06/04). Derbyshire, UK: Health and Safety Laboratories.

Boswell, W. R., and Olson-Buchanan, J. B. (2004) Experiencing mistreatment at work: The role of grievance filing, nature of mistreatment, and employee withdrawal. *Academy of Management Journal, 47,* 129–139.

Bowling, N. A., and Beehr, T. A. (2006) Workplace harassment from the victim's perspective: A theoretical model and meta-analysis. *Journal of Applied Psychology, 91,* 998–1012.

Brun, J., and Lamarche, C. (2006) Assessing the costs of work stress. Université Laval, Quebec, Canada. http://www.cgsst.com/stock/eng/doc272-806.pdf (accessed May 11, 2010).

Caverley, N., Cunningham, J. B., and MacGregor, J. N. (2007) Sickness presenteeism, sickness absenteeism, and health following restructuring in a public sector organization. *Journal of Management Studies, 44,* 304–319.

Cooper, C. L., Liukkonen, P., and Cartwright, S. (1996) *Stress prevention in the workplace: Assessing the costs and benefits for organisations.* Dublin: European Foundation for the Improvement of Living and Working Conditions.

Cortina, L. M., Magley, V. J., Williams, J. H., and Longhout, R. D. (2001) Incivility in the workplace: Incidence and impact. *Journal of Occupational Health Psychology, 6,* 64–80.

Dalton, D. R. (1997) Employee transfer and employee turnover: A theoretical and practical disconnect? *Journal of Organizational Behavior, 18,* 411–413.

Daniels, K., and Harris, C. (2000) Work, psychological well-being and performance. *Occupational Medicine–Oxford, 50,* 304–309.

Di Martino, V., Hoel, H., and Cooper, C. L. (2003) *Preventing violence and harassment in the workplace.* Dublin: European Foundation for the Improvement of Living and Working Conditions.

Djurkovic, N., McCormack, D., and Casimir, G. (2004) The physical and psychological effects of workplace bullying and their relationship to intention to leave: A test of the psychosomatic and disability hypotheses. *International Journal of Organization Theory and Behavior, 7,* 469–497.

——— (2008) Workplace bullying and intention to leave: The moderating effect of perceived organisational support. *Human Resource Management Journal, 18,* 405–422.

Einarsen, S. (1996) Bullying and harassment at work: Epidemiological and psychosocial aspects. PhD thesis, Faculty of Psychology, University of Bergen, Norway.

Einarsen, S. (2000) Harassment and bullying at work: A review of the Scandinavian approach. *Aggression and Violent Behavior, 4,* 371–401.

Einarsen, S., Raknes, B. I., Matthiesen, S. B., and Hellesøy, O. H. (1994) *Mobbing og harde personkonflikter: Helsefarlig samspill på arbeidsplassen* [Bullying and interpersonal conflict: Interaction at work with negative implications for health]. Bergen, Norway: Sigma Forlag.

Ferris, G. R., Zinko, R., Brouer, R. L., Buckley, M. R., and Harvey, M. G. (2007) Strategic bullying as a supplementary, balanced perspective on destructive leadership. *Leadership Quarterly, 18,* 195–206.

Field, T. (1996) *Bully in sight: How to predict, resist, challenge and combat workplace bullying.* Wantage, UK: Wessex Press.

Giga, S., Hoel, H., and Lewis, D. (2008) *Dignity at work: The costs of workplace bullying*. Unite and BERR Partnership Project working together for Dignity at Work. London: Department for Business, Enterprise and Regulatory Reform, Unite the Union.

Gordon, F., and Risley, D. (1999) *The costs to Britain of workplace accidents and work-related ill health in 1995/6*, 2nd ed. London: HSE Books.

Hoel, H., and Cooper, C. L. (2000) Destructive conflict and bullying at work. Unpublished report, University of Manchester Institute of Science and Technology.

Hoel, H., Cooper, C. L., and Faragher, B. (2001) The experience of bullying in Great Britain: The impact of organizational status. *European Journal of Work and Organizational Psychology, 10*, 443–465.

Hoel, H., Einarsen, S., and Cooper, C. L. (2003) Organisational effects of bullying. In S. Einarsen, H. Hoel, D. Zapf, and C. L. Cooper (eds.), *Bullying and emotional abuse in the workplace: International perspectives in research and practice* (pp. 145–161). London: Taylor & Francis.

Hoel, H., Rayner, C., and Cooper, C. L. (1999) Workplace bullying. In C. L. Cooper and I. T. Robertson (eds.), *International review of industrial and organizational psychology* (pp. 195–230), Chichester, UK: John Wiley.

Hoel, H., Sparks, K., and Cooper, C. L. (2001) *The cost of violence/stress at work and the benefits of a violence/stress-free working environment*. Geneva: International Labour Organisation.

Ishmael, A. (1999) *Harassment, bullying and violence at work*. London: The Industrial Society.

Johns, G. (1994) How often were you absent? A review of the use of self-reported absence data. *Journal of Applied Psychology, 79*, 574–591.

———— (1997) Contemporary research on absence from work: Correlates, causes and consequences. In C. L. Cooper and I. T. Robertson (eds.), *International review of industrial and organizational psychology* (pp. 115–173). Chichester, UK: John Wiley.

Keashly, L., and Jagatic, K. (2000) The nature, extent, and impact of emotional abuse in the workplace: Results of a statewide survey. Paper presented at the Academy of Management Conference, Toronto, Canada.

Kile, S. M. (1990) *Helsefarleg leiarskap: Ein explorerande studie* [Leadership with negative health implications: An exploratory study]. Report from the Norwegian General Science Council, Bergen, Norway.

Kivimäki, K., Elovainio, M., and Vathera, J. (2000) Workplace bullying and sickness absence in hospital staff. *Occupational and Environmental Medicine, 57*, 656–660.

Knott, V., Dollard, M. F., and Winefield, A. H. (2004) *Workplace bullying: An assessment of prevalence/nature of bullying behaviours in a correctional services sample*. Adelaide, Australia: Work & Stress Research Group, University of South Australia.

Lee, D. (2000) An analysis of workplace bullying in the UK. *Personnel Review, 29*, 593–612.

Lee, R. T., and Brotheridge, C. M. (2006) When prey turns predatory: Workplace bullying as a predictor of couteraggression/bullying, coping and well-being. *European Journal of Work and Organizational Psychology, 15*, 352–357.

Lee, T. H., Gerhart, B., Weller, I., and Trevor, C. O. (2008) Understanding voluntary turnover: Path-specific job satisfaction effects and the importance of unsolicited job offers. *Academy of Management Journal, 51*, 651–671.

Lee, T. W., Holtom, B. C., McDaniel, L. S., and Hill, J. W. (1999) The unfolding model of voluntary turnover: A replication and extension. *Academy of Management Journal, 42*, 450–462.

Leymann, H. (1990) Mobbing and psychological terror at workplaces. *Violence and Victims, 5*, 119–125.

Leymann, H. (1992) *Vuxenmobbing på svenska arbeidsplatser. Delrapport 1 om frekvenser.* Stockholm: Arbetarskyddstyrelsen.

—— (1996) The content and development of mobbing at work. *European Journal of Work and Organizational Psychology, 5* (2), 165–184.

March, J. G., and Simon, H. A. (1958) *Organizations.* New York: Wiley.

Mathisen, G. E., Einarsen, S., and Mykletun, R. (2008) The occurrences and correlates of bullying and harassment in the restaurant sector. *Scandinavian Journal of Psychology, 49*, 59–68.

McCarthy, P., and Barker, M. (2000) Workplace bullying risk audit. *Journal of Occupational Health and Safety–Australia and New Zealand, 16*, 409–418.

McCarthy, P., Sheehan, M., and Kearns, D. (1995) *Managerial styles and their effects on employees' health and well-being in organizations undergoing restructuring.* Brisbane, Australia: School of Organizational Behaviour and Human Resource Management, Griffith University.

Meschkutat, B., Stackelbeck, M., and Langenhoff, G. (2002) *Der mobbing-report: Repräsentativstudie für die Bundesrepublik Deutschland.* [The mobbing report: Representative study for the Federal Republic of Germany]. Bremerhaven: Wirtschaftsverlag.

Neuman, J. H., and Baron, R. A. (2003) Social antecedents of bullying: A social interactionist perspective. In S. Einarsen, H. Hoel, D. Zapf and C. L. Cooper (eds.), *Bullying and emotional abuse in the workplace: International perspectives in research and practice* (pp. 185–202). London: Taylor & Francis.

Niedl, K. (1996) Mobbing and wellbeing: Economic and personnel development implications. *European Journal of Work and Organizational Psychology, 5*, 239–249.

O'Connell, P. J., Calvert, E., and Watson, D. (2007) *Bullying in the workplace: Survey report.* Dublin: Department of Enterprise Trade and Employment. Economic and Social Research Institute.

Olweus, D. (2003) Bully/victim problems and an effective intervention program. In S. Einarsen, H. Hoel, D. Zapf, and C. L. Cooper (eds.), *Bullying and emotional abuse in the workplace: International perspectives in research and practice* (pp. 62–78). London: Taylor & Francis.

O'Moore, M., Seigne, E., McGuire, L., and Smith, M. (1998) Bullying at work: Victims of bullying at work in Ireland. *Journal of Occupational Health and Safety–Australia and New Zealand, 14*, 569–574.

Price Spratlen, L. (1995) Interpersonal conflict which includes mistreatment in a university workplace. *Violence and Victims, 10*, 285–297.

Quine, L. (1999) Workplace bullying in NHS community trust: Staff questionnaire survey. *British Medical Journal, 318*, 228–232.

—— (2001) Workplace bullying in nurses. *Journal of Health Psychology, 6*, 73–84.

Rayner, C. (1997) The incidence of workplace bullying. *Journal of Community and Applied Social Psychology, 7*, 249–255.

—— (1999) Workplace bullying. PhD thesis, University of Manchester Institute of Science and Technology.

———— (2000) Building a business case for tackling bullying in the workplace: Beyond a cost-benefit analysis. In M. Sheehan, C. Ramsey, and J. Patrick (eds), *Transcending boundaries*. Proceedings of the 2000 Conference, Brisbane, Australia, September.

Rayner, C., Hoel, H., and Cooper, C. L. (2002) *Workplace bullying: What we know, who is to blame, and what can we do?* London: Taylor & Francis.

Seago, J. A. (1996) Work group culture, stress and hostility: Correlations with organisational outcomes. *Journal of Nursing Administration*, 26, 39–47.

Sheehan, M., McCarthy, P., Barker, M., and Henderson, M. (2001) A model for assessing the impact and costs of workplace bullying. Paper presented at the Standing Conference on Organizational Symbolism (SCOS), Trinity College, Dublin, 30 June–4 July.

Sparks, K., Cooper, C. L., Fried, Y., and Shirom, A. (1997) The effects of hours of work on health: A meta-analytical review. *Journal of Occupational and Organizational Psychology*, 70, 391–408.

Steer, S. R., and Rhodes, R. M. (1978) Major influence on employee attendance: A process model. *Journal of Applied Psychology*, 63, 391–407.

Tepper, B. J. (2000) Consequences of abusive supervision. *Academy of Management Journal*, 43, 178–190.

Tepper, B. J. (2007) Abusive supervision in the workplace: Review, synthesis and research agenda. *Journal of Management*, 33, 261–289.

Tepper, B. J., Duffy, M. K., Henle, C. A., and Lambert, L. S. (2006) Procedural injustice, victim precipitation, and abusive supervision. *Personnel Psychology*, 59, 101–123.

Thyholdt, R., Eide, R., and Hellesøy, O. H. (1986) Arbeidsplass Statfjord. Heise og Sykdom på Statfjordfeltet, report no. 6 [Workplace Statfjord: Health and Illness in the Statfjord sector, report no. 6]. Bergen, Norway: Research Centre for Work, Health and Safety, University of Bergen.

Tranøy, T. (2007) *Vallas fall: Et innblikk i den skjulte valgkampen* [The fall of Valla: A look at the hidden powerplay]. Oslo: Forlaget Manifest.

UNISON (1997) *UNISON members' experience of bullying at work*. London: UNISON.

———— (2000) *Police staff bullying report* (No. 1777). London: UNISON.

Vartia, M. (2001) Consequences of workplace bullying with respect to well-being of its targets and the observers of bullying. *Scandinavian Journal of Work, Environment and Health*, 27, 63–69.

World Health Organization (WHO). (2004) *The economic dimension of interpersonal violence*. Geneva: WHO.

Zapf, D., Dormann, C., and Frese, M. (1996) Longitudinal studies in organizational stress research: A review of the literature with reference to methodological issues. *Journal of Occupational Health Psychology*, 1, 145–169.

Zapf, D., and Gross, C. (2001) Conflict escalation and coping with workplace bullying: A replication and extension. *European Journal of Work and Organizational Psychology*, 10, 497–522.

Zapf, D., Knorz, C., and Kulla, M. (1996) On the relationship between mobbing factors, and job content, social work environment, and health outcomes. *European Journal of Work and Organizational Psychology*, 5, 215–237.

6

Measuring Exposure to Workplace Bullying

Morten Birkeland Nielsen, Guy Notelaers, and Ståle Einarsen

CONTENTS

Introduction

During the last 20 years, workplace bullying has been measured and assessed in a range of different studies in order to investigate issues such as the nature, frequency, antecedents, and outcomes of the phenomenon (cf. Einarsen et al., 2003b; Rayner, Hoel, and Cooper, 2002). Despite all this attention on the phenomenon in itself, little is known about how the use of different measurement and estimation methods influences the findings on workplace bullying. Keashly and Harvey (2005) even claim the field has been hurried by a desire to discover such substantive issues as nature and phenomenology at the expense of construct research and research on methodological issues. As a consequence of insufficient research on the development of psychometrically sound measures of workplace bullying, there are reasons to believe that the measurement of the phenomenon has not been as rigorous as one would hope.

In order to shed light on some of the methodological challenges within the field, the aim of this chapter is to present the different quantitative measurement strategies that are used in the assessment of workplace bullying, as well as the strengths and limitations of these different methods. In doing so, this chapter provides an overview of the many options that exist when measuring exposure to workplace bullying in survey research. Hence, the chapter may be seen as a tool that can assist researchers and practitioners in selecting the best measurement methods for assessing workplace bullying and may, in addition, provide guidelines for how results from various studies can best be understood.

Assessing Workplace Bullying

As workplace bullying is a complex phenomenon with many facets, its measurement is not a straightforward task (Cowie et al., 2002; Einarsen et al., 2003a). The difficulties in measuring the phenomenon are reflected through the many different measurement methods and inventories that have been applied in research to this date. Nonetheless, although many different methods have been used, most have either assessed (a) the respondents' overall feeling of being victimised by bullying (often referred to as the self-labelling method), (b) the respondents' perception of being exposed to a range of specific bullying behaviour (often referred to as the behavioural experience method), or (c) a combination of the two methods (Cowie et al., 2002; Mikkelsen and Einarsen, 2001; Salin, 2001). Examples of the self-labelling and the behavioural experience methods are displayed in Table 6.1.

In line with basic theories on measurement, a satisfactory overlap between the theoretical definition and the operational definition of the concept is necessary to achieve valid findings on bullying. That is, the measurement instrument should operationalise the central characteristics of the conceptual definition. Linking back to the theoretical definition presented in the opening chapter of this book, a measurement instrument should therefore assess exposure to negative acts, the regularity and persistency of these acts, the process development of workplace bullying, and the power imbalance between target and perpetrator. Yet, when reviewing the different measurement methods applied in research on workplace bullying, it becomes obvious that these do not necessarily capture all five characterising elements of the definition. In the first parts of this chapter, we will present and discuss strengths and limitations of the methods used in research on workplace bullying and show why differences between methods lead to inconsistent findings that cannot be compared across studies. In order to reduce such methodological issues in future research, the final part of this chapter is devoted to a best practice recommendation for the measurement of workplace bullying.

TABLE 6.1

Examples of the Self-labelling Method and the Behavioural Experience Method

Self-labelling method as presented by Einarsen and Skogstad (1996, pp. 190–191) and as developed by Olweus (1978):

"Bullying takes place when one or more persons systematically and over time feel that they have been subjected to negative treatment on the part of one or more persons, in a situation in which the person(s) exposed to the treatment have difficulty in defending themselves against them. It is not bullying when two equally strong opponents are in conflict with each other."

According to this definition, have you been subjected to bullying at the workplace during the last *six* months?

1. No
2. Yes, once or twice
3. Yes, now and then
4. Yes, about once a week
5. Yes, many times a week

Behavioural experience method. Examples of items from the Negative Acts Questionnaire–Revised (Einarsen et al., 2009):

The following behaviours are often seen as examples of negative behaviour in the workplace. Over the last six months, how often have you been subjected to the following negative acts at work?

Please circle the number that best corresponds with your experience over the last six months:

	Never	Now and then	Monthly	Weekly	Daily
1. Being humiliated or ridiculed in connection with your work	1	2	3	4	5
2. Having key areas of responsibility removed or replaced with more trivial or unpleasant tasks	1	2	3	4	5
3. Persistent criticism of your work and effort	1	2	3	4	5
4. Being the subject of excessive teasing and sarcasm	1	2	3	4	5

The Self-Labelling Method

The self-labelling method is probably the most frequently used approach in research on workplace bullying. In a meta-analysis of the impact of methodological moderators on prevalence rates of workplace bullying, 67% of the 102 included prevalence estimates were based on a self-labelling method (Nielsen et al., 2009a). When applying this method, participants are usually given a single item question asking whether or not they have been bullied within a specific time period. In some studies, the respondents are offered a theoretical definition of bullying before being asked whether or not they have experiences in the workplace that corresponds to the presented definition (e.g., Einarsen and Skogstad, 1996; Olweus, 1989; O'Moore, Lynch, and Niamh, 2003). In other studies, the question about bullying has been asked without a preceding definition (e.g., Lewis, 1999; Rayner, 1997). Response categories also vary between studies. Whilst some studies use a simple yes/ no response alternative, other studies employ a five- to seven-point frequency scale usually ranging from "never bullied" to "bullied daily" (Cowie et al., 2002).

Advantages and Disadvantages of the Self-Labelling Method

A clear advantage of the self-labelling method is that it does not take up much space in a questionnaire while it also is easy to administer (Nielsen, 2009). Furthermore, as the method explicitly asks respondents whether they are exposed to bullying, the face validity of the method is convincing. The method may, in addition, have high construct validity if the respondents are presented with a precise and easy-to-grasp theoretical definition that explains the concept.

Despite these strengths, the self-labelling method does not come without some flaws and difficulties. First of all, the method does not offer any insight in the nature of the behaviours involved. Hence, by using this method one only gets to know whether the respondents perceive themselves as victims of bullying, whereas information on how the bullying took place is ignored. Consequently, because people may have different personal thresholds for labelling themselves as bullied, the self-labelling method is a very subjective approach in which personality, emotional factors, cognitive factors, and mis-perceptions may figure as potential biases. For instance, labelling oneself as a victim of bullying has been found to be associated with feelings of shame that may even last after the bullying has ended (Felblinger, 2008; Lewis, 2004). Because of shame, some targets may also find it threatening to their self-esteem to admit to victimisation, and they will therefore avoid label-ling themselves even though their experiences corresponds with the formal definition of workplace bullying (Nielsen, 2009; Out, 2005). Related to this, perception of masculinity may also affect the thresholds for self-labelling, for research has shown that men label themselves as victims of bullying to a lesser degree than do women (Salin, 2003). Thus, it may be that more men try to avoid disclosing personal information that might make them appear weak or vulnerable.

The subjectivity of the self-labelling method may be especially important to be taken into consideration when the respondents are not offered a definition of the bullying concept. In such cases, some respondents may not self-label simply because the particular situation they experienced does not conform to their personal definition of, and experiences with, bullying (cf. Magley et al., 1999). In a comparison of academic definitions and lay definitions of workplace bullying, it was found that lay definitions excluded central ele-ments found in academic definitions, while at the same time they included elements not found in any of the academic definitions of bullying (Saunders et al., 2007). Although most laypersons included perpetration of negative behaviour as a defining characteristic, only 15.2% of the respondents included power imbalance, and only 14.7 % included the persistency criterion found in the formal definition. Furthermore, many lay definitions also included themes of fairness and respect that are not found in the formal definitions.

Hence, when respondents are not presented with a definition of workplace bullying, one risk is that there will be a large discrepancy between how the researcher and the respondents perceive the phenomenon under investigation. Of course, even when a precise definition is presented, there is no guarantee that respondents actually read and digest the definitions they are provided or that they do not simply stick to and apply their own definition of the bullying concept when answering the self-labelling question. Yet, in a meta-analysis of prevalence rates of workplace bullying, Nielsen et al. (2009a) showed that self-labelling with definition studies yielded far lower estimates of bullying than self-labelling studies without definitions. Consequently, when a definition of the bullying concept is presented for the respondents, it seems that it is taken into consideration.

In sum, the self-labelling method may be described as an approach that is easy to apply in research. Yet, because of the subjectivity bias and the fact that self-labelling does not provide any information on the nature and content of the bullying, the method has important limitations. For instance, the self-labelling with definition approach does not guarantee against respondents who are exposed to a high level of negative and unwanted behaviour but who, for some reason, do not label themselves as victims. Because of such shortcomings, the behavioural experience method has been proposed as an alternative approach when assessing exposure to workplace bullying.

Behavioural Experience Method

When investigating the respondents' exposure to specific bullying behaviours, respondents are usually presented with an inventory that includes various types of unwanted and negative behaviour that may be called bullying if occurring repeatedly over time. The respondents are then asked to report how frequently they have been exposed to the different behaviours listed in the inventory within a given time period.

An array of different inventories has been developed to assess the behaviours involved in workplace bullying. Table 6.2 gives an overview of 27 different inventories that have been used to assess bullying or phenomena similar to workplace bullying. Some of these have been used in only one study, whereas others, such as the Leymann Inventory of Psychological Terror (LIPT; Leymann, 1990a), the Negative Acts Questionnaire (NAQ/NAQ-R; Einarsen, Hoel, and Notelaers, 2009; Einarsen and Raknes, 1997), and the Workplace Aggression Research Questionnaire (Harvey and Keashly, 2003) have been employed in a range of studies. Yet, the NAQ seems to be the

TABLE 6.2

Inventories Used to Assess Behavioural Experiences of Workplace Bullying

Inventory	Country of Origin/Reference	Targets/ Perpetrators	Time Frame	Items (N)	Response Categories	Dimensions	Cronbach's Alpha
Agervold's Questionnaire (based on the NAQ)	Denmark (Agervold, 2007)	Targets	6 months	10	1. Almost daily 2. 2–3 times per week 3. 2–3 times per month 4. never/rarely	One-dimensional	.79
Baron, Neuman, & Geddes's scale	USA (Baron, Neuman, & Geddes, 1999)	Targets	N/A	40	Five-point scale ranging from 1 = Never to 5 = Very often	1. Expression of hostility 2. Obstructionism 3. Overt aggression	.95
Fox & Stallworth's bullying scale	USA (Fox & Stallworth, 2005)	Targets	5 years	25 + 7	Five-point scale ranging from 1 = Never to 5 = Extremely often	1. General bullying 2. Racial/ethnic bullying	.94
Indirect Aggression Scale (IAS)	UK (Forrest, Eatough, & Shevlin, 2005)	Targets and perpetrators	12 months	25	1. Never 2. Once or twice 3. Sometimes 4. Often 5. Regularly	6. Social exclusion 7. Use of malicious humor 8. Guilt induction	Subscales: .81–.89
Interpersonal Workplace Events	Canada (Keashly, Trott, & MacLean, 1994)	Targets	12 months	48	Five-point scale ranging from 1 = Rare to 5 = Always	1. Positive events 2. Abusive events	Subscales: .78–92
Keashly & Jagatic's checklist	Canada (Keashly & Jagatic, 2000)	Targets	12 months	17	Five-point scale ranging from 1 = Never to 5 = Very often	1. Hostile expression 2. Exclusion 3. Reputation	Subscale: 72–82

Lee & Brotheridge's scale	Canada (Lee & Brotheridge, 2006)	Targets and perpetrators	6 months	43 + 43	1. Not at all 2. Once or twice 3. Now and then 4. Once a week 5. Many times a week	Targets: 1. Belittlement 2. Work undermined 3. Verbal abuse Perpetrators: 1. Fall guy/gal 2. Undermined others' work 3. Emotional abuse	Subscales: .60–.94
Leymann Inventory of Interpersonal Terror (LIPT)	Sweden (Leymann, 1990a; Niedhammer, David, Degioanni, & 143 Médecins du travail, 2006; Zapf, Knorz, & Kulla, 1996)	Targets	12 months	45	1. Daily 2. At least once a week 3. At least once a month 4. More seldom 5. Never	1. Attacking the victim with organisational measures 2. Attacking the victim's *social* relationship with *social* isolation 3. Attacking the victim's private life 4. Physical violence 5. Attacking the victim's attitudes 6. Verbal aggression 7. Rumours	Subscales: .79–.86
Mobbing Scale for Academic Nurses	Turkey (Ozturk, Sokmen, Yilmaz, & Cilingir, 2008)	Targets	N/A	82	Five-point scale ranging from 1 = Totally disagree to 5 = Completely agree	1. Effects on psychology and fatigue 2. Effects on the organisation and management 3. Attacks on self-esteem 4. Attacks on personal and professional relationships 5. Effects on social relationships 6. Attacks on showing oneself and communications 7. Attacks on professional practice 8. Effects on health and life	Total scale: .97; Subscales: .62–.86

(continued)

TABLE 6.2 (continued)

Inventories Used to Assess Behavioural Experiences of Workplace Bullying

Inventory	Country of Origin/Reference	Targets/ Perpetrators	Time Frame	Items (N)	Response Categories	Dimensions	Cronbach's Alpha
Negative Acts Questionnaire (NAQ)	Norway (Einarsen & Raknes, 1997)	Targets	6 months	18	1. Never 2. Now and then 3. About monthly 4. About weekly 5. About daily	1. Personal derogation 2. Work-related harassment 3. Social exclusion 4. Social control 5. Physical abuse	Subscales: .33–.85
Negative Acts Questionnaire– Revised 29-item version (NAQ-R 29)	UK/Norway (Hoel, Cooper, & Faragher, 2001)	Targets	6 months	29	1. Never 2. Now and then 3. About monthly 4. About weekly 5. About daily	1. Personal bullying 2. Work-related bullying	
Negative Acts Questionnaire– Revised 22-item version (NAQ-R)	Norway (Einarsen et al., 2009)	Targets	6 months	22	1. Never 2. Now and then 3. Monthly 4. Weekly 5. Daily	1. Personal bullying 2. Work-related bullying 3. Physical intimidating forms of bullying	.90
Negative Acts Questionnaire– Salin's version	Finland/Norway (Salin, 2001)	Targets	12 months	32	1. Never 2. Now and then 3. Monthly 4. Weekly 5. Daily	N/A	N/A
Negative Acts Questionnaire– Reduced Spanish version	Spain/Norway (Jiménez et al., 2007)	Targets	6 months	14	1. Never 2. Now and then 3. Monthly 4. Weekly 5. Daily	1. Work-related bullying 2. Attacks on private and personal life	.85

Instrument	Country (source)	Target/Perpetrator	Time frame	Items	Response scale	Dimensions	Reliability
Negative Workplace Experience Questionnaire	USA/Japan (Kokubun, 2007)	Targets	N/A	95	Frequency: 1. Never 2. Seldom 3. Sometimes 4. Frequently; Intensity: 1. Not at all 2. Slightly 3. Moderately 4. Very abusive	1. Organisational measure 2. Isolation 3. Private information 4. Physical aggression 5. Verbal aggression 6. Attacking attitudes 7. Rumours 8. Teasing or mockery 9. Intimidation 10. Sabotage 11. Unfair treatment 12. Criticised 13. Humiliation 14. Treated unimportant 15. Harassment	N/A
Ólafsson & Jóhannsdóttir's scale	Iceland (Ólafsson & Jóhannsdóttir, 2004)	Targets	12 months	18	1. Never 2. Once or twice 3. Two to three times per month 4. Once a week 5. A few times per week	1. General bullying 2. Work-related bullying	Subscales: .78–.85
Parkins, Fishbein, & Ritchey's scale	USA (Parkins et al., 2006)	Perpetrators	6 months	6	Five-point scale ranging from 0 = Never to 4 = 4 or more times	One-dimensional	N/A
Perceived Victimization Scale	USA (Aquino & Bradfield, 2000)	Targets	6 months	10	1. Never 2. 1–3 times 3. 4–10 times 4. 11–20 times 5. More than 20 times	1. Indirect victimisation 2. Direct victimisation	Subscales: .72–.77

(continued)

TABLE 6.2 (continued)
Inventories Used to Assess Behavioural Experiences of Workplace Bullying

Inventory	Country of Origin/Reference	Targets/ Perpetrators	Time Frame	Items (N)	Response Categories	Dimensions	Cronbach's Alpha
Quine's scale	UK (Quine, 1999, 2003)	Targets	12 months	21	1. No 2. Yes	1. Threat to professional status 2. Threat to personal standing 3. Isolation 4. Overwork 5. Destabilisation	N/A
Rayner's scale	UK (Ayoko, Callan, & Härtel, 2003; Rayner, 1999)	Targets	6 months	15	0. Never 1. Less than once a month 2. Every month 3. Every week 4. Every day	One-dimensional	.95
Vragenlijst Beleving en Beoordeling van de Arbeid (The Assessment and Experience of Work; VBBA)	Netherlands (Hubert & van Veldoven, 2001)	Targets	N/A	4	1. Never 2. Sometimes 3. Often 4. Always	One-dimensional	N/A
Workplace Aggression Research Questionnaire (WAR-Q)	Canada (Harvey & Keashly, 2003; Keashly & Neuman, 2004)	Targets	12 months	60	1. Never 2. Once 3. A few times 4. Several times 5. Monthly 6. Weekly 7. Daily	1. Passive 2. Verbal 3. Indirect 4. Direct 5. Active 6. Physical	N/A

Scale	Country (Source)	Target/Source	Timeframe	Items	Response scale	Dimensions	Reliability
Workplace Aggression Scale	USA (Schat et al., 2006)	Targets		15	1. Never 2. Less than once a month 3. 1–3 days a month 4. 1–2 days a week 5. 3–5 days a week 6. 6–7 days a week	1. Psychological abuse 2. Physical abuse 3. Source of abuse	N/A
Work Harassment Scale (WHS)	Finland (Björkqvist et al., 1994)	Targets	6 months	24	0. Never 1. Seldom 2. Occasionally 3. Often 4. Very often	N/A	.95
Workplace Incivility Scale	USA (Cortina, Magley, Williams, & Langhout, 2001)	Targets	5 years	7	5-point scale ranging from 0 = Never to 4 = Most of the time	One-dimensional	.89
Workplace Ostracism Scale	USA (Ferris, Brown, Berry, & Lian, 2008)	Targets	12 months	10	7-point scale ranging from 1 = Strongly disagree to 7 = Strongly agree	One-dimensional	.89–.96 in four different samples
Workplace Psychologically Violent Behaviours Instrument (WPVB)	Turkey (Yildirim & Yildirim, 2008)	Targets	12 months	33	0. I have never faced 1. I have faced once 2. I have faced several times 3. I face sometimes 4. I frequently face 5. I constantly face	1. Individual's isolation from work 2. Attack on professional status 3. Attack on personality 4. Direct negative behaviours	Total scale: .93 Subscales: .72–.91

Note: N/A = Not answered.

most utilised inventory as a review of studies on the prevalence of workplace bullying showed that 47% of the included behavioural experience studies employed a variation of this instrument (Nielsen et al., 2009a).

A clear limitation of many of the behavioural experience inventories used in research on workplace bullying is that they have not been thoroughly tested and validated in separate validation studies. Hence, little is known about the accuracy, factor structure, and trustworthiness of the different instruments. A notable exception is the NAQ, which has been validated in several studies and countries (e.g., Einarsen et al., 2009; Giorgi, 2008; Jiménez et al., 2007). The most recent version of the NAQ, the NAQ-R, investigates the frequency and persistency of the respondent's exposure to 22 different types of unwanted and negative behaviour that range from such subtle and indirect acts as gossiping to such more direct behaviours as threats of physical abuse. All items are formulated in behavioural terms, with no reference to the word *bullying*. Based on their experiences at the workplace during the last six months, respondents are asked to indicate how often they have been exposed to the 22 negative acts using a response scale ranging from "Never" to "Daily." Hence, the NAQ-R may also be used to identify one-off incidences of psychological aggression and harassment.

In a study on the psychometric properties of the NAQ-R (Einarsen et al., 2009), it was shown that the inventory has high internal stability and excellent criterion and construct validity. The instrument was found to comprise three underlying factors labelled as personal bullying, work-related bullying, and physical intimidating forms of bullying. By using Latent Class Cluster analysis (LCC), it was further shown that the instrument may be used to differentiate between groups of employees with different levels of exposure to bullying, ranging from infrequent exposure to incivility at work to severe exposure to bullying and harassment.

Data based on the behavioural experience inventories may be used in several different ways. The most straightforward approach has been to compute an overall sum score on the basis of the individual items. This sum score may then be applied as a measure of the level of exposure to bullying, which can be further included in correlation analyses, regression analyses, and so on. The behavioural experience approach may also be used to distinguish between different groups of respondents (e.g., targets and nontargets). A common method for separating targets from nontargets is to apply an operational criterion (cf. Solberg and Olweus, 2003). Leymann (1990b), for instance, stated that one has to be exposed to at least one negative act per week over a period of at least six months to be reckoned as a target of bullying. Mikkelsen and Einarsen (2001) claimed that two negative acts are required to classify the experience as bullying, whereas Agervold (2007) proposed that the actions should take place at least three or four times a week and continue for at least six months.

However, the decision that targets are only those respondents who are subjected to a specific number of behaviours with a given frequency (e.g., on a weekly basis) is questionable (Notelaers, De Witte, et al., 2006). First of all, the use of operational criteria is a more or less arbitrary choice that reduces the escalating process of bullying into a strict either-or phenomenon. Second, by using the operational criterion approach, the number of reported targets may actually be a function of the number of items included in the inventory. That is, an inventory that includes a wide range of behaviours may identify more targets than would a shorter inventory. Third, when using a cut-off approach based on the Leymann criterion, for instance, employees exposed to a wide range of specific behaviours, each occurring only now and then, are not regarded as targets of bullying even though they are actually exposed to many negative behaviours regularly. Fourth, the operational criterion approach implies that all kinds of negative acts are equally important when distinguishing between targets and nontargets.

To avoid the limitations with the operational criterion, LCC has been proposed as a method that may overcome many of the earlier mentioned shortcomings of operational criterion method (Nielsen, 2009; Notelaers, Einarsen, et al., 2006). LCC is a statistical method for identifying subtypes of related cases (latent classes) from multivariate categorical data. As a general example, it can be used in larger samples to find distinct diagnostic categories given the presence or absence of a range of experienced symptoms, types of attitude structures from survey responses (Moors, 2003), and consumer segments from demographic and preference variables (Paas, Vermunt, and Bijmolt, 2007) or to examine subpopulations from answers to test items (cf. Vermunt and Magidson, 2002). With regard to bullying, the LCC method can be used to identify different groups of respondents based on their exposure to bullying behaviours.

Compared to the operational criterion method, the LCC method has several advantages. First of all, the method can deal with such violations of statistical assumptions as skewed responses. Second, in resemblance to item-response theory models, the LCC is able to model the discriminatory power of items as well as the difficulty to agree with items (Magidson and Vermunt, 2004; Vermunt, 2001; Vermunt and Magidson, 2002). Third, the LCC method has the ability to classify respondents into several mutually exclusive groups on the basis of the respondents' reported exposure to negative behaviour. Hence, as opposed to the operational criterion method, which only classifies respondent into targets and nontargets, the LCC approach distinguishes empirically between several different groups of respondents regarding the nature and frequency of their exposure to bullying behaviour (Giang and Graham, 2008; Notelaers, Einarsen, et al., 2006). Finally, rather than relying on some arbitrary cut-off value, the number of groups and the prevalence within the different exposure groups are here tested empirically.

The LCC method has been applied in several studies on workplace bullying during the last few years. For instance, in a study among Belgian employees, six different groups of respondents were identified based on their exposure to negative behaviours (Notelaers, Einarsen, et al., 2006). The respondents in the first cluster, which made up 35% of the total sample, reported low exposure to all negative behaviours and were consequently labelled as "not bullied." The respondents in the second cluster also reported relatively low exposure to negative behaviours. However, some negative acts related to the respondents' work performance and occurred more frequently. The cluster was therefore labelled as "limited work criticism." This cluster was the second-largest group and covered some 28% of the respondents. The third cluster was characterised by exposure to a limited amount of work- and person-related criticism. The cluster was labelled as "limited negative encounters," and nearly 17% of the respondents belonged to this group. Respondents in the fourth cluster were subjected to a range of negative acts, yet mainly "now and then." The cluster was labelled as "sometimes bullied," with 9% of the sample belonging to this cluster. The fifth cluster comprised respondents who reported exposure to various work-related negative acts on a frequent basis. The cluster was therefore labelled as "work-related bullying." Eight percent of the respondents were classified in this cluster. The last cluster was characterised by a high probability that the respondents were subjected to negative behaviours at least on a weekly basis. The cluster, which comprised about 3% of the sample, was labelled as "victims of bullying."

A noteworthy aspect of these findings is that the differences in nature and frequency of the exposure to bullying between the six clusters correspond with the view of workplace bullying as a gradually escalating process that involves a set of stages with distinct natures. Similar cluster solutions to the one described earlier have also been obtained in later studies in Norway (Nielsen et al., 2009b) and the United Kingdom (Einarsen et al., 2009).

In some studies, statistically developed cut-off criteria have been proposed as a practical alternative for analysing behavioural experience data (Björkqvist, Österman, and Hjeltbäck, 1994; Ólafsson and Jóhannsdóttir, 2004). For instance, by creating a sum-score of the behavioural experience items and cross-fitting this score with the self-labelling method in a receiver operating characteristics (ROC) curve, Notelaers and Einarsen (2009) showed that it is possible to calculate cut-off values based on the overlap between behavioural experience data and self-labelling data on workplace bullying. The ROC curve, a method used to assess the accuracy of a test, works by giving a plot of the sensitivity of a test versus its false-positive rate for all possible cut points (Obuchowski, 2003). By applying this method on the NAQ-R scale in a representative sample, Notelaers and Einarsen (2009) found that respondents with a score below 33 on the NAQ-R scale cannot be perceived as being bullied. Respondents with a score between 33 and 44 can be classified as being "sometimes" bullied at work, whereas respondents with a score of 45 or higher are classified as victims of workplace bullying. Since these cut-off scores are derived from a Norwegian sample, it should be noted they may not be applicable to other countries, and further research is needed to establish whether these cut-off scores are consistent or differ across cultures.

Advantages and Disadvantages of the Behavioural Experience Method

The behavioural experience method has been considered to be a more objective method than the self-labelling approach because it does not require the respondents to label their experience as bullying and so the decision about whether or not someone is bullied rests with the researcher through the use of an operational criterion or through statistical analysis (e.g., Frese and Zapf, 1988; Notelaers, Einarsen, et al., 2006). Consequently, findings should have a lower risk for being influenced by cognitive and emotional processes. Yet, one may also question the objectivity of the behavioural experience method (Agervold, 2007; Einarsen, 1996).

"Objective bullying" refers to a situation in which actual external evidence of bullying is found (cf. Brodsky, 1976). In other words, to be considered as an objective measurement, the reported unwanted and negative treatment must be confirmed by third parties (Agervold, 2007) or by the perpetrator. Moreover, in resemblance with the self-labelling method, perceptual bias may also affect the reporting of exposure to specific negative behaviour. Since all perceptions are influenced by one's attitudes, personality, and affective states (Bower, Gilligan, and Monterio, 1981; Lazarus, 1982), respondents may experience the same behaviour differently.

Whereas an important strength of the behavioural experience approach is that the method takes the nature, frequency, and duration of the unwanted behaviours into consideration, a limitation is that the power distance between target and perpetrator is not measured (Nielsen, 2009). Consequently, this lack of overlap between the theoretical and operational definition of the bullying concept constrains the construct validity of the method. Moreover, by only asking the respondents about the frequency and duration of the behaviour, we do not know whether the respondents would actually label the behaviour as bullying. In some cases, it may be that behaviour that most people would perceive as potential acts of bullying is seen by other employees as part of their job. An item from the NAQ-R (Einarsen et al., 2009) may serve as an example. Whereas most people would experience threats of violence or physical abuse as potential negative workplace behaviour, such behaviour is something many prison warders or nurses in psychiatric hospitals experience as a part of their jobs on an almost daily basis (cf. Ireland and Snowden, 2002; Maghan, 1999). Hence, in some settings, there is a risk of labelling more or less expected behaviour as bullying when using the behavioural experience method, even though the behaviours may not necessarily be perceived as such by the targets. This could especially be a problem when employing an operational approach such as the Leymann criterion, which does not differentiate between the nature and severity of the various behaviours. Nonetheless, even though many psychiatric wardens and prison guards experience threats of violence and abuse as an unavoidable part of their job, it should, of course, be made clear that such behaviour should not be accepted. Another problem with most of the inventories used to measure exposure to bullying behaviours is that they do not differentiate between different actors of the bullying. Yet, this can easily be solved by adding a question about the perceived perpetrators (see Einarsen, Hoel, and Notelaers, 2009).

Witnesses and Perpetrators

An important limitation of research on workplace bullying is the lack of verification, by peer nominations, for example, of bullying incidents (Coyne et al., 2003; Einarsen, 2000; Hoel, Rayner, and Cooper, 1999). That is, the majority of the research has assessed the phenomenon from the target perspective without obtaining any information to verify the behaviour or without even obtaining the views of other parties, such as the bully or other employees (Coyne et al., 2003). According to Agervold (2007), the assessment of witnesses is the closest one may come to an objective observation of bullying, and school-based research on bullying has often used the peer nomination method to identify both victims and bullies (cf. Cowie et al., 2002; Solberg and Olweus, 2003). Yet, within a workplace setting there are practical and ethical problems of using this approach (Coyne et al., 2003). One problem with the third-party approach is that bullying often takes the form of subtle and indirect behaviour such as withholding information and slander. Hence, third parties may be unable or unwilling to perceive and label something as bullying until it has reached the stage of direct aggression (Einarsen, 1996). Moreover, economic dependency could also prevent people from being honest in their assessment (Björkqvist et al., 1994). This situation would be particularly true in peer nominations where one is asked to assess one's superiors or people in formal positions of power.

Using perpetrators to assess bullying may also be somewhat problematic. One problem is that some perpetrators do not perceive themselves as such because they may not understand or wish to admit that their behaviour can be considered bullying. Social desirability—that is, the tendency of respondents to reply in a manner that will be viewed favourably by others—is another potential problem. In one of the few studies that have investigated the social desirability issue among perpetrators of bullying, it was found that those who admitted to bullying others had low scores on social desirability (Parkins, Fishbein, and Ritchey, 2006). Yet, the fact that those admitting to having bullied others have low scores on social desirability just confirms the presence of this subset of respondents. Those individuals high in social desirability would, of course, be expected to deny being perpetrators.

The Role of Negative Affectivity

Another methodological problem that may influence findings on all the different assessment methods in bullying research is negative affectivity, or NA (cf. Hoel et al., 1999; Spector et al., 2000). Negative affectivity, a frequently cited issue in self-report survey research, is usually defined as a mood-dispositional dimension that reflects pervasive individual differences in negative emotionality and self-concept (Watson and Clark, 1984). Individuals high in NA are usually characterised by distress, unpleasurable engagement, and nervousness, whereas low negative affect is characterised by a state of calmness and serenity. Watson and Clark (1984) conclude that people who express high NA view themselves and a variety of aspects of the world around them in generally negative terms. As a consequence, NA may influence the relationships between variables in organisational research.

For instance, research shows that people with high NA report more stressors and strains (e.g., Burke, Brief, and George, 1993; Watson and Clark, 1984). Hence, there may be a personality-linked predisposition to assign negative meaning and label to events (Hoel et al., 1999). Because of the strong influence of NA on self-reports, it has been recommended that survey researchers statistically control for NA bias in their studies (Burke et al., 1993). Yet, others argue strongly against controlling for NA in every study on work stress because of the potential to remove substantive effects rather than actual bias. For instance, Spector and colleagues (2000) claim that one should only control for a variable when the variable in question has been demonstrated conclusively to be a bias and only a bias. In relation to bullying, negative affect must most definitely be seen as a likely outcome of the experience.

Empirical Findings

The foregoing examination of the different measurement methods clearly shows that the methods are different and that they emphasise different aspects of the bullying phenomenon. The differences between the methods are also reflected through empirical findings. The large variation in reported prevalence rates of bullying may serve as an indicator. In a review of the occurrence of bullying, Zapf and colleagues (2003) reported that studies using self-labelling without a preceding definition yielded prevalence rates between 10% and 25%, while self-labelling with a preceding definition resulted in prevalence rates of 1–4% bullying. For studies using the behavioural method with an operational criterion, prevalence rates seem to vary between 3% and 17%, depending on the cut-off criterion utilised (Nielsen, 2009).

A meta-analysis investigating the impact of methodological moderators (e.g., measurement method) on the prevalence rates of workplace bullying supports the review of Zapf and colleagues (2003). In this meta-analytical study (Nielsen et al., 2009a), which weighted the estimates by sample size, a mean prevalence rate of 11.3% was found for self-labelling with definition estimates, whereas a mean rate of 18.1% was found for self-labelling estimates without a definition. The behavioural experience estimates fell in between with a mean rate of 14.8%. Hence, the results from these reviews show that the self-labelling with the definition method yields the lowest estimates of workplace bullying, whereas the self-labelling without definition provides the highest estimate. In conclusion, findings on workplace bullying are heavily dependent on measurement method, and research findings from different studies should not be directly compared without taking the measurement method into account.

Targets and Victims

Based on the foregoing discussion of strengths and limitations related to the self-labelling method and the behavioural experience method, it can be concluded that the two methods possibly investigate somewhat different characteristics of the bullying phenomenon. Hence, the discrepancy with regard to empirical findings between the two methods may be explained by what the methods actually assess. As mentioned, a main difference between the self-labelling and the behavioural experience method is that the former method includes a cognitive evaluation of whether the respondents feel victimised by the bullying, whereas the latter method mainly investigates the persistency of different negative behaviours without taking the victimisation aspect into consideration. Hence, to avoid mixing the methods together, it may be useful to apply different labels when referring to findings of bullied respondents. As an example of this, Nielsen (2009) proposes that the self-labelling approach assesses self-labelled *victims* of bullying, whereas the behavioural experience method assesses *targets* of bullying behaviour. Although these two concepts have been used interchangeably in the literature on bullying, it may be useful to treat them as separate constructs when investigating workplace bullying. More specifically, a target of bullying is an employee who experiences exposure to systematic and persistent bullying behaviours at the workplace. A victim of bullying, on the other hand, is a person exposed to equivalent systematic and persistent bullying behaviour and who, *in addition*, perceives her- or himself as being victimised by this treatment. This view is supported by prior research and theory that suggest that relational powerlessness is a core determinant of victimisation (cf. Roscigno, Lopez, and Hodson, 2009). Thus, based on such reasoning, all victims are targets of bullying, but all targets are not necessarily victims—a claim also substantiated in empirical research showing that although all victims are targets and, as such, are exposed to a range of bullying behaviours, many targets exist who do not label themselves as victims (Nielsen et al., 2009b).

This distinction between targets and victims seems to correspond to the view of workplace bullying as a process (Zapf and Gross, 2001). During the early phase of the process, when the bullied employees are typically subjected to behaviour that is difficult to pin down because of its occasional, indirect, and discrete nature, one may not yet feel victimised by the bullying. As the treatment gradually develops and escalates to a stage involving more frequent and more direct acts, the nature of the treatment may become clearer. If the target feels that he or she lacks the resources to retaliate in kind against the treatment, a feeling of being victimised may develop. The very subjective nature of the victimisation process may be an advantage of the self-labelling approach in this respect. In any instance, the definitional core of bullying rests on the subjective perception that the target experiences these behaviours as hostile and humiliating and that the behaviours are directed towards him or her, regardless of how the experience is labelled. Anyhow, as it is only the bullied person who knows whether or not he or she feels victimised, the self-labelling method provides unique information about how the bullying is being experienced.

A "Best Practice" Approach for Measuring Workplace Bullying

Having concluded that the self-labelling method and the behavioural experience method assess workplace bullying from somewhat different points of view, we must consider the implication that the methods should be regarded as equally useful, since both approaches seem to provide valid but complementary information on the phenomenon of bullying (Nielsen et al., 2009b). Our recommendation is, therefore, that one should always integrate both methods in the survey when investigating workplace bullying (for exemplifications, see Einarsen et al., 2003a; Mikkelsen and Einarsen, 2001; Salin, 2001). Such a combination has several advantages. First of all, it is in line with the suggestion that more than one operationalisation of a given concept should be used in studies on work stressors (Frese and Zapf, 1988). Second, by combining the two approaches, this cross-fit provides knowledge about different groups of respondents with regard to both their persistency of the exposure to negative behaviours and their subjective interpretation of being victimised by the bullying behaviours (Einarsen et al., 2003a; Nielsen et al., 2009b). Hence, by combining the methods, with both measuring respondents' exposure to persistent bullying behaviour and their perceptions of being victimised by this behaviour, it is possible to capture all characteristics included in the theoretical definition of workplace bullying (cf. Nielsen, 2009).

The next question is, then, which of the different variants of the methods should be used. Clearly, the definitional approach should be preferred to the self-labelling approach without definition. By excluding the definition, one increases the risk that a respondent confirms having been exposed to workplace bullying without actually having experienced such treatment. However, when a definition is present, one can rest assured that there is an agreement between the researcher's and the respondents' understanding of the concept.

With regard to how behavioural experience data should be analysed, we suggest that the LCC method is used rather than the operational criterion. As discussed earlier in this chapter, the operational criterion method has several limitations that may constrict the findings on workplace bullying, whereas the LCC method overcomes most of these difficulties. Compared to the operational criterion method, which separates only between targets and nontargets, the LCC approach has the advantage that it distinguishes between several groups of respondents. Furthermore, the LCC method also distinguishes between different kinds of behaviours, for instance, work-related bullying versus person-related bullying. Finally, by using the LCC method, one obtains a better indication of how workplace bullying develops as a process. Although no longitudinal empirical research using LCC analysis has been conducted to explicitly investigate the dynamics of this process, cross-sectional results clearly show that the different clusters have the potential to complement the self-assessment measure by documenting the emergence of this process (Notelaers, Einarsen, et al., 2006).

Of course, the LCC method is not without challenges. First, the method has yet to be implemented in commonly used statistical programs such as the Statistical Package for the Social Sciences (SPSS), Statistical Analysis Software (SAS), and Statistica. Second, the method requires somewhat advanced statistical knowledge and requires rather large samples, something that may further limit its usability. Consequently, the approach is not directly available to everyone. Furthermore, for practical purposes in organisational settings where the application of such an advanced approach is not feasible, an operational criterion or a statistically developed cut-off value may well be used to classify respondents into targets and nontargets (Einarsen et al., 2009). Hence, because the operational criterion approach—for instance, in the form of the Leymann criterion of exposure to one negative act per week—gives easy access to prevalence estimates of bullying, such an approach may still have some value. Yet, both researchers and practitioners should keep the limitations of the method in mind. Finally, in cases where the aim is to look at the relationship between bullying and other variables, and where it is not necessary to classify respondents into different groups, calculating the sum-score based on different bullying behaviours may be the most useful approach because such a score provides larger variation in the variable than does a dichotomous score.

Conclusion

In this chapter, we have provided an overview of how workplace bullying has been measured and assessed in survey studies to date, as well as the strengths and limitations of the different measurement methods. There are strong methodological, theoretical, and practical reasons why knowledge about the measurement of workplace bullying is valuable: To achieve reliable and valid findings on workplace bullying, one needs to know how and to what extent different methodological factors actually influence the measurement of the phenomenon. Furthermore, by having knowledge about the reliability and validity of the research methods, one can correct, improve, and strengthen the theories in the field. For practical reasons, the use of reliable and valid methods to assess workplace bullying is an important part of the basis for the development and implementation of intervention strategies to prevent workplace aggression (cf. Schat, Frone, and Kelloway, 2006).

Yet, the knowledge about how different measurement methods influence research findings may also represent an ethical dilemma for researchers, practitioners, and organisations (cf. Nielsen, 2009). Because the prevalence rates for bullying vary extensively depending on the operationalisation employed, the measurement method can actually be used to manipulate the observed extent of the bullying phenomenon in a given enterprise or survey. For instance, if one wishes to find a relatively low rate of bullying, the self-labelling with definition method can be chosen; whereas the self-labelling without definition or the behavioural experience method could be used if a higher prevalence rate is desired. Hence, researchers and practitioners need to be cautious when investigating bullying in their surveys.

However, by applying both the self-labelling method and the behavioural experience method in studies on workplace bullying, researchers can decrease the problem of data manipulation, since the two methods used in combination will provide both a "high" and a "low" rate of bullying and will thus present a balanced view.

References

Agervold, M. (2007) Bullying at work: A discussion of definitions and prevalence, based on an empirical study. *Scandinavian Journal of Psychology, 48,* 161–172.

Aquino, K., and Bradfield, M. (2000) Perceived victimization in the workplace: The role of situational factors and victim characteristics. *Organization Science, 11* (5), 525–537.

Ayoko, O. B., Callan, V. J., and Härtel, C. E. J. (2003) Workplace conflict, bullying, and counterproductive behaviors. *International Journal of Organizational Analysis, 11* (4), 283–301.

Baron, R. M., Neuman, J. H., and Geddes, D. (1999) Social and personal determinants of workplace aggression: Evidence for the impact of perceived injustice and the type A behavior pattern. *Aggressive Behavior, 25,* 281–296.

Björkqvist, K., Österman, K., and Hjeltbäck, M. (1994) Aggression among university employees. *Aggressive Behavior, 20,* 173–184.

Bower, G. H., Gilligan, S. G., and Monterio, K. P. (1981) Selectivity of learning caused by affective states. *Journal of Experimental Psychology: General, 11,* 451–473.

Brodsky, C. M. (1976) *The harassed worker.* Lexington, MA: D. C. Heath.

Burke, M. J., Brief, A. P., and George, J. M. (1993) The role of negative affectivity in understanding relations between self-reports of stressors and strains: A comment on the applied psychology literature. *Journal of Applied Psychology, 78* (3), 402–412.

Cortina, L. M., Magley, V. J., Williams, J. H., and Langhout, R. D. (2001) Incivility in the workplace: Incidence and impact. *Journal of Occupational Health Psychology, 6* (1), 64–80.

Cowie, H., Naylor, P., Rivers, I., Smith, P. K., and Pereira, B. (2002) Measuring workplace bullying. *Aggression and Violent Behavior, 7,* 33–51.

Coyne, I., Chong, P. S.-L., Seigne, E., and Randall, P. (2003) Self and peer nominations of bullying: An analysis of incident rates, individual differences, and perceptions of the working environment. *European Journal of Work and Organizational Psychology, 12* (3), 209–228.

Einarsen, S. (1996) *Bullying and harassment at work: Epidemiological and psychosocial aspects.* Bergen: Norway: University of Bergen.

——— (2000) Harassment and bullying at work: A review of the Scandinavian approach. *Aggression and Violent Behavior, 5* (4), 379–401.

Einarsen, S., Hoel, H., and Notelaers, G. (2009) Measuring bullying and harassment at work: Validity, factor structure, and psychometric properties of the Negative Acts Questionnaire–Revised. *Work & Stress, 23* (1), 24–44.

Einarsen, S., Hoel, H., Zapf, D., and Cooper, C. L. (2003a) The concept of bullying at work: The European tradition. In S. Einarsen, H. Hoel, D. Zapf, and C. L. Cooper (eds.), *Bullying and emotional abuse in the workplace. International perspectives in research and practice* (pp. 3–30). London: Taylor & Francis.

——— (eds.). (2003b) *Bullying and emotional abuse in the workplace. International perspectives in research and practice.* London: Taylor & Francis.

Einarsen, S., and Raknes, B. I. (1997) Harassment in the workplace and the victimization of men. *Violence and Victims, 12* (3), 247–263.

Einarsen, S., and Skogstad, A. (1996) Bullying at work: Epidemiological findings in public and private organizations. *European Journal of Work and Organizational Psychology, 5,* 185–201.

Felblinger, D. M. (2008) Incivility and bullying in the workplace and nurses' shame responses. *Journal of Obstetric, Gynecologic, and Neonatal Nursing, 37,* 234–242.

Ferris, D. L., Brown, D. J., Berry, J. W., and Lian, H. (2008) The development and validation of the workplace ostracism scale. *Journal of Applied Psychology, 93* (6), 1348–1366.

Forrest, S., Eatough, V., and Shevlin, M. (2005) Measuring adult indirect aggression: The development and psychometric assessment of the indirect aggression scales. *Aggressive Behavior, 31* (1), 84–97.

Fox, S., and Stallworth, L. E. (2005) Racial/ethnic bullying: Exploring links between bullying and racism in the US workplace. *Journal of Vocational Behavior, 66,* 438–456.

Frese, M., and Zapf, D. (1988) Methodological issues in the study of work stress: Objective vs. subjective measurements and the question of longitudinal studies. In C. L. Cooper and R. Payne (eds.), *Causes, coping and consequences of stress at work* (pp. 371–411). Chichester, UK: John Wiley.

Giang, M. T., and Graham, S. (2008) Using latent class analysis to identify aggressors and victims of peer harassment. *Aggressive Behavior, 34* (2), 203–213.

Giorgi, G. (2008) The Negative Acts Questionnaire Revised (NAQ-R) in Italy. *Prevention Today, 4* (4), 71–83.

Harvey, S., and Keashly, L. (2003) Predicting the risk for aggression in the workplace: Risk factors, self-esteem and time at work. *Social Behavior and Personality, 31* (8), 807–814.

Hoel, H., Cooper, C. L., and Faragher, B. (2001) The experience of bullying in Great Britain: The impact of organizational status. *European Journal of Work and Organizational Psychology, 10,* 443–465.

Hoel, H., Rayner, C., and Cooper, C. L. (1999) Workplace bullying. In C. L. Cooper and I. T. Robertson (eds.), *International review of industrial and organizational psychology* (pp. 195–230). Chichester, UK: John Wiley.

Hubert, A. B., and van Veldoven, M. (2001) Risk sectors for undesirable behaviour and mobbing. *European Journal of Work and Organizational Psychology, 10,* 415–424.

Ireland, J. L., and Snowden, P. (2002) Bullying in secure hospitals. *Journal of Forensic Psychiatry, 13* (3), 538–554.

Jiménez, B. M., Muñoz, A. R., Gamarra, M. M., and Herrer, M. G. (2007) Assessing workplace bullying: Spanish validation of a reduced version of the Negative Acts Questionnaire. *Spanish Journal of Psychology, 10* (2), 449–457.

Keashly, L., and Harvey, S. (2005) Emotional abuse in the workplace. In S. Fox and P. E. Spector (eds.), *Counterproductive behavior: Investigations of actors and targets.* Washington, DC: American Psychological Association.

Keashly, L., and Jagatic, K. (2000) *The nature, extent, and impact of emotional abuse in the workplace: Results of a statewide survey.* Paper presented at the Symposium on Persistent Patterns of Aggressive Behavior at Work, Toronto, Ontario.

Keashly, L., and Neuman, J. H. (2004) Bullying in the workplace: Its impact and management. *Employee Rights and Employment Policy Journal, 8* (2), 335–373.

Keashly, L., Trott, V., and MacLean, L. M. (1994) Abusive behavior in the workplace: A preliminary investigation. *Violence and Victims, 9* (4), 341–357.

Kokubun, S. (2007) *Abusive behavior at work: A cross-cultural comparison between the U.S. and Japan.* San Diego, CA: Alliant International University.

Lazarus, R. S. (1982) Thoughts on the relations between emotion and cognition. *American Psychologist, 37,* 1019–1024.

Lee, R. T., and Brotheridge, C. M. (2006) When prey turns predatory: Workplace bullying as a predictor of counteraggression/bullying, coping, and well-being. *European Journal of Work and Organizational Psychology, 15* (33), 352–377.

Lewis, D. (1999) Workplace bullying—interim findings of a study in further and higher education in Wales. *International Journal of Manpower, 20* (1/2), 106–118.

——— (2004) Bullying at work: The impact of shame among university and college lectures. *British Journal of Guidance and Counselling, 32* (3), 281–299.

Leymann, H. (1990a) *Handbok för användning av LIPT-formuläret för kartläggning av risker för psykologisk vald* [Manual of the LIPT questionnaire for assessing the risk of psychological violence at work]. Stockholm: Violen.

——— (1990b) Mobbing and psychological terror at workplaces. *Violence and Victims, 5* (2), 119–126.

Maghan, J. (1999) Dangerous inmates: Maximum security incarceration in the state prison systems of the United States. *Aggression and Violent Behavior, 4* (1), 1–12.

Magidson, J., and Vermunt, J. K. (2004) Latent Class Models. In D. Kaplan (ed.), *The Sage handbook for quantitative methodology* (pp. 175–198). Thousand Oaks, CA: Sage.

Magley, V. J., Hulin, C. L., Fitzgerald, L. F., and DeNardo, M. (1999) Outcomes of self-labeling sexual harassment. *Journal of Applied Psychology, 84* (3), 390–402.

Mikkelsen, E. G., and Einarsen, S. (2001) Bullying in Danish worklife: Prevalence and health correlates. *European Journal of Work and Organizational Psychology, 10,* 393–414.

Moors, G. (2003) Diagnosing response style behavior by means of a latent-class factor approach: Socio-demographic correlates of gender role attitudes and perceptions of ethnic discrimination reexamined. *Quality & Quantity, 37,* 277–302.

Niedhammer, I., David, S., Degioanni, S., and 143 Médecins du travail (2006) The French version of the Leymann's questionnaire on workplace bullying: The Leymann Inventory of Psychological Terror (LIPT). *Rev Epidemiol Sante Publique, 54* (3), 245–262.

Nielsen, M. B. (2009) Methodological issues in research on workplace bullying: Operationalisations, measurements, and samples. Unpublished doctoral dissertation, University of Bergen, Norway.

Nielsen, M. B., Matthiesen, S. B., and Einarsen, S. (2009a) The impact of methodological moderators on prevalence rates of workplace bullying: A meta-analysis. *Journal of Occupational and Organizational Psychology,* Early online.

Nielsen, M. B., Skogstad, A., Matthiesen, S. B., Glasø, L., Aasland, M. S., Notelaers, G., et al. (2009b) Prevalence of workplace bullying in Norway: Comparisons across time and estimation methods. *European Journal of Work and Organizational Psychology, 18* (1), 81–101.

Notelaers, G., De Witte, H., Vermunt, J., and Einarsen, S. (2006) Pesten op het werk, gewikt en gewogen: Een latente klassen benadering op basis van de negatieve Acts-vragenlijst [How to measure bullying at work? A latent class analysis of the Negative Acts Questionnaire]. *Gedrag en Organisatie, 19* (2), 149–160.

Notelaers, G., and Einarsen, S. (2009) Measuring workplace bullying. Paper presented at the Social Tension at Work and Mental Health Workshop, Berlin.

Notelaers, G., Einarsen, S., De Witte, H., and Vermunt, J. (2006) Measuring exposure to bullying at work: The validity and advantages of the latent class cluster approach. *Work & Stress, 20* (4), 288–301.

Obuchowski, N. A. (2003) Receiver operating characteristic curves and their use in radiology. *Radiology, 229,* 3–8.

Ólafsson, R. F., and Jóhannsdóttir, H. L. (2004) Coping with bullying in the workplace: The effect of gender, age and type of bullying. *British Journal of Guidance and Counselling, 32* (3), 319–333.

Olweus, D. (1989) Prevalence and incidence in the study of antisocial behavior: Definitions and measurement. In M. W. Klein (ed.), *Cross national research in self reported crime and delinquency* (pp. 187–201). Dorndrect: Kluwer.

O'Moore, M., Lynch, J., and Niamh, N. D. (2003) The rates and relative risks of workplace bullying in Ireland, a country of high economic growth. *International Journal of Management and Decision Making, 4* (1), 82–95.

Out, J. (2005) *Meanings of workplace bullying: Labelling versus experiencing and the belief in a just world.* Unpublished dissertation, University of Windsor, Windsor, Ontario, Canada.

Ozturk, H., Sokmen, S., Yilmaz, F., and Cilingir, D. (2008) Measuring mobbing experiences of academic nurses: Development of a mobbing scale. *Journal of the American Academy of Nurse Practitioners, 20* (9), 435–442.

Paas, L. J., Vermunt, J. K., and Bijmolt, T. H. A. (2007) Discrete time, discrete state latent Markov modelling for assessing and predicting household acquisitions of financial products. *Journal of the Royal Statistical Society, 170,* 955–974.

Parkins, I. S., Fishbein, H., and Ritchey, P. N. (2006) The influence of personality on workplace bullying and discrimination. *Journal of Applied Social Psychology, 36* (10), 2554–2577.

Quine, L. (1999) Workplace bullying in NHS community trust: Staff questionnaire survey. *British Medical Journal, 318,* 228–232.

——— (2003) Workplace bullying, psychological distress, and job satisfaction in junior doctors. *Cambridge Quarterly of Healthcare Ethics, 12,* 91–101.

Rayner, C. (1997) The incidence of workplace bullying. *Journal of Community & Applied Social Psychology, 7,* 199–208.

——— (1999) Workplace bullying: Do something! *Journal of Occupational Health and Safety–Australia and New Zealand, 14* (6), 581–586.

Rayner, C., Hoel, H., and Cooper, C. L. (2002) *Workplace bullying: What we know, who is to blame, and what can we do?* London: Taylor & Francis.

Roscigno, V. J., Lopez, S. H., and Hodson, R. (2009) Supervisory bullying, status inequalities and organizational context. *Social Forces, 87* (3), 1561–1589.

Salin, D. (2001) Prevalence and forms of bullying among business professionals: A comparison of two different strategies for measuring bullying. *European Journal of Work and Organizational Psychology, 10* (4), 425–441.

——— (2003) The significance of gender in the prevalence, forms and perceptions of bullying. *Nordiske Organisasjonsstudier, 5* (3), 30–50.

Saunders, P., Huynh, A., and Goodman-Delahunty, J. (2007) Defining workplace bullying behaviour professional lay definitions of workplace bullying. *International Journal of Law and Psychiatry, 30* (4–5), 340–354.

Schat, A. C., Frone, M. R., and Kelloway, E. K. (2006) Prevalence of workplace aggression in the U.S. workforce: Findings from a national study. In E. K. Kelloway, J. Barling, and J. J. Hurrell Jr. (eds.), *Handbook of workplace violence* (pp. 47–89). Thousand Oaks, CA: Sage.

Solberg, M. E., and Olweus, D. (2003) Prevalence estimation of school bullying with the Olweus Bully/Victim Questionnaire. *Aggressive Behavior, 29,* 239–268.

Spector, P. E., Zapf, D., Chen, P. Y., and Frese, M. (2000) Why negative affectivity should not be controlled for in job stress research: Don't throw out the baby with the bath water. *Journal of Organizational Behavior, 21* (1), 79–95.

Vermunt, J. K. (2001) The use of restricted latent class models for defining and testing nonparametric and parametric IRT models. *Applied Psychological Measurement*, 25, 283–294.

Vermunt, J. K., and Magidson, J. (2002) Latent class cluster analysis. In J. Hagenaars and A. L. McCutcheon (eds.), *Applied latent class analysis* (pp. 89–106). Cambridge: Cambridge University Press.

Watson, D., and Clark, L. A. (1984) Negative affectivity: The disposition to experience aversive emotional states. *Psychological Bulletin*, 96, 465–490.

Yildirim, D., and Yildirim, A. (2008) Development and psychometric properties of workplace psychologically violent behaviours instrument. *Journal of Clinical Nursing*, 17, 1361–1370.

Zapf, D., Einarsen, S., Hoel, H., and Vartia, M. (2003) Empirical findings on bullying in the workplace. In S. Einarsen, H. Hoel, D. Zapf, and C. L. Cooper (eds.), *Bullying and emotional abuse in the workplace: International perspectives in research and practice* (pp. 103–126). London: Taylor & Francis.

Zapf, D., and Gross, C. (2001) Conflict escalation and coping with workplace bullying: A replication and extension. *European Journal of Work and Organizational Psychology*, 10, 497–522.

Zapf, D., Knorz, C., and Kulla, M. (1996) On the relationship between mobbing factors, and job content, social work environment, and health outcomes. *European Journal of Work and Organizational Psychology*, 5, 215–237.

Section III

Explaining the Problem

7

Individual Antecedents of Bullying: Victims and Perpetrators

Dieter Zapf and Ståle Einarsen

CONTENTS

Introduction

Since the beginning of bullying research, the causes of bullying at work have been a hot issue of debate in both the popular press and the scientific community. While some argue that such individual antecedents as the personality of bullies and victims may indeed be involved as causes of bullying (e.g., Coyne et al., 2000), others have totally disregarded the role of individual characteristics. Heinz Leymann (1993, 1996), one of the founders of bullying research, categorically claimed that organisational factors relating to the organisation of work and the quality of leadership behaviour were the main causes of bullying. He rejected the idea that the personal characteristics of the victim are capable of playing any part in the development of bullying at work. This standpoint is also strongly advocated by some victims of bullying and their organisational networks. Other victims and their spokespersons have claimed that bullying is mainly caused by the psychopathic personality of the bully (e.g., Field, 1996).

On no account do we deny that organisational issues have to be considered in the discussion of bullying causes (see Salin and Hoel, this volume). However, our own standpoint is that no comprehensive model of workplace bullying would be satisfactory without also including personality and other individual factors of both perpetrators and victims, and their contributing effects to the onset, escalation, and consequences of the bullying process (Einarsen, 2000; Hoel et al., 1999; Zapf, 1999b). However, one has to tread carefully with respect to these issues, as one might easily be accused of blaming the victim, on the one hand, and witch-hunting, on the other. Yet, against the background of communication theory (Watzlawik et al., 1967) and the psychology of interpersonal conflict (e.g., van de Vliert, 1998), any one-sided and monocausal explanations are highly unlikely. Rather, one may have to take a broad range of potential causes of bullying into account, which may lie within the organisation, the perpetrator (the bully), the social psychology of the work group, and also the victim (Einarsen, 1999; Zapf, 1999b). Personality factors may also be more or less relevant depending on the level of intensity of the bullying, as shown in a study by Nielsen et al. (2008). In addition, subgroups of targets may exist portraying quite different personality characteristics, as shown in some studies (e.g., Glaso et al., 2007, 2009).

Furthermore, the personality of the victim may be relevant in explaining perceptions of and reactions to workplace bullying as shown in the study by Nielsen and colleagues (2008) on the effect of sense of coherence on the targets' stress reactions when being exposed to bullying but not necessarily as relevant in explaining the behaviour of the bully or even who are singled out as targets in the first place (Einarsen, 2000; Glaso et al., 2009). It is also likely that the personalities of perpetrator and victim may be of more relevance in some cases than in others. Empirical evidence (Glaso et al., 2007; Matthiesen and Einarsen, 2001; Zapf, 1999b) indicates that bullying cases differ in the degree to which personality is involved as a potential cause. There is certainly no such victim personality (e.g., the "notorious complainer") that can explain bullying in general (Glaso et al., 2007, 2009). Rather, specific explanations may be valid for specific subgroups but not for every case of bullying. Moreover, it is likely that several antecedents together contribute to the development of bullying, although one antecedent may sometimes play a dominating role (see also Hoel and Cooper, 2001).

Bearing these precautions in mind, the aim of this chapter is to discuss individual antecedents of workplace bullying and to review the empirical evidence. We will focus on individual antecedents of bullying that can be related to the perpetrator, the victim, or both. Causes of bullying lying within perpetrators may, of course, overlap with factors relevant to both the social group and the organisation, for example, in the case of a supervisor being the bully. Supervisors and managers are often made responsible for organisational circumstances—such as organisational culture and the organisation of work—that may contribute to the development of bullying.

This chapter aims at discussing individual antecedents involved in severe cases of bullying, but it does not intend to cover all aspects of interpersonal conflict and social stress at work. Applying a relatively stringent definition of bullying, we will focus on those severe cases of highly escalated conflicts that usually last longer than one year, and in many cases even several years (Zapf, 1999a). The following definition of bullying, or mobbing, will be the basis for this chapter:

> Bullying at work means harassing, offending, or socially excluding someone or negatively affecting someone's work tasks. In order for the label bullying (or mobbing) to be applied, a particular activity, interaction, or process has to occur repeatedly and regularly (e.g., weekly) and over a period of time (e.g., about six months). Bullying is an escalating process in the course of which the person confronted ends up in an inferior position and becomes the target of systematic negative social acts. A conflict cannot be called bullying if the incident is an isolated event or if two parties of approximately equal strength are in conflict. (Einarsen et al., this volume, p. 22)

Bullying research is still in need of studies that allow clear cause–effect analyses (Einarsen, 2000; Zapf, Dormann, et al., 1996; Zapf and Einarsen, 2005). Bullying is supposed to lead to health complaints on the part of the victim. Leymann (1996; Leymann and Gustafsson, 1996) even assumed that bullying can change the personality of the victim. However, one can equally assume, for example, that anxious, depressive, or obsessive behaviour produces a negative reaction in a group, which leads to bullying after some time. Careful language is therefore necessary when talking about cause and effect. However, although single cases can be questioned, we do believe that the arguments in this chapter will at least apply to some cases of bullying.

When talking about individual antecedents of bullying, it should be noted that a distinction has to be made between cause and guilt or responsibility and that in the case of the victim, for example, a cause "within the victim" may as much point to the social group as to the individual. An example may be a group that for some reason is not able to integrate a person who is different in one respect or another. However, being perceived as an antecedent may also mean that something is to be seen as a cause or, in some cases, even a responsibility. A manager who feels threatened by a subordinate may start to bully this person. A socially incompetent and narrow-minded person may, by her behaviour, create constant hassle for her colleagues, thus provoking an aggressive response. In other cases, a person, either bully or victim, may not be responsible for triggering a conflict but he or she may be the main reason behind the escalation of the conflict. Finally, personal characteristics may take the role of a moderator. That is, although neither cause nor mediator forces escalation of the conflict, bullying might happen to some people but not to others. There is, for example, evidence that simply being significantly different from the rest of the group—for example, being the only woman among men—increases one's risk of becoming an outsider and thus the victim of bullying (Schuster, 1996; Zapf, 1999a). In a study of assistant nurses where men held only some 4% of the jobs, men were found to be bullied more than twice as often as the women (Eriksen and Einarsen, 2004).

In conclusion, given the present state of research, empirical results are often compatible with more than one of the mechanisms just described. These constraints should be kept in mind when considering the following discussion.

Individual Antecedents of Bullying: The Perpetrator

Bullying research has revealed that bullies seem to be male more often than female, and supervisors and managers more often than colleagues (see Zapf et al., this volume). Yet we do not know much about the characteristics of the perpetrators of bullying. For obvious reasons, it is difficult to collect valid information about them. Most of the available information comes from studies based on victim reports. In an interview study among 30 Irish victims of bullying, all victims blamed the difficult personality of the bully (Seigne, 1998). Some victims reported that bullying was related to the bully's transfer to a position of power. Zapf (1999b) found that the most frequent reasons reported by victims were "They wanted to push me out of the company" and "A hostile person influenced others." A weak, unsure superior; competition for tasks, status, or advancement; and competition for the supervisor's favour were other motives perceived by victims in the studies of Björkqvist et al. (1994) and Vartia (1996). That is, at least from the victims' perspective, the cause of bullying is quite frequently identified with a particular perpetrator.

Summarising empirical findings on the bully as a cause of workplace bullying, we suggest three main types of bullying related to certain perpetrator characteristics: (1) self-regulatory processes with regard to threatened self-esteem, (2) lack of social competencies, and (3) bullying as a result of micropolitical behaviour.

Bullying as Protection of Self-Esteem

Many theorists (e.g., Baumeister, 1994; Baumeister et al., 1993, 1996; Dauenheimer et al., 2002) assume that protecting or enhancing one's self-esteem is a basic human motive that influences and controls human behaviour in many social situations. Self-esteem can be understood as having a favourable global evaluation of oneself (Baumeister et al., 1996). In every social interaction, mutual recognition of one's status is a core issue. Social interactions go smoothly as long as the interaction partners feel respected and recognised in their position, which means that self-evaluation and external evaluation are in agreement. Conflicts arise where agreement is not the case (Dauenheimer et al., 2002; Watzlawik et al., 1967). It is often assumed that people with low self-esteem tend to become aggressive and start to bully other people. Parkins et al. (2006), for example, hypothesised that self-esteem is negatively related to active workplace bullying behaviour. However, the researchers did not find the expected significant relationship. In contrast, Baumeister et al. (1996), based on a literature review on self-esteem, violence, and aggression, proposed that it is *high* rather than low self-esteem that is related to aggressive behaviour. People with low self-esteem are usually not aggressive because they fear losing the battle. Rather, they present depressive reactions and withdrawal. In contrast, one major cause of aggressive response is threatened egotism, that is, a favourable self-appraisal that encounters an external unfavourable evaluation. When favourable views about oneself are questioned, contradicted, or impugned, people may aggress. This is particularly so if unrealistically positive or inflated views of self prevail, especially if these self-appraisals are uncertain, unstable, or heavily dependent on external validation.

> People who regard themselves as superior beings might feel entitled to help themselves to the resources of other, seemingly lesser beings, and indeed they might even aggress against these beings without compunction, just as people kill insects or mice without remorse. (Baumeister et al., 1996, p. 8)

Aggression is most commonly directed at the source of the negative evaluation and serves to refute and prevent negative evaluations as well as to constitute a means of symbolic dominance and superiority over the other person. In terms of self-regulation, aggression is used to defend positive self-appraisals, instead of adapting to the more negative appraisal of oneself proposed by others, because the resulting decrease in self-esteem is aversive for nearly everyone. Thus, (too) high self-esteem can lead to tyrannical behaviour because it may be related to perfectionism, arrogance, and narcissism (Ashforth, 1994; Baumeister et al., 1993, 1996; Kets de Vries and Miller, 1985).

In summarising their review of empirical research, Baumeister et al. concluded:

> In all spheres we examined, we found that violence emerged from threatened egotism, whether this was labelled as wounded pride, disrespect, verbal abuse, insults, anger manipulations, status inconsistency, or something else. For huge nationalities, medium and small groups, and lone individuals, the same pattern was found: Violence resulted most commonly from feeling that one's superiority was somehow being undermined, jeopardised, or contradicted by current circumstances. (Baumeister et al., 1996, p. 26)

People with unstable high self-esteem may well become aggressive in response to even seemingly minor or trivial threats to self-esteem. Kernis et al. (1993) found, for example, that people with unstable self-esteem were most prone to respond defensively to unfavourable feedback. Kernis et al. (1989) reported that the highest levels of self-reported angry and hostile responses were associated with individuals who had high but unstable self-esteem scores. Similar results were found by Stucke (2002) in a study on workplace bullying. In her study, employees indicated first whether or not they actively used different kinds of bullying behaviours and whether or not they were the receivers of such behaviour. Second, narcissism and stability of self-esteem were measured. Narcissism implies high self-esteem along with the disregard of others. Active bullying behaviour was highest for a group high in narcissism but low in self-esteem stability. Obviously, these individuals had to stabilise their high but unstable self-esteem by treating other individuals negatively. Matthiesen and Einarsen's findings (2007) also support this view. In their study, the perpetrator's self-esteem was not different from that of a control group, but there were indirect indications that the self-esteem was less stable. Moreover, in line with Baumeister et al., the perpetrators showed more aggression after provocation against both supervisors and friends compared to a group who were neither bullies nor victims.

Qualitative data collected by the authors confirm these findings. For example, a typical constellation for bullying is that the department supervisor retires and a highly qualified employee of the department hopes to become his or her successor. Instead, for some reason, a person from another department is appointed whose qualification for the job is not obvious. Right from the beginning a conflict arises, and the new supervisor, who feels threatened by this person, uses all possible means to defend his or her own position and starts bullying the potential rival. Accordingly, in the study by O'Moore et al. (1998) in Ireland, many victims claimed that the bully had just recently entered a position of power.

Moreover, Baumeister et al. (1996) suggested that various negative emotions such as frustration, anger, or anxiety play a mediating role between self-esteem and aggression. Envy can also play a significant role in this context. Envy arises when someone else has what the envious person wants, which can imply that the envious person feels less worthy and less deserving than the other (Salovey, 1991). Smith et al. (1994) found that envy leads to hostility only if the person retains a favourable view of self as deserving a particular positive outcome, in which case the envied person's advantage is seen as unjust and unfair. In several studies on workplace bullying using reports of victims, envy on the part of the bullies is considered one main reason for bullying (Björkqvist et al., 1994; Einarsen et al., 1994a; Seigne, 1998; Vartia, 1996). In Vartia's study, this was the case in 68% of all incidents. Two-thirds of the victims in the study by Seigne (1998) identified the bully's envy, especially with regard to certain qualifications, as a main reason for bullying. Similar numbers were reported by Zapf (1999b).

In addition, bullying as a kind of personal retaliation can be explained as a self-regulatory mechanism of self-esteem protection (cf. the section on the provocative victim). A typical case relates to romantic affairs between bosses and their secretaries. The secretary's unexpected termination of the relationship leads to severely hurt feelings on the part of the manager, who, because everybody knows about the "secret" relationship, feels that not only his or her self-esteem but also his or her reputation as a manager are at stake. Thus, he or she starts to bully the secretary with the goal of expelling him or her from the firm. Bullying as a result of self-esteem protection may occur especially frequently if the bullies are managers, since being dominant and self-assertive, having high self-esteem, and protecting this self-esteem are characteristics normally expected from this group.

Bullying as a Lack of Social Competencies

In other cases, a lack of social competencies seems to be a dominating factor. One aspect is lack of emotional control. Supervisors might vent their anger by regularly yelling at one of their subordinates. There are many published cases that correspond to such a pattern (e.g., Adams, 1992) and "being shouted at" is a typical item in many instruments to measure bullying (e.g., Einarsen et al., 2009; Zapf, Knorz, et al., 1996). Again, this behaviour may just be an indication of a general style where aggressive outlets typically come as a response of provocation, as found in the study by Matthiesen and Einarsen (2007). Bullying might also be a consequence of a lack of self-reflection and perspective taking. This implies that some bullies are not fully aware of what they are doing and how their behaviour affects the victims. In the study by Einarsen et al. (1994b), "thoughtlessness" was seen as a cause of bullying in 46% of all cases. In the study of Jenkins et al. (2010), several bullies played down their deeds and claimed that they were not aware of the consequences of their behaviour. This is particularly so if several individuals bully someone. The personal contribution of a bully may then be small—such a relatively insignificant event as gossiping, not greeting, not passing on information, or horseplay, events that may occur, for example, every two or three weeks. If such behaviour is acted out by four or five individuals, this implies that the victim is bullied approximately twice a week. The bullies may be unaware of this situation because of the little communication between perpetrators and victims, as well as because perpetrators may not receive realistic feedback about their behaviour. Baumeister et al. (1990) analysed narrative interviews and found that victims of aggression, anger, and harassment interpreted such experiences as a series of events. The victims experienced frequent and single acts as systematic and intentional behaviour directed against them. They tolerated the bullying behaviour for some time, but then they tended suddenly to overreact. The perpetrators, by contrast, perceived the behaviours as isolated and therefore tolerable events. Thus they were surprised by the reaction of the victims, finding the victims' behaviour exaggerated and difficult to understand.

Thus, to be able to reflect on the consequences of one's own behaviour, a bully must be able to reflect on the overall situation and even take the perspective of the victim, which may show that the incidental behaviour of various bullies can be perceived by the victim as systematic and frequent harassment. However, it is assumed that some bullies may lack this ability of perspective taking. In a study of 2,200 members of seven Norwegian labour unions, some 5% of the respondents admitted that they had bullied others at work (Einarsen et al., 1994b). The self-reported bullies differed from other employees in many respects, describing themselves as high on social anxiety, low on social competence, low on self-esteem, but generally high on aggressiveness. These bullies reacted with aggressiveness in a wide range of situations and towards a wide range of perceived provocations. That bullies are low in emotional stability, as in the study of Coyne et al. (2003), adds to this picture. This study suggests that self- and peer-reported bullies tended to have difficulties coping with personal criticism, to be easily upset, and to view the world as threatening. Apart from cases where bullies admit "light" forms of bullying (e.g., Krum, 1995, who reported that the bullies admitted they had spread rumours and had refused to talk to a certain colleague), bullies will normally not be prepared to report their behaviour, probably because of social desirability and even for legal reasons. In addition, many bullies will view their own behaviour not as bullying but, rather, as a reasonable reaction to a difficult and tense situation. In the study by Jenkins et al. (2010) of a sample of managers who had been accused of bullying, the managers claimed to have a reasonable excuse for their behaviour. Hence, the results presented earlier must be interpreted with caution and must probably not be generalised to all alleged bullies. However, these characteristics may at least apply to a subgroup of bullies. Since bullying may come in many forms and shapes and evolve in a range of different situations (see Einarsen, 1999), that a single personality profile would be common to all bullies is highly unlikely.

Bullying as Micropolitical Behaviour

It has been suggested that some cases of bullying follow the logic of micropolitical behaviour in organisations (Neuberger, 1999, 2006; Salin, 2003). The concept of micropolitics is based on the premise that organisations do not consist of fully determined structures and processes. Rather, organisations require members of the organisation to assist and fill the gaps in the formal structure. That is, organisations both require individuals to take part in decision making and offer them possibilities for influence and decision making. A second premise is that members in organisations try to protect and improve their status in the organisation, which may correspond to the self-esteem protection discussed earlier. Thus, they normally use their possibilities for decision making not only in the interests of the organisation (i.e., to reach organisational goals) but also in their own interest (i.e., to reach personal goals in the sense of status protection and improvement). These decision processes are assumed to be political in nature. In order to be able to have influence on decisions, individuals may deem it necessary to build coalitions and sometimes to plot against competitors.

Such micropolitical behaviour cannot be equated with bullying, however, for micropolitical behaviour aims at protecting one's own interests and improving one's own position (Neuberger, 2006). This behaviour may, of course, include actions that negatively affect other persons and correspond to the negative view of organisational politics that dominated the literature for a long time (Neuberger, 2006). But overall the focus is on one's own interests and not on the destruction of others. This is particularly so because coalitions vary and current competitors may become allies later. Nevertheless, such micropolitical behaviour may occasionally take the form of bullying, and there may be a thin line between the acceptable use of power and bullying. Bullying as micropolitical behaviour indicates harassment of another person in order to protect or improve one's own position in the organisation, or it may be done in order to weaken the position and reputation of a potential rival. From an outside perspective, micropolitical bullying may appear as the most "rational" of all forms of bullying because "reasonable" motives (striving to get a position, influence, resources, etc.) can often be identified (cf. Salin, 2003). Moreover, such bullying behaviour may sometimes only slightly transgress such organisational norms and values as being dominant, competitive, or high-achieving. Social dominance was found to be related to active bullying in the study of Parkins et al. (2006). Individuals high in social dominance might more easily justify that their advantages resulting from micropolitical behaviour are based on disadvantages for others. Some data on victim perceptions may support the concept of micropolitical bullying. Using the Perceptions of Organizational Politics Scale (POPS) of Kacmar and Ferris (1991), and a modified version of the Negative Acts Questionnaire (NAQ) as a measure of bullying (see Einarsen et al., 2009), Salin (2003) found a correlation of .30 between these variables. Similar results were found by Jenderek et al. (2008). Micropolitical behaviour has been described as a phenomenon occurring mainly at the middle and higher hierarchical levels of an organisation. One may conclude that managers do profit more by using bullying as a form of micropolitical behaviour, which may be one of the explanations as to why supervisors and managers are so often among the bullies (Hoel et al., 2001; Zapf et al., this volume).

So far, we have presented various explanations for how individual antecedents of bullying could lie within the perpetrator. However, looking at data that support the view that certain perpetrators are responsible for bullying, we have to take into consideration that potential errors of attribution may play a part in this evidence: It is easier to attribute unpleasant feelings to a person than to invisible circumstances (cf. Neuberger, 1999). It is not unlikely that where the real causes are to be found in organisational circumstances or in the social system, they are instead attributed to certain persons, especially if the source of information is the victim.

Finally, Brodsky (1976) concludes that even if perpetrators indeed have some common personality characteristics making them prone to bullying, they will not exhibit such behaviour unless they are in an organisational culture that rewards, or at least is permissive of, such behaviours. Also, research in the field of sexual harassment has shown that even though some men have stronger proclivities for sexually harassing behaviours than do others, they will portray such behaviours only if they perceive the social climate to encourage them (Schneider et al., this volume).

Individual Antecedents of Bullying: The Victim

Among the reasons for bullying offered by Zapf (1999b) were several items that suggest the reasons may lie, at least partly, within the victim him- or herself. It is interesting to note that these items were endorsed by few victims. Altogether, 37% of the victims agreed with one of these items, whereas 63% did not see any personal involvement in the emergence of the bullying episode. Notably, only 2% of the victims admitted that their performance was below average. Other examples of personal reasons were deficits in social skills or "being difficult," for example, being overly accurate, being aggressive, or moaning.

Several researchers have analysed the relation between personality and bullying, most of them using the big-five personality framework (McCrae and Costa, 1990) and analysing the relation among low emotional stability (neuroticism), extraversion, agreeableness, openness for experience, and conscientiousness. Almost all studies found a relationship between bullying and neuroticism or low emotional stability (Coyne et al., 2000, 2003; Glaso et al., 2007; Matthiesen and Einarsen, 2001; O'Moore et al., 1998; Rammsayer and Schmiga 2003; Schwickerath, 2009; Vartia, 1996; von Holzen-Beusch, 1999), an exception being one study by Djurkovic et al. (2006), whose study was based on a sample of students, and a study by Lind and colleagues (Lind et al., 2009), who studied female targets working in nursing homes. Effects for other variables such as extraversion were sometimes found (victims being less extraverted: Coyne et al., 2000; Glaso et al., 2007; Schwickerath, 2009), but sometimes not (Coyne et al., 2003; Rammsayer and Schmiga, 2003; von Holzen-Beusch, 1999). Less agreeableness was found in the study of Glaso et al. (2007); however, Coyne et al. (2003), Rammsayer and Schmiga (2003), and von Holzen-Beusch (1999) found no effects. Contradictory results occurred for conscientiousness. Glaso et al. (2007) found victims to be less conscientious; Coyne et al. (2000) and von Holzen-Beusch (1999) found victims to be more conscientious; and Coyne et al. (2003), Rammsayer and Schmiga (2003), and

Schwickerath (2009) found no effects. Finally, more openness for experience was found by Rammsayer and Schmiga (2003); however, no effect was found in the studies of Glaso et al. (2007), Schwickerath (2009), and von Holzen-Beusch (1999). The studies by Glaso et al. (2007, 2009) indicate that these contradictory results may be explained by the fact that there are subgroups of victims with quite different types of personality. In one study, they found that although a group consisting of some 30% of the sample were characterised by low openness to experience, the majority of the victims were actually higher on openness to experience compared to a control group (Glaso et al., 2007). Furthermore, Glaso et al. (2009) showed that although some 50% of all victims portrayed themselves as persons with substantial problems in interpersonal relations, so did some 40% of the general Norwegian working population. Hence, those working with victims (e.g., counsellors, human resources personnel, or safety representatives) will undoubtedly meet many victims who are indeed difficult to handle. However, so are actually many nonvictims.

Yet, in the study by Glaso et al. (2009), being exploitable and vulnerable in interpersonal relations did seem to be typical for many targets. Niedl (1995) claims that targets of bullying will be victimised only if they are unable to defend themselves for any reason or if they are unable to escape the situation as a result of any dependency on their part. This dependency may either be of a social nature (hierarchical position, power relations, group membership), a physical nature (physical strength), an economic nature (labour market, private economy), or a psychological nature (self-esteem, personality, cognitive capacity). Hence, the issue of individual antecedents located within the victim relates to many issues. Primarily, some individuals may generally or in specific situations be at risk because of social, demographic, or personal factors, which increase their chances of experiencing bullying. Second, a victim's personality and behaviours may be possible factors in eliciting aggressive behaviours in others. Third, psychological factors may be involved in the ability to defend oneself in highly escalated conflicts with peers and superiors. In a Norwegian survey (Einarsen et al., 1994b), many victims felt that their lack of coping resources and self-efficacy—such as low self-esteem, shyness, and lack of conflict management skills—contributed to the problem. Fourth, individual factors may also be involved as potential moderating factors explaining why some more than others develop stress reactions and health problems after exposure to bullying (see also Einarsen, 2000). As in the case of the perpetrators, there seem to be various groups and mechanisms in which victim characteristics dominate in the development of bullying: (1) the exposed position of the victim, (2) social incompetence and self-esteem deficits, and (3) overachievement and conflict with group norms.

The Salience and Outsider Position of the Victim

Research on groups suggests that people who are outsiders in some respect and who differ from the rest of the group carry a risk of getting in trouble with others and may even be forced into the role of a scapegoat (Thylefors, 1987). Social psychologists have repeatedly demonstrated that individuals who do not belong to the group are devaluated, whereas group members are much more positively evaluated (e.g., Brown, 1997). According to social identity theory (Tajfel and Turner, 1986), being different may cause others to see the persons as "one of them" and not "one of us," which again may, in certain circumstances, lead to displaced aggression towards the person seen as the outsider.

In addition to the negative consequences of being categorised as an out-group member, processes operating within groups may be related to mobbing as well. In general, groups have more or less well-defined norms or proto-types that define who they are in comparison with other groups. Ingroup members who deviate from ingroup norms in an antinormative direction tend to be derogated more strongly than outgroup members who perform the same behaviour (Marques et al., 1988). A large body of research has fol-lowed up and replicated this so-called black-sheep effect. There is some evidence that derogation of deviants is enhanced when the status (i.e., the superiority) of the ingroup is uncertain (Marques et al., 2001). In contrast, the benefits accruing to pro-norm deviants or highly prototypical group mem-bers appear to generalise very well to organisational contexts. For instance, organisational members in leadership roles are evaluated more positively and are more effective as leaders the higher their perceived prototypicality (e.g., Ullrich et al., 2009).

Outsiders have a weaker social network and receive less social support (Cohen and Wills, 1985). According to labelling theory (Neuberger, 1999), deviant behaviour may be escalated when small peculiarities that are in themselves unimportant are used to label (as a mischief-maker, moaner, fail-ure, etc.) and socially exclude someone. Leymann (1993) also saw a socially exposed position of the victim as a risk factor for becoming a victim. In one of his studies, male kindergarten teachers who were a minority were more often among the victims than females. Identical findings came in a study among assistant nurses (Eriksen and Einarsen, 2004). In another study (Lindroth and Leymann, 1993), 21.6% of the handicapped employees in a nonprofit organisation were bullied, but only 4.4% of the nonhandicapped were so treated. In the study by Zapf (1999a), 14% of the victims claimed to be different from the other members of the work group according to such visible characteristics as age, gender, or physical handicap, whereas only 8% of the control group said so.

The Vulnerable Victim: Social Competence and Self-Esteem

Among laypeople, the most common view with regard to victim character-istics is certainly a belief that some people are more vulnerable than others because they are low on self-assertiveness, unable to defend themselves, and unable to manage the inevitable conflicts constructively. Therefore, they are seen as natural victims of bullying. A few studies have tried to find some evidence for this assumption. Together with professionals and bullying victims, Zapf and Bühler (in Zapf, 1999a) developed a list of 45 items according to which victims of bullying may see themselves as being different from the rest of their work group, including such items as a lack of social skills and unassertive behaviour. One item, for example, prompted, "Compared with my colleagues...I do not recognise conflicts as quickly as others." The results showed that victims more often than a control group saw themselves as different from their colleagues. Most interestingly, the results showed that there were obviously very heterogeneous groups of victims. There was one group that did indeed correspond to what most people would expect: individuals low in social competencies, bad as conflict managers, and with unassertive and weak personalities. Averaging the results of two samples reported in Zapf (1999a), one concludes that 33% of the victims saw themselves as more unassertive and as worse conflict managers than their colleagues, compared to only 16% of a control group. In a study of Schwickerath (2009), an inpatient sample of bullying victims in comparison to a control group showed lower self-confidence and lower assertiveness Also, a study among 2,200 Norwegian employees in seven organisational settings showed that victims of bullying were characterised by being low on self-esteem, high on social anxiety, and low on social competence. However, they did not differ from the average in terms of aggressiveness (Matthiesen and Einarsen, 2007). Vartia (1996) found a relationship between bullying and neuroticism and a negative self-image in a sample of municipal employees. Using personality tests based on the five-factor model in a study of 60 Irish victims of bullying, Coyne et al. (2000) found that in comparison with a control sample, victims were more anxious and suspicious and had more problems coping with difficult situations. Moreover, victims were less assertive, competitive, and outspoken than nonvictims. Lindemeier (1996) carried out psychiatric analyses of 87 victims of bullying. Thirty-one percent of the patients reported a general tendency to avoid conflict, and 27% reported low self-esteem problems before bullying began. Moreover, 23% reported that they had always been emotionally labile and had taken everything very seriously. In line with these findings are the results on the relationship between bullying and neuroticism presented in the earlier cited studies. Yet, looking at all the studies, one conclusion may be that vulnerability characteristics are more typical in studies employing victims in clinical samples than in studies employing representative samples where all respondents are still at work (Lind et al., 2009).

These findings on the vulnerable victim also largely overlap with the description of victims of bullying in school (see also Olweus, 1978). In addition, in a study among all nurses and assistant nurses working in three Norwegian nursing homes, victims of bullying were shown to have little sense of humour (Einarsen, 1997); that is, they portrayed a negative attitude towards humour and the use of humour at work. On the basis of interviews with American victims of bullying, Brodsky (1976) claimed that many victims are of a humourless nature. On meeting a notorious teaser, they may feel they are being victimised and bullied by becoming the laughingstock of the department, by the use of practical jokes, or by real or perceived excessive teasing.

Using the Minnesota Multiphasic Personality Inventory–2 (MMPI-2), a personality test developed for clinical purposes, Matthiesen and Einarsen (2001) investigated a group of 85 Norwegian victims of bullying. When reaching a certain threshold, the MMPI-2's scales are assumed to indicate levels of psychological problems and disturbances that require psychological treatment. The authors found elevated levels on various scales for this sample of victims. As a group, the victims were described as being oversensitive, suspicious, depressive, and tending to convert psychological distress into psychosomatic symptoms. The group members were also seen as possibly having problems with understanding more subtle psychological explanations for their own problems. Recently, Matthiesen and Einarsen's findings were replicated in an Italian sample of bullying victims (Girardi et al., 2007). Interestingly enough, these results were identical to those of Gandolfo (1995), who used the same instrument on American victims claiming compensations from employers because of bullying. However, while the victims in Gandolfo's study were mostly younger men, the Norwegian study consisted of mainly older females.

On the application of cluster analysis, a statistical method to group individuals, thereby maximising similarity within the group and maximising differences across groups based on a given set of variables (here, the MMPI-2 scales), three stable clusters could be identified: the "seriously affected," the "normals," and the "disappointed and depressed" (Matthiesen and Einarsen, 2001). While the first group reported a wide range of psychological and emotional problems and symptoms, the second group did not portray any particular psychological symptoms of a neurotic or psychotic nature. However, while the second group reported exposure to a wide range of specific bullying behaviours, the former reported exposure to fewer acts of bullying. The last group, being depressed and somewhat paranoid, consisted of those victims who were bullied at present. Matthiesen and Einarsen (2001) interpret these results to be indicative of a vulnerability factor in a specific group of victims.

Depue and Monroe (1986) and Dohrenwend et al. (1984) suggested that people high in negative affect (NA, including anxious, depressive, and neurotic symptoms) create or enact adverse circumstances through their own behaviour. They may, therefore, create or contribute to the development of problems and conflicts at work. For example, people high in NA might more often be involved in conflicts with others at work, be less able to manage their own workflow, and perform less well in their jobs compared to people with low NA. In addition, there is a large literature showing that other people may respond negatively to depressed individuals (e.g., Sacco et al., 1993). All these behaviours going along with high NA may increase the base rate of conflicts and may thus increase the likelihood of conflict escalation. Applying these findings to bullying, it may be concluded that individuals low in self-esteem, self-assertion, and social competencies but high in anxiety and depression may be bullied not just because they are defenceless and thus easy targets. Rather, because of their own behaviour, the victims may actively produce conflicts that may cause them to become the targets of aggression and harassment.

One may ask whether there is some relationship between bullying at school and workplace bullying, which would also point to individual antecedents of bullying. Smith et al. (2003) report some data based on retrospective reports on school bullying of individuals having taken part in a study on workplace bullying. They found a small correlation: victims of bullying at school were slightly more bullied in the workplace. The effect was somewhat larger for individuals having been both bullies and victims at school. Moreover, those who reported that as victims they "did not really cope" with school bullying had an increased risk of becoming victims of workplace bullying. These results may be interpreted as further support that individuals with a general lack of good conflict management skills have a somewhat higher risk of being bullied at work.

In the study by Zapf (1999b), victims high on unassertiveness/avoidance more often claimed that their performance was below average and that they were bullied because of their nationality or because of a physical disability. Victims high in unassertiveness/avoidance also portrayed the worst conflict management behaviours, being high in conflict avoiding and obliging, low in compromising and integrating. The clearest effect appeared for those avoiding conflict. Moreover, those victims who were low on unassertiveness/avoidance showed results comparable with those of a control group.

Individual antecedents may play a role not only in the onset of a bullying conflict but also in the process of conflict escalation. Knorz and Zapf (1996) and Zapf and Gross (2001) compared the coping behaviour of successful victims (those individuals whose overall situation substantially improved) and unsuccessful victims (those whose overall situation became worse and worse in spite of their coping trials). The successful victims more frequently avoided mistakes or such behaviours as frequent absenteeism, which could be turned against them, and they were obviously better in recognising and avoiding behaviours that escalate rather than de-escalate the conflict.

Schuster (1996) related the literature on school and workplace bullying to the literature on social status and peer rejection among schoolchildren. According to Schuster, research consistently shows that compared with accepted schoolchildren, children rejected by their peers are perceived as less pro-social and more aggressive by both external observers and peers. Rejected children assume more hostile intentions in others and generate fewer and less effective solutions for social problems or conflicts. Moreover, they use less successful strategies to be accepted by a new group. If this line of research can be transferred to workplace bullying, it would underline that being socially incompetent as well as nonassertive may contribute to rejection on behalf of colleagues and superiors and thus explain why some individuals may easily become a target of workplace bullying.

Overachievement and Clash with Group Norms

Various studies seem to include a further behavioural pattern related to overachievement and conscientiousness. In the study by Zapf (1999a), 69% of the victims reported being more conscientious than their colleagues, whereas 40% of the control group reported being so. Moreover, 62% of the victims reported being more achievement oriented than their colleagues, compared to 41% of the control group. These data fit with Brodsky's qualitative observations:

> The harassed victim generally tends to be conscientious, literal-minded, and somewhat unsophisticated. Often, he is an overachiever who tends to have an unrealistic view of himself and of the situation in which he finds himself. He may believe he is an ideal worker and that the job he is going to get will be the ideal job. As a result, he has great difficulty in adjusting not only to the imperfections of the situation but to the imperfections of his own functioning as well. (Brodsky, 1976, p. 89)

Similarly, Coyne et al. (2000) found that victims in comparison with a control group were generally more rule bound, honest, punctual, and accurate. Such persons may be highly annoying to others, which again may contribute to frustration and aggression outlets in their colleagues. Following the earlier line of argument on high but unstable self-esteem as a cause of aggression in perpetrators, feelings of being unjustly treated and belittled may, of course, also easily be generated among this subgroup of victims.

Victims showing these characteristics may also not be very high on empathy. In some cases, they may actually be highly qualified and experienced workers. The problem is that they clash with the norms of the work group to which they belong because they often "know it better," tend to be legalistic, and insist on their own view, as well as have difficulties in taking the perspective of others. By being overcritical, they may be a constant threat to the self-esteem of their colleagues and superiors. Being at odds with group norms may imply that they challenge low performance standards, informal rules, and privileges. In some cases, they may actually be the "good guys" as seen from an employer's point of view. However, in practice, the management is more dependent on the group than on the victim. Therefore, and because information about the conflict situation is likely to be biased in favour of the group, the management tends to take the view of the group rather than that of the victim, thus leaving the victim in a hopeless position (cf. Leymann, 1993).

The Provocative Victim

Although in public the picture of the defenceless victim may prevail, there are victims who are constantly striking back and who may even have started the conflict. When individuals perceive that norms of acceptable behaviours have been violated by aggressive behaviour, they often have the need to retaliate to reestablish justice; this action and reaction, in turn, may spiral into further aggression from both sides (Andersson and Pearson, 1999). The need to retaliate is fuelled by the need to protect oneself. Andersson and Pearson argued that personal identity becomes salient in interpersonal conflicts and that aggressive responses are more likely the more individuals perceive their identity to be threatened or attacked. In the earlier section on bullies, we described aggression as a means of self-esteem protection. There is empirical evidence that active bullying is positively correlated with being bullied (Lee and Brotheridge, 2006). Victims who are active bullies at the same time may be called *provocative victims* (Matthiesen and Einarsen, 2007), a concept borrowed from schoolyard bullying (Olweus, 1978). These individuals are victims who show active bullying behaviour and can be seen as perpetrators and victims at the same time. According to Olweus (1978), these persons are characterised by a combination of both anxious and aggressive reaction patterns. In the study of Matthiesen and Einarsen, 2.1% of the sample defined themselves as both perpetrators and victims. In line with findings on schoolyard bullying, the provocative victims were low in self-esteem and social competence but high in aggression. Interestingly, almost 50% of this group said that they bullied other children at school. Moreover, one-third of them admitted that they were bullied earlier in their career. The provocative victims may possess what Levinson (1978) called an "abrasive personality" (Matthiesen and Einarsen, 2007), which is characterised by insensitive and ruthless behaviours, especially when under social pressure. Revenge and retaliation may be motives for counteraggression (Bies and Tripp, 1996). Yet, counteraggression may sometimes be successful in helping to stop bullying. However, victims may sometimes use aggression against their perpetrators in their fight for justice, but in doing so they may only further escalate the conflict (Zapf and Gross, 2001). There may also be displaced aggression (Marcus-Newhall et al., 2000); that is, aggressive acts may be directed against other persons to ameliorate frustration (Hoel et al., 1999).

Chiming Victim Status

A final question is whether there can be any positive effects involved in declaring oneself a victim. When interpreting the frequencies of bullying in the various studies (cf. Zapf et al., this volume), the question occurs whether bullying will be overreported or underreported (see also Nielsen et al., this volume). There is some evidence that victims of bullying would hesitate to label themselves so, especially if the bullying is subtle, of low intensity, and consisting of indirect forms of aggression (Zapf, 1999a). Given that being a victim implies not being accepted among superiors and colleagues, and even being unable to solve the major problems of one's working life, one might assume that bullying tends to be underreported. In anecdotes, it is repeatedly reported that victims tried to hide their problems as long as they could (e.g., Leymann, 1993).

However, there are also reasons to believe that labelling oneself a victim may have positive implications. In many European countries, the dominant public understanding of bullying is that an innocent victim is harassed by unfair bullies or by an unfair organisation. First of all, labelling oneself as a victim of bullying may then be used to obtain personal goals, for example, to receive an early pension or to win a case of unfair dismissal in court. Moreover, victim status may be used as a justification of oneself in several respects: a victim neither initiates nor escalates a conflict. A victim is fair and innocent, whereas the bullies are unfair and guilty. A victim suffers and hence earns sympathy. A victim is in a powerless position and cannot do anything against the superior strength of the alleged bully. Because the victim is innocent in every respect, he or she is not responsible for solving the problem, which is seen as the responsibility of management (cf. Neuberger, 1999). In this sense, victim status can be used for the protection of self-esteem. This self-concept as a victim is one of the central problems in some of the therapeutic approaches in rehabilitation hospitals caring for victims of bullying (Schwickerath, 2001; Schwickerath and Zapf, this volume). Therapists report that victims are unwilling to become active in therapeutic meetings because they believe that as innocent victims it is not their task to change anything.

In sum, there is the epidemiological problem that some victims of bullying tend to hide their victim status. However, there is also the practical problem that some individuals have started to use the term *bullying* or *mobbing* to achieve personal goals, which means that not every person who says so is really a victim of bullying at work (see also Hoel and Einarsen, this volume).

Conclusion

In this chapter we have discussed the theoretical and empirical evidence of individual antecedents of bullying from the perspective of both the perpetrator and the victim. We have come to the conclusion that on the part of the perpetrator, protection of self-esteem, lack of social competence, and micropolitically motivated behaviour and, on the part of the victim, being in a salient position, being low on social competence and self-assertiveness, and showing high achievement and high conscientiousness are likely individual antecedents that may contribute to the occurrence of workplace bullying.

However, several studies have concluded that personal characteristics are *not* a general explanation of bullying (Coyne et al., 2003; Glaso et al., 2007, 2009; Lind et al, 2009; Matthiesen and Einarsen, 2001; Zapf, 1999b). Rather, these studies identified *subgroups* where such an explanation is likely, while also identifying individuals who had normal competence, personality, and health profiles but who were at the same time targets of serious and frequent bullying at work. Moreover, these studies also identified a group of victims who were even rather conscientious, achievement oriented, creative, resourceful, and open to experience (Glaso et al., 2007; Rammsayer and Schmiga, 2003; Zapf, 1999a). Thus, it is suggested that not all victims of bullying have little social competencies and a neurotic personality profile. Rather, it is likely that there are specific subgroups to whom these characteristics apply, and there are other victims who may not differ from respective control groups and in whose cases the main reason for bullying may lie elsewhere. However, even if personality factors do play a role in the development of workplace bullying, this finding does not undermine or change the responsibility of managers and employers in the prevention and management of such problems.

Professionals dealing with bullying have to keep in mind that they typically meet self-selected groups of victims. Hence, psychiatrists and psychotherapists may overestimate the degree to which the causes of bullying lie within the victim him- or herself. Victims of bullying clearly caused by organisational factors or factors within a certain bully are often extremely happy when the situation is over. They tend to cope with their problem alone, and they may even leave the company or seek a new position elsewhere. They may consult their family doctor, but they will seldom approach a psychiatrist or psychotherapist. On the other hand, the overachievers tend to begin battles for their rights. They may try to get every assistance needed to demand their legal rights, even if their cases continue for months or even years. They may consult trade union representatives and lawyers. Hence, personal experiences with bullying at work should not be overgeneralised.

To summarise: So far there are few hard facts regarding the causes of bullying. However, taking all the existing empirical data together, there is sufficient evidence that there are many possible causes and probably often multiple causes of bullying, be they causes within the organisation, within the perpetrator, within the social system, or within the victim. One-sided and simplistic discussions are in this respect usually misleading. One should consider carefully the circumstances of each bullying case, as in our experience they can be extremely different.

References

Adams, A. (1992) *Bullying at work: How to confront and overcome it.* London: Virago Press.

Andersson, L. M., and Pearson, C. M. (1999) Tit for tat? The spiralling effect of incivility in the workplace. *Academy of Management Review, 24,* 452–471.

Ashforth, B. (1994) Petty tyranny in organizations. *Human Relations, 47,* 755–778.

Baumeister, R. F. (1994) *Losing control: How and why people fail at self-regulation.* San Diego, CA: Academic Press.

Baumeister, R. F., Heatherton, T. F., and Tice, D. M. (1993) When ego threats lead to self-regulation failure: Negative consequences of high self-esteem. *Journal of Personality and Social Psychology, 64,* 141–156.

Baumeister, R. F., Smart, L., and Boden, J. M. (1996) Relation of threatened egotism to violence and aggression: The dark side of high self-esteem. *Psychological Review, 103,* 5–33.

Baumeister, R. F., Stillwell, A., and Wotman, S. R. (1990) Victim and perpetrator accounts of interpersonal conflicts: Autobiographical narratives about anger. *Journal of Personality and Social Psychology, 59,* 994–1005.

Bies, R. J., and Tripp, T. M. (1996) Beyond distrust: "Getting even" and the need for revenge. In R. M. Kramer and T. Tyler (eds.), *Trust in organizations* (pp. 246–260). Thousand Oaks, CA: Sage.

Björkqvist, K., Österman, K., and Hjelt-Bäck, M. (1994) Aggression among university employees. *Aggressive Behavior, 20,* 173–184.

Brodsky, C. M. (1976) *The harassed worker.* Lexington, MA: D. C. Heath.

Brown, R. (1997) Beziehungen zwischen Gruppen [Relationships between groups]. In W. Stroebe, M. Hewstone, and G. M. Stephenson (eds.), *Sozialpsychologie: Eine Einführung* (pp. 545–576). Berlin: Springer.

Cohen, S., and Wills, T. A. (1985) Stress, social support, and the buffering hypothesis. *Psychological Bulletin, 98,* 310–357.

Coyne, I., Seigne, E., and Randall, P. (2000) Predicting workplace victim status from personality. *European Journal of Work and Organizational Psychology, 9,* 335–349.

Coyne, I., Smith-Lee Chong, P., Seigne, E., and Randall, P. (2003) Self and peer nominations of bullying: An analysis of incident rates, individual differences, and perceptions of the working environment. *European Journal of Work and Organizational Psychology, 12,* 209–228.

Dauenheimer, D., Stahlberg, D., Frey, D., and Petersen, L.-E. (2002) Die Theorie des Selbstwertschutzes und der Selbstwerterhöhung [The theory of self-worth protection and self-worth enhancement]. In D. Frey and M. Irle (eds.), *Theorien der Sozialpsychologie: Motivation-, Selbst- und Informationsverarbeitungstheorien,* vol. 3 (pp. 159–190). Bern, Switzerland: Huber.

Depue, R. A., and Monroe, S. M. (1986) Conceptualization and measurement of human disorder in life stress research: The problem of chronic disturbance. *Psychological Bulletin, 99,* 36–51.

Djurkovic, N., McCormack, D., and Casimir, G. (2006) Neuroticism and the psychosomatic model of workplace bullying. *Journal of Managerial Psychology, 21,* 73–88.

Dohrenwend, B. S., Dohrenwend, B. P., Dodson, M., and Shrout, P. E. (1984) Symptoms, hassles, social supports, and life events: Problems of confounded measures. *Journal of Abnormal Psychology, 93,* 222–230.

Einarsen, S. (1997) Bullying among females and their attitudes towards humour. Paper presented at the Eighth European Congress of Work and Organizational Psychology, April 2–5, Verona, Italy.

—— (1999) The nature and causes of bullying. *International Journal of Manpower, 20,* 16–27.

———— (2000) Harassment and bullying at work: A review of the Scandinavian approach. *Aggression and Violent Behavior*, *4*, 371–401.

Einarsen, S., Hoel, H., and Notelaers, G. (2009) Measuring exposure to bullying and harassment at work: Validity, factor structure and psychometric properties of the Negative Acts Questionnaire–Revised. *Work and Stress*, *23*, 24–44.

Einarsen, S., Raknes, B. I., and Matthiesen, S. B. (1994a) Bullying and harassment at work and their relationships to work environment quality: An exploratory study. *European Journal of Work and Organizational Psychology*, *4*, 381–401.

Einarsen, S., Raknes, B. I., Matthiesen, S. B., and Hellesøy, O. H. (1994b) *Mobbing og harde personkonftikter. Helsefarlig samspill pa arbeidsplassen* [Bullying and severe interpersonal conflicts: Unhealthy interaction at work]. Soreidgrend, Norway: Sigma Forlag.

Eriksen, W., and Einarsen, S. (2004) Gender minority as a risk factor for exposure to bullying at work: The case of male assistant nurses. *European Journal of Work and Organizational Psychology*, *13*, 473–492.

Field, T. (1996) *Bullying in sight*. Wantage, UK: Success Unlimited.

Gandolfo, R. (1995) MMPI-2 profiles of worker's compensation claimants who present with claimant of harassment. *Journal of Clinical Psychology*, *51*, 711–715.

Girardi, P., Monaco, E., Prestigiacomo, C., Talamo, A., Ruberto, A., and Tatarelli, R. (2007) Personality and psychopathological profiles in individuals exposed to mobbing. *Violence and Victims*, *22*, 172–188.

Glaso, L., Matthiesen, S. B., Nielsen, M. B., and Einarsen, S. (2007) Do targets of workplace bullying portray a general victim personality profile? *Scandinavian Journal of Psychology*, *48*, 313–319.

Glaso, L, Nielsen, M. B., and Einarsen, S. (2009) Interpersonal problems among perpetrators and targets of workplace bullying. *Journal of Applied Social Psychology*, *39*, 1316–1333.

Hoel, H., and Cooper, C. L. (2001) The origins of bullying. In N. Tehrani (ed.), *Building a culture of respect: Managing bullying at work* (pp. 3–19). London: Taylor & Francis.

Hoel, H., Cooper, C. L., and Faragher, B. (2001) The experience of bullying in Great Britain: The impact of organisational status. *European Journal of Work and Organizational Psychology*, *10*, 443–465.

Hoel, H., Rayner, C., and Cooper, C. L. (1999) Workplace bullying. In C. L. Cooper and I. T. Robertson (eds.), *International review of industrial and organizational psychology*, vol. 14 (pp. 195–230). Chichester, UK: Wiley.

Jenderek, K., Schwickerath, J., and Zapf, D. (2008) Victims of workplace bullying in a psychosomatic hospital. Symposium on Workplace Bullying and Health: Organizational and Personal Interventions. Eighth Conference of the European Academy of Occupational Health Psychology, November 12–14, Valencia, Spain.

Jenkins, M., Winefield, H., Zapf, D., and Sarris, A. (2010) Listening to the bullies: An exploratory study of managers accused of workplace bullying. Paper submitted for publication.

Kacmar, K. M., and Ferris, G. R. (1991) Perceptions of organizational politics scale (POPS): Development and construct validation. *Educational and Psychological Measurement*, *51*, 193–205.

Kernis, M. H., Cornell, D. P., Sun, C. R., Berry, A., and Harlow, T. (1993) There is more to self-esteem than whether it is high or low: The importance of stability of self-esteem. *Journal of Personality and Social Psychology*, *65*, 1190–1204.

Kernis, M. H., Grannemann, B. D., and Barclay, L. C. (1989) Stability and level of self-esteem as predictors of anger arousal and hostility. *Journal of Personality and Social Psychology, 56,* 1013–1022.

Kets de Vries, M. F. R., and Miller, D. (1985) Narcissism and leadership. *Human Relations, 38,* 583–601.

Knorz, C., and Zapf, D. (1996) Mobbing—eine extreme Form sozialer Stressoren am Arbeitsplatz [Mobbing—an extreme form of social stressors at work]. *Zeitschrift für Arbeits- und Organisationspsychologie, 40,* 12–21.

Krum, H. (1995) Mobbing—eine unethische Form der Kommunikation am Arbeitsplatz [Mobbing—an unethical form of communication at work]. Unpublished diploma thesis. Technical University of Darmstadt, Germany.

Lee, R. T., and Brotheridge, C. M. (2006) When prey turns predatory: Workplace bullying as a predictor of counter-aggression/bullying, coping, and well-being. *European Journal of Work and Organizational Psychology, 15,* 352–377.

Levinson, H. (1978) The abrasive personality. *Harvard Business Review, 56,* 86–94.

Leymann, H. (1993) *Mobbing—Psychoterror am Arbeitsplatz und wie man sich dagegen wehren kann* [Mobbing—psychoterror in the workplace and how one can defend oneself]. Reinbeck, Germany: Rowohlt.

——— (1996) The content and development of mobbing at work. *European Journal of Work and Organizational Psychology, 5,* 165–184.

Leymann, H., and Gustafsson, A. (1996) Mobbing and the development of post-traumatic stress disorders. *European Journal of Work and Organizational Psychology, 5,* 251–276.

Lind, K., Glaso, L., Pallesen, S., and Einarsen, S. (2009) Personality profiles among targets and non-targets of bullying. *European Psychologist, 14,* 231–237.

Lindemeier, B. (1996) Mobbing: Krankheitsbild und Intervention des Betriebsarztes [Bullying: Symptoms and intervention of the company physician]. *Die Berufsgenossenschaft,* June, 428–431.

Lindroth S., and Leymann, H. (1993) *Vuxenmobbning mot en minoritetsgrupp av män inom barnomsorgen: Om mäns jämställdhet i ett kvinnodominerat yrke* [Bullying of a male minority group within child-care: On men's equality in a female-dominated occupation]. Stockholm: Arbetarskyddstyrelsen.

Marcus-Newhall, A., Pedersen, W. C., Carlson, M., and Miller, N. (2000) Displaced aggression is alive and well: A meta-analytic review. *Journal of Personality and Social Psychology, 78,* 670–689.

Marques, J. M., Abrams, D., and Serôdio, R. G. (2001) Being better by being right: Subjective group dynamics and derogation of in-group deviants when generic norms are undermined. *Journal of Personality and Social Psychology, 81,* 436–447.

Marques, J. M., Yzerbyt, V. Y., and Leyens, J.-P. (1988) The black sheep effect: Judgmental extremity towards ingroup members as a function of ingroup identification. *European Journal of Social Psychology, 18,* 1–16.

Matthiesen, S. B., and Einarsen, S. (2001) MMPI-2-configurations among victims of bullying at work. *European Journal of Work and Organizational Psychology, 10,* 467–484.

——— (2007) Perpetrators and targets of bullying at work: Role stress and individual differences. *Violence and Victims, 22,* 735–753.

McCrae, R. R., and Costa, P. T. (1990) *Personality in adulthood.* New York: Guilford.

Neuberger, O. (1999) *Mobbing: Übel mitspielen in Organisationen* [Mobbing: Bad games in organisations], 3rd ed. Munich: Rainer Hampp.

———— (2006) Mikropolitik: Stand der Forschung und Reflexion [Micropolitics: State of the art and reflection]. *Zeitschrift für Arbeits- und Orgnisationspsychologie, 50*, 189–202.

Niedl, K. (1995) *Mobbing/bullying am Arbeitsplatz: Eine empirische Analyse zum Phänomen sowie zu personalwirtschaftlich relevanten Effekten von systematischen Feindseligkeiten* [Mobbing/bullying at work: An empirical analysis of the phenomenon and of the effects of systematic harassment on human resources management]. Munich: Hampp.

Nielsen, M. B., Matthiesen, S. B., and Einarsen, S. (2008) Sense of coherence as a protective mechanism among targets of workplace bullying. *Journal of Occupational Health Psychology, 13* (2), 128–136.

Olweus, D. (1978) *Aggression in the schools: Bullies and whipping boys.* Washington, DC: Hemisphere, Wiley.

O'Moore, M., Seigne, E., McGuire, L., and Smith, M. (1998) Victims of bullying at work in Ireland. *Journal of Occupational Health and Safety–Australia and New Zealand, 14*, 568–574.

Parkins, I. S., Fishbein, H. D., and Ritchey, P. N. (2006) The influence of personality on workplace bullying and discrimination. *Journal of Applied Social Psychology, 36*, 2554–2577.

Rammsayer, T., and Schmiga, K. (2003) Mobbing und Persönlichkeit: Unterschiede in grundlegenden Persönlichkeitsdimensionen zwischen Mobbing-Betroffenen und Nicht-Betroffenen [Bullying and personality: Differences with regard to basic personality traits between bullying victims and non-victims]. *Wirtschaftspsychologie, 5* (2), 3–11.

Sacco, W. P., Dumont, C. P., and Dow, M. G. (1993) Attributional, perceptual, and affective responses to depressed and nondepressed marital partners. *Journal of Consulting and Clinical Psychology, 61*, 1076–1082.

Salin, D. (2003) Bullying and organisational politics in competitive and rapidly changing work environments. *International Journal of Management and Decision Making, 4*, 35–46.

Salovey, P. (1991) Social comparison processes in envy and jealousy. In J. Suls and T. A. Wills (eds.), *Social comparison: Contemporary theory and research* (pp. 261–285). Hillsdale, NJ: Erlbaum.

Schuster, B. (1996) Rejection, exclusion, and harassment at work and in schools. *European Psychologist, 1*, 293–317.

Schwickerath, J. (2001) Mobbing am Arbeitsplatz: Aktuelle Konzepte zu Theorie, Diagnostik und Verhaltenstherapie [Bullying in the workplace: Current concepts on theory, diagnostics and behaviour therapy]. *Psychotherapeut, 46*, 199–213.

———— (2009) Mobbing am arbeitsplatz—Stationäre verhaltenstherapie von patienten mit mobbingerfahrungen [Bullying at the workplace: Inpatient treatment of victims of bullying]. Lengerich, Germany: Pabst Science.

Seigne, E. (1998) Bullying at work in Ireland. In C. Rayner, M. Shcehan, and M. Barker (eds.), *Bullying at work. 1998 research update conference proceedings.* Stafford, UK: Staffordshire University.

Smith, P. K., Singer, M., Hoel, H., and Cooper, C. L. (2003) Victimisation in the school and the workplace: Are there any links? *British Journal of Psychology, 94*, 175–188.

Smith, R. H., Parrott, W. G., Ozer, D., and Moniz, A. (1994) Subjective injustice and inferiority as predictors of hostile and depressive feelings in envy. *Personality and Social Psychology Bulletin, 20*, 717–723.

Stucke, T. (2002) Narzismus und Selbstkonzeptklarheit als Persönlichkeitskorrelate bei Mobbingtätern und Mobbingopfern [Narcissism and clearness of the self-concept as personality correlates of perpetrators and victims of mobbing]. *Zeitschrift für Arbeits- und Organisationspsychologie, 46*, 216–221.

Tajfel, H., and Turner, J. (1986) The social identity theory of intergroup behavior. In S. Worchel and W. G. Austin (eds.), *Psychology of intergroup relations* (pp. 7–24). Chicago: Nelson.

Thylefors, I. (1987) *Syndabockar: Om utstötning och mobbning i arbetslivet* [Scapegoats: On expulsion and bullying in working life]. Stockholm: Natur och Kulture.

Ullrich, J., Christ, O., and van Dick, R. (2009) Substitutes for fairness: Prototypical leaders are endorsed whether they are fair or not. *Journal of Applied Psychology, 94*, 235–244.

van de Vliert, E. (1998) Conflict and conflict management. In P. J. D. Drenth, H. Thierry, and C. J. J. Wolff (eds.), *Handbook of work and organizational psychology*, vol. 3: *Personnel Psychology*, 2nd ed. (pp. 351–376). Hove, UK: Psychology Press.

Vartia, M. (1996) The sources of bullying: Psychological work environment and organizational climate. *European Journal of Work and Organizational Psychology, 5*, 203–214.

von Holzen-Beusch, E. (1999) Warum Mobbingbetroffene ihren Arbeitsplatz nicht wechseln. Eine explorative Studie [Why mobbing victims don't change their jobs. An exploratory study]. Unpublished licentiate thesis, Department of Psychology, Zurich University, Switzerland.

Watzlawick, P., Beavin, J. H., and Jackson, D. D. (1967) *Pragmatics of human communication: A study of interactional patterns, pathologies, and paradoxes*. New York: Norton.

Zapf, D. (1999a) Mobbing in Organisationen: Ein Überblick zum Stand der Forschung [Mobbing in organisations: A state-of-the-art review]. *Zeitschrift für Arbeits- und Organisationspsychologie, 43*, 1–25.

——— (1999b) Organizational, work group related and personal causes of mobbing/bullying at work. *International Journal of Manpower, 20*, 70–85.

Zapf, D., Dormann, C., and Frese, M. (1996) Longitudinal studies in organizational stress research: A review of the literature with reference to methodological issues. *Journal of Occupational Health Psychology, 1*, 145–169.

Zapf, D., and Einarsen, S. (2005) Mobbing at work: Escalated conflicts in organizations. In S. Fox and P. E. Spector (eds.), *Counterproductive work behavior: Investigations of actors and targets* (pp. 237–270). Washington, DC: American Psychological Association.

Zapf, D., and Gross, C. (2001) Conflict escalation and coping with workplace bullying: A replication and extension. *European Journal of Work and Organizational Psychology, 10*, 497–522.

Zapf, D., Knorz, C., and Kulla, M. (1996) On the relationship between mobbing factors and job content, the social work environment and health outcomes. *European Journal of Work and Organizational Psychology, 5*, 215–237.

8

Social Antecedents of Bullying:
A Social Interactionist Perspective

Joel H. Neuman and Robert A. Baron

CONTENTS

As described in previous chapters, workplace bullying involves persistent patterns of behavior in which one or more individuals engage in actions intended to harm others (e.g., Hoel et al., 1999). It is our contention that bullying, although not always identified as such, involves acts of interpersonal aggression—any form of behavior directed toward the goal of harming or injuring another living being who is motivated to avoid such treatment (Baron and Richardson, 1994). While both phenomena involve actions that are intentional in nature, the persistence of aggression over time, evidenced in episodes of bullying, serves as a distinguishing characteristic. That is, while a single act of intentional harm-doing constitutes an act of aggression, it would not, by definition, constitute bullying. In short, and of central importance to the present chapter, we believe that workplace bullying involves repeated acts of interpersonal aggression directed against specific targets in work settings, or what we would refer to as *workplace aggression*— efforts by individuals to harm others with whom they work (Neuman and Baron, 1997a). Furthermore, we propose that anything that serves as an antecedent to aggression may contribute to—and increase the likelihood of—workplace bullying.

Having said all this, we do recognize that bullying in workplaces, like bullying in other contexts, represents a special or unique form of aggression in certain respects. For instance, the persons involved in bullying episodes are generally participants in ongoing, long-term relationships: they may work together for months, years, or even decades. Second, "bullying and harassment imply a difference in actual or perceived power and 'strength' between the persecutor and the victim" (Einarsen and Skogstad, 1996, p. 187). Third, since bullying often occurs openly, in front of many observers, it is clear that norms concerning such behavior differ from the societal norms that regulate aggression generally and—in most instances—condemn aggression, especially repeated aggression against weak or helpless victims, as inappropriate. Thus, a key question to be addressed is this: why do societal norms against aggression fail to apply, or apply only weakly, where workplace bullying is concerned? Related to this, we also ask, in what additional ways is workplace bullying different from aggression in many other contexts, and why is this so?

Building on these basic assertions, we draw on a substantial literature devoted to interpersonal aggression and examine what we believe to be important social antecedents to bullying. These *social factors* are distinct from *individual* causes of aggression, which center on the characteristics (e.g., personality traits) of the persons who engage in workplace bullying. Social factors, in contrast, involve the words and/or deeds of individuals—that is, actions that elicit or condone aggression and the context in which these actions occur. In addition, we will also focus on the aforementioned social norms, for these, too, often exert strong effects on the nature, form, and frequency of overt acts of aggression. Specifically, we will consider important social norms that serve to shape and reinforce aggression as well as the process by which norm violations (injustice perceptions) elicit retaliation or predispose individuals toward aggression and bullying. Then, employing modern theories of aggression, we discuss the mediating variables through which norm violations and injustice perceptions may lead to aggression; specifically, we consider situations that produce frustration and stress, generate assaults on individual dignity and self-worth, and elicit negative affect, physiological arousal, and hostile thoughts in the persons involved. Finally, we conclude by identifying several contemporary business practices that produce such mediating factors and create social conditions that are ripe for bullying—business practices, we might add, that are ubiquitous in today's work settings.

A Social Interactionist Perspective

The following elements are central to the social interactionist approach and the arguments we put forth in this chapter. First, this perspective treats interpersonal and situational factors as critical in instigating aggression. Second, it recognizes that aggressors "often view their own behaviour as legitimate and even moralistic. Thus, beliefs about justice and equity, the assignment of blame, and the accounts that people give to excuse or justify their behaviour are central" (Felson and Tedeschi, 1993, p. 2). Third, it focuses on the interactions of actors, targets, and third parties and views aggression as both instrumental (proactive) and hostile (reactive) in nature (Neuman and Baron, 2003, 2005; Tedeschi and Felson, 1994). With respect to instrumental aggression, we refer to an act in which aggression is employed to obtain some valued outcome. In such instances, the harm inflicted on the victim is merely a means to an end. Reactive (hostile) aggression, on the other hand, is generally viewed as being impulsive, thoughtless, driven by anger, having the ultimate objective of harming the target, and often coming in reaction to some perceived provocation (Anderson and Bushman, 2002).

In their seminal work on the social interactionist approach, Felson and Tedeschi interpret aggression as instrumental behavior used to influence (coerce) others, to establish and protect valued social identities (toughness, competence), or to achieve justice or retribution. In their view, much aggression can be motivated by a variety of goals, and so many instances of hostile aggression may be viewed as instrumental in nature (Felson and Tedeschi, 1993; Tedeschi and Felson, 1994). Recently, scholars have proposed that the instrumental–reactive dichotomy has outlived its usefulness (e.g., Anderson and Bushman, 2002; Bushman and Anderson, 2001). In truth, human behavior (including bullying) is frequently instigated by mixed motives, and it is difficult to distinguish between purely instrumental or reactive motivations. As an example, a bully might want something possessed by the target while, at the same time, feeling hate or contempt for that individual. Consequently, some acts may involve both hostile *and* instrumental motives and encompass both short- and long-term objectives. Having said this, we will employ the proactive–reactive dichotomy for ease of presentation and organization as well as to emphasize the interactive and temporal nature of bullying episodes (i.e., actions and reactions that play out over time).

The Social Antecedents of Aggression

When asked to describe situations that made them angry, most individuals refer to something another person said or did, something that caused them to become upset and view aggression against this person as justified (Harris, 1993). In short, the things that make people most angry are the words and deeds of other people—the social causes of aggression. As a point of departure for this chapter, we examine an important norm that serves to help shape social interaction.

The Norm of Reciprocity

Contrary to advice proffered in the "golden rule," people tend to do unto others as others have actually done unto them. This behavioral norm has been found to "exert a powerful influence upon various social behaviors ranging from altruism and assistance on the one hand through aggression and violence on the other" (Baron et al., 1974, p. 374). Recognition of the importance of reciprocity in interpersonal relationships has a long history. "There is no duty more indispensable than that of returning a kindness," says Cicero, adding, "all men distrust one forgetful of a benefit" (as cited in Gouldner, 1960, p. 161). Cicero's admonition speaks to the obligation one owes a benefactor and the distrust that accrues to a person who fails to repay such an obligation. In a similar but opposite vein, when people feel attacked, they often respond with an attack of comparable severity (Geen, 1968). Gouldner (1960) recognized these negative forms of reciprocity as "sentiments of retaliation where the emphasis is placed not on the return of benefits but on the return of injuries" (p. 172). The act of "exacting justice," "evening the score," "righting the wrong," and "balancing the scales" is also prominently featured in the biblical injunction to exact "an eye for an eye and a tooth for a tooth." Several years ago, Stuckless and Goranson (1992) noted that "the terms revenge and vengeance [were] not even designated as keywords in the American Psychological Abstracts but [were] subsumed under the classification of reciprocity" (p. 26). To summarize the importance of reciprocity in Simmel's words (1950), "All contacts among men rest on the schema of giving and returning the equivalence" (p. 387). This dynamic is so common to social experience that it led Becker (1956) to view the human species as "Homo Reciprocus" (p. 1).

As beliefs are antecedent to the formation of attitudes, intentions, and, ultimately, the enactment of behavior (Ajzen and Fishbein, 1980), recent work on reciprocity has focused on reciprocity beliefs. In contentious work climates, individuals are probably justified in developing "a generalised expectancy or belief regarding the lack of trustworthiness of particular individuals, groups, or institutions that is predicated upon a specific history of interaction with them" (Kramer, 1994, p. 200). The *rational distrust* that develops in such situations may lead to a negative reciprocity orientation—a belief in people's general malevolence and cruelty that encourages negative reciprocity as a strategy to prevent exploitation (Eisenberger et al., 2004). Recently, Mitchell and Ambrose (2007) found that such beliefs increased the likelihood of supervisor-directed aggression in response to abusive supervision. From a social interactionist perspective, this may create a self-fulfilling prophecy in which supervisors mistreat subordinates, who in turn retaliate (overtly or covertly), thus providing justification for the bullying. A recent meta-analysis by Bowling and Beehr (2006) found support for a reciprocity-attribution model in which targets retaliate against the perpetrators of aggression as well as the organizations for allowing the situation to exist. Similarly, Glomb and Liao (2003) found that aggressive behavior exhibited by members of an individual's work group is a significant predictor of that individual's tendency towards aggressive behavior. In sum, beliefs suggesting that others do not or will not follow the reciprocity principle can sometime serve as an important elicitor of aggression, in a wide range of contexts.

According to Gouldner (1960), reciprocity is accounted for as a result of the development of a beneficial cycle of mutual reinforcement between parties in social exchange. But what constitutes *benefit* as opposed to *detriment*? Clearly, individuals must recognize that they are being advantaged or disadvantaged in any given situation, and this recognition would presuppose a calculus by which justice perceptions are made. In the following sections, we examine these justice perceptions in some detail and discuss the manner in which justice judgments are made as well as the process by which justice is restored.

Injustice Perceptions

The relationship between perceptions of unfair treatment and the restoration of equity is the motivation underlying a significant portion of the research on social justice. As suggested by Homans (1974), "We should be much less interested in injustice if it did not lead so often to anger and aggression" (p. 257). When we consider that fairness perceptions are manifest in almost every aspect of the employment relationship (e.g., personnel selection, performance evaluations, promotions, raises, merit pay, or allocation of work assignments and office space), the importance of incorporating social justice theories in work settings seems obvious. Accordingly, there has been a burgeoning interest in the area of organizational justice—research related to people's perceptions of fairness on the job (Greenberg and Colquitt, 2005).

In reviewing the literatures related to social justice and human aggression, we find that four classes of variables emerge as central to both research streams (Neuman and Baron, 1998). These factors include "unjust" situations that (1) violate norms, (2) produce frustration and stress, (3) induce negative affect (emotion), and (4) assault individual dignity and self-worth. In the sections that follow, we examine each of these factors in turn.

Norm Violations

Justice refers to an appropriate correspondence between a person's fate and that to which he or she is entitled—that is, what is deserved. Rule and

Ferguson (1984) discuss norm violations as an instance of "is–ought discrepancy" and suggest that anger, blaming, and retaliation might ensue when people see their partly idiosyncratic norms of proper conduct (the "oughts") violated. In a similar vein, referent cognitions theory (Folger, 1986) suggests that with respect to outcome allocation, "resentment is maximized when people believe they *would* have obtained better outcomes if the decision maker had used other procedures that *should* have been implemented" (Cropanzano and Folger, 1989, pp. 293–294). A more recent formulation of this reasoning (now referred to as *fairness theory*) suggests that in order for an event to be perceived as unfair, it must violate ethical standards to which people are expected to adhere, and the target must assign blame and accountability for the violation (Folger and Cropanzano, 1998, 2001). Finally, Tedeschi and Felson (1994) have suggested that any factor that increases the likelihood that norm violations will be committed should lead to grievances and coercive interactions. But under what circumstances are norm violations likely to result in the assignment of blame and accountability, the formulation of grievances, the elicitation of anger, and coercive interactions and retaliation? In attempting to answer this question, researchers have focused on issues related to distributive, procedural, and interactional forms of (in)justice.

Distributive Justice

The earliest research on social justice focused almost exclusively on the issue of distributive justice—that is, the perceived fairness of the outcomes or allocations that an individual receives. In short, this research was concerned with how individuals make fairness judgments about their outcomes and how they react when they perceive inequity in the distribution of those outcomes. With respect to inequity, individuals may perceive that they have unfairly benefited or that they have been unfairly disadvantaged in a particular situation. In the language of equity theory (Adams, 1965), this would represent instances of overpayment and underpayment, respectively. With respect to the present chapter, our focus will be on underpayment inequity (i.e., perceptions of relative deprivation).

In the seminal research on distributive justice and relative deprivation, the connection between perceived injustice and aggression is clearly evident. As noted by Homans (1974), if a state of injustice exists and it is to a person's disadvantage—that is, the person experiences deprivation—he or she will display anger. Adams (1965) echoed this position in his original formulation of equity theory, when he observed that "men do not simply become dissatisfied with conditions they perceive to be unjust. They usually do something about them" (p. 276). But what do they do? Fortunately, reactions to injustice do not always involve aggression. For example, in response to perceptions of underpayment inequity, individuals may simply put forth a greater effort in the hope of increasing their outcomes or rewards. Unfortunately, the organizational justice literature is replete with examples of less acceptable responses, and many of these may be aggressive in nature.

From the organization's perspective, theft would certainly constitute an unacceptable response to employee perceptions of distributive injustice. In a study designed to explore this issue, Greenberg (1990) compared theft rates within three manufacturing plants belonging to the same organization during a period in which pay cuts were administered. In response to an ongoing financial crisis, the company imposed a pay cut of 15% over a 10-week period in two plants. In a third and demographically similar plant, no pay cuts were administered (this served as the control condition). Although theft rates, as measured by shrinkage, were traditionally low in all three plants, Greenberg found significantly greater theft rates in the two plants where the pay cuts were administered, in comparison to the control condition.

Other researchers have found similar examples of "getting even" for perceived inequity in outcome distribution in actual work settings. In a study by Altheide et al. (1978), involving bread delivery drivers, stealing was generally viewed by the drivers as a way of compensating for a pay system that unfairly penalized them for making mistakes. Similarly, Zeitlin (1971) studied employees in a retail clothing store and found that theft was generally accepted as an appropriate means by which employees could "get something additional for [their] work since [they weren't] getting paid enough" (p. 26). Beyond theft of merchandise, employees may compensate for relative deprivation by stealing time, or engaging in non-work-related activities during working hours (Atkinson, 2006).

The purpose of the preceding discussion is not to focus attention on theft, which is clearly not the central issue of this chapter or text; rather, it is meant to provide empirical support for the connection between distributive injustice and retaliation. Having said this, it is important to note that the motive for theft is not always greed, or the desire to profit financially. It is often the case that the target of an "injustice" merely wants to inflict harm on the source of that injustice or, as we shall demonstrate, simply lash out against any convenient target. This response is most clearly evident with respect to instances of sabotage, in which company or individual property is damaged as an act of revenge for unfair treatment, and there is ample evidence of this in work settings (cf. Analoui, 1995).

Before concluding this section, it is important to note that outcomes are not synonymous with wages, salaries, or employee benefits. Other important outcomes relate to quality of work life, social support, and opportunities for growth and development, to name just a few. In a study by Neuman and Baron (1997b), employees indicated the extent to which they were satisfied with their opportunities for growth and development and the social conditions in which they worked. Dissatisfaction with these outcomes was significantly correlated with workplace aggression. Specifically, respondents who were dissatisfied with these outcomes were significantly more likely to engage in workplace aggression against the perceived source of the injustice. More recently, distributive injustice was found to be significantly related to hostility and indirect, obstructional forms of aggression (Chory and Hubbell, 2008).

As noted at the beginning of this section, reactions to perceived inequity may involve behaviors that are prosocial or antisocial in nature. Unfortunately, the research related to distributive justice is of little help in predicting the likelihood of one type of response over another. In part, the reason for this dilemma centers on the fact that distributive justice research has tended to focus on the outcomes but not the underlying causes (or perceived causes) for those outcomes. As it turns out, just as individuals are concerned with the fairness of the outcomes that they receive, they also are sensitive to the process used to determine those outcomes (Thibaut and Walker, 1975) and the nature of the interactions that characterize those transactions (Bies, 1987). Folger and Greenberg (1985) were the first researchers to consider procedural justice issues in work settings, and since their initial efforts a considerable amount of research has been conducted. To date, this line of inquiry strongly suggests that (1) fair process can mitigate against an aggressive response to unfavorable outcomes and (2) unfair process may be more strongly linked to aggression than unfair outcomes. Now we turn to some evidence supporting our assertion that unfair treatment is an important antecedent to aggression and bullying.

The Link between Unfair Treatment and Workplace Violence and Aggression

There is a substantial and growing literature suggesting that perceptions of unfair (insensitive) treatment on the part of management and/or cowork-ers often serve as antecedents to, and mediators of, workplace aggression and violence (Barling et al., 2009; Dupré and Barling, 2006). For example, the perpetrators of workplace homicide often point to what they believe was unfair treatment at the hands of a supervisor or coworker as a justification for their actions. In Hoad's summary of the most common causes of work-place violence in the United Kingdom (1993), feeling aggrieved—that is, hav-ing a sense of being treated unfairly, whether real or imagined— was ranked as the most common cause of aggression. Similarly, in a study reported by Weide and Abbott (1994), over 80% of the cases of workplace homicide they examined involved employees who "wanted to get even for what they per-ceived as [their] organizations' unfair or unjust treatment of them" (p. 139). All of these findings are consistent with empirical research demonstrating a link between perceptions of unfair treatment and interpersonal conflict (Cropanzano and Baron, 1991).

In another related study (Neuman and Baron, 1997b), individuals who reported that they had been treated unfairly by their supervisors were significantly more likely than those who were satisfied with their treatment to indicate that they engaged in some form of workplace aggression. Additionally, this study provided evidence that this aggression was directed against the source of that perceived injustice. For example, when individuals expressed dissatisfaction with organizationally controlled outcomes (e.g., job security, pay and fringe benefits, or social conditions), they were more likely to aggress against the entire organization. However, when dissatisfaction was associated with a particular manager (e.g., respect, fair treatment, or support and guidance received from their boss), they were more likely to indicate that they aggressed against that particular individual. Data from this study also suggest that aggression is directed against targets other than the source of the perceived injustice; that is, there is evidence of displaced aggression.

In a six-year longitudinal study on workplace sabotage, Analoui (1995) found that 65% of all acts of sabotage stemmed from discontent with management and its behavior toward workers. Similarly, Crino and Leap (1989) suggest that in response to violations of the psychological employment contract, employer–employee loyalty can be severely damaged. Consequently, "once that loyalty has been destroyed, an employee is more likely to commit an act of sabotage" (p. 32).

Finally, as part of a large-scale project within the U.S. Department of Veterans' Affairs (VA), in two questionnaire administrations (occurring in 2000 and 2002), being treated in a rude or disrespectful manner was significantly correlated with workplace aggression, $r(3844) = .53$ and $r(3087) = .55$, respectively. The greater the degree of interactional or interpersonal injustice, the greater the degree to which respondents reported experiencing and observing workplace aggression (Neuman, 2004). Also, related to this study, aggregated data from 26 VA facilities revealed that injustice perceptions were significantly related to independent measures of discrimination, stress, and violence.

To summarize, recent theorizing on organizational justice suggests that aggression is more likely to occur when unfavorable outcomes result from an "unfair" process, thus highlighting the importance of interpersonal sensitivity in mitigating against aggression. But what is it about norm violations, whether distributive, procedural, or interpersonal, that may elicit an aggressive response? To answer this question, we briefly turn our attention to modern theories of aggression and the role of some important mediating variables.

The General Affective Aggression Model: A Modern Perspective on Human Aggression

Early views of human aggression tended to emphasize the influence of one, or at most, a few variables. The most famous of these, perhaps, was the *frustration–aggression hypothesis,* a theoretical perspective that attached central importance to the role of frustration, or the thwarting of ongoing, goal-directed behavior (Dollard et al., 1939). As we will note later, although modern views of aggression include frustration as one potential cause of such behavior, they assign far less importance to this variable than do the sweeping suggestions that were part of the original frustration–aggression hypothesis. The original formulation proposed that (1) *all* aggression stems from frustration and (2) frustration *always* produces aggression. Subsequent research demonstrated clearly that these assertions were false in that aggression stems from many sources other than frustration and frustration does not always lead to increased aggression. In fact, aggression often generates feelings of resignation or despair rather than overt assaults against the perceived causes of such thwarting—reactions that often characterize a target's response to bullying.

Modern perspectives on aggression, in contrast, recognize that such behavior stems from a wide range of social, situational, and personal factors (e.g., Baron and Richardson, 1994). One such model that has attained widespread acceptance is the *General Affective Aggression Model* (GAAM) proposed by Anderson and his colleagues (Anderson et al., 1996). According to this theory, aggression is triggered or elicited by a wide range of *input variables*—aspects of the current situation (the focus of this chapter) and/or tendencies or predispositions that individuals bring with them to a given context. Among variables included in the first category are frustration, provocation or attack from another person, exposure to aggressive models, the presence of cues or stimuli associated with aggression (e.g., guns or other weapons), and virtually anything that causes individuals to experience negative affect (e.g., harsh and unfair criticism, unpleasant environmental conditions, or pain or discomfort produced by physical injuries). Variables in the second category (individual difference factors) include traits that predispose individuals toward aggression (e.g., high irritability, the type A behavior pattern, or negative affectivity), certain attitudes and beliefs about violence (e.g., the view that violence is acceptable or a demonstration of one's masculinity), and specific skills related to aggression (e.g., knowledge of how to use various weapons or having skills useful in attacking others physically or verbally).

Central to the present chapter, the GAAM further suggests that these situational and individual difference variables lead to aggression through their impact on three basic intervening processes, which we refer to as *critical internal states.* First, they may increase physiological *arousal* or excitement. Second, they elicit *negative affect*—that is, feelings of anger and other hostile emotions along with outward signs of these emotions (e.g., angry facial expressions). Finally, they elicit hostile *cognitions*; in other words, they induce individuals to bring hostile thoughts to mind, to remember aggression-related experiences, and so on. Depending on an individual's appraisals (interpretations) of the current situation and possible restraining factors (e.g., threat of retaliation or strict disciplinary and enforcement policies), aggression then occurs or does not occur.

The foregoing description is necessarily brief but captures, we believe, the sophisticated nature of this and related views of human aggression. Such views are, indeed, much more complex than the suggestion that aggression stems from frustration or any other single factor. These contemporary views are also more accurate in reflecting the multifaceted nature of human aggression and the wide array of variables and processes that influence its occurrence, form, and targets. An overview of the General Affective Aggression Model is presented in Figure 8.1. As shown in this figure, the model also suggests the occurrence of interactions between arousal, affective states, and aggression-related cognitions. This interaction is extremely important in that the elicitation of any single critical internal state (physiological, affective, or cognitive) tends to evoke the others, thereby initiating a potential cascade of hostile thoughts and feelings.

While the GAAM is valuable in itself, it is especially pertinent here because it calls attention to variables and processes that may influence aggression in any social context. In the section that follows, we suggest such specific variables, ones that are ubiquitous in today's work settings.

Frustration, Stress, and Aggression

As suggested, norms represent entitlements, that is, what is deserved or "ought to be." To the extent that norm violations block the attainment of some desired goal, a state of relative deprivation may result. This relative deprivation may, in turn, lead to a sense of frustration experienced as injustice. In fact, Brown and Herrnstein (1975) concluded that frustration "may not be fundamentally different from injustice, inequity, and relative deprivation" (p. 271), a sentiment echoed by Crosby (1976), who observed that "by definition, the sense of injustice is a part of relative deprivation" (p. 91). Additionally, propositions derived from cognitive dissonance theory suggest that the presence of inequity creates a state of tension (physiological and psychological) and that this tension is proportional to the magnitude of the inequity (Adams, 1965).

While the assertion that frustration always precedes aggression was dismissed long ago, there is evidence to suggest that under certain limited conditions (e.g., when the frustration is perceived as intentional or unjustified), frustration may produce a state of readiness or instigation to aggress (Geen, 1991). In organizational settings, frustration has been found to be positively correlated with aggression against others, interpersonal hostility, sabotage, strikes, work slowdowns, stealing, and employee withdrawal (Heacox and Sorenson, 2005; Spector, 1975).

The fact that frustration and aggression often stem from *relative* deprivation is particularly pertinent in an era in which many people feel deprived. After decades of excellent economic growth and prosperity, there has been a

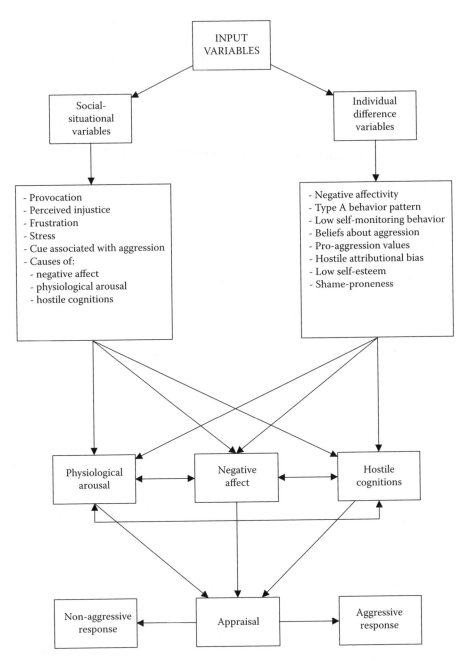

FIGURE 8.1
The General Affective Aggression Model.

sharp rise in expectations and sense of entitlement. Paradoxically, these rising employee expectations are running headlong into increasingly callous attitudes on the part of managers and executives who seem more focused on "bottom line" issues rather than the welfare of their employees. In spite of the current serious global economic downturn, CEOs in the United States continue to receive extremely large (some would say "excessive") pay packages. Last year, S&P 500 CEOs averaged $10.5 million in compensation, which is 344 times the pay of a typical American worker, and compensation levels for the top 50 hedge and private equity fund managers averaged $588 million each, more than 19,000 times as much as typical U.S. workers earned (Anderson et al., 2008). Given the current circumstances and the dire predictions for the immediate future, frustration and injustice—more real than imagined—will likely contribute more toxicity (Frost, 2003, 2004) and stress to already contentious work settings.

Stress and Aggression

Research, as well as everyday experience, suggests a relationship between stress and physical violence. With respect to work settings, Chen and Spector (1992) found a relationship between work-related stressors and interpersonal aggression, hostility, sabotage, complaints, and intentions to quit. They also reported a modest relationship between work stressors and theft and employee absence. Recent research has supported the relationship between workplace stressors and aggressive behavior (Dollard et al., 2007; Vigoda, 2002). Anecdotally, workplace stressors have been cited as a causal factor in many high-profile cases of coworker-involved homicides in the United States.

With respect to the relationship between norm violations and stress, Tedeschi and Felson (1994) suggest that stress has negative effects on performance and increases violations of politeness norms. Consequently, these effects on behavior often contribute to the development of coercive episodes and aggression. This would be very consistent with recent research demonstrating what appears to be an increase in "workplace incivility" in the United States (Pearson et al., 2001), a trend that has been observed in academic settings as well (Desrochers, 2009).

Individual Dignity and Self-Worth

Blows to one's self-esteem (i.e., an inability to maintain a positive self-image) may, on occasion, lead to aggression. For example, Averill's examination of everyday episodes of anger among adults (1982) suggests that a common cause of anger is a loss of pride or a loss of self-esteem. The experience of shame among college students has been associated with a desire to punish others (Wicker et al., 1983) as well as anger arousal, suspiciousness, resentment, irritability, a tendency to blame others for negative events, and indirect expressions of hostility (Tangney et al., 1992). Among fifth-grade boys, shame proneness was positively correlated with both self-reports of anger and teacher reports of aggression (Tangney et al., 1991).

Performance evaluations perceived to be unfairly negative; failure to obtain the raise, promotion, choice work assignment, or office to which people believe they are entitled; and being unfairly criticized by a supervisor in the presence of other employees are all blows to one's self-esteem. In fact, *all* injustices are, by their very nature, personal evaluations in which individuals have been judged to be worth less (or *worthless*) in comparison to some referent. To the extent that such an evaluation results in feelings of shame and embarrassment, it may elicit defensive, retaliative anger and a tendency to project blame outward. This tendency has also been referred to as "humiliated fury," a situation in which the acute pain of shame may elicit a rage that is directed toward the self and a real or imagined disapproving other.

When we consider the importance of a job in defining one's sense of identity, it is clear that threats to our employment relationship, hierarchical status, professional reputation, and/or chances for advancement pose significant threats to our feelings of self-worth. Unfortunately, even the most casual review of the management literature (both popular and academic) reveals a workplace environment that is fraught with ego-crushing practices. In an era of downsizing, rightsizing, layoffs, restructuring, reengineering, outsourcing, and mergers and acquisitions, the need for productivity increases is creating an environment where, according to Abraham Zaleznik, "tough is passé. Today you're dealing with a variety of head games. That's where the cruelty is" (Dumaine, 1993, p. 39). Finally, predatory bullying (Einarsen, 1999: Glasø et al., 2007) often focuses, specifically, on damaging the self-esteem of the target.

Instrumental Forms of Aggression

The preceding discussion has focused almost exclusively on hostile and affective (or reactive) aggression, where the ultimate objective is to intentionally inflict harm on others. Understandably, much of the literature on workplace bullying has focused on this form of aggression. As noted previously, many acts of aggression and bullying may be instrumental in nature, where the harm inflicted is not the ultimate aim but merely a means to a desired end.

In a dog-eat-dog business world where "no good deed goes unpunished," a world filled with "hostile takeovers," "poison pills," and "creative tension," and where you are either the victor or the vanquished, there appear to be a sufficient number of opportunities to learn the so-called benefits of being aggressive. Stories of emotionally abusive CEOs of highly successful companies are legion. For example, Linda Wachner of giant apparel maker Warnaco advised an executive to "fire people if he wanted to be taken seriously," and Herbert Haft of the Dart Group fired his own wife and son because he thought they were usurping his power. The rants of Steve Jobs, of Apple Computer, and the "Jack Attacks" of Jack Connors, founder of the Boston advertising firm Hill Holliday, represent just a few visible examples of the kinds of emotional abuse (Keashly, 1998) that are rampant in today's work settings.

The lessons learned in such hostile environments center on the importance of obtaining, maintaining, and projecting power. The Machiavellian dictum that it is better to be feared than loved and the notion that the ends justify the means—or in the case of bullying, justify the *meanness*—have been widely embraced in contemporary work settings (e.g., Bing, 2002; Griffin, 1991). Consequently, these behaviors are likely to be adopted through a social learning process (Bandura, 1973). This is especially true given that abuse of power is seldom punished and more often rewarded (Salin, 2003).

As noted by Salin (2003), workplace bullying can also be viewed as political behavior, as "actions by individuals which are directed toward the goal of furthering their own self-interests without regard for the well-being of others or their organization" (Kacmar and Baron, 1999, p. 5). In fact, victims often report that office politics and favoritism are the causes of the bullying they have experienced. This explanation is particularly true as it relates to a number of decisions that are often associated with office politics, such decisions as those relating to promotions, task assignments, budget allocations, performance appraisals, allocations of resources, and dismissal (Drory, 1993). With regard to dismissals, Zapf and Warth (1997) note that workplace bullying is often employed as a means of expelling unwanted employees who might be difficult to terminate through more formal mechanisms. They refer to this tactic as "personnel management by other means" (as cited in Salin, 2003; Salin, 2005). In an analysis of questionnaire data from 377 members of the Finnish Association of Graduates in Economics and Business Administration, Salin (2005) found support for the relationship between workplace bullying and organizational politics. Using the Negative Acts Questionnaire to measure bullying (Einarsen and Hoel, 2001; Einarsen and Raknes, 1997) and the Perceptions of Organizational Politics scale (Ferris and Kacmar, 1992), Salin (2005) found a significant correlation between bullying and perceptions of politics, $r(377) = .30, p < .01$. Beyond victim self-reports, individuals having witnessed bullying reported significantly greater perceptions of organizational politics than those individuals who had not reported seeing bullying. Qualitative analyses of narrative reports from respondents supported the use of bullying as a tactic to expel employees perceived to be threats or burdens. On a related note, perceptions of organizational politics have recently been linked to employee emotional reactions, frustration, dissatisfaction, and reductions in organizational citizenship behavior (Rosen et al., 2009).

Finally, in research on the underlying dimensionality of bullying, two prominent factors involve social isolation (ostracism) and assaults on the targets' professional reputation. Similarly, research on organizational politics reveals that *marginalization* and *scapegoating* are frequently used tactics employed to make victims look bad while enhancing the image and political position of the perpetrators.

Regardless of the immediate or ultimate objectives, today's hostile cultures and climates are fertile environments for the social antecedents of aggression and bullying.

From Perceived Injustice to Workplace Bullying: The Role of Norms, Effect/Danger Ratio, and Displacement

So far, we have argued that social factors play a key role in workplace aggression and that in this respect, violations of the norms of reciprocity and fairness are especially crucial. When individuals perceive that they have been treated unfairly, they experience stress, negative affect, and other critical internal states that together set the stage for increased workplace aggression. Further, we have suggested that perceptions of unfairness are increasingly common in modern workplaces because of changes that have occurred within them in recent years. But how, precisely, does workplace bullying fit into this picture? We believe that the answer lies in three interrelated factors: (1) efforts by individuals to maximize what has been termed the *effect/danger ratio* (Bjorkqvist et al., 1994); (2) strong tendencies toward displaced aggression, which result from such efforts; and (3) norms that encourage, or simply condone, workplace bullying and thus assure the silence or active complicity of witnesses to such behavior.

Effect/Danger Ratio

When adults aggress against others, they generally seek to maximize the harm produced, while minimizing the danger to themselves through retaliation, social censure, or other potential consequences (e.g., Bjorkqvist et al., 1994). Previous research suggests that workplaces are no exception to this general rule and that, in fact, individuals often prefer disguised forms of workplace aggression for this very reason: such forms permit them to harm the victim while making it difficult for the victim to identify them as the source of such harm (Baron et al., 1999). Even when perpetrators occupy higher-level positions than their subordinate targets and thus do not necessarily fear direct retaliation from their victims, they may engage in covert forms of aggression to avoid reprisals from superiors or groups of individuals. In general, when the use of covert (i.e., passive and indirect) forms of reactive aggression against particular individuals proves too dangerous, perpetrators may redirect their aggression against other, more convenient targets.

Displaced Aggression

Because of such pressures to maximize the effect/danger ratio, we believe that there are equally strong pressures toward *displacement of aggression* in the workplace. Displacement of aggression—recently found to be a reliable and powerful phenomenon (e.g., Marcus-Newhall et al., 2000)—refers to the tendency to aggress against someone other than the source of strong provocation because aggressing against the source of such provocation is too dangerous. Since perceived unfairness often stems from treatment by supervisors or by the entire organization, and since aggression against such targets is dangerous and would violate the desire to maximize the effect/danger ratio, we believe that tendencies toward displaced aggression are frequently strong in many workplaces. As a result, persons "made ready to aggress" by conditions within their workplaces often select targets who are relatively weak and defenseless, and a pattern of bullying against such individuals may then arise. Recent research confirms such displacement of aggression within and beyond work settings (Barling et al., 2009; Bushman et al., 2005; Hoobler and Brass, 2006).

Norms Supporting Workplace Bullying

These tendencies are further strengthened in modern organizations by the existence of norms emphasizing toughness and survival of the fittest. Such norms, in turn, reflect the weakening of the traditional perceived contract between organizations and their employees: no longer do organizations guarantee continued employment, even after long and faithful service. This norm is increasingly true given the present state of the global economy, which often involves relocation of manufacturing and other company units to off-shore locations and promotes the use of virtual offices and organizations, thereby reducing face time and the ability to build and maintain relationships. Current conditions suggest that it is perfectly appropriate for organizations to close even profitable operations and dismiss loyal and hard-working employees if greater profit is to be made elsewhere. Norms of toughness, in turn, tend to reduce the likelihood that witnesses to workplace bullying will take action against it. On the contrary, such norms tend to increase the odds that witnesses will join in and even applaud the actions of workplace bullies. From the perspective of the effect/danger ratio, risk is reduced and the benefits are more likely to outweigh the potential costs.

Projecting an air of toughness, or simply being "difficult," pays other dividends, too—at least from the bully's perspective. Whatever the underlying motivations for such behavior, people are hesitant to approach others who are known to be volatile or difficult to deal with. Rather than expose themselves to such behavior, managers and peers will often choose to take the path of least resistance to avoid dealing with unpleasant individuals. Unfortunately, this avoidance comes at the expense of others who are forced to do the work that should have been assigned to the bully. In addition, the bully achieves his or her desired objectives (e.g., less work or the exercise of power over others) and is, therefore, reinforced in the use of bullying behavior. People then begin to resent the unfair distribution of work, and this resentment, in turn, may lead to displaced aggression against others (and the organization), resulting in reactive (hostile) aggression.

In sum, from a social interactionist perspective, we suggest that many instances of workplace bullying can be understood in the following terms. Bullying behavior is initiated for some instrumental purpose, as a result of one or more individual predispositions, or in reaction to social situations. From the perspective of one or more targets, such behavior violates the basic norm of reciprocity, and resulting perceptions of injustice produce feelings of frustration, negative affect, and stress on the part of many individuals. Since overt or direct aggression against the perceived sources of such feelings (i.e., the perceived causes of unfairness) is often dangerous and therefore violates the desire to maximize the effect/danger ratio, aggression is displaced onto relatively weak and defenseless victims. Norms in modern organizations emphasizing toughness make it unlikely that witnesses to workplace bullying will intervene on behalf of the victims. On the contrary, such witnesses may actively support bullying or become willing participants; at the very least, they tend to ignore the bullying or to view it as a normal feature of their workplace. Importantly, norms of toughness make it less likely that victims will complain, since doing so would, no doubt, be viewed as a sign of weakness. As we argued at the outset, then, workplace bullying may indeed stem from social factors relating to the real or perceived treatment individuals receive from others at work.

Conclusion

In this chapter, we have provided some examples of the many ways in which social factors may serve as antecedents to aggression. We began with a discussion about the norm of reciprocity and how people are predisposed to repay in kind the type of treatment they receive. Following this, we discussed aggression as a response to the violation of important social norms and perceptions of distributive, procedural, and interpersonal injustice and focused on the specific roles of frustration and stress, negative affect, and assaults on individual dignity and self-worth. Then, drawing on contemporary theories of aggression, we demonstrated how various critical internal states might initiate a process that often concludes with aggression. Finally, we demonstrated how the effect/danger ratio, displaced aggression, and organizational norms emphasizing toughness serve as a bridge between perceptions of injustice and aggression and bullying.

While we have touched on several important social antecedents and described a process by which these antecedents may lead to aggression and bullying, our discussion has been necessarily brief and far from exhaustive. For example, although we have touched on aspects of the interplay between perpetrators and victims, we have not provided a detailed discussion of the role of victims as antecedents to episodes of bullying. There is literature suggesting that bullies may engage in aggression as a response to perceived provocation by a victim. While this possibility is implicit in our previous discussion about perceived injustice, we never specifically addressed this issue nor several other dynamics associated with bully–victim interaction. Also, we did not explicitly discuss the fact that people are seldom merely bullies or victims but may, on occasion, occupy both roles (Ma, 2001; Unnever, 2005)—another important social dynamic. While we certainly believe that these issues are important to a full understanding of bullying from a social interactionist perspective, space constraints do not permit such a discussion. Finally, with respect to the General Affective Aggression Model presented earlier, this framework is not meant to capture every aspect of social interaction but, rather, to represent the mediating process by which situational and dispositional variables may lead to individual acts of aggression. As such, it focuses heavily on affective (i.e., hostile, angry, or reactive) aggression and is not meant to describe all forms of instrumental (goal-directed) aggression. Nevertheless, the model describes situations present in a large proportion of cases involving both individual acts of aggression and persistent episodes of bullying.

Before concluding, it is important to highlight the cumulative effects of norm violations and injustices in creating a hostile work environment. As noted, hostile cognitions represent a critical internal state that may provide *instigation toward aggression.* As also noted previously, this situation results when people's expectations of trust have systematically and repeatedly been violated, leading to suspicion and mistrust, especially in ambiguous situations (a common occurrence in human interactions). Research demonstrates that individuals are more likely to make sinister or hostile attributions about the behavior of others in such circumstances (Dodge et al., 1990; Kramer, 1994). This response predisposes people toward perceptions of injustice even when no real injustice exists. Also, as noted in our discussion of the GAAM, these cognitions are likely to elicit negative affect and physiological arousal, thereby contributing to an escalating and potentially volatile dynamic. Recently, increased attention has been focused on the role of emotions in work settings (Ashkanasy and Daus, 2002) and the role of cognitive and affective drivers of behavior (Baron, 2008; Weiss and Beal, 2005; Weiss and Cropanzano, 1996). As noted by Amabile and Kramer (2007), a critical driver of behavior is the "stream of emotions, perceptions, and motivations [that people experience] as they react to and make sense of the events of the workday" (p. 74). This view, of course, is central to a social interactionist analysis of bullying episodes. Unfortunately, with few exceptions (Glomb,

2002; Keashly and Neuman, 2008), little work has focused on experiences within the ongoing social dynamics between perpetrators and victims and the forces that exacerbate or ameliorate hostility. We strongly recommend that more qualitative research focus on this process (interviews, random events or experience sampling, diary methodology, etc.).

In conclusion, although it is often easy (and occasionally accurate) to attribute workplace bullying to internal, dispositional causes (e.g., he or she is just a bully), we have attempted to demonstrate here that aggression and bullying derive from a wide variety of factors, many of which are social in nature. The bad news is that these factors are all too common in everyday experience. The good news is that they are often amenable to change.

References

Adams, J. S. (1965) Inequity in social exchange. In L. Berkowitz (ed.), *Advances in experimental social psychology*, vol. 2 (pp. 267–299). New York: Academic Press.

Ajzen, I., and Fishbein, M. (1980) *Understanding attitudes and predicting social behavior*. Englewood Cliffs, NJ: Prentice-Hall.

Altheide, D. L., Adler, P. A., Adler, P., and Altheide, D. A. (1978) The social meanings of employee theft. In J. M. Johnson and J. D. Douglas (eds.), *Crime at the tip: Deviance in business and the professions* (pp. 90–124). Philadelphia: Lippincott.

Amabile, T. M., and Kramer, S. J. (2007) Understanding the subtext of business performance. *Harvard Business Review, 85*, 72–83.

Analoui, F. (1995) Workplace sabotage: Its styles, motives and management. *Journal of Management Development, 14*, 48–65.

Anderson, C. A., Anderson, K. B., and Deuser, W. E. (1996) Examining an affective aggression framework: Weapon and temperature effects on aggressive thoughts, affect, and attitudes. *Personality and Social Psychology Bulletin, 22*, 366–376.

Anderson, C. A., and Bushman, B. J. (2002) Human aggression. *Annual Review of Psychology, 53*, 27–51.

Anderson, S., Cavanagh, J., Collins, C., Pizzigati, S., and Lapham, M. (2008). Executive excess 2008: How average taxpayers subsidize runaway pay. Fifteenth annual CEO compensation survey. http://www.faireconomy.org/files/executive_ excess_2008.pdf (accessed February 11, 2009).

Ashkanasy, N. M., and Daus, C. S. (2002) Emotion in the workplace: The new challenge for managers. *Academy of Management Executive, 16*, 76–86.

Atkinson, W. (2006) Stealing time. *Risk Management, 53*, 48–52.

Averill, J. R. (1982) *Anger and aggression: An essay on emotion*. New York: Springer Verlag.

Bandura, A. (1973) *Aggression: A social learning analysis*. Englewood Cliffs, NJ: Prentice-Hall.

Barling, J., Dupré, K. E., and Kelloway, E. K. (2009) Predicting workplace aggression and violence. *Annual Review of Psychology, 60*, 671–692.

Baron, R. A. (2008) The role of affect in the entrepreneurial process. *Academy of Management Review, 33*, 328–340.

Baron, R. A., Byrne, D., and Griffitt, W. (1974) *Social psychology: Understanding human behavior*. Boston: Allyn and Bacon.

Baron, R. A., Neuman, J. H., and Geddes, D. (1999) Social and personal determinants of workplace aggression: Evidence for the impact of perceived injustice and the type A behavior pattern. *Aggressive Behavior, 25,* 281–296.

Baron, R. A., and Richardson, D. R. (1994) *Human aggression,* 2nd ed. New York: Plenum.

Becker, H. (1956) *Man in reciprocity*. New York: Praeger.

Bies, R. J. (1987) The predicament of injustice: The management of moral outrage. In L. L. Cummings and B. M. Staw (eds.), *Research in organizational behavior,* vol. 9 (pp. 289–319). Greenwich, CT: JAI Press.

Bing, S. (2002) *What would Machiavelli do? The ends justify the meanness*. New York: HarperCollins.

Bjorkqvist, K., Osterman, K., and Lagerspetz, K. M. J. (1994) Sex differences in covert aggression among adults. *Aggressive Behavior, 20,* 27–33.

Bowling, N. A., and Beehr, T. A. (2006) Workplace harassment from the victim's perspective: A theoretical model and meta-analysis. *Journal of Applied Psychology, 91,* 998–1012.

Brown, R., and Herrnstein, R. J. (1975) *Psychology*. Boston: Little, Brown.

Bushman, B. J., and Anderson, C. A. (2001) Is it time to pull the plug on the hostile versus instrumental aggression dichotomy? *Psychological Review, 108,* 273–279.

Bushman, B. J., Bonacci, A. M., Pedersen, W. C., Vasquez, E. A., and Miller, N. (2005) Chewing on it can chew you up: Effects of rumination on triggered displaced aggression. *Journal of Personality and Social Psychology, 88,* 969–983.

Chen, P. Y., and Spector, P. E. (1992) Relationships of work stressors with aggression, withdrawal, theft and substance use: An exploratory study. *Journal of Occupational and Organizational Psychology, 65,* 177–184.

Chory, R. M., and Hubbell, A. P. (2008) Organizational justice and managerial trust as predictors of antisocial employee responses. *Communication Quarterly, 56,* 357–375.

Crino, M. D., and Leap, T. L. (1989) What HR managers must know about employee sabotage. *Personnel,* May, 31–32, 34–36, 38.

Cropanzano, R., and Baron, R. A. (1991) Injustice and organizational conflict: The moderating effect of power restoration and task type. *International Journal of Conflict Management, 2,* 5–26.

Cropanzano, R., and Folger, R. (1989) Referent cognitions and task decision autonomy: Beyond equity theory. *Journal of Applied Psychology, 74,* 293–299.

Crosby, F. (1976) A model of egoistical relative deprivation. *Psychological Review, 83,* 85–113.

Desrochers, C. (2009) Thriving in academe: Classroom civility. Is it just me? http://www2.nea.org/he/advo-new/front.html (accessed February 7, 2009).

Dodge, K. A., Price, J. M., Bachorowski, J. A., and Newman, J. P. (1990) Hostile attributional biases in severely aggressive adolescents. *Journal of Abnormal Psychology, 99,* 385–392.

Dollard, J., Doob, L., Miller, N., Mowrer, O. H., and Sears, R. R. (1939) *Frustration and aggression*. New Haven, CT: Yale University Press.

Dollard, M. F., Skinner, N., Tuckey, M. R., and Bailey, T. (2007) National surveillance of psychosocial risk factors in the workplace: An international overview. *Work & Stress, 21,* 1–29.

Drory, A. (1993) Perceived organizational climate and job attitudes. *Organization Studies, 14*, 59–71.

Dumaine, B. (1993) America's toughest bosses. *Fortune*, October 18, 39–42, 44, 48, 58.

Dupré, K. E., and Barling, J. (2006) Predicting and preventing supervisory workplace aggression. *Journal of Occupational Health Psychology, 11*, 13–26.

Einarsen, S. (1999) The nature and causes of bullying at work. *International Journal of Manpower, 20*, 16–27.

Einarsen, S., and Hoel, H. (2001) The Negative Acts Questionnaire: Development, validation and revision of a measure of bullying at work. Paper presented at the Ninth European Congress of Work and Organizational Psychology, Prague, May.

Einarsen, S., and Raknes, B. I. (1997) Harassment in the workplace and the victimization of men. *Violence and Victims, 12*, 247–263.

Einarsen, S., and Skogstad, A. (1996) Bullying at work: Epidemiological findings in public and private organizations. *European Journal of Work and Organizational Psychology, 5*, 185–201.

Eisenberger, R., Lynch, P., Aselage, J., and Rohdieck, S. (2004) Who takes the most revenge? Individual differences in negative reciprocity norm endorsement. *Personality and Social Psychology Bulletin, 30*, 787–799.

Felson, R. B., and Tedeschi, J. T. (eds.) (1993) *Aggression and violence: Social interactionist perspectives*. Washington, DC: American Psychological Association.

Ferris, G. R., and Kacmar, K. M. (1992) Perceptions of organizational politics. *Journal of Management, 18*, 93–116.

Folger, R. (1986) Rethinking equity theory: A referent cognitions model. In H. W. Bierhoff, R. L. Cohen, and J. Greenberg (eds.), *Justice in social relations* (pp. 145–162). New York: Plenum.

Folger, R., and Cropanzano, R. (1998) *Organizational justice and human resource management*. Thousand Oaks, CA: Sage.

——— (2001) Fairness theory: Justice as accountability. In J. Greenberg and R. Cropanzano (eds.), *Advances in organizational justice* (pp. 1–55). Palo Alto, CA: Stanford Press.

Folger, R., and Greenberg, J. (1985) Procedural justice: An interpretive analysis of personnel systems. In K. Rowland and G. Ferris (eds.), *Research in personnel and human resource management*, vol. 3 (pp. 141–183). Greenwich, CT: JAI Press.

Frost, P. J. (2003) *Toxic emotions at work*. Boston: Harvard Business School Press.

——— (2004) Handling toxic emotions: New challenges for leaders and their organization. *Organizational Dynamics, 33*, 111–127.

Geen, R. G. (1968) Effects of frustration, attack and prior training in aggressiveness on aggressive behavior. *Journal of Personality and Social Psychology, 9*, 316–321.

——— (1991) *Human aggression*. Pacific Grove, CA: Brooks/Cole.

Glasø, L., Matthiesen, S. B., Birkeland Nielsen, M., and Einarsen, S. (2007) Do targets of workplace bullying portray a general victim personality profile? *Scandinavian Journal of Psychology, 48*, 313–319.

Glomb, T. M. (2002) Workplace anger and aggression: Informing conceptual models with data from specific encounters. *Journal of Occupational Health Psychology, 7*, 20–36.

Glomb, T. M., and Liao, H. (2003) Interpersonal aggression in work groups: Social influence, reciprocal, and individual effects. *Academy of Management Journal, 46*, 486–496.

Gouldner, A. W. (1960) The norm of reciprocity. *American Sociological Review, 25,* 161–178.

Greenberg, J. (1990) Employee theft as a reaction to underpayment inequity: The hidden cost of pay cuts. *Journal of Applied Psychology, 75,* 561–568.

Greenberg, J., and Colquitt, J. A. (eds.) (2005) *Handbook of organizational justice.* Mahwah, NJ: Lawrence Erlbaum.

Griffin, G. R. (1991) *Machiavelli on management: Playing and winning the corporate power game.* New York: Praeger.

Harris, M. B. (1993) How provoking! What makes men and women angry? *Aggressive Behavior, 19,* 199–211.

Heacox, N. J., and Sorenson, R. C. (2005) Organizational frustration and aggressive behaviors. *Journal of Emotional Abuse, 4,* 95–118.

Hoad, C. D. (1993) Violence at work: Perspectives from research among 20 British employers. *Security Journal, 4,* 64–86.

Hoel, H., Rayner, S., and Cooper, C. L. (1999) Workplace bullying. In C. L. Cooper and I. T. Robertson (eds.), *International review of industrial and organizational psychology,* vol. 14 (pp. 195–229). New York: John Wiley.

Homans, G. C. (1974) *Social behavior: Its elementary forms,* rev. ed. New York: Harcourt Brace.

Hoobler, J. M., and Brass, D. J. (2006) Abusive supervision and family undermining as displaced aggression. *Journal of Applied Psychology, 91,* 1125–1133.

Kacmar, K. M., and Baron, R. A. (1999) Organizational politics: The state of the field, links to related processes, and an agenda for future research. In G. Ferris (ed.), *Research in personnel and human resources management,* vol. 17 (pp. 1–39). Greenwich, CT: JAI Press.

Keashly, L. (1998) Emotional abuse in the workplace: Conceptual and empirical issues. *Journal of Emotional Abuse, 1,* 85–117.

Keashly, L., and Neuman, J. H. (2008) Aggression at the service delivery interface: Do you see what I see? *Journal of Management & Organization, 14,* 180–192.

Kramer, R. M. (1994) The sinister attribution error: Paranoid cognition and collective distrust in organizations. *Motivation and Emotion, 18,* 199–230.

Ma, X. (2001) Bullying and being bullied: To what extent are bullies also victims? *American Educational Research Journal, 38,* 351–370.

Marcus-Newhall, A., Pedersen, W. C., and Miller, N. (2000) Displaced aggression is alive and well: A meta-analytic review. *Journal of Personality and Social Psychology, 78,* 670–689.

Mitchell, M. S., and Ambrose, M. L. (2007) Abusive supervision and workplace deviance and the moderating effects of negative reciprocity beliefs. *Journal of Applied Psychology, 92,* 1159–1168.

Neuman, J. H. (2004) Injustice, stress, and aggression in organizations. In R. W. Griffin and A. M. O'Leary-Kelly (eds.), *The dark side of organizational behavior* (pp. 62–102). San Francisco: Jossey-Bass.

Neuman, J. H., and Baron, R. A. (1997a) Aggression in the workplace. In R. Giacalone and J. Greenberg (eds.), *Antisocial behavior in organizations* (pp. 37–67). Thousand Oaks, CA: Sage.

——— (1997b) Type A behavior pattern, self-monitoring, and job satisfaction as predictors of aggression in the workplace. In G. Chao (ed.), *Counterproductive job performance and organizational dysfunction.* St. Louis, MO: Society for Industrial and Organizational Psychology.

—— (1998) Perceived injustice as a cause of—and justification for—workplace aggression and violence. In A. M. O'Leary-Kelly and D. P. Skarlicki (eds.), *Advances in organizational justice theories: The motivation to engage in dysfunctional behavior*. San Diego, CA: Academy of Management.

—— (2003) Social antecedents of bullying: A social interactionist perspective. In S. Einarsen, H. Hoel, D. Zapf, and C. L. Cooper (eds.), *Bullying and emotional abuse in the workplace: International perspectives in research and practice* (pp. 185–202). London: Taylor & Francis.

—— (2005) Aggression in the workplace: A social psychological perspective. In S. Fox and P. E. Spector (eds.), *Counterproductive work behavior: Investigations of actors and targets* (pp. 13–40). Washington, DC: American Psychological Association.

Olweus, D. (1978). *Aggressions in the schools; Bullies and whipping boys*. Washington DC; Hemisphere, Wiley.

Pearson, C. M., Andersson, L. M., and Wegner, J. W. (2001) When workers flout convention: A study of workplace incivility. *Human Relations, 54*, 1387–1420.

Rosen, C. C., Harris, K. J., and Kacmar, K. M. (2009) The emotional implications of organizational politics: A process model. *Human Relations, 62*, 27–58.

Rule, B. G., and Ferguson, T. J. (1984) The relation among attribution, moral evaluation, anger, and aggression in children and adults. In A. Mummendey (ed.), *Social psychology of aggression: From individual behavior to social interaction* (pp. 143–155). Berlin: Springer Verlag.

Salin, D. (2003) Bullying and organisational politics in competitive and rapidly changing work environments. *International Journal of Management and Decision Making, 4*, 35–46.

—— (2005) Workplace bullying among business professionals: Prevalence, gender differences and the role of organizational politics. http://www.pistes.uqam.ca/v7n3/pdf/v7n3a2en.pdf (accessed January 30, 2009).

Simmel, G. (1950) *The sociology of Georg Simmel*. Trans. K. H. Wolff. Glencoe, IL: Free Press.

Spector, P. E. (1975) Relationship of organizational frustration with reported behavioral reactions of employees. *Journal of Applied Psychology, 60*, 635–637.

Stuckless, N., and Goranson, R. (1992) The vengeance scale: Development of a measure of attitudes toward revenge. *Journal of Social Behavior and Personality, 7*, 25–42.

Tangney, J. P., Wagner, P. E., Burggraf, S. A., Gramzow, R., and Fletcher, C. (1991) Children's shame-proneness, but not guilt-proneness, is related to emotional and behavioral maladjustment. Paper presented at the Society for Research in Child Development in Seattle, Washington, June.

Tangney, J. P., Wagner, P., Fletcher, C., and Gramzow, R. (1992) Shamed into anger? The relation of shame and guilt to anger and self-reported aggression. *Journal of Personality and Social Psychology, 62*, 669–675.

Tedeschi, J. T., and Felson, R. B. (1994) *Violence, aggression, and coercive actions*. Washington, DC: American Psychological Association.

Thibaut, J., and Walker, L. (1975) *Procedural justice: A psychological analysis*. Hillsdale, NJ: Lawrence Erlbaum.

Unnever, J. D. (2005) Bullies, aggressive victims, and victims: Are they distinct groups? *Aggressive Behavior, 31*, 153–171.

Vigoda, E. (2002) Stress-related aftermaths to workplace politics: The relationships among politics, job distress, and aggressive behavior in organizations. *Journal of Organizational Behavior, 23,* 571–592.

Weide, S., and Abbott, G. E. (1994) Murder at work: Managing the crisis. *Employment Relations Today, 21,* 139–151.

Weiss, H. M., and Beal, D. J. (2005) Reflections on affective events theory research on emotion in organizations. In N. M. Ashkanasy, C. Härtel, and W. J. Zerbe (eds.), *The effect of affect in organizational settings,* vol. 1 (pp. 1–21). Amsterdam: Elsevier.

Weiss, H. M., and Cropanzano, R. (1996) Affective events theory: A theoretical discussion of the structure, causes and consequences of affective experiences at work. In B. M. Staw and L. L. Cummings (eds.), *Research in Organizational Behavior,* vol. 18 (pp. 1–74). Greenwich, CT: JAI Press.

Wicker, F. W., Payne, G. C., and Morgan, R. D. (1983) Participant descriptions of guilt and shame. *Motivation and Emotion, 7,* 25–39.

Zapf, D., and Worth, K. (1997, August) Mobbing: Subtile Kriegsführung am Arbeitsplatz. *Psychologie Heute,* 20–25, 28–29.

Zeitlin, L. R. (1971, June) A little larceny can do a lot for employee morale. *Psychology Today,* 22, 24, 26, 64.

9

Organisational Causes of Workplace Bullying

Denise Salin and Helge Hoel

CONTENTS

Introduction

Already in the 1980s Heinz Leymann, a pioneer of workplace bullying research, emphasised the importance of organisational antecedents of bullying. He strongly rejected the idea of a specific "victim personality." Instead, he forcefully argued that anybody could become a target of bullying under the right circumstances (e.g., Leymann, 1996). In this way, he emphasised the important link between a poor work environment and bullying.

Despite this clear hypothesis, only a limited number of studies in the 1990s and early 2000s actually sought to empirically identify which factors in the work environment may contribute to bullying and the mechanisms through which a poor environment results in bullying (e.g., Einarsen, Raknes, and Matthiesen, 1994; Vartia, 1996; Zapf, Knorz, and Kulla, 1996b). Nevertheless, in the past few years the interest in organisational antecedents of bullying, or what is referred to as "the work environment hypothesis," has grown and new studies have been carried out, increasing our understanding of this important and detrimental phenomenon. A general finding from these studies is that both victims and observers of bullying report a more negative work environment than those who were not bullied (Baillien, Neyens, and De Witte, 2008; Einarsen et al., 1994; Hauge, Skogstad, and Einarsen, 2007;

Vartia, 1996) and that the worst work environment is associated with those most severely bullied (Zapf et al., 1996b). However, it is important to bear in mind that bullying is a complex and multicausal phenomenon and can seldom be explained by one factor alone (Salin, 2003a; Zapf, 1999).

The aim of this chapter is to summarise research findings on organisational causes or antecedents of workplace bullying. Building on our previous contributions (Hoel and Salin, 2003; Salin, 2003a), this chapter updates and incorporates the most important new findings and insights obtained over the last few years. We will emphasise that we understand workplace bullying as a complex and dynamic process, where both action and reaction should be understood within the social context in which they take place (see also Neuman and Baron, this volume). Thus, situational factors may increase the vulnerability of targets or recipients of bullying behaviour and contribute to their response to such acts. Whilst acknowledging this reality, the present investigation will limit itself to an exploration of the influence of various organisational antecedents on the behaviour of perpetrators.

Our review of organisational causes of bullying examines and synthesises empirical findings and new theoretical insights about the relationship between work environment factors and workplace bullying. For this purpose, we have structured the chapter under five separate subheadings: (1) job design and work organisation, (2) organisational cultures and climate, (3) leadership, (4) reward systems, and (5) organisational change. For the sake of clarity, we discuss the factors separately, although we acknowledge that in reality they are highly intertwined. Towards the end of the chapter, we examine the limitations of the studies undertaken so far in the field.

Work Organisation and Job Design

Bullying has frequently been associated with a negative and stressful working environment (Einarsen et al., 1994; Leymann, 1996). To account for such a relationship, various work environment factors can be considered to produce or elicit occupational stress, which may increase the risk of conflict and bullying. In some cases, it can also be argued, the presence of work stressors by themselves may be perceived as harassment, particularly when they are attributed to hostile intentions (Brodsky, 1976; Einarsen et al., 1994).

Looking at the evidence, role conflict, role ambiguity, and lack of clear goals are features of work organisations that as early as the 1990s were found to be linked to bullying (Einarsen et al., 1994; Vartia, 1996). More recent studies from a variety of European countries, including Norway, Belgium, and Spain, have confirmed these findings, all reporting role conflict and role ambiguity to be among the strongest predictors of workplace bullying (Baillien and De Witte, 2009; Baillien, Neyens, and De Witte, 2008; Einarsen et al., 1994; Hauge, Skogstad, and Einarsen, 2007; Moreno-Jiménez , Rodríguez-Muñoz, Pastor, Sanz-Vergel, and Garrosa, 2009). Thus, bullying seems to thrive where employees perceive contradictory expectations, demands, and values in their jobs and where expectations are perceived as unclear or unpredictable. This conclusion is confirmed by a meta-analysis of empirical studies undertaken between 1987 and 2005, which included 90 separate samples and which concluded that role conflict and role ambiguity were among the strongest predictors of harassment (Bowling and Beehr, 2006).

Work intensification and increasing pressure have often been suggested as precursors of bullying. However, early research failed to unequivocally demonstrate such a relationship. So, whilst Appelberg, Romanov, Honkasalo, and Koskenvuo (1991) identified time pressure as a source of interpersonal conflict, and Hoel and Cooper (2000) reported significant differences in terms of workload between those who were currently bullied, those who had been previously bullied, those who had witnessed bullying, and those who had neither witnessed nor experienced bullying, some other studies failed to support their findings (e.g., Vartia, 1996). However, recent studies seem to provide more support for such a relationship (Baillien et al., 2008; Baillien and De Witte, 2009; Bowling and Beehr, 2006; Hauge et al., 2007; Moreno-Jimenez et al., 2009).

It has also been argued that the problem of bullying comes to the fore when a high degree of pressure is present in a work environment that offers individuals little control over their own work (Einarsen et al., 1994). Such an interpretation would be in line with Karasek's "job demand–control model of stress" (1979), in which strain is seen as the likely outcome of a combination of high demands and low decision latitude. Yet, Zapf et al. (1996b), in a German study of victims of "mobbing," fail to find support for such a hypothesis. Still, the authors argue that time pressure may indirectly affect bullying by undermining the opportunity to resolve conflicts. In line with this finding, they also rejected an earlier view, put forward by Thylefors (1987) and Sjøtveit (1992), that bullying may be the result of boredom.

Certain physical aspects of the work environment may also act as antecedents of aggressive behaviour and bullying. Hence, work undertaken under noisy, hot (or cold) circumstances or in cramped conditions has been found to be associated with increased feelings and attitudes of hostility (Anderson, Anderson, and Deuser, 1996; Einarsen, 1996). Similarly, in a qualitative study, Baillien et al. (2008) found that working in high temperatures, crowded spaces, or in otherwise unpleasant and irritating environments and relying on sharing tools and equipment were all associated with higher risk for bullying. This finding could also help to explain the high levels of bullying often reported in restaurant kitchens, frequently described as cramped, hot, and noisy work environments (e.g., Einarsen and Skogstad, 1996; Mathisen, Einarsen, and Mykletun, 2008).

Organisational Culture and Climate

Studies of workplace bullying discussing the role and impact of the culture of the organisation have often emphasised that in many organisations with high levels of bullying, negative and abusive acts were indirectly "permitted," whether or not the behaviour was actually an integrated part of the culture. This finding supports the view of Brodsky (1976) that for harassment to occur, there needs to be a culture that permits and rewards it (p. 83). Accordingly, bullying is seen to be prevalent in organisations where employees and managers feel that they have the support, or at least implicitly the blessing, of senior managers to carry on their abusive and bullying behaviour (Einarsen, 1999). Furthermore, new managers will quickly come to view this form of behaviour as acceptable and normal if they see others get away with it and are even rewarded for it (Rayner, Hoel, and Cooper, 2002). The findings of Neyens, Baillien, De Witte, and Notelaers (2007) that bullying was more common in organisations with no antibullying policies seems to support such a view, albeit indirectly or by implication. Nevertheless, Salin (2009) found that having a policy did not predict what kind of action organisations take in cases of bullying, thus indicating that policies may have a preventive effect rather than necessarily influencing what interventions, if any, organisations may choose to take.

The importance of socialisation processes, whereby new members gradually adopt shared, destructive norms conducive to bullying, has been demonstrated in other settings. Thus, in a study of bullying in the fire service, Archer (1999) explored how bullying in a "paramilitary' setting may become institutionalised and passed on as tradition, thus reinforcing the dominance of white males. In this respect, Archer identified the training process as a powerful source of socialisation of such behaviour. Some parallels can be drawn to a seemingly very different work environment, the luxury restaurant kitchen, where there exists a widespread view that top-quality food can be produced only in an atmosphere of blind subordination, resembling the discipline and subordination found in military environments. This view has been reinforced by the media and some celebrity chefs, whose aggressive and bullying behaviour, at times exhibited on air in programmes like *Hell's Kitchen*, has often been excused with reference to their artistry and creativity (Johns and Menzel, 1999). Although the presence of abusive behaviour has in fact been shown to have a negative relationship to external evaluations of restaurant (Mathisen et al., 2008), the normality of abusive behaviour may become internalised and reproduced by means of processes of learning and socialisation, handed down from one generation of chefs to the next, and in turn affecting all kitchen staff (Bloisi and Hoel, 2008).

Similarly, several studies of student or trainee nurses (e.g., Hoel, Giga, and Davidson, 2007; Randle, 2003) reveal that socialisation processes account for the widespread presence of bullying within the nursing occupation, whilst

Randle (2003) found that student nurses were exposed to bullying on a large scale. Hoel et al. (2007) found less evidence for that, arguing that their status, high visibility (wearing separate uniforms), and presumed vulnerability actually seemed to protect against more sustained abuse and bullying but not against exploitation and negative behaviour per se. To cope with their situation, nursing students responded by becoming harder, gradually changing their professional outlook and expectations, thus indirectly contributing to normalisation and reproduction of abuse and bullying in the health service.

In the preceding examples, bullying can be considered as being built into or becoming part of the culture. Organisations characterised by an extreme degree of conformity and group pressure seem to be particularly prone to bullying. Consequently, bullying seems to flourish in institutions such as prisons, hospitals for the mentally ill, and the armed forces, where compliance and discipline are of overriding importance (e.g., Ashforth, 1994).

While organisations characterised by a strict focus on power relations, a very formal atmosphere, and extreme goal-orientation may be associated with bullying, Baillien et al. (2008) showed that the opposite—that is, a power vacuum, a too informal atmosphere, and a manager who is too tolerant— were also risk factors. Their research thus demonstrated that leaders who are "too much people oriented" or "too tolerant" caused a lot of frustration. The Baillen et al. study (2008) thus demonstrated the need for a "healthy" balance between extremes in order to decrease the risk of bullying.

Partly linked to a very informal atmosphere is the issue of workplace humour. Humiliating jokes, surprises, and insults can sometimes be part of the socialisation process whereby new members are tested. However, this kind of humour can easily go sour and turn into bullying if the target, for some reason, cannot defend him- or herself or does not take it as a joke (Einarsen and Raknes, 1997). In some cases, bullying disguised as harsh humour can also be used to punish colleagues who do not conform to shared norms (Collinson, 1988). Hence, practical jokes were reported as one of the most common negative acts among male industrial workers at a Norwegian marine engineering plant (Einarsen and Raknes, 1997).

More generally, the social climate and the communication climate in workplaces have been shown to be related to workplace bullying (Einarsen et al., 1994; Vartia, 1996). For example, Vartia (1996) found that the general atmosphere in workplaces where bullying occurred more often was described as quarrelsome and sullen or strained and competitive. Problems there were also less likely to be discussed openly. Zapf et al. (1996b), in turn, found that bullied employees reported less social support from colleagues and supervisors. The importance of the social climate has also been supported by several more recent studies (Baillien et al., 2008; Baillien and De Witte, 2009), and the frequency of interpersonal conflicts, in particular, has been highlighted (Baillien, Neyens, De Witte, and De Cuyper, 2009; Hauge et al., 2007; Skogstad, Matthiesen, and Einarsen, 2007b). Strandmark and Hallberg (2007), in their qualitative studies in the Swedish public-service sector, have particularly highlighted the importance of professional and value conflicts, which might give rise to a struggle for power that may in turn evolve into bullying. The evidence from survey studies suggests that conflicts with the immediate supervisor seem to be a stronger predictor of bullying than conflicts with colleagues (Skogstad et al., 2007b), although the national context seems to impact the prevailing mode of conflict (Beale and Hoel, in press).

Leadership

With managers in positions of power often identified as perpetrators, a scrutiny of the impact of different leadership styles on bullying appears to be essential. However, given that most studies of leadership have focused on the effectiveness of leaders, where negative leadership has been seen as equivalent to ineffective leadership, little attention has traditionally been paid to more destructive aspects of leadership (cf. Einarsen, Aasland, and Skogstad, 2007; Hoel, Glasø, Hetland, and Einarsen, in press; Skogstad, Einarsen, Torsheim, Aasland, and Hetland, 2007a). Recent research seems to support the importance of leadership styles. For example, in a representative, large-scale study of the Norwegian workforce, Hauge et al. (2007) (N = 2,539) found both tyrannical and laissez-faire leadership behaviour to be among the strongest predictors of bullying.

Accordingly, several other studies have confirmed a relationship between bullying, on the one hand, and an autocratic leadership and an authoritarian way of settling conflicts or dealing with disagreements, on the other (O'Moore and Lynch, 2007; Vartia, 1996). An authoritarian style of leadership may also create a climate of fear, where there is little or no room for dialogue and where complaining may be considered futile. Such a form of autocratic or coercive leadership seems to come close to what Einarsen et al. (2007) label as tyrannical leadership and Ashforth (1994) referred to as "petty tyranny."

In a study of more than 5,000 British employees, Hoel et al. (in press) found that bullying was positively associated with noncontingent punishment (NCP), a leadership style where punishment is used arbitrarily (Podsakoff, Todor, and Skov, 1982), with an autocratic leadership as well as with a laissez-faire style of leadership. By contrast, bullying was negatively associated with a participative leadership style. However, the study revealed that targets and observers of bullying did not entirely share perceptions in terms of leadership and bullying. Thus, whilst the strongest association between leadership and bullying was found for NCP, an autocratic leadership styled emerged as the strongest predictor for observers. Thus, where punishment is meted out arbitrarily, its unpredictability may contribute to its being labelled as bullying. For observers, however, bullying is primarily associated with an autocratic or coercive style of management. Although unpleasant, being exposed

□ See Ferris ref, below

to such a style of management may be more predictable and thus easier to make sense of and to protect against.

So far our focus has been on seemingly abusive styles of management. However, abdication of leadership or a so-called laissez-faire style of management also appears to provide fertile ground for bullying between peers or colleagues. There are also a number of empirical studies, both surveys (e.g., Einarsen et al., 1994; Hauge et al., 2007; Hoel et al., in press) and qualitative studies (e.g., Strandmark and Hallberg, 2007) that support this. Strandmark and Hallberg (2007) identified the emergence of "unofficial leaders" as one of the risks of weak or indistinct official leadership (cf. Leymann, 1996). The mechanisms through which laissez-faire leadership leads to bullying were studied in a large-scale Norwegian study (Skogstad et al., 2007a). The authors found that experiencing laissez-faire leadership by one's immediate supervisor was associated with high levels of role conflict and role ambiguity and with increased conflict levels with coworkers. Path-analysis showed that the association between laissez-faire leadership and bullying was mainly mediated through these three work stressors. It follows that laissez-faire leadership primarily affects bullying by giving rise to a stressful work environment. Nonetheless, Skogstad et al. (2007a) also found a direct effect between laissez-faire leadership and bullying. To account for such a relationship, they argued, sometimes the absence of adequate leadership itself may be experienced by subordinates as rejection and expulsion.

Reward Systems and Competition

Following a model proposed by Salin (2003a), bullying is typically an interplay among enabling, motivating, and precipitating or triggering structures and processes. Whilst many of the factors discussed so far might be classified as enabling factors—that is, as factors that create fertile ground and allow bullying to occur in the first place—Salin's model (2003a) also emphasised the role of motivating factors—that is, factors that create an incentive for a perpetrator to engage in such behaviour. Although Björkqvist, Österman, and Hjelt-Bäck (1994) used the term *danger-ratio effect* to refer to the cost–benefit analysis applied by perpetrators, there have been few studies examining the possible benefits of bullying for perpetrators and contributing organisational factors. An exception is the controversial work by Ferris, Zinko, Brouer, Buckley, and Harvey (2007) on "strategic bullying." Drawing upon organisational politics literature, the authors argue that bullying can have positive effects for the bully as well as for the organisation. Thus, strategic bullying might be seen to contribute to enhancing the reputation and the power of the bully. In terms of positive organisational outcomes, it is argued that bullying can be used to ensure compliance or "getting the job done."

Salin's model might that with this situation perpetrator

prevalence in military/civ in workplace environs?

bullying as strategic/rational

According to the authors, the fact that bullying is likely to cause distress could be seen as advantageous to the organisation in the sense that it might help the organisation get rid of an unproductive or otherwise unwanted employee, where other approaches have been seen to fail, albeit at a substantial cost to the individual at the receiving end. Whilst being at odds with most other contributions in the field, the Ferris et al. study (2008) highlights that certain organisational practices, like those of the cultures previously described, may in fact encourage bullying. The study reveals that from a management or even an organisational perspective, bullying can be highly rational, a fact that has often been overlooked in the bullying literature (see Beale, this volume).

politics

competitive climate

Some organisational practices related to particular aspects of reward systems and high internal competition may also act as antecedents of bullying. For example, in a study among business professionals, Salin (2003b) found a significant, positive correlation between perceptions of organisational politics and workplace bullying. This finding supports the view that bullying may sometimes be used as a micropolitical strategy to enhance the perpetrator's own position. In terms of the presence of bullying, she also reported a positive relationship with high internal competition, with a competitive climate also being identified as something typical by other bullying researchers (O'Moore, Seigne, McGuire, and Smith, 1998; Vartia, 1996). In very competitive work environments, bullying may thus be used strategically to punish and get rid of over- or underachieving colleagues or subordinates who are considered either threats or burdens.

Few studies have examined empirically the effect of different reward systems on workplace bullying. However, in a Finnish study, Sutela and Lehto (1998) reported that performance-based reward systems were associated with increased risk for bullying. This seems to support our foregoing argument. In this respect, bullying may be used not only to sabotage the work performance of others but also to achieve compliance in order to meet departmental goals and objectives. Furthermore, according to Collinson (1988), the introduction of collective bonus systems may reinforce some workers' concerns about controlling their colleagues, possibly resulting in bullying in extreme cases. This notion is supported by a study by Zapf et al. (1996b), who found that victims reported greater requirement for cooperation or teamwork than did nonvictims. Enforced teamwork may, therefore, be considered a possible antecedent of bullying as a fertile ground for conflict development, particularly if linked to interteam competition for limited rewards.

In terms of motivating factors, workplace bullying has also been referred to as "personnel work by other means" (Zapf and Warth, 1997). For instance, in sectors and countries characterised by strict employment laws, bullying may be a way of trying to expel "undesired" employees (*International Herald Tribune*, 2004). Similarly, it might be argued that a higher degree of bureaucracy and stricter rules for laying off workers in the public sector may increase the value of using bullying as a strategy for circumventing

bullying & workforce shaping

rules and eliminating unwanted persons. This argument can account for the higher prevalence rate of bullying often reported in the public sector (cf. Salin, 2001; Zapf et al., 2003). By contrast, Hodson, Roscigno, and Lopez (2006) found that bullying was particularly strongly linked to organisational chaos and job insecurity. Using ethnographic evidence from nearly 150 book-length organisational ethnographies, representing a range of workplaces and organisations, they arrived at a somewhat different conclusion with respect to the impact of bureaucracy on bullying. In this, they made a distinction between two forms of bureaucracies, or what they referred to as "coercion" and "facilitation," which they found to be related to bullying in opposite ways. Whilst coercive bureaucracy is associated with lack of autonomy, leading to feelings of powerlessness and often associated with bullying, facilitative bureaucracy was seen to contribute to reducing role conflict and role ambiguity and was negatively associated with bullying.

Organisational Change and Bullying

In Salin's framework (2003a), precipitating factors were described as factors that involved changes of various kinds, whether internal or external to the organisation. These factors were considered likely triggers of bullying.

The potential impact of organisational change on bullying has received attention from a number of writers in the field. One of the early studies on the relationship between a range of organisational changes and aggression was conducted by Baron and Neuman (1996). Backed up by confirmatory statements from observers, the strongest predictors of aggression emerging from their study were found to be use of part-time workers, change in management, and pay cuts or pay freezes. In a large-scale survey in the United Kingdom, Hoel and Cooper (2000) also reported significant differences between bullied and nonbullied workers when asking about different changes taking place within the past six months. Those who had been bullied reported more cases of "major organisational change," budget cuts, "major technological change," and "major internal restructuring." In addition, change of management has also frequently been found to be associated with bullying (Hoel and Cooper, 2000; UNISON, 1997). For example, Rayner (1997) found that targets most frequently mentioned a recent change in job (51%) or change in manager (31%) as the event coinciding with the onset of bullying. As far as the impact of change is concerned, according to the conflict literature conflicts are likely to be more prevalent in the formative and normative period of group development, when groups are less stable (Tuckman, 1965). Hence, as argued by Zapf and Einarsen (2005), if the probability of conflict is high, the prevalence of bullying is similarly likely to be elevated.

Although several authors have reported that there is a relationship between organisational change and bullying, the mechanisms explaining these dynamics have received little attention and conclusions have often been hypothetical or speculative. Many of the early articles suggested that change may increase the risk of bullying because managers tend to adopt more autocratic practices to bring about change. Similarly, downsizing has been argued to lead to harsher internal competition, thereby increasing the risk that employees and managers may rely on any means deemed necessary to "get rid of competitors" (Salin, 2003b).

Nevertheless, recent studies have provided a more detailed and nuanced picture of the relationship between organisational change and bullying. In this respect, Skogstad et al. (2007b) found a significant association, in their Norwegian representative sample, between exposure to bullying and organisational changes, albeit a low to moderate relationship. Work environment changes and reduction in staff and pay significantly predicted both task- and person-related bullying, but with a stronger relationship found for the former. The same study also examined whether this relationship was mediated by interpersonal conflicts. The results indicated that there was a direct relationship rather than a mediated one and that organisational changes and interpersonal conflicts are separate, but mainly independent, precursors of bullying at work. Furthermore, interpersonal conflicts with one's immediate superior were by far the strongest predictor of bullying in that study.

Similarly, in a Belgian study, Baillien and De Witte (2009) examined the mediating effect of several different stressors on bullying. They found that the relationship between organisational change and bullying was fully mediated by role conflict and job insecurity. Similar to Skogstad et al.'s Norwegian study (2007b), it revealed only a modest correlation between bullying and organisational change, questioning the assumption of organisational change as one of the most important triggers of bullying. They further concluded that organisational change seems to increase the risk of bullying only when it actually results in a poorer or worse work environment (e.g., role conflict and insecurity) for the individual.

It is worth noting that several survey studies (Hoel and Cooper, 2000; Skogstad et al., 2007b) have questioned the relationship between restructuring and bullying, a relationship that has been proposed by some authors (McCarthy et al., 1995; Sheehan, 1996). Also, when specifically investigating the relationship between layoffs (redundancies) and aggression, Barling, Dupres, and Kelloway (2009), with reference to perceptions of justice, argue that it is not the layoffs per se that represent the problem in terms of workplace aggression but the way in which the layoffs are carried out.

Organisational change is often the result of larger societal change and cannot always be clearly separated from such. Any discussion of organisational changes would, therefore, be incomplete without acknowledging how these are affected by larger trends in society. Economic globalisation has increased competition, and in order to survive in the current economic environment, organisations are restructuring and downsizing with the aim of cutting costs, with greater pressures on everyone in work as a result (Cooper, 1999). It has also been argued that by introducing market philosophies into areas previously unaffected by such pressures, for example, within the health and the education sectors, the relationship between managers and staff has changed, with work intensification and increased managerial discretion and control as a result (Beale, this volume; Ironside and Seifert, 2003; Lee, 2000).

Limitations: A Critical Review

During the past few years, our knowledge of workplace bullying has increased and the methods for studying have become increasingly sophisticated. Despite this advance, there are several limitations that need to be kept in mind when interpreting the results. Hauge et al. (2007) highlight that only a handful of studies have analysed more than a few predictors of bullying associated with the work environment and that most use correlational analysis rather than a multivariate regression that would allow for exploring the relative strength of different job stressors. They also point out that most studies have used additive models, rather than acknowledging interaction effects.

Moreover, with a few exceptions, the findings discussed earlier are the results of cross-sectional studies that do not allow for robust conclusions with regard to causality (Zapf, Dormann, and Frese, 1996a). Thus, whilst a poor working environment may directly or indirectly give rise to bullying, alternative interpretations may be suggested. For example, anxious or depressed individuals may create tension and elicit negative reactions from colleagues and managers alike (Einarsen et al., 1994). Moreover, bullying may itself have a negative effect on the work environment, for example by negatively affecting internal communication, thereby giving rise to more stress and further organisational problems (Zapf, 1999). To what extent bullying is the result of a poor work environment, or whether victims only perceive their own environment as more negative in general, also remains open to debate, as discussed earlier.

Despite calls for more longitudinal studies, few studies have applied such designs. Similarly, there is a paucity of multilevel studies. Thus whilst the work environment hypothesis operates, strictly speaking, on the group level, in terms of risk factors of bullying, our current knowledge is primarily drawn from studies focusing on the individual level. Most studies have also used quantitative methods, although a few recent studies have applied qualitative research methods, possibly introducing alternative perspectives of the role of organisational antecedents in the bullying process (e.g., Baillien et al., 2008; Hoel et al., 2007; Strandmark and Hallberg, 2007).

Studies of bullying have often varied in their operationalisation of bullying, some relying on self-labelling, others on an operational classification method (e.g., NAQ–R) (Einarsen, Hoel, and Notelaers, 2009). This disparity, too, may give rise to contradictory results. As shown by Hauge et al. (2007), who used both methods simultaneously, the relationship between work environment stressors and bullying was stronger when measuring exposure to specific negative acts than when measuring it in terms of self-labelling. This result seems to indicate that whilst a poor work environment is a good predictor of negative acts, the actual labelling process is affected by many other factors, as well. Moreover, they also studied the predictors of work-related and person-related bullying separately. Whilst their overall result showed that the same stressors were relevant for both forms of bullying, the magnitude and relative strengths of the predictors varied for each type.

When discussing risk factors of bullying, it is worth noting that the impact of specific antecedents is likely to vary between occupational settings. This was shown, for example, in a study by Einarsen et al. (1994), where 14 different subgroups of employees and employers were studied. Different antecedents of bullying of a situational or organisational nature are also unlikely to affect different demographic groups uniformly (Hoel, Cooper, and Faragher, 2001). In the same way as organisational antecedents may combine and interact with particular characteristics of the persons involved, the effects and influence of individual antecedents are likely to vary across organisational contexts and between demographic groups. In this context, one would expect that different antecedents may take on different meanings in different settings and that factors such as gender, age, and ethnicity would impinge upon such meanings. Interestingly enough, existing research also seems to indicate that women are more likely to emphasise the role of organisational factors, whereas men emphasise victim characteristics (Salin, 2003c; Salin, in press). However, whether this is only a matter of perception or just reflects different risk factors among men and women remains open.

Similar findings have been reported in aggression research, where both individual and situational factors were predicting workplace aggression (Hershcovic et al., 2007). Such a finding can be seen as giving support to ideas about an interactionist approach to workplace aggression and bullying; thus, situational factors are seen as necessary but not sufficient for aggression to occur. To uncover such connections and the relative impact of various antecedents in different contexts would require a greater emphasis on more qualitative research methods, as well as the development and application of more sophisticated research instruments. In response, Inness, Barling, and Turner (2005) carried out a study to explore the relative impact of individual and situational factors in supervisor-targeted aggression. For that purpose, they randomly enlisted Canadians who held down two separate jobs, one official and one unofficial (the latter, so-called moonlighting). The authors concluded that "employees' aggressive behaviour is contingent on the quality of their experiences in that particular workplace, rather than experiences outside of that workplace" (p. 736), thus giving support to the importance of organisational causes in explaining bullying and aggression.

Conclusion

In this chapter we have explored a large number of factors at the level of the organisation, factors that may, on their own or in combination, give rise to bullying behaviour and the escalation of bullying processes. Whilst these antecedents have largely been discussed one by one, their relationship and interconnectivity have been emphasised. However, in real-life scenarios, the interaction between features of the organisation and the larger environment within which it is located is likely to be far more complex than research has so far uncovered. One may also anticipate that where a number of antecedents may be present at the same time, synergetic effects may occur, increasing the risk of bullying scenarios emerging. However, in order to reveal patterns of interaction, more sophisticated research methods need to be developed and applied, among them multilevel and longitudinal designs.

References

Anderson, C. A., Anderson, K. B., and Deuser, W. E. (1996) Examining an affective aggression framework: Weapon and temperature effects on aggressive thoughts, affects and attitudes. *Personality and Social Psychology Bulletin, 22*, 366–376.

Appelberg, K., Romanov, K., Honkasalo, M-L., and Koskenvuo, M. (1991) Interpersonal conflicts at work and psychosocial characteristics of employees. *Social Science Medicine, 32*, 1051–1056.

Archer, D. (1999) Exploring "bullying" culture in the para-military organisation. *International Journal of Manpower, 20* (1/2), 94–105.

Ashforth, B. (1994) Petty tyranny in organizations. *Human Relations, 47* (7), 755–778.

Baillien, E., and De Witte, H. (2009) Why is organizational change related to workplace bullying? Role conflict and job insecurity as mediators. *Economic and Industrial Democracy, 30* (3), 348–371.

Baillien, E., Neyens, I., and De Witte, H. (2008) Organizational, team related and job related risk factors for bullying, violence and sexual harassment in the workplace: A qualitative study. *International Journal of Organisational Behaviour, 13* (2), 132–146.

Baillien, E., Neyens, I., De Witte, H., and De Cuyper, N. (2009) A qualitative study on the development of workplace bullying: Towards a three way model. *Journal of Community & Applied Social Psychology, 19*, 1–16.

Barling, J. Duprè, K. E., and Kelloway, E. K. (2009) Predicting workplace aggression and violence. *Annual Review of Psychology, 60*, 671–692.

Baron, R. A., and Neuman, J. H. (1996) Workplace violence and workplace aggression: Evidence on their relative frequency and potential causes. *Aggressive Behavior*, 22, 161–173.

Beale, D., and Hoel, H. (in press) Workplace bullying, industrial relations and the challenge for management: Britain and Sweden compared. *European Journal of Industrial Relations*. doi: 10.1177/0959680110364826.

Björkqvist, K., Österman, K., and Hjelt-Bäck, M. (1994) Aggression among university employees. *Aggressive Behavior*, 20, 173–184.

Bloisi, W., and Hoel, H. (2008) The expectation of abusive work practices and bullying among chefs: A review of the literature. *International Journal of Hospitality Management*, 27, 649–656.

Bowling, N., and Beehr, T. (2006) Workplace harassment from the victim's perspective: A theoretical model and meta-analysis. *Journal of Applied Psychology*, 91 (5), 998–1012.

Brodsky, C. M. (1976) *The harassed worker*. Lexington, MA: D. C. Heath.

Collinson, D. L. (1988) "Engineering humour": Masculinity, joking and conflict in shop-floor relations. *Organization Studies*, 9 (2), 181–199.

Cooper, C. L. (1999) The changing psychological contract at work. *European Business Journal*, 11, 115–118.

Einarsen, S. (1996) Bullying and harassment at work: Epidemiological and psychosocial aspects. Doctoral dissertation, Faculty of Psychology, Department of Psychosocial Science, University of Bergen, Norway.

——— (1999) The nature and causes of bullying at work. *International Journal of Manpower*, 20 (1/2), 16–27.

Einarsen, S., Aasland, M., and Skogstad, A. (2007) Destructive leadership behaviour: Definition and conceptual model. *Leadership Quarterly*, 18, 207–216.

Einarsen, S., Hoel, H., and Notelaers, G. (2009). Measuring exposure to bullying and harassment at work: Validity, factor structure and psychometric properties of the Negative Acts Questionnaire–Revised. *Work & Stress*, 23 (1), 24–44.

Einarsen, S., and Raknes, B. I. (1997) Harassment in the workplace and the victimization of men. *Violence and Victims*, 12, 247–263.

Einarsen, S., Raknes, B. I., and Matthiesen, S. B. (1994) Bullying and harassment at work and their relationships to work environment quality: An exploratory study. *European Work and Organizational Psychologist*, 4, 381–401.

Einarsen, S., and Skogstad, A. (1996) Bullying at work: Epidemiological findings in public and private organizations. *European Journal of Work and Organizational Psychology*, 5 (2), 185–201.

Ferris, G., Zinko, R., Brouer, R., Buckley, R., and Harvey, M. (2007) Strategic bullying as a supplementary, balanced perspective on destructive leadership. *Leadership Quarterly*, 18 (3), 195–206.

Hauge, L. J., Skogstad, A., and Einarsen, S. (2007) Relationships between stressful work environments and bullying: Results of a large representative study. *Work & Stress*, 21 (3), 220–242.

Hershcovic, S., Turner, N., Barling, J., Arnold, K., Dupré, K., Inness, M., LeBlanc, M., and Sivanathan, N. (2007) Predicting workplace aggression: A meta-analysis. *Journal of Applied Psychology*, 92 (1), 228–238.

Hodson, R., Reocigno, V. J., and Lopez, S. H. (2006) Chaos and the abuse of power: Workplace bullying in organizational and interactional context. *Work and Occupations*, 33, 382–416.

Hoel, H., and Cooper, C. L. (2000) *Destructive conflict and bullying at work*. Manchester, UK: Manchester School of Management, University of Manchester Institute of Science and Technology.

Hoel, H., Cooper, C. L., and Faragher, B. (2001) The experience of bullying in Great Britain: The impact of organizational status. *European Journal of Work and Organizational Psychology, 10* (4), 443–465.

Hoel, H., Giga, S. I., and Davidson, M. J. (2007) Expectations and realities of student nurses' experiences of negative behaviour and bullying in clinical placement and the influences of socialisation processes. *Health Services Management Research, 20,* 270–278.

Hoel, H., Glasø, L., Hetland, J., Cooper, C. L., and Einarsen. S. (in press) Leadership as predictor of self-reported and observed workplace bullying. *British Journal of Management.* doi: 10.1111/j.1467-8551.2009.00664.x.

Hoel, H., and Salin, D. (2003) Organisational antecedents of workplace bullying. In S. Einarsen, H. Hoel, D. Zapf, and C. L. Cooper (eds.), *Bullying and emotional abuse in the workplace: International perspectives in research and practice* (pp. 203–218). London: Taylor & Francis.

Inness, M., Barling, J., and Turner, N. (2005) Understanding supervisor-targeted aggression: A within person, between jobs design. *Journal of Applied Psychology, 90,* 731–739.

International Herald Tribune (2004) The workplace: Bully them to make them leave, September 8, 11.

Ironside, M., and Seifert, R. (2003) Tackling bullying in the workplace: The collective dimension. In S. Einarsen, H. Hoel, D. Zapf, and C. L. Cooper (eds.), *Bullying and emotional abuse in the workplace: International perspectives in research and practice* (pp. 383–398). London: Taylor & Francis.

Johns, N., and Menzel, P. J. (1999) "If you can't stand the heat!" . . . kitchen violence and culinary art. *Hospitality Management, 18,* 99–109.

Karasek, R. A. (1979) Job demands, job decision latitude and mental strain: Implications for job redesign. *Administrative Science Quarterly, 24,* 285–308.

Lee, D. (2000) An analysis of workplace bullying in the UK. *Personnel Review, 29* (5), 593–612.

Leymann, H. (1996) The content and development of mobbing at work. *European Journal of Work and Organizational Psychology, 5* (2), 165–184.

Mathisen, G. E., Einarsen, S., and Mykletun, R. (2008) The occurrences and correlates of bullying and harassment in the restaurant sector. *Scandinavian Journal of Psychology, 49,* 59–68.

McCarthy, P., Sheehan, M., and Kearns, D. (1995) Managerial styles and their effect on employees health and well-being in organizations undergoing restructuring. Report for Worksafe Australia, Griffith University, Brisbane.

Moreno-Jiménez, B., Rodríguez-Muñoz, A., Pastor, J. C., Sanz-Vergel, A. I., and Garrosa, E. (2009) The moderating effects of psychological detachment and thoughts of revenge in workplace bullying. *Personality and Individual Differences, 46,* 359–364.

Neyens, I., Baillien, E., De Witte, H., and Notelaers, G. (2007) Kwantitatieve studie van taak-, team- en organisatorische risicofactoren voor pesten op het werk [A quantitative study of task- team- and organizational risk factors of bullying at work]. *Tijdschrift voor Arbeidsvraagstukken, 23* (4), 306–320.

O'Moore, M., and Lynch, J. (2007) Leadership, working environment and workplace bullying. *International Journal of Organization Theory and Behavior, 10,* 95–117.

O'Moore, M., Seigne, E., McGuire, L., and Smith, M. (1998) Victims of bullying at work in Ireland. *Journal of Occupational Health and Safety–Australia and New Zealand, 14* (6), 569–574.

Podsakoff, P. M., Todor, W. D., and Skov R. (1982) Effects of leader performance contingent and non-contingent reward and punishment behaviors on subordinate performance and satisfaction. *Academy of Management Journal, 25,* 812–821.

Randle, J. (2003) Bullying in the nursing profession. *Journal of Advanced Nursing, 43,* 395–401.

Rayner, C. (1997) The incidence of workplace bullying. *Journal of Community and Applied Social Psychology, 7* (3), 199–208.

Rayner, C., Hoel, H., and Cooper, C. L. (2002) *Workplace bullying: What we know, who is to blame, and what can we do?* London: Taylor & Francis.

Salin, D. (2001) Prevalence and forms of bullying among business professionals: A comparison of two different strategies for measuring bullying. *European Journal of Work and Organizational Psychology, 10* (4), 425–441.

Salin, D. (2003a) Ways of explaining workplace bullying: A review of enabling, motivating and precipitating structures and processes in the work environment. *Human Relations, 56* (10), 1213–1232.

Salin, D. (2003b) Bullying and organisational politics in competitive and rapidly changing work environments. *International Journal of Management and Decision-Making, 4* (1), 35–46.

Salin, D. (2003c) The significance of gender in the prevalence, forms, and perceptions of workplace bullying. *Nordiske Organisasjonsstudier, 5* (3), 30–50.

Salin, D. (2009) Organisational responses to workplace harassment: An exploratory study. *Personnel Review, 38* (1), 26–44.

Salin, D. (in press) The significance of gender for third parties' perceptions of negative interpersonal behaviour: Labelling and explaining negative acts. *Gender, Work, and Organization.* doi:10.1111/j.1468-0432.2009.00465.x.

Sheehan, M. (1996) Case studies in organisational restructuring. In P. McCarthy, M. Sheehan, and W. Wilkie (eds.), *Bullying: From backyard to boardroom.* Alexandria, NSW, Australia: Millenium Books.

Sjøtveit, J. (1992) *Når veven rakner: Om samhold og mobbning på arbeidsplassen* [When the social fabric disintegrates: About solidarity and mobbing at work]. Oslo: Folkets Brevskole.

Skogstad, A., Einarsen, S., Torsheim, T., Aasland, M., and Hetland, H. (2007a) The destructiveness of laissez-faire leadership behaviour. *Journal of Occupational Health Psychology, 12* (1), 80–92.

Skogstad, A., Matthiesen, S. B., and Einarsen, S. (2007b) Organizational changes: A precursor of bullying at work? *International Journal of Organizational Theory and Behavior, 10* (1), 58–94.

Strandmark, M., and Hallberg, L. (2007) The origin of workplace bullying: Experiences from the perspective of bully victims in the public service sector. *Journal of Nursing Management, 15* (3), 332–341.

Sutela, H., and Lehto, A. M. (1998) Henkinen väkivalta on koko työyhteisön ongelma. [Psychological violence is a problem of the whole work community]. *Hyvinvointikatsaus, 3,* 18–24.

Thylefors, I. (1987) *Syndabockar: Om utstötning och mobbning i arbetslivet* [Scapegoats: About expulsion and bullying in working life]. Stockholm: Natur och Kultur.

Tuckman, B. W. (1965) Developmental sequence in small groups. *Psychological Bulletin, 63,* 384–399.

UNISON (1997) *UNISON members' experience of bullying at work.* London: UNISON.

Vartia, M. (1996) The sources of bullying: Psychological work environment and organizational climate. *European Journal of Work and Organizational Psychology, 5* (2), 203–214.

Zapf, D. (1999) Organisational, work group related and personal causes of mobbing/ bullying at work. *International Journal of Manpower, 20* (1/2), 70–85.

Zapf, D., Dormann, C., and Frese, M. (1996a) Longitudinal studies in organisational stress research: A review of the literature with reference to methodological issues. *Journal of Occupational Health Psychology, 1,* 145–169.

Zapf, D., and Einarsen, S. (2005) Mobbing at work: Escalated conflict in organizations. In S. Fox and P. E. Spector (eds.), *Counterproductive work behaviour: Investigations of actors and targets* (pp. 237–270). Washington, DC: American Psychological Association.

Zapf, D., Einarsen, S., Hoel, H., and Vartia, M. (2003) Empirical findings on bullying in the workplace. In S. Einarsen, H. Hoel, D. Zapf, and C. L. Cooper (eds.), *Bullying and emotional abuse in the workplace: International perspectives in research and practice* (pp. 103–126). London: Taylor & Francis.

Zapf, D., Knorz, C., and Kulla, M. (1996b) On the relationship between mobbing factors, and job content, social work environment, and health outcomes. *European Journal of Work and Organizational Psychology, 5* (2), 215–237.

Zapf, D., and Warth, K. (1997) Mobbing: Subtile Kriegsführung am Arbeitsplatz. *Psychologie Heute,* August, 20–25, 28–29.

10

Sexual Harassment Research in the United States

Kimberly T. Schneider, John B. Pryor, and Louise F. Fitzgerald

CONTENTS

What Is Sexual Harassment?

In the United States, the concept of workplace sexual harassment is considered a form of gender-based discrimination under federal law. Sexual harassment developed first as a legal concept and subsequently as a concept empirically studied by social scientists. In 1980, the U.S. Equal Employment Opportunity Commission (EEOC) issued guidelines that have become the cornerstone of legal and policy definitions of sexual harassment throughout the United States. These guidelines describe two general types of sexual harassment: (1) unwelcome sex- or gender-related behavior that creates a *hostile environment* and (2) *quid pro quo* behaviors, where the unwelcome behavior becomes a term or condition of employment or advancement. The EEOC Guidelines emphasize the importance of examining contextual factors in each case to determine whether sexual harassment has taken place. Many subsequent legal cases have since refined the legal understanding of sexual harassment and have specified criteria indicating that the unwelcome sexual or gender-related behavior must be sufficiently severe or pervasive so as to alter the work environment. While sexual harassment often represents a pattern of behavior in the workplace, it is possible for single episodes to cross a threshold of severity to be considered sexual harassment in a legal sense—particularly instances of *quid pro quo* sexual harassment.

From a legal standpoint, intentions for sexually harassing behavior are superfluous. The U.S. legal understanding of sexual harassment focuses upon potential employment consequences for victims. Sexual harassment is illegal because it is thought to represent a barrier for equal employment. Whether the perpetrator intended the behavior to be offensive or not is irrelevant. That the behavior occurred and was unwelcome are the main considerations, although there are individual, gender, and contextual differences in what is considered unwelcome behavior (*Ellison v. Brady*, 1991). While psychological damage may sometimes result from sexual harassment, U.S. courts do not require plaintiffs to prove psychological damage to establish that they have been sexually harassed. Plaintiffs are only required to show that the behavior would be considered offensive or abusive by a reasonable person who shares the perspective of the victim (*Harris v. Forklift Systems, Inc.*, 1993). A hostile work environment created by sexual or gender-related behavior can be considered sexual harassment even for coworkers who are not directly targeted (*Broderick v. Ruder*, 1990). Same-sex and other-sex harassment are both possible under federal law (*Oncale v. Sundowner Offshore Services, Inc.*, 1998).

Two important cases appearing in U.S. federal courts in the late 1990s refined the legal understanding of employer liability in sexual harassment cases (*Burlington Industries v. Ellerth*, 1998; and *Faragher v. City of Boca Raton*, 1998). Employers are always legally liable for supervisor harassment that involves tangible employment actions (e.g, firing or demotion), whereas employers may or may not be liable for supervisor and coworker harassment that did not lead to tangible employment action (hostile environment harassment). An employer's liability depends upon whether the employer judged to have taken reasonable care to correct or prevent such harassment (e.g., offer accessible channels for filing complaints) and whether the employee disregarded preventive or corrective strategies put in place by the employer (see EEOC, 1999).

Perhaps the key legal issue that distinguishes sexual harassment from bullying is that plaintiffs in sexual harassment lawsuits in the United States must prove that the harassment is based upon gender. In practice, this is difficult to prove because actual differential comparisons of how men and women are treated are often lacking in sexual harassment cases. Attorneys often appeal to people's general beliefs about how men and women are typically treated to try to argue that a behavior was gender based.

Sexual Harassment from a Social Scientific Perspective

The developing legal understanding of sexual harassment has influenced not only how sexual harassment has been studied but also what constructs are studied, with social scientists being influenced by the legal literature and vice versa. The legal understanding of a "reasonable person standard" has been informed by research in the social sciences on gender differences in the interpretation of potentially sexually harassing behaviors. As in studies of bullying, much of the research concerning sexual harassment is based upon victim surveys. Some of the earliest systematic surveys of sexual harassment were conducted by the U.S. Merit Systems Protection Board (USMSPB, 1981, 1988, 1995). Scientific samples of federal workers were asked to indicate the frequency with which they had experienced several kinds of "unwanted sexual attention" on the job in the last 24 months. Remarkably, these three MSPB surveys found very similar incidence rates. A breakdown of the percentages of women and men reporting each of the specific behaviors is presented in Table 10.1.

One finding that was replicated in all MSPB surveys concerns the role of organizational power. Contrary to popular stereotypes about sexual harassment involving supervisors targeting underlings, a large majority of the respondents reported that the perpetrators were coworkers or other employees with no supervisory authority. Notably, the behaviors most frequently described as the most significant incidents were "unwanted sexual teasing,

TABLE 10.1

Percentages of Women and Men Reporting Sexual Harassment Experiences

Women	1980	1987	1994
Sexual teasing, jokes, remarks	33	35	37
Sexual looks, gestures	28	28	29
Deliberate touching, cornering	26	26	24
Pressure for dates	15	15	13
Suggestive letters, calls, materials	9	12	10
Pressure for sexual favours	9	9	7
Stalking	NA	NA	7
Actual/attempted rape, assault	1	0.8	4
Any type	42	42	44
Men	**1980**	**1987**	**1994**
Sexual teasing, jokes, remarks	10	12	14
Sexual looks, gestures	8	9	9
Deliberate touching, cornering	7	8	8
Pressure for dates	3	4	4
Suggestive letters, calls, materials	3	4	4
Pressure for sexual favours	2	3	2
Stalking	NA	NA	2
Actual/attempted rape, assault	0.3	0.3	2
Any type	15	14	19

Source: From the 1981, 1988, and 1995 U.S. Merit Systems Protection Board Surveys of Federal workers. With permission.

jokes, remarks, or questions" and "unwanted sexually suggestive looks or gestures."

Another early survey of sexual harassment was conducted by Gutek (1985) using Los Angeles County employees. Gutek asked respondents whether they had ever experienced (1) sexual comments from a man (or a woman if the respondent was male) that was meant to be insulting (reported by 12.2% of women and 12.6% of men); (2) sexual looks or gestures from a man that were meant to be insulting (reported by 9.1% of women and 12.3% of men); (3) sexual touching from a man (reported by 15.3% of women and 20.9% of men); (4) requests to go out with a man as part of a job (reported by 2.8% of women and 2.7% of men); and (5) requests to engage in sexual relations with a man as part of a job (reported by 2.7% of women and 1% of men).

Obviously, the MSPB surveys found more sexually harassing behaviors reported than did Gutek's survey (1985). Such differences could have been the result of differences in the samples or wording differences in the survey questions. For example, the MSPB survey asked about "unwanted sexually suggestive looks or gestures," whereas the Gutek survey asked about "sexual looks or gestures from a man that he meant to be insulting," the latter specifying an intentional motive. Another potentially crucial difference is that Gutek's survey explicitly asked about opposite-sex sexual harassment, whereas the MSPB surveys allowed reports of behaviors perpetrated by either sex. While the MSPB surveys consistently found more women reporting unwanted sexual behaviors at work than men, Gutek (1985) found slightly more men than women reporting three out of five of the behaviors in her survey. There are many other points of methodological divergence between these two surveys that make them difficult to compare, illustrating a general problem in the sexual harassment literature: the absence of a common method of measurement. Fortunately, some progress has been made on this issue.

One problem common to the MSPB surveys and the Gutek (1985) survey is the use of discrete behavioral questions without reference to underlying constructs. This approach makes it difficult to pose fundamental psychometric questions related to reliability and validity. There have been several attempts to develop more psychometrically sound instruments for measuring self-reported sexual harassment incidents (Dekker and Barling, 1998; Fitzgerald et al., 1995a). Perhaps the most extensive research on instrument development concerns the Sexual Experiences Questionnaire (SEQ). The SEQ was originally developed to assess sexual harassment experiences in an academic context (Fitzgerald et al., 1988), but it has subsequently been adapted for use in business settings (Fitzgerald et al., 1997), military settings (Fitzgerald et al., 1999; Stark et al., 2002), and cross-cultural settings (Gelfand et al., 1995; Wasti et al., 2000).

Early versions of the SEQ assess victims' experiences of three forms of sexually harassing behavior: sexual coercion, unwanted sexual attention, and gender harassment. Sexual coercion includes the use of threats or bribes to solicit sexual involvement. Unwanted sexual attention includes repeated requests for dates and persistent attempts to establish unwanted sexual relationships. With regard to gender harassment, behaviors are focused on verbal and nonverbal behavior that conveys insulting and hostile attitudes toward women (Gelfand et al., 1995). In the SEQ, each of these three types of sexually harassing behaviors is measured by multiple behavioral items on a frequency scale referencing the past 24 months. Another important methodological aspect of the SEQ is that the term *sexual harassment* is not mentioned in any of the behavioral questions but appears only at the survey's end to avoid potentially biased responses resulting from variations in respondents' definitions of sexual harassment.

Research has shown consistently in many organizational and cultural settings that gender harassment is the most common form of sexual harassment, followed by unwanted sexual attention, and then by sexual coercion. In comparing this to a legal analysis of sexual harassment, sexual coercion often constitutes *quid pro quo* sexual harassment, whereas unwanted sexual attention and gender harassment constitute what is termed hostile environment sexual harassment. Sometimes unwanted sexual attention may be argued to constitute a form of *quid pro quo* sexual harassment if toleration of such behavior becomes a term or condition of employment (Fitzgerald et al., 1995a). Recent revisions of the SEQ also distinguish two subtypes of gender harassment: sexist hostility (e.g., insulting behavior based on gender) and sexual hostility (insulting and explicit sexual behavior; Fitzgerald et al., 1999; Stark et al., 2002).

Research has found that most targets of sexually harassing behavior are women and most perpetrators are men. A recent meta-analysis indicates that 58% of women will likely encounter sexually harassing behavior in the workplace during their careers (Ilies et al., 2003), whereas incidence rates for men are generally much lower. Research by Waldo and his colleagues (Waldo et al., 1998) examined a revision of the SEQ that is more gender neutral in content and applicable to men and included additional items that focus on harassment regarding the enforcement of traditional gender roles, a violation that Stockdale et al. (1999) found resulted in frequent male-male harassment. One conclusion drawn from research using this revised SEQ (i.e., the Sexual Harassment of Men scale; Waldo et al., 1998) is that men may be exposed to behavior that is potentially harassing far more than previously estimated (however, they may also be more likely than women to find it flattering; Berdahl and Aquino, 2009). Another conclusion is that men are much more likely to be the targets of same-sex harassment than women.

Much of the previous research on sexual harassment has used a behavioral item-counting method in establishing whether people have experienced sexual harassment. This approach relies heavily on researchers' conceptualizations of what behaviors constitute harassment (e.g., Fitzgerald et al., 1988; Martindale, 1992; USMSPB, 1981, 1988, 1995). One problem with this method of determining percentages of harassment experiences is that some behaviors might be relatively innocuous—for example, the occasional sexual joke. One might speculate that a solution to this conundrum would simply be to ask people whether they considered the behaviors they experienced to be sexual harassment (Ilies et al., 2003). However, many researchers have argued that this self-labeling approach is inherently unreliable because of the fuzzy understanding that lay people have of sexual harassment. Also, there is empirical evidence that people who do not label their experiences as sexual harassment nevertheless experience psychological distress, negative job attitudes, and other consequences at rates comparable to those who do label them as sexual harassment (Magley et al., 1999). An alternative approach couples an assessment of the occurrence of potentially harassing behaviors with respondents' appraisals of the behaviors they experienced (Fitzgerald et al., 1999). This approach emphasizes the similarities between harassment experiences and other forms of stress (Lazarus, 1993). Palmieri et al. (2000) suggest that the two-step approach of asking (1) Did a respondent experience any SEQ behaviors? coupled with (2) Were these behaviors negatively appraised (i.e., were respondents upset)? is useful in discerning whether a survey respondent actually experienced sexual harassment. Berdahl (2007) also emphasized the importance of including an evaluative assessment of potentially harassing behaviors along with frequency counts. In their meta-analysis, Ilies et al. (2003) found that alternate surveying methods do affect estimates of incidence rates. Asking participants whether experiences constitute sexual harassment resulted in a 24% incidence rate, whereas asking whether respondents appraised the behaviors they experienced negatively resulted in a 58% incidence rate. Thus, the two-step approach, by counting only experiences that negatively affected respondents, seems to hold the promise of converging social scientific and legal understandings.

Social Antecedents of Sexual Harassment in the Workplace

Research reveals several "risk factors" associated with the occurrence of sexual harassment. One set of risk factors involves the nature of the job, specifically the "gender context," or the numbers of men and women who interact at work. In addition, traditionally masculine jobs or jobs where women are "gender pioneers" (i.e., among the first women in that job) pose a higher risk for sexual harassment for women (Neibuhr, 1997; Ragins and Scanduri, 1995). In an examination of surveys from five organizations, Berdahl (2007a) found that women working in male-dominated organizations experienced more frequent harassment than those working in female-dominated organizations. More specifically, women with relatively masculine personalities working in contexts dominated by men experienced the highest levels of harassment, suggesting an interaction between context and the target's personality.

Another set of risk factors centers around social norms or organizational climate factors (Gruber, 1998). Bowling and Beehr's recent meta-analysis (2006) indicated that workplace harassment levels are predicted by stressful work environments. They propose that the ambient stress in such environments may encourage potential harassers to enact their frustration by targeting others. Some organizations, through their stated policies, enforced sanctions, and effective leadership, actively and successfully discourage sexual harassment. Other organizations, through their vague policies, indifference, or even implicit acceptance of harassment, may condone sexual harassment. Men are more likely to harass women when management is perceived as tolerating or condoning such behavior. A study conducted at a U.S. federal agency (Pryor et al., 1993; Pryor et al., 1995) found that women were more likely to experience sexual harassment in offices where the men believed that local management discouraged complaints. In studies of military personnel also reported by Pryor et al. (1995), the degree to which women on a base reported experiences of sexual harassment was related to the degree to which the men perceived the local commander as tolerating or condoning such behavior. In a U.S. Department of Defense (DoD)–wide survey (Williams et al., 1999), perceived military efforts related to the implementation of sexual harassment policies that could be viewed as climate indicators (e.g., thoroughly investigating complaints and enforcing penalties) were more strongly related to the incidence of harassment than training efforts or providing resources to victims. Fitzgerald et al. (1997) found that individual perceptions of organizational tolerance (i.e., degree of risk for reporting harassment, likelihood complaints would be taken seriously, and likelihood the harasser would be punished) were significantly related to harassment incidence. Again, these social climate factors seem to emanate from managerial stances on sexual harassment, implying that leadership within an organization can influence the incidence of sexual harassment. Tolerant climates may send the message to perpetrators that it is acceptable to harass less powerful people, a connection supported by evidence linking climate perceptions and the sexual harassment experiences of gender and ethnic minorities (Bergman and Henning, 2008).

Other recent research expands the focus of harassment to work teams and indicates that the ambient level of sexual harassment in a work group is related to higher levels of team conflict (Raver and Gelfand, 2005). Raver and Gelfand even found an indirect link between ambient sexual hostility and team financial performance, a relationship mediated by team conflict and cohesion. Hitlan et al. (2006) focused on how the experience of being a bystander to coworkers' harassment may influence women's attributions regarding their own harassment experiences. In their sample of predominantly Hispanic working students, those students who were aware of more frequent gender harassment of their coworkers were less upset by their own gender harassment experiences. Hitlan et al. (2006) suggested that rampant gender harassment in one's work group may allow women to more easily make external attributions regarding the harassment. Similar results were reported for direct and bystander experiences of ethnic harassment (Low et al., 2007).

Gruber (1998) and others have demonstrated that organizational policies can play an important role in preventing sexual harassment. Fairly soon after the U.S. Equal Employment Opportunity Commission issued guidelines on sexual harassment (EEOC, 1980), many organizations and universities adopted sexual harassment policies (Kelley, 2000; National Council for Research on Women, 1995). Research on the introduction of sexual harassment policies in institutions of higher education found that while official complaints about sexual harassment typically increased following a policy introduction (Robertson et al., 1988), prevalence surveys showed that reporting rates declined (Williams et al., 1992). It is not surprising that complaints would rise following the introduction of a sexual harassment policy, since most policies outline the procedures for filing complaints. As Williams et al. (1999) found, even when a policy purports zero tolerance of sexual harassment, as in the U.S. armed forces, the failure of local management to implement the policy can create a climate tolerant of sexual harassment. From a social norm standpoint, policies and their implementation are inputs to an overall organizational climate and communicate information about desired social norms for employee behavior. When management fails to enforce policies against sexual harassment, this communicates that the "real" norms are different from the "stated" norms.

Research on social norms indicates that men who want to harass sexually are more likely to do so in social settings where they feel they can get away with it (Dekker and Barling, 1998; Pryor and Meyers, 2000). A recent study by Bacharach et al. (2007) focused on local norms related to workplace drinking. Although Gutek (1985) identified drinking behavior as a potential risk factor for sexual harassment, Bacharach et al.'s study of union members was the first to find empirical evidence of the importance of this specific norm in predicting gender harassment incidents. Heavy workplace drinking during breaks was positively associated with the level of gender harassment reported by female coworkers, and the relationship was even stronger in groups with more heavily embedded and permissive norms related to drinking.

In summary, when we observe sexual harassment that has taken place repeatedly over a long period of time, it is reasonable to assume that a climate of tolerance has existed or the organization has shown indifference. When victims complain and nothing is done, the harasser is encouraged to continue such behavior. Finally, there is also good evidence that not all men are would-be harassers. Individual proclivities for sexual harassment are more likely to translate into sexually harassing behaviors when those individuals with such proclivities find themselves in a climate of organizational tolerance (Dekker and Barling, 1998; Pryor and Meyers, 2000).

Characteristics of Harassers and Harassing Behaviors

The lion's share of research and theory on harassers has focused on the most common perpetrators and victims—men who sexually harass women. In examining the psychological characteristics of men who sexually harass women, two things are important to keep in mind. First, the term *sexual harassment* is used to refer to a variety of different types of behavior, so it is unlikely that a single psychological profile would characterize all harassers. Second, sexually harassing behavior has been shown to be more likely to occur in some social settings than others (Dekker and Barling, 1998; Pryor et al., 1995). So even though some men may have stronger proclivities for sexually harassing behaviors than others, they may not act upon these proclivities when the social environment encourages professional behavior. This framework for understanding the antecedents of sexually harassing behavior is called a Person X Situation analysis. Pryor et al. (1993) have suggested that both person factors (relatively stable personality and attitudinal characteristics) and situational factors (organizational climate) may contribute to repeated sexually harassing behavior. Discussed next are the potential proclivities for different forms of sexual harassment.

Gender Harassment

Empirical studies show that gender harassment is the type of sexual harassment that occurs most often and is most likely to occur in isolation of the others (Fitzgerald et al., 1999). Gender harassment is not sexual behavior, per se, in that its aim does not seem to be to gain sexual access to the target. Rather, it may be better construed as sexist behavior or behavior that is intended to put down or offend those targeted. For example, women who are perceived as gender role deviants are often punished through gender harassment (Berdahl, 2007). Pryor and Whalen (1996) theorized that individual differences in sexism may contribute to male propensities for gender harassment. Recent research (Hitlan et al., 2009) has confirmed this prediction. Manipulated situational factors, such as having one's sense of masculinity threatened by being outperformed by a woman on a masculine task and being exposed to a film depicting the sexist treatment of women, also contributed to men's propensities to exhibit sexist verbal behaviors in these studies.

Dekker and Barling (1998) applied a similar analysis of person and situation factors in a study of Canadian university faculty and staff. Male respondents were asked whether they had engaged in gender harassment behaviors with a member of the opposite sex at work in the past three months. Men who admitted to having performed various forms of gender harassment were likely to possess what the authors called *inappropriate sexual harassment beliefs*—for example, the belief that women exaggerate sexual harassment problems (a person factor). Men who admitted to having engaged in gender harassment were also likely to see the organizational climates where they worked as supporting such behavior. Thus, the tendency to admit gender harassment was related to perceptions that there were few company sanctions for such behavior and that the workplace was highly sexualized (situational factors).

Finally, Maass et al. (2003) found that both person and situation factors predicted male participants' choosing to expose women to pornographic computer images—a behavior that might be considered a form of gender harassment. Maass et al. found that some person factors, such as *gender identification* (i.e., the degree to which men's self-concepts were identified with the social category of male), not only produced "main effects" in predicting sexually harassing behavior but also interacted with situational factors, such as *threat legitimacy* (i.e., whether a woman described herself as traditional or nontraditional) such that those men who more strongly self-identified as males were even more likely to harass when the woman was nontraditional.

Unwanted Sexual Attention

Pryor and Whalen (1996) theorized that a common contributor to men's persistence in making unwanted sexual advances towards women might simply be a lack of social sensitivity. Research by Abbey (1982, 1987) and others (see Stockdale, 1993) suggests that men often interpret women's friendly behaviors as conveying more sexual intent than women intend to convey. Ridge and Reber (2002) suggest that men's erroneous initial expectations about sexual intentions may lead them to engage in a dynamic behavioral confirmation process that exacerbates their misperceptions of women's mutual sexual intent.

Dekker and Barling (1998) also found evidence for person and situation factors as potential contributors to unwanted sexual behaviors. Men with weak perspective-taking skills, adversarial sexual beliefs, or inappropriate sexual harassment beliefs were likely to engage in sexualized harassment. In addition, the presence of fewer company sanctions tended to exacerbate men's proclivities to engage in sexualized harassment such that these men were even more likely to engage in sexualized harassment when company sanctions were perceived as few.

Sexual Coercion

Since 1987, a great deal of research has explored male proclivities for such sexually exploitative or coercive behaviors as a willingness to use a social power differential to enlist sexual cooperation (for reviews, see Pryor and Meyers, 2000, or Pryor and Wesselmann, in press). Based on a variety of empirical studies, a psychological profile of men who might be likely to use social power for sexually exploiting women includes the following: (1) proclivities for rape and other forms of sexual aggression, (2) a strong desire to dominate women, (3) psychological justification of male dominance, (4) a view of sex in nonrelationship terms (e.g., to escape boredom, achieve social recognition, or indulge appetites), (5) a traditional view of gender roles, (6) weak perspective-taking skills, and (7) a personality that combines low conscientiousness (especially high impulsiveness) and low openness (narrow-mindedness) (Gutek and Done, 2001; Pryor and Whalen, 1996).

Research has found that whether men with proclivities for sexually exploitative behavior choose to behave in this way in situations where they have power over women seems to be related to situation factors including organizational climate and local social norms (Pryor et al., 1995). For example, research has found that such men are strongly influenced by the behaviors of their peers and people in authority. When such sexually coercive behaviors are modeled or condoned by peers and authority figures, these men are also more likely to enact them.

In summary, there is a building body of research that supports the Person X Situation Model for understanding the occurrence of different types of sexually harassing behavior. Certain personal characteristics of men seem to present risks for sexually harassing behavior. Social situation factors can enable or disable, and encourage or discourage, such behavior. So far, this analysis has been applied only to male-perpetrator and female-target behaviors.

Victim Characteristics

There has been less research on personal characteristics of victims of sexual harassment than research related to harassers' characteristics. Some demographic differences may be operating. Black women, for example, report experiencing more frequent unwanted sexual attention and sexual coercion than do white women (Berdahl and Moore, 2006; Bergman and Drasgow, 2003). Rosen and Martin (1998) found that victims of sexual harassment in the U.S. military were more likely than nonvictims to have a history of childhood sexual abuse. Whether would-be harassers target these people because they appear more vulnerable or whether the victims of previous abuse are more likely to report sexual harassment on surveys is not known. Bowling and Beehr's meta-analysis (2006) indicated that victims' individual differences had a much weaker relationship with harassment experiences than with environment conditions, with the exception of victims' levels of negative affect (NA). However, it is important to note that it is unclear whether NA predisposes one to harassment or is a result of the harassment. Experimental research by Dall'Ara and Maass (1999) has found that male harassers may target women more for gender harassment when the women are perceived to be egalitarian than when they are perceived to have traditional gender role orientations. Men may thus harass women they see as threatening male hegemony.

Consequences of Sexual Harassment in the Workplace

Most research on the consequences of sexual harassment has focused upon female targets of this behavior. Research has found evidence of negative psychosocial reactions relevant to women's personal and professional lives (Gutek and Koss, 1993). Pryor (1995) found that the degree to which harassed women experience negative psychosocial reactions was related to various factors. Harassed women who thought that complaining to management was ineffective were more likely to experience both of these types of negative psychosocial reactions. This characteristic sense of futility or helplessness when one's complaints are unheeded seems to add to the stress harassed women experience. Pryor (1995) also found that the degree to which harassed women experienced negative emotional reactions was related to the duration of the harassment and the organizational status of the harasser. Long-term harassment from someone who has the power to do you harm seems particularly stressful. A recent study of women sexually harassed by customers or clients high in power supports this finding (Gettman and Gelfand, 2007). Because the negative impact of stress is cumulative to some degree, long-term sexual harassment is likely to be particularly harmful (Hitlan et al., 2006).

The experience of sexual harassment at work is considered a form of *work-related stress* (Fitzgerald et al., 1995b). Many forms of sexual harassment, including exposure to sexist put-downs or unwanted sexual remarks, are correlated with subsequent psychological problems for women (e.g., Glomb et al., 1997). Across studies, there appears to be a consistent association between experiencing sexual harassment and symptoms of post-traumatic stress disorder (PTSD), anxiety and depression, and alcohol misuse (Berdahl and Aquino, 2009; Dansky and Kilpatrick, 1997; Rospenda et al., 2000, 2008; Schneider et al., 1997). Research by Schneider et al. (2001) has established a causal connection between experiences of gender harassment and physiological measures of stress. Women who were subjected to sexist remarks from a male coworker showed significant levels of autonomic physiological activity consistent with the pattern of cardiac and vascular reactions that people generally display in threat situations. Such cardiovascular reactivity to stress has been related to coronary heart disease and depressed immune functioning. This finding suggests that women who are continually exposed to sexist behavior may be at a greater risk for long-term health consequences.

Research by Fitzgerald and her colleagues (Fitzgerald et al., 1997) examined the work-related outcomes of sexual harassment for women in the context of an integrated model of the antecedents and consequences of sexual harassment in an organizational setting. A structural equation analysis indicated that sexual harassment was related to work withdrawal (e.g., absenteeism and tardiness) through its impact on job satisfaction and was related to job withdrawal (e.g., turnover) through its impact on health conditions and health satisfaction. Similar findings were reported by Schneider et al. (1997) using different samples and indicated that even relatively mild, but frequent, sexual harassment had a detrimental impact. There are clear negative job-related attitudinal correlates of harassment as well (Berdahl and Aquino, 2009; Lapierre et al., 2005; Schneider et al., 1997). Lapierre et al.'s meta-analysis (2005) estimated a coefficient of −.32 between sexual aggression and job satisfaction. This negative job-related impact appears to exist even among employees who report finding the sexual behavior somewhat enjoyable (Berdahl and Aquino, 2009).

Training and Education Programs and Sexual Harassment

There seems to be a general perception that training/education programs have some positive impact upon sexual harassment. For example, all the federal agencies in the MSPB surveys provided training programs, and 63% of the respondents (USMSPB, 1995) indicated that the training provided by their agencies helped reduce or prevent sexual harassment. However, one gap seems to be in the scientific evaluation of these training programs. Evaluations often consist almost entirely of post-training course evaluations that are, in essence, gauges of how participants feel about the training experience. No federal agency in the MSPB studies conducted evaluations of whether the sexual harassment training offered in their organizations had any short- or long-term effects. An attempt was made to explore the possible connections between the perceptions of training efforts of various federal agencies and their sexual harassment rates (USMSPB, 1988). These analyses revealed no consistent correlations between training efforts and agency incidence of sexual harassment, although there is evidence that women who experienced more unwanted sexual attention perceived training as being less effective (Newman et al., 2003). The 1995 DoD survey of U.S. military personnel indicated that although the majority of military personnel rated sexual harassment training as at least moderately effective, the thoroughness of the specific training experiences was not predictive of sexual harassment rates (Bastian et al., 1996). In a secondary analysis of the MSPB data, Antecol and Cobb-Clark (2003) found that training was associated with an increased labeling among men and women of potentially harassing behaviors as sexually harassing. This may be one effective goal of training: to reduce gender differences in definitions of harassing behaviors.

A few experimental or quasi-experimental studies, mainly using under-graduate samples, have examined the effectiveness of short-term training upon harassment-related attitudes and perceptions (for a review, see Pryor and Meyers, 2000). Some of these studies appear to suffer potential experimenter demand biases (e.g., Beauvais, 1986). York et al. (1997) found that a frequently used training technique, that of asking participants to analyze written cases, can affect undergraduates' sensitivity to sexually harassing behaviors portrayed in training films. Training has also been shown to have an impact on gender differences in reactions to sexual harassment. Moyer and Nath (1998) examined whether and how men and women differed in their ability to discriminate descriptions of sexually harassing and nonharassing behavior, and whether any gender differences in judgments could be reduced by training. Without training, women were found to be both more biased (they more often judged sexual harassment when it was not there) and more accurate (they better discriminated harassment from nonharassment) than men. After training, Moyer and Nath found, women's and men's judgments converged, with men adopting a response bias similar to that of women.

One goal of training might be to reduce the likelihood that men at risk for sexually harassing behaviors would actually perform them. Research by Perry et al. (1998) represents one of the only studies that sought to establish that training can have such an impact. Perry and her colleagues exposed undergraduate volunteers to a popular, commercially available sexual harassment training film. This research found that men with strong proclivities for sexually harassing behaviors as measured by the Likelihood to Sexually Harass (LSH) scale (Pryor and Wesselmann, in press) in a pretraining test showed gains in knowledge about sexual harassment after viewing the film and displayed fewer sexual advances to a woman to whom they were assigned to teach a golf lesson. The Perry et al. (1998) study suggests that these cognitive changes based on education may be followed by short-term changes in behavior. Interestingly, when Perry and her colleagues remeasured LSH scores in a post-training test, they found that men's scores did not differ significantly from pretraining measures. This result suggests that exposure to the training film might have had an impact upon participants by making salient antiharassment norms rather than by changing men's inherent proclivities for sexual harassment. Additional research supports the idea that short-term workplace training programs have little effect in changing the pro-harassment attitudes of high-LSH men (Robb and Doverspike, 2001).

Other training and education studies have focused primarily on samples of potential victims. For example, Barak (1994) found that participating in a cognitive-behavioral workshop can increase women's understanding of sexual harassment and enhance their coping skills. Whereas Barak's workshops included only women, most training does not segregate workers by gender. The practicality of segregating workers into groups by gender for sexual harassment training while maintaining a climate of gender equality remains to be seen. Indeed, there seems to be a potential to intensify a "we/ they" sense among men and women in the workplace that could actually exacerbate harassment problems (see Pryor and Whalen, 1996).

In summary, while workers who attend training and education programs seem to feel that such programs have a positive impact upon sexual harassment in the workplace, research is mixed about such a connection. The various experimental studies reviewed showed that in the short term, training and education programs can have demonstrable effects upon how people, especially men, think about sexual harassment and how men with harassment proclivities behave. While limited in scope, these experimental results are encouraging. Certainly, such short-term changes are necessary if training is ultimately to have an impact on workplace behaviors.

Focus on a New Direction in Research: Multiple Types of Harassment

A recently emerging research focus has examined targets' overlapping social identities as bases for harassment. People may be targeted for harassment because of their gender, their race or ethnicity, their sexual orientation, or a variety of other social dimensions. Salient overlapping identities—for example, being a black woman—could conceivably exacerbate the risk of harassment (Bergman and Henning, 2008). Interestingly, it is not yet clear whether the impact of experiencing harassment based on multiple dimensions or identities is worse than experiencing only gender or sexual harassment (Buchanan and Fitzgerald, 2008). Some studies comparing women who experienced harassment based on both gender and ethnicity have found additive but not multiplicative effects (Berdahl and Moore, 2006; Bergman and Henning, 2008), whereas one study found multiplicative effects (Buchanan and Fitzgerald, 2008). Bergman and Henning (2008) explained that models of general stress may explain the lack of strong multiplicative effects; that is, stress responses may be universal given stressors of a particular strength. This view is supported by studies comparing job-related and psychological well-being correlates of women who experienced multiple types of ethnic harassment (Schneider et al., 2000) and women who were both direct targets and bystanders to sexual harassment or incivility (Hitlan et al., 2006; Miner-Rubino and Cortina, 2007).

There are many interesting social psychological phenomena relevant to an examination of harassment based on an intersection of women's identities, including differential appraisals of such behavior based on the source, power differences, and suspected motives (Woods et al., 2009). Woods et al. (2009) focused on black women's experiences and found that cross-racial sexual harassment was appraised more negatively than intraracial harassment and included more frequent racial content within the incident. Such *racialized sexual harassment* may represent a unique construct that has differential antecedents and consequences, and such experiences certainly merit increased attention. There is also a great need for research exploring the intersection of sexual orientation harassment and sexual harassment.

Conclusions

For U.S. researchers, sexual harassment is considered to be a form of gender-based discrimination. Sexually harassing behaviors are more likely in organizational environments that tolerate or condone such behavior. While some individuals seem more likely to commit these behaviors than others, even those with proclivities for sexually harassing behavior seem reluctant to harass when local social norms discourage such behavior. Experiences of sexually harassing behavior are potentially sources of stress for workers. Such negative behaviors undermine the physical and mental health of workers and reduce their productivity. Future research should lead to a better understanding of how to encourage a healthy and productive workplace where each person is treated with dignity and respect, a workplace where work is enjoyed and is perceived as the fulfillment of human potential and not as an ordeal to be dreaded.

References

Abbey, A. (1982) Sex differences in attributions for friendly behavior: Do males misperceive females' friendliness? *Journal of Personality and Social Psychology, 42,* 830–838.

———— (1987) Misperceptions of friendly behavior as sexual interest: A survey of naturally occurring incidents. *Psychology of Women Quarterly, 11,* 173–194.

Antecol, H., and Cobb-Clark, D. (2003) Does sexual harassment training change attitudes? A view from the federal level. *Social Science Quarterly, 84,* 826–842.

Bacharach, S. B., Bamberger, P. A., and McKinney, V. M. (2007) Harassing under the influence: The prevalence of male heavy drinking, the embeddedness of permissive workplace drinking norms, and the gender harassment of female coworkers. *Journal of Occupational Health Psychology, 12,* 232–250.

Barak, A. (1994) A cognitive-behavioral educational workshop to combat sexual harassment in the workplace. *Journal of Counseling and Development, 72,* 595–602.

Bastian, L. D., Lancaster, A. R., and Reyst, H. E. (1996) *Department of Defense 1995 sexual harassment survey.* Arlington, VA: Defense Manpower Data Center.

Beauvais, K. (1986) Workshops to combat sexual harassment: A case of studying changing attitudes. *Signs, 12,* 130–145.

Berdahl, J. L. (2007) The sexual harassment of uppity women. *Journal of Applied Psychology, 92,* 425–437.

Berdahl, J. L., and Aquino, K. (2009) Sexual behavior at work: Fun or folly? *Journal of Applied Psychology, 94,* 34–47.

Berdahl, J. L., and Moore, C. (2006) Workplace harassment: Double jeopardy for minority women. *Journal of Applied Psychology, 91,* 426–436.

Bergman, M. E., and Drasgow, F. (2003) Race as a moderator in a model of sexual harassment: An empirical test. *Journal of Occupational Health Psychology, 8,* 131–145.

Bergman, M. E., and Henning, J. B. (2008) Sex and ethnicity as moderators in the sexual harassment phenomenon: A revision and test of Fitzgerald et al. (1994). *Journal of Occupational Health Psychology, 13,* 152–167.

Bowling, N. A., and Beehr, T. A. (2006) Workplace harassment from the victim's perspective: A theoretical model and meta-analysis. *Journal of Applied Psychology, 91,* 998–1012.

Broderick v. Ruder, 685 F. Supp. 1269, 46 EPD 37,963 (DDC 1988).

Buchanan, N. T., and Fitzgerald, L. F. (2008) Effects of racial and sexual harassment on work and the psychological well-being of African American women. *Journal of Occupational Health Psychology, 13,* 137–151.

Burlington Industries, Inc. v. Ellereth, 123 F.3d 490 (1998).

Dall'Ara, E., and Maass, A. (1999) Studying sexual harassment in the laboratory: Are egalitarian women at higher risk? *Sex Roles, 41,* 681–704.

Dansky, B. S., and Kilpatrick, D. G. (1997) Effects of sexual harassment. In W. O'Donohue (ed.), *Sexual harassment: Theory, research, and treatment* (pp. 152–174). Boston: Allyn and Bacon.

Dekker, I., and Barling, J. (1998) Personal and organizational predictors of workplace sexual harassment of women by men. *Journal of Occupational Health Psychology, 3,* 7–18.

Ellison v. Brady, 924 F.2d 872 (CA9 1991).

EEOC (Equal Employment Opportunity Commission) (1980) Guidelines and discrimination because of sex (Sec. 1604.11) *Federal Register, 45,* 74676–74677.

———— (1999) Enforcement guidance: Vicarious employer liability for unlawful harassment by supervisors. http://www.eeoc.gov/policy/docs/harassment.html (accessed March 15, 2009).

Faragher v. City of Boca Raton, 111 F.3d 1530 (1998).

Fitzgerald, L. F., Drasgow, F., and Magley, V. J. (1999) Sexual harassment in the armed forces: A test of an integrated model. *Military Psychology, 11* (30), 329–343.

Fitzgerald, L. F., Gelfand, M. J., and Drasgow, F. (1995a) Measuring sexual harassment: Theoretical and psychometric advances. *Basic and Applied Social Psychology, 17* (4), 425–445.

Fitzgerald, L. F., Hulin, C., and Drasgow, F. (1995b) The antecedents and consequences of sexual harassment in organizations: An integrated model. In G. Keita and J. J. Hurrell Jr. (eds.), *Job stress in a changing workforce: Investigating gender, diversity, and family issues* (pp. 55–73). Washington, DC: American Psychological Association.

Fitzgerald, L. F., Drasgow, F., Hulin, C. L., Gelfand, M. J., and Magley, V. (1997) Antecedents and consequences of sexual harassment in organizations: A test of an integrated model. *Journal of Applied Psychology, 82,* 578–589.

Fitzgerald, L. F., Magley, V. J., Drasgow, F., and Waldo, C. R. (1999) Measuring sexual harassment in the military: The Sexual Experiences Questionnaire (SEQ–DoD). *Military Psychology, 11* (3), 243–263.

Fitzgerald, L. F., Shullman, S. L., Bailey, N., Richards, M., Swecker, J., Gold, Y., Ormerod, A. J., and Weitzman, L. (1988) The incidence and dimensions of sexual harassment in academia and the workplace. *Journal of Vocational Behavior, 32*, 152–175.

Gelfand, M. J., Fitzgerald, L. F., and Drasgow, F. (1995) The structure of sexual harassment: A confirmatory analysis across cultures and settings. *Journal of Vocational Behavior, 47* (2), 164–177.

Gettman, H. J., and Gelfand, M. J. (2007) When the customer shouldn't be king: Antecedents and consequences of sexual harassment by clients and customers. *Journal of Applied Psychology, 92*, 757–770.

Glomb, T. M., Richman, W. L., Hulin, C. L., Drasgow, F., Schneider, K. T., and Fitzgerald, L. F. (1997) Ambient sexual harassment: An integrated model of antecedents and consequences. *Organizational Behavior and Human Decision Processes, 71* (3), 309–328.

Gruber, J. E. (1998) The impact of male work environments and organizational policies on women's experiences of sexual harassment. *Gender and Society, 12* (3), 301–320.

Gutek, B. A. (1985) *Sex and the workplace.* San Francisco: Jossey-Bass.

Gutek, B. A., and Done, R. S. (2001) Sexual harassment. In R. Unger (ed.), *Handbook of the psychology of women and gender* (pp. 367–387). New York: John Wiley.

Gutek, B. A., and Koss, M. P. (1993) Changed women and changed organizations: Consequences of and coping with sexual harassment. *Journal of Vocational Behavior, 42* (1), 28–48.

Harris v. Forklift Systems, Inc., 115 S. Ct. 367 (1993).

Hitlan, R. T., Pryor, J. B., Olson, M., and Hesson-McInnis, M. (2009) Antecedents of gender harassment: An analysis of person and situation factors. *Sex Roles, 61*, 794–807.

Hitlan, R. T., Schneider, K. T., and Walsh, B. M. (2006) Upsetting behavior: Reactions to personal and bystander sexual harassment experiences. *Sex Roles, 55*, 187–195.

Ilies, R., Hauserman, N., Schwochau, S., and Stibal, J. (2003) Reported incidence rates of work-related sexual harassment in the United States: Using meta-analysis to explain reported rate disparities. *Personnel Psychology, 56*, 607–631.

Kelley, M. (2000) Sexual harassment in the 1990s. *Journal of Higher Education, 71*, 548–568.

Lapierre, L. M., Spector, P. E., and Leck, J. D. (2005) Sexual versus nonsexual workplace aggression and victims' overall job satisfaction: A meta-analysis. *Journal of Occupational Health Psychology, 10*, 155–169.

Lazarus, R. S. (1993) Coping theory and research: Past, present, and future. *Psychosomatic Medicine, 55*, 234–247.

Low, K. S. D., Radhakrishnan, P., Schneider, K. T., and Rounds, J. (2007) The experiences of bystanders of workplace ethnic harassment. *Journal of Applied Social Psychology, 37*, 2261–2297.

Maass, A., Cadinu, M., Guarnieri, G., and Grasselli, A. (2003) Sexual harassment under social identity threat: The computer harassment paradigm. *Journal of Personality and Social Psychology, 85*, 853–870.

Magley, V. J., Hulin, C. L., Fitzgerald, L. E., and DeNardo, M. (1999) Outcomes of self-labeling sexual harassment. *Journal of Applied Psychology, 84*, 390–402.

Martindale, M. (1992) Sexual harassment in the military, 1988. *Sociological Practice Review, 2*, August, 200–216.

Meritor Savings Bank v. Vinson, 106 S. Ct. 2399 (1986).

Miner-Rubino, K., and Cortina, L. M. (2007) Beyond targets: Consequences of vicarious exposure to misogyny at work. *Journal of Applied Psychology, 92*, 1254–1269.

Moyer, R. S., and Nath, A. (1998) Some effects of brief training interventions on perceptions of sexual harassment. *Journal of Applied Social Psychology, 28* (4), 333–356.

National Council for Research on Women (1995) *Sexual harassment: Research and resources*. New York: National Council for Research on Women.

Newman, M. A., Jackson, R. A., and Baker, D. D. (2003) Sexual harassment in the federal workplace. *Public Administration Review, 63*, 472–483.

Niebuhr, R. E. (1997) Sexual harassment in the military. In W. O'Donohue (ed.), *Sexual harassment: Theory, research, and treatment* (pp. 250–262). Boston: Allyn and Bacon.

Oncale v. Sundowner Offshore Services, Inc., et al., 118 S. Ct. 998 (1998).

Palmieri, P., Harned, M., Collinsworth, L., Fitzgerald, L. F., and Lancaster, A. (2000) *Who counts? A rational-empirical algorithm for determining the incidence of sexual harassment in the military*. DMDC Technical Report. Urbana: University of Illinois–Urbana/Champaign.

Perry, E. L., Kulik, T., and Schmidtke, J. M. (1998) Individual differences in the effectiveness of sexual harassment awareness training. *Journal of Applied Social Psychology, 28* (8), 698–723.

Pryor, J. B. (1995) The psychosocial impact of sexual harassment on women in the U.S. military. *Basic and Applied Social Psychology, 17*, 581–603.

Pryor, J. B., Giedd, J. L., and Williams, K. B. (1995) A social psychological model for predicting sexual harassment. *Journal of Social Issues, 51*, 69–84.

Pryor, J. B., LaVite, C., and Stoller, L. (1993) A social psychological analysis of sexual harassment: The person/situation interaction. *Journal of Vocational Behavior, 42* (Special issue), 68–83.

Pryor, J. B., and Meyers, A. B. (2000) Men who sexually harass women. In L. B. Schlesinger (ed.), *Serial offenders: Current thought, recent findings, unusual syndromes*. Boca Raton, FL: CRC Press.

Pryor, J. B., and Wesselmann, E. D. (in press) An updated account of the Likelihood to Sexually Harass Scale. In T. D. Fisher, C. M. Davis, W. H. Yarber, and S. L. Davis (eds.), *Sexuality-related measures: A compendium*. Beverly Hills, CA: Sage.

Pryor, J. B., and Whalen, N. J. (1996) A typology of sexual harassment: Characteristics of harassers and the social circumstances under which sexual harassment occurs. In W. O'Donohue (ed.), *Sexual harassment: Theory, research, and treatment* (pp. 130–151). Needham Heights, MA: Allyn and Bacon.

Ragins, B. R., and Scandura, T. A. (1995) Antecedents and work-related correlates of reported sexual harassment: An empirical investigation of competing hypotheses. *Sex Roles, 32* (7–8), 429–455.

Raver, J. L., and Gelfand, M. J. (2005) Beyond the individual victim: Linking sexual harassment, team processes, and team performance. *Academy of Management Journal, 48*, 387–400.

Ridge, R. D., and Reber, J. S. (2002) I think she's attracted to me: The effect of men's beliefs on women's behavior in a job interview scenario. *Basic and Applied Social Psychology, 24*, 1–14.

Robb, L. A., and Doverspike, D. (2001) Self-reported proclivity to harass as a moderator of the effectiveness of sexual harassment prevention training. *Psychological
Reports, 88*, 85–88.

Robertson, C., Dyer, C. E., and Campbell, D. (1988) Campus harassment: Sexual
harassment policies and procedures at institutions of higher learning. *Signs:
Journal of Women in Culture and Society, 13*, 792–812.

Rosen, L. N., and Martin, L. (1998) Childhood maltreatment history as a risk factor
for sexual harassment among U.S. Army soldiers. *Violence and Victims, 13* (3),
269–286.

Rospenda, K. M., Fujishiro, K., Shannon, C. A., and Richman, J. A. (2008) Workplace
harassment, stress, and drinking behavior over time: Gender differences in a
national sample. *Addictive Behaviors, 33*, 964–967.

Rospenda, K. M., Richman, J. A., Wislar, J. S., and Flaherty, J. A. (2000) Chronicity of
sexual harassment and generalized work-place abuse: Effects on drinking outcomes. *Addiction, 95*, 1805–1820.

Schneider, K. T., Hitlan, R. T., and Radhakrishnan, P. (2000) An examination of the
nature and correlates of ethnic harassment experiences in multiple contexts.
Journal of Applied Psychology, 85, 3–12.

Schneider, K. T., Swan, S., and Fitzgerald, L. F. (1997) Job-related and psychological effects of sexual harassment in the workplace: Empirical evidence from two
organizations. *Journal of Applied Psychology, 82* (3), 401–415.

Schneider, K. T., Tomaka, J., Palacios-Esquivel, R., and Goldsmith, S. D. (2001)
Women's cognitive, affective, and physiological reactions to a male co-worker's
sexist behavior. *Journal of Applied Social Psychology, 31*, 1995–2018.

Stark, S., Chernyshenko, O. S., Lancaster, A. R., Drasgow, F., and Fitzgerald, L. F.
(2002) Toward standardized measurement of sexual harassment: Shortening the
SEQ-DoD using item response theory. *Military Psychology, 14*, 49–72.

Stockdale, M. S. (1993) The role of sexual misperceptions of women's friendliness
in an emerging theory of sexual harassment. *Journal of Vocational Behavior, 42*,
84–101.

Stockdale, M. S., Visio, M., and Batra, L. (1999) The sexual harassment of men:
Evidence for a broader theory of sexual harassment and sex discrimination.
Psychology, Public Policy, and Law, 5, 630–664.

USMSPB (U.S. Merit Systems Protection Board) (1981) *Sexual harassment in the Federal
workplace: Is it a problem?* Washington, DC: U.S. Government Printing Office.

——— (1988) *Sexual harassment in the federal government: An update.* Washington, DC:
U.S. Government Printing Office.

——— (1995) *Sexual harassment in the federal government: Trends, progress, continuing
challenges.* Washington, DC: U.S. Government Printing Office.

Waldo, C. R., Berdahl, J. L., and Fitzgerald, L. F. (1998) Are men sexually harassed?
If so by whom? *Law and Human Behavior, 22*, 59–79.

Wasti, S. A., Bergman, M. E., Glomb, T. M., and Drasgow, F. (2000) Test of the cross-
cultural generalizability of a model of sexual harassment. *Journal of Applied
Psychology, 85*, 766–778.

Williams, J. H., Fitzgerald, L. F., and Drasgow, F. (1999) The effects of organizational
practices on sexual harassment and individual outcomes in the military. *Military
Psychology, 11*, 303–328.

Williams, J. H., Lam, J. A., and Shivery, M. (1992) The impact of a university policy on
the sexual harassment of female students. *Journal of Higher Education, 63*, 50–64.

Woods, K. C., Buchanan, N. T., and Settles, I. H. (2009) Sexual harassment across the color line: Experiences and outcomes of cross- versus intraracial sexual harassment among black women. *Cultural Diversity and Ethnic Minority Psychology, 15,* 67–76.

York, K. M., Barblay, L., and Zajack, A. (1997) Preventing sexual harassment: The effect of multiple training methods. *Employee Responsibilities and Rights Journal, 10,* 277–289.

11

Discrimination and Bullying

Duncan Lewis, Sabir Giga, and Helge Hoel

CONTENTS

Introduction

While other chapters have debated and explored boundaries and questions around bullying, its antecedents, and its precedents, this chapter discusses how bullying and discrimination can be so readily colocated. Our position on why these two concepts are linked needs to be made clear at the outset. If we consider organisations as a microcosm of society, then it is our belief that many of the economic, social, political, historical, and global issues affecting intergroup relationships and experiences of discrimination outside work are just as likely to have an impact on experiences in work.

The forms of discrimination we are concerned with involve experiences similar to those associated with workplace bullying as described in other chapters, including a level of oppression or detriment through acts of force or deprivation by individuals or groups in positions of power against others in relatively less powerful positions. Thompson (2006) refers to this form of discrimination as "unfair" discrimination whereby the perception of being "different" is a factor in being treated inequitably or unjustly. This chapter, therefore, does not concern itself with concepts of positive discrimination (see, e.g., Russell and O'Cinneide, 2003) or justifiable discrimination on the grounds of skills or experiences but instead focuses on the negative aspects and consequences of discrimination.

The first part of this chapter will explore conventional equality strands or risk groups of discrimination and look at the evidence to see how, and to what extent, discrimination equates to bullying. In doing so, we also explore why these strands are seen as risk groups. Specifically, how might behaviours and standards that we know as bullying manifest as discrimination and vice versa? And do repeat experiences of exposure to discrimination end up as or equate to a form of bullying? In the second part of the chapter, we explore some organisational interventions that we believe are necessary to begin to address the interconnectivity between bullying and discrimination.

Throughout this chapter we will, in particular, draw on experiences from a UK context. Readers will therefore need to reflect on our circumstances with that of their own. The rationale for setting our position so clearly and unequivocally is that although there are common pieces of legislation affecting European Union (EU) states, the United Kingdom, as most other countries, has its own unique context in terms of its history, laws, codes of conduct, and practices.

Discrimination and Bullying: An Overview

The process of dealing with discrimination is as much a social model as it is an organisational one, and the process itself is long and demanding in terms of time and effort. Finding solutions to address specific forms of discrimination may not be universally acceptable, particularly as in the short term, interventions may require targeting or focusing on a specific group and, at the same time, potentially excluding others. As such, attitudes to equality initiatives relating to gender, race, sexual orientation, religion and belief, and age and disability can differ significantly. Similarly, these equality strands are not universally viewed as equal because lobby groups have differing power bases, with some possessing more influence than others. This situation might be explained by the maturity of the legislation implemented to support each of these areas. It may also be attributed to the historical period of a social and political campaign. For example, the African American Civil Rights movement of Dr. Martin Luther King Jr., Rosa Parks, Malcolm X, and others has its roots in the emancipation of slaves first raised by President Lincoln in 1849, while much later, in 2007, the EU declared the year to be one of "Equal Opportunities for All." What this discrepancy tells us is that the history of addressing discrimination goes back centuries yet is as current today as it has ever been. In particular, there is much discussion presently around the changing demographics of many countries, which is likely to have a dramatic impact on the nature of the workforce—for example, a longer lifespan and an ageing population requiring employees to work beyond the statutory retirement age. Therefore, although much discrimination has historically focused predominantly on gender and race, time and place is showing us a constant shift in the focus of discriminatory practices and as such is, in our opinion, likely to remain an issue that is never fully resolved.

As already indicated, the term *discrimination* is frequently used to refer to unfair or perceived unfair treatment based on social group membership, for example, sex, ethnicity, age, or other personal factors such as being pregnant (e.g., Fiske, 1998; Grainger and Fitzner, 2007; Harris, Lievens, and Van Hoye, 2004). However, a distinction should be made between discrimination and "unfair decisions," which are based on other factors that are not necessarily related to some form of prejudice (Harris et al., 2004, p. 55). In legal terms, discrimination has often been associated with intentionality, with early legislation focusing on deliberate actions. However, as intent is difficult to prove, perhaps because people for reasons of prejudice or stereotyping may be unaware of their unequal treatment of others or that they could plead ignorance as a form of defence, most areas of UK employment discrimination law consider the end effect, and so individuals cannot simply argue that it was not their intention to discriminate. Similarly, U.S. law has introduced the term *disparate impact* to refer to the unequal effect of decisions and behaviour on social groups (Goldman et al., 2008).

Risk groups, or what U.S. legislation refers to as "protected groups," are broadly similar across the EU with the dominant focus on age, gender, sexual orientation, disability, religion or belief, and ethnicity. In Europe, these categories are commonly referred to as the "central equality strands" for which the EU currently offers legal protection against discrimination (*Discrimination in the European Union*, 2008). From a UK legal perspective, in addition to the protection offered to the foregoing strands, it is also unlawful to discriminate on the grounds of marital status, gender reassignment or sex change, trade union membership, and employment status (fixed or part time). By contrast, in the United States, federal discrimination legislation does not include discrimination on the grounds of sexuality, although protection for this category may vary by state, and therefore it is potentially not illegal in some states to discriminate on this basis.

Within the UK legislative framework (see www.direct.gov.uk, 2009), four types of employment discrimination are highlighted: (1) *direct discrimination*, for example, treating someone less favourably on the basis of their belonging to one of the aforementioned protected groups; (2) *indirect discrimination*, for example, applying a general rule, such as pay, that is disadvantageous to one of the groups; (3) *harassment*, which is any unwanted conduct that can be linked to one of the aforementioned protected groups and violates an individual's dignity or creates an intimidating, hostile, degrading, humiliating, or offensive environment for an individual; and (4) *victimisation*, or treating individuals unfairly because they have made a complaint or raised a grievance.

Although there is a lack of consensus on a definition of workplace bullying, researchers are increasingly agreeing on some core features or dimensions, among them persistency and long-term duration of exposure to negative or unwanted behaviour (Di Martino et al., 2003; Einarsen et al., this volume). In this respect, most academics distinguish themselves from more general definitions that simply focus on the hostile working environment created by the presence of unwanted behaviour (e.g., Grainger and Fitzner, 2007). Another core feature of bullying is the power imbalance between targets and perpetrators, whether this emerges from asymmetrical positions of power within organisational hierarchies, from others sources of social power, or from the social dynamics of conflict situations. The preceding argument also implies that bullying may be a dynamic and often escalating process. For the purpose of this chapter, we subscribe to the following definition:

> Bullying at work means harassing, offending, socially excluding someone or negatively affecting someone's work tasks. In order for the label bullying (or mobbing) to be applied to a particular activity, interaction or process it has to occur repeatedly and regularly (e.g., weekly) and over a period of time (e.g., about six months). Bullying is an escalating process in the course of which the person confronted ends up in an inferior position and becomes the target of systematic negative social acts. A conflict cannot be called bullying if the incident is an isolated event or if two parties of approximately equal "strength" are in conflict. (Einarsen et al., 2003)

To us, bullying is rooted in principles of unreasonableness (e.g., Keashly, 1998) and may include aspects of all the aforementioned forms of discrimination. However, we argue that not all forms of discrimination can be labelled as bullying and vice versa for three main reasons. First, discrimination is predominantly based on being treated unfairly because of membership to a protected group. We argue that this would not necessarily have to be the case for bullying. Second, in line with the dominant view, one of the key aspects of workplace bullying is persistency of negative experiences. Persistent discrimination or situations where the presence of discrimination appears to form a pattern can therefore be perceived as bullying, particularly when it involves negative or unfair acts of the type often associated with bullying; however, a singular experience of discrimination, such as being denied access to a training programme, would not normally qualify as bullying. As discussed earlier, under most definitions of bullying and areas of UK discrimination law, the motives or intentions behind a person's actions are irrelevant. It therefore becomes imperative to develop an understanding of the contextual factors and implications of each situation. However, first let us explore what is already known about the connection between bullying and discrimination.

Discrimination and Bullying: An Exploration of the Evidence from an Equalities Perspective

Although it may be false to suggest that all minorities experience discrimination and bullying, there is growing evidence suggesting that minority status could be a contributing factor to receiving differential treatment in the workplace (see, e.g., Hoel and Giga 2006; Lewis and Gunn 2007). Furthermore, because of inherent problems associated with data collection when working with minorities, including problems of access and smaller numbers, along with the sensitive nature of the subject matter, difficulties often occur with sampling. As a result, convenience samples are frequently used, giving rise to issues such as self-selection bias (Croteau, 1996).

In one of the earliest studies of bullying and discrimination, Archer (1999) highlighted a culture of white male dominance in a UK fire service by referring to the negative experiences and bullying of individuals because of their gender and race in a predominantly white male environment. Discrimination came in the form of maintenance of majority interests to ensure continued dominance. Bullying was used to reinforce supremacy.

From recent UK studies on workplace bullying, differences in prevalence rates vary significantly when comparing black and minority ethnic (BME) figures with the wider population. In Hoel and Giga's study (2006), 25.2% of all BME respondents indicated they had been bullied compared to 11.8% of white participants. Similarly, in Lewis and Gunn's study (2007), 35% of the BME sample indicated they had been bullied compared to 9% of white respondents. Fox and Stallworth (2005) demonstrated how Hispanics and Latinos reported higher levels of general bullying than did white and Asians counterparts. African Americans and Hispanics and Latinos all reported higher levels of ethnic bullying. This pattern is repeated in a UK study on behalf of the Royal College of Nursing (Ball and Pike, 2006); participants in this study were asked whether they thought their experiences of bullying were directly linked to their ethnic minority status. Sixty-one percent of BME nurses sampled said that the bullying and harassment they experienced was race related. On a cautionary note, however, there is evidence to suggest that experiences of bullying vary among minority groups (see, e.g., Fox and Stallworth, 2005).

We can explore the boundaries between discrimination and bullying from the case of *Yeboah v. Crofton*, where a West African man employed by a local council in London was continuously subjected to false allegations made by a fellow employee. It was found that the accusations were not based on any evidence of underperformance at work but on an individual's prejudice and belief that all people from West Africa were corrupt.

The lines between race and religious discrimination are blurred legislative and practice perspectives. Religious conflict has been a major issue in the United Kingdom, particularly the historical tensions between Catholics and Protestants in Northern Ireland and some parts of Scotland. Increasing levels of immigration, particularly from countries previously governed by the United Kingdom, have resulted in greater ethnic and religious diversity. Religious belief has received increased prominence recently through the mounting worldwide debate around religious freedom, particularly after the terrorist attacks in New York in 2001 and in London in 2005, which have resulted in reported increases in Islamaphobia. It could be argued that there may, at least in the short term, be a reported increase of bullying on the grounds of religion particularly if the individual's religious association is visibly obvious to others.

In regard to sexual orientation, in 2007 a study by the gay rights campaign group Stonewall reported that nearly 20% of lesbians and gay men were experiencing bullying because of their sexual orientation, with 13% of the British population (close to 4 million people) reporting that they had witnessed verbal bullying of lesbian, gays, and bisexuals (LGB) in the workplace, whilst nearly 4% (1.2 million) reported witnessing physical anti-gay bullying at work (Stonewall, 2007). Compared to most other groups, with the exception of people with disabilities, levels of discrimination and bullying for LGB employees are higher (Grainger and Fitzner, 2007). Verbal abuse, physical attacks, sexual harassment, unfair treatment, and a homophobic work culture are among the most commonly reported forms of bullying (Acas, 2007). Moreover, compared to other minorities, LGBs tend to focus on the experience as bullying rather than as discrimination (Di Martino, Hoel, and Cooper, 2003). This finding is mirrored in the types of sexual orientation–based claims taken to industrial tribunals (Acas, 2007). In *Majrowski v. Guys and St Thomas' NHS Trust* (House of Lords, 2006), a gay employee succeeded with his complaint against his line managers for repeatedly subjecting him to verbal abuse, public humiliation, unreasonable deadlines, and being ignored.

European Union figures show that of those respondents self-identified as members of a minority, disabled respondents (31%) reported experiencing the highest levels of discrimination (*Discrimination in the European Union*, 2008). Evidence from the United Kingdom (Fevre, Robinson, Jones, and Lewis, 2008; Grainger and Fitzner, 2007) shows that disabled employees and those with long-term health conditions are more likely to report bullying, or exposure to the types of negative behaviour that constitute bullying, than are their nondisabled counterparts. Fevre et al. (2008) illustrate that individuals with learning, psychological, or emotional disabilities, rather than those with physical disabilities, are the most likely to report exposure to negative behaviours.

Although there may be a perception that women are bullied in higher numbers than men, most studies suggest that men and women are bullied in approximately equal numbers (Hoel, Cooper, and Faragher, 2001; Zapf et al., this volume). That women are found to be overrepresented among targets might reflect female participation rates in high-risk occupations, for example, in health, retailing, and other service occupations (Di Martino et al., 2003). Alternative explanations include entry into such traditional male-dominated occupations as the uniformed sectors (Archer, 1999) and business professionals (Salin, 2001) or into other positions historically occupied by men (Hoel and Cooper, 2001).

According to Salin (2003), women's apparently overrepresentativeness among targets may also be explained by a possible interaction between higher exposure to negative behaviour and bullying and greater acceptance of labelling their experience as harassment and bullying. In some cases, however, men employed in primarily female occupations have reported higher prevalence rates than have women (Eriksen and Einarsen, 2004; Leymann and Lindroth, 1993). In explaining their findings, Eriksen and Einarsen refer to Pryor and Fitzgerald (2003), who argue that women are more likely to experience sexual harassment when they are numerically in a minority. This finding also appears to tie in with Leymann's view that social exposure in a work group leads the more vulnerable to be targeted for bullying (1996). In this respect, and in line with Archer's fire service study (1999), Eriksen and Einarsen (2004) suggest that gender minorities could be seen as a threat to the prevailing culture.

Most studies of age and harassment and bullying suggest that being young is associated with elevated risk of exposure, with younger employees generally experiencing more unwanted negative behaviour and harassment than older counterparts (see, e.g., Di Martino et al., 2003). According to Hoel et al. (1999), this hazard might mirror cultural as well as labour market differences in terms of age of entry into the labour market, retirement age, and employment protection regulations. Nevertheless, any discrepancy between exposure to negative behaviour and perceptions of discrimination and bullying could also be explained by different thresholds for various age groups, possibly linked to expectations of treatment at work. However, as reported elsewhere in this book (Zapf et al., this volume), the findings on bullying as it relates to age remain inconclusive.

Suggesting that different forms of oppression in the workplace, such as sexual harassment, racial harassment, and bullying, are interconnected, Hoel and Beale (2006) refer to Lee's argument that these forms of oppression should be considered interactive rather than seen as additive experiences (2002). In other words, they do not necessarily exist in isolation but might, rather, reinforce coexistence. This relationship could also imply that where bullying is prevalent, there is a greater risk of discrimination and vice versa. Although our discussions focus on individual equality strands, there is, of course, the possibility that individuals belong to more than one minority group—for example, age and disability—in which case they face the potential of experiencing discrimination on multiple fronts.

The foregoing accounts show that protected groups often associate their experience of bullying with their minority status. Of course, it could be that this is a result of a degree of hypersensitivity, or what Salin (2003), in explaining women's overrepresentation among targets of bullying, refers to as a lower threshold compared to men experiencing similar behaviours. The same could be said for ethnic minorities, although several bullying studies have reported higher levels of exposure to negative behaviour by ethnic minorities (e.g., Fox and Stallworth, 2005; Lewis and Gunn 2007).

It has also been argued that minorities' experiences of discrimination in one situation may result in a tendency to blame negative outcomes on discrimination in other situations (James, Lovato, and Cropanzano, 1994). This resembles the "vigilante hypothesis" (e.g., Allport, 1954), which suggests that the frequent exposure to prejudice and discrimination that members of low-status groups experience makes them more alert, with thoughts of discrimination stimulated or prompted in ambiguous circumstances. However, a study by Major, Gramzow, McCoy, and Lewin (2002) found no support for the vigilante hypothesis, with the rate of reported discrimination by members of low-status groups precisely mirroring their enhanced exposure to discrimination as well as the severity of their experience. Others have suggested that membership in a low-status group may actually prompt a person to seek to reduce cognitions of discrimination because of the psychological costs involved (Branscombe, 1998, cited in Major et al., 2002), a finding that could equally apply to perceptions of bullying.

Dealing with Discrimination and Bullying in the Workplace

The Case for Dealing with Workplace Bullying

The evidence for bullying having a detrimental impact at individual, organisational, and societal levels is well established and explored elsewhere in this volume (see also Hoel et al. and Høgh et al., this volume).

From a national economic and organisational development perspective, it makes sense that the skills set of the whole workforce is utilised fully rather than creating barriers for individuals because of differences in gender, ethnicity, age, sexual orientation, religion, and so on. As a result, there is much discussion on the business case for diversity with arguments that a diverse workforce brings with it the development of a wider range of skills and experiences, improvement in understanding the needs of diverse populations, and a general improvement in the image of organisations through their association with diverse groups (Noon, 2007). Similarly, it may also be argued that bullying is likely to become more intense with the increasing globalizing trend for work and workers (Hoel and Cooper, 2001). This evolution of economic activity has seen greater movements in skilled workers geographically or through embedding overseas employees into organisations that operate globally (Lin, 1999).

Operations that encourage multiple variations of culture and subculture, particularly within a short time span, must be aware that such variation can have an unsettling impact on the cultural norms of groups and affect inter-group relations (Baron and Neuman, 1998; Harvey, Treadway, Heames, and Duke, 2009). Of particular concern here are the power relationships between, and sometimes even within, minority groups, with the conflicts that often arise between ethnicity or religion and sexual orientation serving as a case in point. Thus, as Yamada (2003) commented, this situation has created complex legal and moral issues associated with tackling bullying. If national and global spheres of operation are likely to create tensions around cultural norms, the likelihood makes a compelling case for tackling bullying and discrimination together.

Whilst the cost argument is compelling, we caution against an overemphasis on an economically rational argument for tackling bullying and discrimination. If diversity and equality of opportunity are human rights based on moral legitimacy (Noon, 2007), then there appears to be a strong case for placing bullying within the same category. If managers and organisations simply adopt an economic rationality for tackling bullying, the moral imperative could be lost behind a business veneer. Furthermore, if managers alone "own" the problem, then tackling bullying becomes a managerial agenda rather than a moral one. Under such circumstances, the long-term benefits of tackling bullying might be seen by managers as not offering "quick wins," specifically where managers are often driven by short-term targets. It therefore seems to be more sensible to argue that linking bullying to discrimination on moral grounds is likely to have stronger foundations for dealing with both.

Interventions

Utilising Existing Legislation and Developing a New Legal Framework

Even though through organisational policies most organisations could demonstrate a fairly developed aptitude on the issue of bullying, one of the main challenges of addressing bullying in workplaces in many countries is the absence of custom-made legislative frameworks. Recognising the weaknesses and shortcomings of existing legislation (e.g., Guerrero, 2004; Hoel and Einarsen, 2010), this absence may mean that individuals who wish to pursue a case of bullying may develop a false sense of security if they consider themselves to be protected by internal organisational policies or by the law. Moreover, in terms of a legal perspective, individuals who want to take a bullying case forward may be forced to seek legal redress through other avenues such as antidiscrimination, health and safety or whistleblowing legislation, or even constructive dismissal routes. This absence of specific antibullying legislation also suggests that workers from protected or minority groups may have more scope to pursue their bullying claims through antidiscrimination legislation, whereas employees who are not from a minority group may have limited options to take such a course of action (Porteous, 2002). In this scenario, employers may also be aware that what may initially be seen as a case of bullying may quickly turn into one of discrimination if the parties involved have characteristics that may warrant such an approach, hence creating a situation where the employer is faced with an altogether more serious accusation. Of course, the potential for such an outcome could also result in organisations trying to conceal cases of bullying much earlier in case there is a risk of linking them to discriminatory practice.

A clear challenge is, therefore, to find more effective ways of grappling with complex constructs. Many managers should be well versed in bullying policies and antidiscrimination principles. However, an absence of specific antibullying legislation in many countries could have negative consequences for tackling bullying simply because that absence suggests that bullying is not seen as important. In the case of the United Kingdom, nongovernment agencies such as Acas and the Equalities and Human Rights Commission (EHRC) are pivotal in promoting awareness of rights and responsibilities to both employers and employees by offering documentation and training to managers and organisations when any new legislation comes into force. An absence of specific antibullying legislation means that tackling bullying at an organisational level consequently becomes difficult, since there is no external agency to drive training and raise awareness. Thus, we argue that bridging awareness of the link between bullying and discrimination, particularly through a wider agenda for dignity and respect at work, could become a critical focus in helping to address bullying at work.

Development of Organisational Policies

Alongside legislative practices, organisations have constructed policies and processes (see Rayner and Lewis, this volume) to attempt to deal with bullying and other forms of unfairness at work. However, these policies and processes have not always been implemented as intended. Salin (2008) demonstrated that success with antibullying policies is often attributable to young and enthusiastic human resources managers keen to drive new initiatives. We contend that making explicit interconnections between bullying and discrimination policies provides mutual synergies. The evidence is reasonably incontrovertible in demonstrating how ethnic minorities (Fox and Stallworth, 2005; Lewis and Gunn, 2007) are more likely to report high levels of exposure to bullying behaviours. Similar evidence exists for disabled employees and those with long-term health conditions (Fevre et al., 2008; Grainger and Fitzner, 2007). We also know, at least in the United Kingdom, that gay and lesbian employees are more likely to label their negative workplace experiences as bullying rather than discrimination (Di Martino et al., 2003). With this clear body of evidence linking minorities to bullying behaviours, it seems prudent to link them at the policy level if we are to minimize the at-risk groups and negative impact to all.

The challenge still exists, however, to broaden the involvement of organisational constituents in the policy arena. Such broadening is possibly best achieved by utilising the aforementioned enthusiasm of human resources professionals as evidenced by Salin (2008) and drawing in wider participation from general employees, not simply those with an equalities agenda. Such an approach could benefit efforts to fight discrimination and bullying, since the latter group tends to be populated by individuals and sub-groups with personalised agendas. Although bullying and discrimination should be identified and discussed as two separate issues within policies, their interconnectivity should be addressed. Organisational understanding of the groups at risk for bullying and discrimination and the interconnectivity between the phenomena should also be reflected in regular monitoring and evaluation of their effectiveness. Such monitoring may be achieved, for example, by assessing the number and types of complaints received and by incorporating relevant questions in quantitative (e.g., staff surveys) or qualitative (e.g., focus groups) assessments.

Harvey et al. (2009) advocate that any demonstration of tolerance of bullying would send a counterproductive message to the establishment of other good practices, and at the very least it would result in an erosion of trust and confidence in the organisation (Fox and Stallworth, 2005). Whilst we endorse the principle of zero tolerance for bullying, consideration should be given to the potential detrimental impact on the complainant if sanctions are draconian. If bringing forward a milder, albeit genuine grievance leaves a complainant fearful of the punitive impact on the perpetrator—for example, being dismissed—then the complainant may be reluctant to take matters further, thus nullifying the zero tolerance policy. This result would be of particular worry for members of vulnerable groups whose position may already be precarious.

Development of Management Practice

Managers should be well versed in antidiscrimination practices because although there has been UK and EU legislation for decades, bullying as such is relatively new as a subject for legislation, having surfaced in discussions only in the last 10–15 years. Is the management treatment of bullying therefore happening only at a superficial level currently, whereas antidiscrimination practices are more embedded? Given this question, we advocate that it is critical to reeducate employees, and specifically managers, on the obvious links between bullying and discrimination; neither practice is a fringe issue that affects only minorities.

We contend that it is therefore critically important to better educate and understand what bullying is and is not. Acas (2007) demonstrates that prior to seeking advice, managers are often blinded by their own perceptions and biases on such issues as religion and sexual orientation. This finding suggests that some fundamental reappraisal is necessary to educate managers to think and act in nonprejudicial and nonjudgmental ways not only in personal actions but also in seeking solutions. This change might be best achieved by training managers and supervisors in skills associated with social psychology such as stereotyping and conflict awareness and resolution. If managers are sightless to certain types of negative behaviours simply because they perceive such behaviours as within normal boundaries, solutions to bullying and discrimination will not be sought. Bullying can be a proxy for discrimination and vice versa. It is therefore critical that those individuals with responsibility for management, regardless of level, are reconnected with the subtle and sometimes not-so-subtle behaviours being enacted.

Conclusions

The issues of bullying and discrimination are not only complex but also, as we have demonstrated, unavoidably colocated, especially given that the workplace is likely to be one of the main environments for social interaction (Estlund 2003). As we increasingly function in diverse societies and in globalised workplaces, interactions between workers of different backgrounds become inevitable; consequently, the need to have appropriate skills to manage intergroup relations becomes more essential. Whilst it cannot fall to organisations alone to address discrimination and bullying, the workplace is likely to see clear demonstrations of discriminatory behaviour in ways that are both obvious and subtle. Although some forms of discrimination—in particular, harassment as defined within the UK legislative framework—are extremely personal and damaging to the individual and organisation, we argue that whereas the former is easiest to deal with, the latter is more difficult to uncover and tackle.

We advocate a return to basic principles of not only understanding diversity but also colocating it within a policy domain. Between-group differences (Modood, Berthoud, Lakey, Nazroo, Smith, Virdee, and Beishon, 1997) and within-group differences (Dale, Fieldhouse, Shaheen, and Kalra, 2002) show us that diversity requires us to recognise difference and that seeing employees as a homogeneous whole is not what diversity or good human resources management is about. Difference must be celebrated, not treated in a uniform way. For this approach to become a reality, however, organisations need to openly address the problems of communication, training, and policy frameworks as well as the way the workforce is managed on a day-to-day basis.

Similarly, fairness and ensuring equality of opportunity are regarded as universal rights, but as Noon (2007) notes, this precept is followed only so long as it meets the needs of the organisation and its business first. If our rationality to tackle bullying is first and foremost economic, we should expect nothing less than a managerially economic response. We therefore advocate that moral rationality must also be a key element in approaching discrimination and bullying.

As we have highlighted earlier, the process of change is a slow one but ultimately one to which we should all yield; bullying and discrimination can be clearly colocated. It is some 30 years since the United Kingdom elected its first female prime minister, and only in 2009 did the United States of America elect its first black president. These events are noteworthy for their signals of the longevity taken for gender and race transformations. As Marian Wright Edelman (1939–) tells us, change is best achieved one step at a time, but sometimes our steps are too slow and ponderous.

References

Acas (2007) *Sexual orientation and religion or belief discrimination in the workplace.* London: Advisory, Conciliation and Arbitration Service (Acas).

Allport, G. (1954) *The nature of prejudice.* Reading, MA: Addison Wesley.

Archer, D. (1999) Exploring "bullying" culture in the para-military organisation. *International Journal of Manpower, 20* (1/2), 94–105.

Ball, J., and Pike, G. (2006) *At breaking point: A survey of the wellbeing and working lives of nurses in 2005.* London: Royal College of Nursing.

Baron, R. A., and Neuman, J. H. (1998) Workplace aggression—the iceberg beneath the tip of workplace violence: Evidence on its forms, frequency and targets. *Public Administration Quarterly, 21,* 446–464.

Croteau, J. M. (1996) Research on the work experiences of lesbian, gay and bisexual people: An integrative review of methodology and findings. *Journal of Vocational Behaviour, 48,* 195–209.

Dale, A., Fieldhouse, E., Shaheen, N., and Kalra, V. (2002) The labour market prospects for Pakistani and Bangladeshi women. *Work, Employment and Society, 16* (1), 5–26.

Di Martino, V., Hoel, H., and Cooper, C. L. (2003) *Preventing violence and harassment in the workplace.* European Foundation for the Improvement of Living and Working Conditions. Luxembourg: Office for Official Publications of the European Communities.

Directgov (2009) Discrimination in the workplace. http://www.direct.gov.uk/en/Employment/ResolvingWorkplaceDisputes/DiscriminationAtWork/DG_10026557 (accessed September 16, 2009).

Discrimination in the European Union: Perceptions, experiences and attitudes. 2008. Special Eurobarometer 296 / Wave 69.1.

Einarsen, S., Hoel, H., Zapf, D., and Cooper, C. L. (2003) The concept of bullying at work: The European tradition. In S. Einarsen, H. Hoel, D. Zapf, and C. L. Cooper (eds.), *Bullying and emotional abuse in the workplace: International perspectives in research and practice* (pp. 3–30). London: Taylor & Francis.

Eriksen, W., and Einarsen, S. (2004) Gender minority as a risk factor of exposure to bullying at work: The case of male assistant nurses. *European Journal of Work and Organisational Psychology, 13* (4), 473–492.

Estlund, G. (2003) *Working together: How workplace bonds strengthen a diverse democracy.* Oxford, UK: Oxford University Press.

Fevre, R., Robinson, A., Jones T., and Lewis, D. (2008) *Disability: Work fit for all.* London: Equalities and Human Rights Commission, Insight Report No. 1.

Fiske, S. T. (1998) Stereotyping, prejudice and discrimination. In D. T. Gilbert, S. T. Fiske, and G. Lindzey (eds.), *The handbook of social psychology,* vol. 2, 4th ed. (pp. 457–411). New York: McGraw-Hill.

Fox, S., and Stallworth, I., E. (2005) Racial/ethnic bullying: Exploring links between bullying and racism in the U.S. workplace. *Journal of Vocational Behavior, 66,* 438–456.

Goldman, B. M., Gutek, B., Stein, J. H., and Lewis, K. (2008) Employment discrimination in organizations: Antecedents and consequences. *Journal of Management, 32,* 786–830.

Grainger, H., and Fitzner, G. (2007) *The first fair treatment at work survey.* London: Department of Trade and Industry.

Guerrero, M. (2004) The development of moral harassment (or mobbing) law in Sweden and France as a step towards EU legislation. *Boston College International and Comparative Law Review, 27,* 477–500.

Harris, M. M., Lievens, F., and Van Hoye, G. (2004) "I think they discriminated against me": Using prototype theory for understanding perceived discrimination in selection and promotion situations. *International Journal of Selection and Assessment, 12,* 54–65.

Harvey, M., Treadway, D., Heames, J. T., and Duke, A. (2009) Bullying in the 21st century global organization: An ethical perspective. *Journal of Business Ethics, 85,* 27–40.

Hoel, H., and Beale, D. (2006) Workplace bullying, psychological perspectives and industrial relations: Towards a contextualised and interdisciplinary approach. *British Journal of Industrial Relations, 44,* 239–262.

Hoel, H., and Cooper, C. L. (2001) Origins of bullying: Theoretical frameworks for explaining workplace bullying. In N. Tehrani (ed.), *Building a culture of respect: Managing bullying at work* (pp. 3–19). London: Taylor & Francis.

Hoel, H., Cooper, C. L., and Faragher, B. (2001) The experience of bullying in Great Britain: The impact of organizational status. *European Journal of Work and Organizational Psychology, 10* (4), 443–465.

Hoel, H., and Einarsen, S. (2010) Shortcomings of anti-bullying regulations: The case of Sweden. *European Journal of Work and Organizational Psychology, 19* (1), 30–50.

Hoel, H., and Giga, S. (2006) *Destructive interpersonal conflict in the workplace: The effectiveness of management interventions.* Manchester Business School, University of Manchester.

Hoel, H., Rayner, C., and Cooper, C. L. (1999) Workplace bullying. *International Review of Industrial and Organizational Psychology, 14,* 195–230.

House of Lords (2006) Judgments, *Majrowski (respondent) v. Guy's and St. Thomas' NHS Trust (appellants).* http://www.publications.parliament.uk/pa/ld200506/ldjudgmt/jd060712/majro-1.htm (accessed August 11, 2007).

James, K., Lovato, C., and Cropanzano, R. (1994) Correlational and know-group comparison validation of a workplace prejudice/discriminatory inventory. *Journal of Applied Social Psychology, 24,* 1573–1592.

Keashly, L. (1998) Emotional abuse in the workplace: Conceptual and empirical issue. *Journal of Emotional Abuse, 1,* 85–117.

Lee, D. (2002) Gendered workplace bullying in the restructured UK Civil Service. *Personnel Review, 31,* 205–227.

Lewis, D., and Gunn, R. W. (2007) Workplace bullying in the public sector: Understanding the racial dimension. *Public Administration, 83* (3), 641–665.

Leymann, H. (1996) The content and development of mobbing at work. *European Journal of Work and Organizational Psychology, 5* (2), 165–184.

Leymann, H., and Lindroth, S. (1993) Vuxenmobnning mot manliga for-skollarare [Mobbing of male teachers at kindergartens]. Stockholm: Arbetarskyddsstyrelsen.

Lin, N. (1999) Social networks and status attainment. *Annual Review of Sociology, 15,* 467–487.

Major, B., Gramzow, R. H., McCoy, S., and Lewin, S. (2002) Perceiving personal discrimination: The role of groups status and legitimising ideology. *Journal of Personality and Social Psychology, 82,* 269–282.

Modood, T., Berthoud, R., Lakey, J., Nazroo, J., Smith, P., Virdee, S., and Beishon, S. (1997) *Ethnic minorities in Britain.* London: PSI.

Noon, M. (2007) The fatal flaws of diversity and the business case for ethnic minorities. *Work, Employment and Society, 21* (4), 773.

Porteous, J. (2002) Bullying at work: The legal position. *Managerial Law, 44* (4), 77–90.

Pryor, J. B., and Fitzgerald, L. F. (2003) Sexual harassment research in the United States. In S. Einarsen, H. Hoel, D. Zapf, and C. L. Cooper (eds.), *Bullying and emotional abuse in the workplace: International perspectives in research and practice* (pp. 79–100). London: Taylor & Francis.

Russell, M., and O'Cinneide, C. (2003) Positive action to promote women in politics: Some European comparisons. *International and Comparative Law Quarterly, 52,* 587–614.

Salin, D. (2001) Prevalence and forms of bullying among business professionals: A comparison of two different strategies for measuring bullying. *European Journal of Work and Organizational Psychology, 10* (4), 425–441.

——— (2003). The significance of gender in the prevalence, forms and perceptions of bullying. *Nordiske Organisasjonsstudier, 5,* 30–50.

——— (2008) The prevention of workplace bullying as a question of human resource management: Measures adopted and underlying organizational factors. *Scandinavian Journal of Management, 24,* 221–231.

Stonewall (2007). *Living together: British attitudes to lesbians and gay people.* London: Stonewall.

Thompson, N. (2006) *Anti-discriminatory practice.* Basingstoke, UK: Palgrave Macmillan.

Yamada, D. (2003) Workplace bullying and the law: Towards a transnational consensus. In S. Einarsen, H. Hoel, D. Zapf, and C. L. Cooper (eds.), *Bullying and emotional abuse in the workplace: International perspectives in research and practice.* London: Taylor & Francis.

Yeboah v. Crofton, IRLR 634, CA (2002). http://www.hmcourts-service.gov.uk/judgmentsfiles/j1209/Yeboah_v_Crofton.htm (accessed October 2009).

12

An Industrial Relations Perspective of Workplace Bullying

David Beale

CONTENTS

Introduction

Workplace bullying research shares some common ground with the study of industrial relations. Obviously, both are concerned with employment. Both are related to some form and degree of conflict at work, and whether and how it might be resolved. Both recognise an imbalance of power as a central element of their analyses; and both, at least in Britain—where the bullying of subordinates by managers is the predominant pattern—are primarily focused on the relationship between managers and workers. Yet there has been little attempt to bring the two fields of study together. There are, of course, important differences. Workplace bullying is a much newer and obviously much more specific area of study. Whilst this has drawn heavily on psychological and social psychological perspectives with attention concentrated on the micro-level processes, industrial relations has had workplace, sectoral, national, and international concerns and has drawn on different academic disciplines to inform its approach. However, these differences in conjunction with the similarities add interest to the question of whether new areas of understanding could be opened up in both fields by exploring workplace bullying issues within the context of industrial relations. This particular chapter attempts to answer part of that question by indicating some of the ways in which industrial relations might contribute to the workplace bullying debate.

The chapter starts by outlining the basic elements of industrial relations as the subject has been studied in Britain. Next are discussed the pioneering but somewhat neglected work of the Norwegian Jon Sjøtveit (1992a, 1992b, 1994) and the explicit attempt by British academics Ironside and Seifert (2003) to see bullying in industrial relations terms. These were the first publications to indicate a possible bridge between the two fields of study, though Sjøtveit's work did this implicitly. Then comes a summary of the potential for a more contextualised and interdisciplinary approach to the study of workplace bullying that draws on aspects of industrial relations (Hoel and Beale, 2006); and in relation to this possible approach, Kelly's focus on mobilisation theory in the context of industrial relations (Kelly, 1998) is highlighted as particularly significant for its potential to establish such linkages Finally, there follows a clarification of this author's own position, along with tentative conclusions and suggestions for new directions in workplace bullying research.

Industrial Relations

Industrial relations is something of a misnomer, as the field is not limited to the study of industrial employment sectors, and similarly—as restructuring of the economy and changes to labour markets have demanded—it is not exclusively concerned today with unionised sectors and workplaces. Whilst *labour relations* is sometimes used as an alternative, in Britain the terms *employee relations* and *employment relations* are now in common usage, with each suggesting a somewhat different emphasis and perspective intended to reflect these changes. However, many researchers still retain the term *industrial relations*—as I do here—not least in recognition of the field's tradition and its continuing relevance, though whilst recognising its actual and potential development.

Whilst industrial relations as a term was in common usage in Britain and North America in the 1920s (Edwards, 2003, p. 1), the pioneering work 20 years earlier of British labour historians Sidney and Beatrice Webb (1897) can reasonably be cited as the starting point for academic study of the subject. However, it was not until after World War II that strong industrial relations academic groups began to emerge in half a dozen leading British universities, and industrial relations established itself as a distinct and fully recognised field of study (or as a discipline, as some prefer to call it) (Ackers and Wilkinson, 2003, p. 1). Whilst in Anglo-Saxon countries industrial relations established itself as such in universities, this has not been the case in continental Europe, where it was studied within various social sciences and mainly as part of industrial sociology (Frege, 2003, p. 242). In Britain, institutional approaches and an empirical emphasis were central to its early postwar development (Flanders and Clegg, 1954); whilst Fox's articulation of unitary, pluralist, and Marxist perspectives, as a loose theoretical framework for industrial relations enquiry, continues to influence work in the subject to this day (Fox, 1966, 1974). Academically, industrial relations has drawn traditionally on history, sociology, law, politics, economics, and to some extent social psychology to develop its approach and analysis.

At the heart of the study of industrial relations is the central relationship that arises from paid employment, ultimately a relationship between employer and worker but one that commonly also involves management, unions, government, and the state. Four defining aspects of the relationship have been suggested: (1) "the creation of an economic surplus"; (2) labour as an uncertain purchase for employers in that when they hire workers, they purchase not the *actuality* of labour but the *potential* of workers to work; (3) an imbalance of power in the relationship between worker and employer; and (4) the presence of both cooperation and conflict in the situation (Blyton and Turnbull, 1994, p. 31, as cited in Kelly, 1998, p. 4).

How pay is determined, whether unilaterally by employers and management or through some form of joint regulation by management and unions (i.e., collective bargaining), as well as the outcomes of pay determination processes, are central concerns of industrial relations. But so too is the regulation of the job itself, including the terms and conditions of employment in addition to pay, the organisation of work, the tasks prescribed, the intensity and quality of work required, and the general process by which management endeavours to maximise productivity. With regard to the latter—which is commonly referred to as the *labour process*—however prescriptive and detailed the job description and the written terms of the contract of employment may be, a significant degree of uncertainty inevitably persists in respect of what work is actually done, and alternative and competing management and workers' interpretations of what is reasonably required of particular workers are certain to occur to varying degrees. This variability in turn raises key questions for industrial relations not only of how and to what extent management exercises control over its workforce—and thus maximises its output—but also of the extent and ways in which workers might resist this. Thus, the labour process is intertwined with the power relations in the workplace, with conflict and cooperation, with the generation of an economic surplus, and with the boundaries of managerial prerogative that are likely to be perceived differently by management and union, to be contested by both, and to shift in either direction over time.

Yet industrial relations does not operate in a vacuum, and its legal, political, economic, technological, social, historical, and international contexts are all important in helping to determine its *particular* nature and dynamics. Also, whilst a minority Marxist body of literature has been established in the industrial relations field and continues to be developed (e.g., Hyman, 1975; Kelly, 1998), institutional and descriptive approaches and pluralist perspectives have tended to persist; in addition, industrial relations has come under considerable pressure from the development of human resource management as an academic alternative. Traditionally, the collective dimension has been central to the study of industrial relations, although in the last decade or two increased attention has been paid within it to individualism at work and to industrial relations in nonunion workplaces.

Much British industrial relations analysis in the 1980s and 1990s focused on the demise of trade union membership and influence and attempted to explain the dramatic decline in strike activity. At the same time, other industrial relations academics responded to the substantial shift in the balance of power in favour of management and employers by adopting a more overt managerialist agenda in their research. Thus, they made management the primary focus of study, with their primary concern the solution of managerial problems rather than analysis of management per se, and many therefore switched their allegiance from the social science orientation of industrial relations to the relatively prescriptive ethos of human resource management. However, industrial relations can arguably embrace the concerns of human resource management within its remit, as Edwards's leading text (2003) admirably demonstrates. Thus, with its analysis of *all* the parties in the employment relationship, the processes, dynamics, and changing context of that relationship, its relatively strong interdisciplinary traditions, and its empirical and to some extent theoretical foundations, industrial relations continues to have much to say about the world of work, in spite of the decline in power and influence of trade unions in Western countries and probably much more so now in the context of the current global economic crisis.

In this chapter, it is suggested that seeing workplace bullying through the eyes of industrial relations can suggest a new perspective that may contribute significantly to contextualising the workplace bullying debate; that is, an industrial relations perspective can relate workplace bullying more systematically to worker and management interests in the political, economic, historical, legal, and social contexts of particular countries, employment sectors, work organisations, and occupations.

This is not to say in any sense that the strong psychological and social psychological analyses that have been central to the workplace bullying debate to date are somehow invalid, nor that they will not continue to drive much of the workplace bullying debate. Rather, just as some important new developments in the field of industrial relations have begun to make use of social psychological approaches, for example, in exploring the ways in which workers may or may not mobilise and act collectively to resist management, so too is there arguably much that industrial relations might offer to the study of workplace bullying through both its traditional and more recent interests and its interdisciplinary approaches—an approach that this author is keen to advocate.

Workplace Bullying and the Pioneering Work of Jon Sjøtveit

To progress the argument, this section looks at how Jon Sjøtveit (1992a, 1992b, 1994) effectively paved the way for such an approach. Sjøtveit was a well-known Norwegian trade union activist and researcher, having an industrial background and substantial experience in health and safety at work. Thus, his starting point with regard to workplace bullying was as a trade unionist, but one who sought a deeper theoretical explanation than then available. He adopted a radical perspective, influenced by a Marxist interpretation of the nature of the employment relationship. The quality of his analysis is impressive: he was awarded an MPhil degree for his work from the University of Oslo in 1992, and in 1993 he won a prestigious Fafo Institute for Labour and Social Research prize for the best social science thesis. However, he remained true to his trade union roots, and his book (Sjøtveit, 1992a) was aimed primarily at workers themselves, written in a very accessible style, and intended to promote discussion and change practice by means of workers' education study circles. Unfortunately, his work has been somewhat neglected in more recent years, as sadly he died shortly after this initiative. In addition, his writing has not been previously available in English, although an unpublished translation of some of his work has recently come to light (Dobson, 2000).

An important aspect of Sjøtveit's analysis was to consider workplace bullying in terms of the labour process (1992a, 1992b, 1994). Thus, Sjøtveit accepted that management under capitalist conditions had to consider how best it might control labour and maximise the output of workers. At the outset, here is an important link between Sjøtveit and the study of industrial relations. Whilst most of the early industrial relations academics in Britain did not recognise the significance of Marx in this regard, nevertheless central to their perspective was the contested "wage-effort" bargain between workers and employers, as embodied in their pluralist assumptions. Arguably, this was essentially the same issue but seen in different terms, from a different perspective, and with different implications, not least in that the resulting conflict could be managed and processed effectively within the capitalist system. However, with the publication of Braverman's seminal work (1974) on the labour process and a growing Marxist challenge to the industrial relations academic orthodoxy in the 1970s (Hyman, 1975), many began to conceptualise this aspect of the employment relationship more closely in Marxist terms, even though some did not see themselves as Marxists in the full sense of the term (Fox, 1974).

Sjøtveit also placed considerable emphasis on the significance of collectivism and individualism in the workplace and on the role that attitudes, perceptions, and norms played in the extent to which a particular workplace culture and the perspectives of the workforce were characterised as such. However, he did not simply equate collectivism with trade unionism but also saw the latter as the formal organisation of the former. He focused on the way in which workers perceived their situation and behaved in the workplace, whilst recognising that workers' collectivism could obviously provide the basis for effective trade unionism. For example, Sjøtveit saw banter (the *cheeky remark*) in the workplace as a common source of strengthening workers' collectivism, as an expression of community—but also, when it transgressed certain social boundaries, as a potentially destructive factor that could isolate individuals and lead to bullying (Sjøtveit, 1992a).

Whilst the debate about Britain has focused primarily on downward bullying, that is, bullying by managers of subordinates, Sjøtveit's work in the Norwegian context was also concerned with horizontal bullying, that is, bullying by fellow workers, which might often take the form of a group victimising an individual. He saw not only vertical but also horizontal forms of bullying as very serious challenges to workers' collectivism, reflected in the title of his book, *Når Veven Rakner* (1992), translated as *When the Social Web Breaks* (Dobson, 2000). "Bullying has an effect on not only those who are its victims, but also strikes at our sense of community. If the community is unable to protect the individual, the individual will be reluctant to take an interest in the community. The social web breaks down" (Dobson, 2000, p. 7). Thus, Sjøtveit advocated fellow workers' intervention to stop all forms of bullying as essential for the protection of the workers' collectivism, therefore suggesting that it should be taken up as an important trade union issue. He was interested especially in those workers who observed bullying incidents and in the kind of response of such third-party observers as the bullying process unfolded.

He argued that observer intervention at an early stage was critical in prevention, and he saw this in terms of whether observers were motivated to challenge the perpetrator out of a sense of workers' solidarity with their victimised colleague. Seeing observer intervention in these terms effectively politicises the act. However, if there was no third-party intervention by workers at an early stage, then—as the bullying escalated and the victim's behaviour and possibly perception of the situation changed—such intervention later would be much more difficult and less likely to occur. This approach poses a challenge not only to management and to workers themselves but also, in the context of bullying in Norwegian workplaces, to many trade union activists. Although aspects of the situation may differ, Sjøtveit's notion of observer intervention is relevant not only to horizontal but also to downward bullying.

Sjøtveit did not relate this perspective explicitly to the field of industrial relations, at least as we know it in Britain and the Anglo-Saxon countries, but this is not surprising since industrial relations is not a distinct field of academic study in Norway. Instead, he placed it firmly within a sociological perspective and its interpretations of the employment relationship. However, several of the key elements of Sjøtveit's approach clearly share common ground with the British study of industrial relations. His focus on the labour process and the dimensions of collectivism and individualism that cut through his analysis of bullying are obvious elements here. So too is the fact that whilst adopting an essentially sociological perspective, he was also very concerned about the micro-level processes, not least with regard to the intervention or otherwise of observers of bullying. In the context of industrial relations literature in Britain, Kelly's work especially (1998) has looked at the ways in which workers' collectivism and mobilisation might or might not occur in particular contexts, and he has emphasised the value of exploring the micro-level process involved in this, thus introducing social psychological concepts to his industrial relations analysis. However, before explaining and discussing the relevance of Kelly's work, Ironside and Seifert's contribution (2003) must be examined.

Bullying and Ironside and Seifert's Industrial Relations Perspective

In the previous edition of this volume (Einarsen et al., 2003), the chapter by British writers Ironside and Seifert (2003) was a new departure in how the issue of workplace bullying might be seen, for they looked at it almost exclusively from the viewpoint of industrial relations, focusing on downward bullying within the British context. Mike Ironside and Roger Seifert are well known in the industrial relations academic field in Britain and for the trade union orientation of their work over many years with industrial relations colleagues at Keele University. Much of their published work has focused on public-sector industrial relations per se, and their study of workplace bullying (Ironside and Seifert, 2003) has been their first and only publication in the workplace bullying field. Arguably, inadequate attention has been paid to their contribution to the workplace bullying debate. This is no doubt because as industrial relations specialists, they are not widely known to workplace bullying specialists and also because they have not followed up with further publications in this area of work. However, theirs is a key contribution in demonstrating the potential of a rather different and thought-provoking industrial relations perspective of bullying.

Ironside and Seifert (2003) not only placed the issue within the debates of industrial relations as a field of study but also related it to the concerns of industrial relations practitioners and especially trade unionists. They emphasised the significance of a historical perspective, as well as the role and roots of workplace bullying in the birth and development of capitalism. Although Ironside and Seifert make no reference to Sjøtveit's work and instead draw on the concerns of British industrial relations, they do make the same assumptions about the nature of the labour process and the question of managerial control of labour, as well as similar assumptions about the prospect of resistance to it by workers. Essentially, they see bullying as a rational ingredient of management, arising from the essential nature of the capitalist employment relationship, and as a method that can be expected to be employed when other forms of control do not have the intended effect. Therefore, they do not see bullying as an accident or primarily the work of misguided managers. Also, implicit in this perspective are the possibilities and limits of bullying as means of managerial control.

They highlighted three key features of the labour market in particular as being significant in contributing to the potential for workplace bullying by managers of subordinates. First, they emphasised the fundamental power imbalance between workers and employers (Ironside and Seifert, 2003, pp. 384–385). This feature relates to their second point that workers are numerically much greater than employers and have to compete with each other to gain employment. It also relates to their third point that the consequences for workers who fail to find employment are much more damaging to them than the consequences for employers who fail to hire labour. Thus, the fear of unemployment—the degree of which may vary but is never entirely absent for workers—affects workers' behaviour and attitudes at work, dismissal being the ultimate sanction that employers can implement. Ironside and Seifert therefore suggest that workers, unlike employers, have serious limitations to their choices regarding how they earn their living and the particular job they do in that for them, the alternatives to their current job might be unemployment or another job that is no better or perhaps worse than the one they hold. These labour market factors and the imbalance of power associated with them affect not only the attitudes and behaviour of workers themselves but also those of management and employer, and the way in which the relationship between them is constructed and develops.

Ironside and Seifert follow through their historical line of argument by drawing attention to the changes to British industrial relations in the context of Thatcherism, suggesting the possible significance of these changes for the apparent increase in bullying and harassment in recent years. In doing so, they highlight the considerable shift in the balance of power that has favoured employers and the intensification of profit, cost, and performance pressures for both private- and public-sector organisations. It has been suggested that the incidence of workplace bullying is considerably greater in the public than in the private sector in Britain (Hoel and Cooper, 2000; Hoel et al., 2004; UNISON, 1997; Zapf and Einarsen, 2003); appropriately, Ironside and Seifert have suggested that a number of key public-sector developments may be significant here. They draw attention to government initiatives to impose financial restraints on the sector, to enforce quasi-business restructuring and culture, and to implement complex performance measurement systems, all leading to work intensification and increasingly Taylorised managerial methods (Ironside and Seifert, 2003, pp. 387–389). They suggest that this set of initiatives has led to more rigid managerial control of labour and had a major impact on workplace industrial relations. These developments are of particular significance in the case of public-service workers, since they challenge the traditional levels of relative autonomy associated with such skilled and professional occupations and can be expected to provoke a sense of grievance amongst these workers (Hoel and Beale, 2006).

It is difficult to establish with confidence, on the basis of existing research and in light of the short history of workplace bullying as a distinct subject of study, whether there actually has been a significant increase in workplace bullying and harassment over the last 30 years in Britain, but certainly there are grounds to suspect this, and serious concern about the issue by trade unions, employers, government, and nongovernmental organisations (NGOs) has clearly emerged over the last 10 years or so. In addition, Ironside and Seifert present a credible argument regarding the emergence of industrial relations factors that are probably conducive to an increase in workplace bullying over the last two or three decades in Britain.

In terms of contesting and preventing bullying, Ironside and Seifert advocate a collectivist approach. However, rather than exploring the micro-level processes as Sjøtveit did, they focus instead on industrial relations practice and promote the use of workplace grievance and disciplinary procedures as means of workers pursuing bullying cases, rather than through distinct anti-bullying procedures (Ironside and Seifert, 2003, pp. 389–397). In conjunction with this approach, they emphasise the role of trade unions in the promotion of a collective challenge to workplace bullying.

Context, Agency, Bullying, and Industrial Relations

Both Sjøtveit and Ironside and Seifert are to be commended for taking the first steps in indicating the potential for employing industrial relations approaches in the study of workplace bullying. A key strength of Ironside and Seifert is that they highlight the industrial relations context in which bullying occurs. Some other studies of workplace bullying in Britain have made valuable contributions to placing the issue within employment contexts, including studies of bullying in the Civil Service, further education, and the National Health Service (Lee, 2002; Lewis, 1999; Quine, 1999). However, this work does not engage with the more fundamental concepts about the employment relationship in the way that Ironside and Seifert do, and thus arguably it falls short of embracing a more comprehensive industrial relations and sociological perspective in terms of context. On the other hand, Ironside and Seifert might be criticised for their emphasis *exclusively* on context at the expense of the subjective element, of social agency in terms of the individual. Whilst it is fair to say that they highlight the role of trade unions in respect of intervention to contest bullying, and there is therefore an element of social agency in this sense, nevertheless they fail to look at the issue at a micro-level in terms of the perceptions and interaction of the individual. As a result, it has been suggested that this aspect can be appropriately addressed by drawing together Ironside and Seifert's work with aspects of Sjøtveit's approach (Hoel and Beale, 2006).

To phrase this in sociological terms, it could be argued that the established workplace bullying debate has paid inadequate attention to the significance of social structure in its explanations. On the contrary, Ironside and Seifert attempt to explain workplace bullying essentially in social structural terms, giving little attention to the social actor. Whilst Sjøtveit focuses particularly on the social actor and his or her interaction, he also evidently recognises the significance of social structure, thus embracing a more comprehensive sociological analysis. At the same time, and understandably in the Norwegian context, his work is divorced from the British industrial relations debates and perspectives, and the latter might also contribute significantly to the workplace bullying field. Hence, there is a good case for creating linkages between Sjøtveit, and Ironside and Seifert, as a basis for a more interdisciplinary approach to the issue.

With regard to Sjøtveit—as a trade union activist, whose starting point for analysis of bullying embraced the concept of the labour process—his approach was collectivist, sociological, and essentially political. But at the same time he was especially interested in the detailed processes of how and why particular workers did or did not respond to bullying in collectivist terms. He was therefore concerned with the social psychological processes, the subjective element, whilst recognising that neither psychological nor social psychological perspectives *alone* could answer the questions he posed. There are important parallels here with the work that Kelly (1998) has pioneered in Britain, with regard to the application of mobilisation theory to industrial relations.

John Kelly is a professor of industrial relations, formerly of the London School of Economics and currently at Birkbeck, University of London. Very well known in the British industrial relations field, he has published extensively, and his work is within the Marxist tradition. His early background was in psychology, and his insights into the psychological and social psychological aspects of industrial relations have been important in his development of an approach to the field, which has typically related the micro-level processes of the employment relationship and workers' attitudes to their collective behaviour to broader social structural factors and the context and dynamics of capitalism. This insight was demonstrated particularly in his critical overview and suggestions for the development of industrial relations as a field of study (Kelly, 1998). Herein, he proposed the application of mobilisation theory—an approach developed quite extensively by sociologists in the study of social movements—to the question of workers' attitudes, behaviour, and, in particular, willingness to act individually or collectively in particular circumstances. In doing so, he provoked considerable debate and interest amongst industrial relations specialists, particularly from those who were keen to look in detail at behaviour in the workplace itself. Much previous industrial relations research that was concerned with workers as a party to industrial relations focused on trade unions per se, whilst Kelly's starting points here were the individual worker, the worker's perception and response to the demands of the employment relationship, and whether this would or would not lead to a collective pattern of behaviour, that is, to the organisation and mobilisation of workers in the form of trade union action. This approach has very much in common with Sjøtveit's focus, although Sjøtveit did not relate the issues explicitly to mobilisation theory, either its particular concepts or literature.

Mobilisation theory highlights perceptions of injustice, the attribution of blame, social identification with the issue, collectivist or individualist interpretations of the situation, and whether and why this might or might not translate into a collective response (Gamson, 1992, 1995; Kelly, 1998, pp. 24–65; McAdam, 1988; Tilly, 1978). In applying this approach to industrial relations, Kelly does not discuss the question of workplace bullying explicitly, but he does raise issues about oppressive workplace behaviour more generally (Kelly, 1998), and there is considerable potential to relate his ideas about workers' mobilisation to workplace bullying.

In addition, some industrial relations writers have emphasised social agency in terms of the role of union leadership, albeit in the context of trade union renewal and union militancy (notably Darlington, e.g., 2002, for a summary of his argument). Not only is union leadership relevant to an analysis of workers' mobilisation and militancy but also it strengthens Sjøtveit's argument about observer intervention to contest bullying at its early stage and the role that workplace union activists could play in this. Thus, unions and sectors that have strong workplace union organisation and effective local union leaders who are proactive in challenging management, provide encouragement to their wider workforce to act similarly, and thus amongst other things increase the likelihood of intervention by both union activists and workers themselves to contest bullying by managers. Of course, the combination of such workplace union leadership and militancy with managerially led radical and rapid change in the workplace that threatens to worsen established employment conditions may create a cycle of conflict focused on the "frontier of control," of which bullying by managers and intervention by union activists would probably be a fairly central feature. In such circumstances, bullying managers may make examples of more vulnerable and isolated workers to convey a message to the wider workforce and union, whilst union activists may attempt to establish a more disciplined union membership to ensure solidarity, with any union "dissidents" and nonunion members placed under increased pressure by the union and possibly ostracised.

Turning to Ironside and Seifert's work, first, their industrial relations perspective suggests considerable opportunities for empirical work that could explore the relationship among workplace bullying and economic restructuring, the impact of unemployment and recession on workplace attitudes, the decline in union power and influence, the reduction in coverage of collective bargaining and its weakness in terms of outcomes favourable to unions, and shifts in the relative balance of power between workers and management.

Second, they emphasise key features of the labour market in terms of the imbalance of power between workers and employers, the limits on workers' choice regarding their employment, and the relative vulnerability and financial insecurity of workers. Whilst the question of power cuts through so much of both the industrial relations and the workplace bullying literatures, its conceptualisation is not unproblematic. Kelly (1998) has argued that there is scope for a clearer theoretical approach to power in industrial relations, and perhaps this is the case in the workplace bullying field too, or arguably at least more might be done to state the assumptions made about the concept in particular studies in both fields. Power in this chapter refers to *distributive power*, that is, power secured at the expense of another party (Hyman, 1975, pp. 26–27; Martin, 1992, as cited in Hoel and Beale, 2006, pp. 249–250). Of course, there is an alternative poststructuralist view of the concept; and although it is not one shared by this author and little interest in it has been shown in industrial relations, a few researchers have made such assumptions in the workplace bullying literature (e.g., Liefooghe, 2003).

These key features suggested by Ironside and Seifert clearly have a central impact on the labour process, especially regarding the nature of managerial control of labour in the workplace and of workers' resistance to it. They open up a wealth of interesting and important questions for bullying research, through both quantitative and qualitative work that could directly enquire about workers' perceptions of the relationship between the form and extent of bullying and about victims' and observers' responses to it. For example, the periodic Workplace Employee Relations Surveys in Britain (the largest survey exercises of their kind globally) suggest important possibilities for the collection of invaluable quantitative data that can link workplace bullying and harassment to some of these industrial relations issues.

Some Conclusions

This chapter has suggested some new perspectives and issues for the workplace bullying agenda, linking some of the approaches, concepts, and subjects of study of industrial relations to the established body of theory and research associated with bullying, and making use particularly of the insights of Sjøtveit, Ironside and Seifert, and Kelly (and to some extent Darlington) in doing so. Here, the main emphasis has been placed on ways in which the analytical framework of the workplace bullying debate might be developed, rather than on the ethical dimensions of the issue. However, I am by no means indifferent to the latter but believe that developing the analytical framework more fundamentally in terms of sociological and particularly industrial relations dimensions can, amongst other things, lay stronger foundations for the ethical and prescriptive aspects of the debate. In spite of this emphasis on analysis of the issue, I am in no sense a detached or disinterested observer but an active trade unionist who wishes to see workplace bullying contested and prevented wherever and whenever possible, but who believes that a more fundamental analysis of the issue in terms of industrial relations equips us more effectively to tackle it. Thus, as a workplace union representative, I also believe in a praxis of contesting workplace bullying of the kind embraced by Sjøtveit (1992a). At the same time, my current work and forthcoming publications in the bullying field are deliberately focused on a series of broad theoretical questions in an attempt to establish some new theoretical parameters, suggesting a new framework that might be taken into account in future empirical research and in ethical and prescriptive considerations about workplace bullying.

As was suggested near the start of this chapter, industrial relations is concerned not least with conflict and cooperation between workers and management, in respect of the amount and determination of pay for particular jobs, the type and amount of work done in return, the effort involved, the organisation and periodic reorganisation of work, the way it might be measured and/or intensified, and various managerial attempts to maximise productivity. An important part of the argument here is that analyses of bullying need to engage in the particular way in which these issues are played out in particular workplaces, occupations, and employment sectors. This is primarily because the industrial relations aspects may have a very direct bearing on management's propensity to bully workers; in addition, these aspects may have a similar bearing on the prospects of effective resistance to and prevention of bullying; and finally, they will also vary considerably among workplaces, occupations, and sectors—and indeed among national contexts. However, there are also other important industrial relations factors that cut across workplaces, occupations, and employment sectors. These other factors suggest some interesting questions to be explored, for example, about possible relationships between patterns of bullying and workplace unionism, union recognition, and nonunion workplaces; different methods of pay determination; human resource management models and "high performance work practices"; management, union and/or workers' attitudes to equal opportunities and diversity; employee involvement and participation; company size; the introduction of new work systems; technological change; and, of course, job security, the threat of redundancy, and the impact of economic recession. There is also the very important question of trade union approaches to bullying prevention and the advantages and possible disadvantages for unions in the way they have commonly made use of the bullying rhetoric—at least in Britain—in industrial relations practice (Hoel and Beale, 2006).

On a more theoretical basis, this chapter has suggested that workplace bullying analysis might engage more fully in the notion of the wider power relationship between workers and employers as indicated by industrial relations and sociological approaches; and in perceiving the employment relationship as a labour process with labour as an uncertain purchase, with its consequences in terms of control and resistance. This view poses the question of bullying as a potential managerial control mechanism, with implications for the issue of managerial legitimacy (Beale and Hoel, in press); it also poses important issues about autocratic control and conflict versus cooperation as management strategies, as well as about their significance for workers' compliance and commitment.

If a more interdisciplinary approach were adopted with regard to the study of workplace bullying, an approach that made more use of industrial relations and sociological perspectives and explanations alongside established psychological and social psychological analyses, then potentially both the common characteristics of the bullying behaviour, the associated interaction, consequences, and perceptions of it, *and* analysis of it in terms of its wider context and of its contextual significance might be given appropriately similar weight in attempting to understand the issue. There is potential for more studies that compare workplace bullying trends and processes—for example, between countries or perhaps between occupations or between small and large workplaces within the same employment sector and country—and these could employ some of the established perspectives of industrial relations and explore the ideas of Sjøtveit, Ironside and Seifert, and Kelly. Such studies could do much both to contextualise the analysis more fundamentally and to identify the similarities and differences between such case studies. In this way, the significant variables might be identified more comprehensively through an incorporation of the industrial relations dimensions, and, one hopes, we could begin to develop a deeper understanding of the reasons for the different incidences of bullying and the presence or absence of effective intervention to contest it. A problem with the established approaches is, arguably, that some of the disciplinary restrictions imposed by psychological parameters are by definition problematic in making adequate sense of the contextual dimension. It is true that the debate to date has to some extent incorporated some sociological ways of seeing, but nevertheless the argument here is that it could go much further in embracing fundamental sociological and especially industrial relations approaches and literature.

Whether we like it or not and whatever they may be, as workers, managers, employers, or researchers and academics, we all make assumptions about the nature of the employment relationship. As much of my previous published work in the industrial relations field suggests, I have a Marxist perspective of the employment relationship (e.g., Beale, 2003, 2004, 2005a, 2005b). My position is based on an acceptance of the key Marxist notions of labour power, surplus value, and the fundamental contradictions of capitalist development. It is also based on my agreement with much of Braverman's analysis of the labour process (Braverman, 1974), with its Marxist assumptions and its emphasis on employers' need to drive down labour costs, on the issue of managerial control of labour arising from the purchase of labour power, on the inevitability of conflict between capital and labour (whether overt or covert), on the significance of social class, and on the importance of placing one's analysis in the wider context of capitalism. However, unlike Braverman, I would also emphasise workers' resistance to these developments in analysis of the labour process.

This combination of assumptions underlies my approach to the issue of workplace bullying. If one accepts such a view of the employment relationship, or at least the essential part of it, it follows that the nature of capitalism *inevitably* generates a degree of workplace bullying and that changes to the particular character and dynamics of capitalism in particular countries, employment sectors, organisations, and occupations need to be explored and tracked to help to explain the variations in the incidence of bullying. However, first, managers are also social actors themselves and do not *always and necessarily* serve their employer's best interest in terms of the basic tenets of the labour process analysis. In short, managers sometimes bully workers for their own reasons regardless of the employer's interests, and therefore it follows that not all bullying is an inevitable product of capitalist relations. Second, whilst a degree of bullying might be generated inevitably, it can also be contested—sometimes effectively—but the assumptions about the employment relationship and capitalism that have been outlined here inform, affect, and limit the ways in which this might occur. With these factors in mind, collective workers' resistance in the terms suggested by Sjøtveit and Kelly is arguably the way forward. This is not intended to suggest that the policies of government or employers aimed to challenge bullying are irrelevant or that unions cannot play a role in promoting them, but these policies have significant limits, and the solidarity of workers in the workplace needs to be central. Third, it also means that if a degree of bullying is inevitably generated by the nature of capitalism, then workers' solidarity to contest it is an indefinite process under this social system; furthermore, that once one bullying problem is solved, another will probably emerge sooner or later; or that market, political, or organisational restructuring factors will override previous "solutions" in particular workplaces. Fourth, of course, not *all* labour movement activists accept the ongoing inevitability of capitalism; for those who do not, challenging the root causes of workplace bullying that are related to the nature of the labour process can be linked to wider anticapitalist, socialist, and militant trade union campaigns. These four key points outline my own position.

A significant proportion of British industrial relations academics accept all or many of the basic assumptions that have been outlined here about the nature of the capitalist employment relationship. However, not all do so, and for those working in the workplace bullying field who do not, the basic issues considered in this chapter still raise important questions and points that are relevant to them and to all those who reject the essential tenets of a Marxist interpretation of the employment relationship. First, much of what Sjøtveit, Ironside and Seifert, and Kelly have to say raises questions of wider relevance to all. For example, it is difficult to deny that workers think and behave both as individuals and collectively at different times, in different circumstances, and that when they behave collectively, this is significant in redressing the prevailing balance of power between workers and management. All previous literature accepts that the balance of power is fundamental to the experience of bullying, and therefore it is difficult to deny that in these terms both Sjøtveit and Kelly raise important questions about when and how collective workers' behaviour might challenge bullying. Ironside and Seifert place the balance of power within the context of the labour market, suggesting that the ultimate threat of unemployment is fundamental to the issue. This notion might encourage managers to believe that they can bully some workers with impunity, with workers just putting up with it as a result and looking for a job elsewhere if they can. Second, industrial relations academics who adopt a pluralist rather than a Marxist or Marxist-influenced interpretation of the employment relationship still recognise that there is a conflict of interest between managers and workers. Thus, regardless of Marxist assumptions, the study of industrial relations by its nature raises basic questions that workplace bullying research to date has not really embraced as a core feature of its analysis. By doing so, the view of this author is that workplace bullying research could be broadened and deepened significantly, whilst continuing to recognise and develop the fundamental contribution that psychological and social psychological contributions to the workplace bullying field have and will continue to make.

References

Ackers, P., and Wilkinson, A. (2003) Introduction: The British industrial relations tradition—formation, breakdown and salvage. In P. Ackers and A. Wilkinson (eds.), *Understanding work and employment: Industrial relations in transition* (pp. 1–27). Oxford: Oxford University Press.

Beale, D. (2003) Engaged in battle: Exploring the sources of workplace union militancy at Royal Mail. *Industrial Relations Journal, 34* (1), 82–95.

———— (2004) The impact of restructuring in further education colleges, *Employee Relations, 26* (5), 465–479.

———— (2005a) The promotion and prospects of partnership in Inland Revenue. In M. Stuart and M. Martinez Lucio (eds.), *Partnership and modernisation in employment relations.* London: Routledge.

———— (2005b) Miners' support groups and the significance of geography, identity and leadership in the 1984–85 British miners' strike. *Capital and Class, 87,* 125–150.

Beale, D., and Hoel, H. (in press) Workplace bullying and the employment relationship: Questions of prevention, control and context. *Work, Employment and Society.*

Blyton, P., and Turnbull, P. (2004) *The dynamics of employee relations*, 3rd ed. Basingstoke, UK: Palgrave Macmillan.

Braverman, H. (1974) *Labor and monopoly capital: The degradation of work in the twentieth century.* New York: Monthly Review Press.

Darlington, R. (2002) Shop stewards' leadership, left-wing activism and collective workplace union organisation. *Capital and Class, 76,* 95–126.

Dobson, S. (2000) When the social web breaks: Community and bullying in the workplace. Unpublished translation of J. Sjøtveit (1992), *Når veven rakner: Om samhold og mobbing på arbeidsplassen.* Oslo: Folkets Brevskole.

Edwards, P. (ed.) (2003) *Industrial relations: Theory and practice*, 2nd ed. Oxford: Blackwell.

Einarsen, S., Hoel, H., Zapf, D., and Cooper, C. L. (eds.) (2003) *Bullying and emotional abuse in the workplace: International perspectives in research and practice.* London: Taylor & Francis.

Flanders, A., and Clegg. H. A. (eds.) (1954) *The system of industrial relations in Great Britain: Its history, law and institutions.* Oxford: Blackwell.

Fox, A. (1966) Industrial sociology and industrial relations. Royal Commission on Trade Unions and Employers' Associations Research Paper 3. London: HMSO.

——— (1974) *Beyond contract: Work, power and trust relations.* London: Faber.

Frege, C. (2003) Industrial relations in continental Europe. In P. Ackers and A. Wilkinson (eds.), *Understanding work and employment: Industrial relations in transition* (pp. 242–262). Oxford: Oxford University Press.

Gamson, W. A. (1992) *Talking politics.* Cambridge: Cambridge University Press.

Gamson, W. A. (1995) Constructing social protest. In H. Johnston and B. Klandermans (eds.), *Social movements and culture* (pp. 85–106). London: UCL Press.

Hoel, H., and Beale, D. (2006) Workplace bullying, psychological perspectives and industrial relations: Towards a contextualised and interdisciplinary approach. *British Journal of Industrial Relations, 44* (2), 239–261.

Hoel, H., and Cooper, C. L. (2000) Destructive conflict and bullying at work. Unpublished report. Manchester, UK: Manchester School of Management, Manchester Institute of Science and Technology.

Hoel, H., Faragher, B., and Cooper, C. L. (2004) Bullying is detrimental to health, but all bullying behaviours are not necessarily equally damaging. *British Journal of Guidance and Counselling, 32,* 367–387.

Hyman, R. (1975) *Industrial relations: A Marxist introduction.* London: Macmillan.

Ironside, M., and Seifert, R. (2003) Tackling bullying in the workplace: The collective dimension. In S. Einarsen, H. Hoel, D. Zapf, and C. L. Cooper (eds.), *Bullying and emotional abuse in the workplace: International perspectives in research and practice* (pp. 383–398). London: Taylor & Francis.

Kelly, J. (1998) *Rethinking industrial relations: Mobilization, collectivism and long waves.* London: Routledge.

Lee, D. (2002) Gendered workplace bullying in the restructured UK Civil Service. *Personnel Review, 31,* 205–227.

Lewis, D. (1999) Workplace bullying: Interim findings of a study in further and higher education in Wales. *International Journal of Manpower, 20,* 106–118.

Liefooghe, A. P. D. (2003) Employee accounts of bullying at work. *International Journal of Management and Decision Making*, 4 (1), 24–34.

Martin, R. M. (1992) *Bargaining power*. Oxford, UK: Clarendon Press.

McAdam, D. (1988) Micromobilization contexts and recruitment to activism. *International Social Movements Research*, 1, 125–154.

Quine, L. (1999) Workplace bullying in NHS community trust: Staff questionnaire survey. *British Medical Journal*, 3, 228–232.

Sjøtveit, J. (1992a) *Når veven rakner: Om samhold og mobbing på arbeidsplassen*. Oslo: Folkets Brevskole.

——— (1992b) Mobbing på arbeidsplassen som diskurs og som sosialt phenomen. [Bullying in the workplace as discourse and as a social phenomenon], MPhil dissertation, University of Oslo.

——— (1994) A trade union approach to the problem of bullying at work. Unpublished manuscript. Workers' Educational Association of Norway.

Tilly, C. (1978) *From mobilization to revolution*. New York: McGraw-Hill.

UNISON (1997) *UNISON members' experience of bullying at work*. London: UNISON.

Webb, B., and Webb, S. (1897) *Industrial democracy*. London: Longman.

Zapf, D., and Einarsen, S. (2003) Individual antecedents of bullying: Victims and perpetrators. In S. Einarsen, H. Hoel, D. Zapf, and C. L. Cooper (eds.), *Bullying and emotional abuse in the workplace: International perspectives in research and practice* (pp. 165–184). London: Taylor & Francis.

13

Workplace Bullying as the Dark Side of Whistleblowing

Stig Berge Matthiesen, Brita Bjørkelo, and Ronald J. Burke

CONTENTS

Introduction

Over 10 years ago, Lennane and De Maria (1998) wrote an article in the *Medical Journal of Australia* entitled "The Downside of Whistleblowing" in which they described the potential devastating experiences that employees reporting some kind of wrongdoing may be confronted with. Whistleblowing occurs when an employee witnesses wrongdoing of some kind and reports it. Those who are informed of the wrongdoing may be leaders in the organisation or external authorities. The voicing of concerns may terminate the wrongdoing. But whistleblowing can also backfire. A whistleblower often risks punishment in return for his or her initiative to stop the acts of wrongdoing.

Box 13.1 illustrates the punishment risks for whistleblowing. The three cases reviewed in the box received considerable media interest in Norway from 2000 and a few years later on. Two of the cases concern whistleblowing

BOX 13.1   THREE NORWEGIAN WHISTLEBLOWER CASES WITH HUGE MEDIA COVERAGE

Case 1: The director of finance's story. After repeatedly having reported internally about Siemens Business Systems' systematic over-billing of their client, the Norwegian military, director of finance Per Yngve Monsen anonymously, following the corporation's own internal guidelines, reported to the main office in Germany about the malpractice. This report backfired on Monsen. A witch hunt to detect the identity of the whistleblower was then launched, and he was met with both formal and informal sanctions. The whistleblower Monsen was later discharged as the only employee in an organisational downsizing process. Monsen sued his employer and won. Although Monsen won his case, the costs of reporting, psychologically and socially, were huge. An anxiety problem arose. Monsen admits that because of the devastating aftereffects, in retrospect he now regrets his initiative to blow the whistle within his former company. After several years without a new job, he now works as an administrative executive for a branch of the Salvation Army, Fretex, which runs a large number of secondhand stores all across Norway.

Case 2: The lawyer's story. Lawyer Kari Breirem worked as a director in one of Norway's biggest law firms, BA-HR, when she was asked to sign an invoice addressed to a Norwegian politician. The 1.5 million Norwegian Crowns (NOK) fee (about US$2.1 million) was set up to cover expenses the previous minister of health had had during a time period where he also received pay from the state for his previous work as a politician. The fee was paid by the company Aker RGI for work in association with the takeover of the Norwegian company Kværner. The transaction was made through the lawyer's company to hide that the politician received the fee in the same time span as he was receiving payment from the Norwegian state. Breirem copied the invoice and sought legal advice. Without Breirem's knowledge, her lawyer decided to hand the invoice over to the Norwegian national authority for investigation and prosecution of financial and environmental crime. Although today Breirem is employed as the director of a court of appeal, she also experienced life threats that required her to have police protection. She also faced ostracism by her previous colleagues, who treated her as a persona non grata when encountering her either within judicial circles or in private settings.

Case 3: The physician's story. The chief physician Carl-Magnus Edenbrandt reported about potential active euthanasia in the treatment of terminally ill cancer patients at a Norwegian hospital. Edenbrandt first reported through internal channels before he notified the Norwegian health authorities. The perceived wrongdoer was a doctor who was on vacation at the same time as he was in charge of this type of treatment. The doctor was later reprimanded for not having followed the reporting procedures for a given treatment and was given a warning. Because of this incident, with its aftermath whistleblowing report, the procedures for medical practice and decision making concerning terminally ill patients were considerably enforced. Edenbrandt himself, however, found that the possibility of continuing his Norwegian career as a physician came to a stop and that also his fellow colleagues who had supported him were given the silent treatment at his workplace. Luckily, the Swedish-born whistleblower Carl-Magnus Edenbrandt got a new job after returning back to his homeland, Sweden. He now works with palliative care patients.

in response to financial fraud, with the whistleblowers being a director of finance in a private company and a lawyer in a reputable law practice. The third case is from the public sector, a hospital. Here, a doctor reports a colleague to the health authorities on suspicion that a fellow doctor has committed euthanasia (mercy killing). What the three cases have in common is that the whistleblowers can be seen as holding middle-management positions. They all suffered retaliation, and all lost their jobs as a result of having blown the whistle. The whistleblowers also developed health problems afterwards, as reported in the literature describing their cases (Breirem, 2007; Lundquist, 2001; Monsen, 2008).

The research field of workplace bullying and whistleblowing have coexisted for the last two decades, with few cross-references between them. Workplace bullying research can be seen as use of inquiry mostly conducted by researchers, typically psychologists, in the Nordic countries and Northern Europe, whereas whistleblowing research has been dominated by social scientists and law specialists from the United States. The modest association between the two research fields can be seen as a paradox, as bullying is probably one of the major downsides and negative consequences after whistleblowing at work.

Our research interest in whistleblowing started by coincidence, as one of the authors of the present chapter acted as an expert witness in a Norwegian workplace bullying court trial in 2001. The employee exposed to bullying was stripped of his job obligations and relocated to another unit as a consequence of his having voiced his concern about weak management and unprofessional conduct among his fellow prison colleagues. Neither the bullying target himself nor the rest of the court seemed to realize the whistleblowing aspects of the prison case.

Back in 2001, the focus of the court was largely concerned with the issue of workplace bullying, whether this had taken place or not, for legal protection of whistleblowers was not yet acknowledged in the Norwegian Work Environment Act (Directorate of Labour Inspection, 2007). The employee exposed to bullying, gradually realizing that he was actually a whistleblower, lost the court case. The court ruled that the former prison officer, the plaintiff, now a recipient of disability benefits, had provoked the retaliation and the subsequent workplace bullying he was met with. This verdict ended this particular whistleblower case.

We will now outline how this chapter is built up. First, we clarify the concept of whistleblowing. Then, we outline whistleblowing as a process, using a simplified model that consists of three different scenarios, one of these being the retaliation scenario. The next section provides different definitions of retaliation, together with a review of some central empirical findings, and pinpoints some methodological problems within the retaliation research field. There follows a section on how whistleblowers may become victimised. Because the concept of retaliation comprises a distinct link between whistleblowing and workplace bullying, a brief concept clarification of workplace bullying is given before a further theoretical elaboration of whistleblowing is presented, including the distinction between short-term whistleblower retaliation and long-term retaliation in the form of workplace bullying. The last section of the chapter sketches possible interventions that may be implemented to avoid reprisals against whistleblowers or bullying of those who have voiced their concern about organisational misconduct in the workplace. There follow some concluding remarks.

Whistleblowing

How can whistleblowing be linked to workplace bullying? Before we investigate this relationship more thoroughly, the relevant concepts will be clarified. The view on whistleblowing in general and whistleblowers in particular has been very mixed throughout the years, not the least among lay people. Whistleblowing has been described as a selfish and spiteful act motivated by greed and personal interests, but equally as a selfless, altruistic act that is undertaken only at extraordinary personal costs (Miethe, 1999). Thus, whistleblowers have been labelled as "snitches," "rats," "moles," "finks," and "blabbermouths," but also as ethical resisters and people of conscience (Hersh, 2002). Whistleblowing is also regarded as a kind of civil courage, be it in the workplace or in other sectors of the society (Alford, 2008; Glazer and Glazer, 1988; Zapf, 2007).

Also, among social scientists there has been obvious disagreement over whether egoism or altruism is the ultimate motivation for whistleblowing, and whether the term *whistleblower* should be restricted to describe only employees or others conducting altruistic acts (Miceli and Near, 1992). The most influential definition of whistleblowing (cf. King, 1999) was expressed by Near and Miceli a quarter of a century ago: "The disclosure by organization members (former or current) of illegal, immoral or illegitimate practices under the control of their employers, to persons or organisations that may be able to effect action" (1985, p. 4). Thus, whistleblowing is a dynamic process involving at least three social actors, each of whom takes actions in response to the others: wrongdoers, who commit the alleged wrongdoing; whistleblowers, who observe the wrongdoing, define it as such, and report it; and recipients of the report of wrongdoing.

Most studies of whistleblowing have used the Near and Miceli (1992) definition as a starting point for their inquiries even if some alternative views and definitions do exist (Jubb, 1999). This definition implicitly states that a whistleblower can voice his or her concern both externally and internally. Still, the definition continues to generate controversy, since some researchers argue that a whistleblower who reports wrongdoing within the organisation, but with no information given to outside authorities, is not really a whistleblower (cf. Miceli, Near, and Dworkin, 2008). Generally, one may claim that there are two main types of whistleblowers (Miethe, 1999): "Internal" whistleblowers report misconduct to another person within the organisation, who can then take corrective action. The internal contact person may be an immediate supervisor, "ombudsman," union representative, or company executive. "External" whistleblowers, in contrast, expose fraud, waste, and abuse to external agents. Common sources for external reporting are law enforcement officials, lawyers, media reporters, and an assortment of local, state, and federal agencies, according to Miethe.

Whistleblowing Process

Several models of the whistleblowing process exist (Miceli and Near, 1992). In Figure 13.1, these models are synthesised into a simplified model of a typical whistleblowing process (scenario A or B). Some organisational wrongdoing or misconduct occurs within an organisation. This wrongdoing is witnessed by someone (relation 1 in scenario A or scenario B). An important recognition is that many of those individuals who are observing an illegal or unethical conduct stay passive (relation 2 in scenario A), which means that nothing happens; the wrongdoing conduct can continue to damage the organisation further (relation 3 in scenario A).

A successful whistleblowing action is described in scenario B. After witnessing the wrongdoing, the individual chooses to report it to a superior in

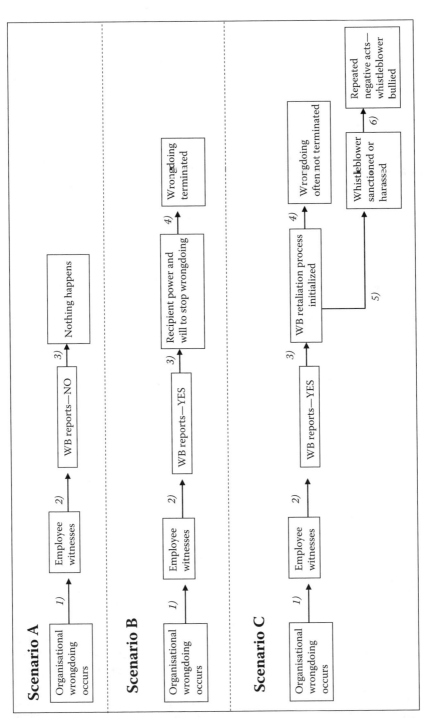

FIGURE 13.1
Three possible scenarios for those who blow the whistle.
Note: WB = whistleblowing.

the organisation or to someone else with the power to stop the misconduct (relation 3 in scenario B). This person has the skill and awareness to terminate the wrongdoing behaviour. The illegal or unethical conduct ceases (relation 4 in scenario B). Usually, the wrongdoers are sanctioned or punished; they may, for example, lose their jobs. Alternatively, depending on how the misconduct is handled, a warning and strong message to never repeat the misconduct may be given.

Scenario C in Figure 13.1 illustrates the retaliatory process in some detail. After unethical or illegal behaviour has occurred, once or several times, the employee witnessing it (relation 1 in scenario C) takes action as a messenger to inform someone to stop the misconduct (relation 2). However, what happens next is that the messengers themselves become targets; that is, some kind of retaliatory process is initiated (relation 3). Weak leadership, an organisational culture that remains defensive and protective, and the need for a scapegoat are some of the many possible factors that may trigger reprisals against the whistleblower (Near and Miceli, 1986; Rehg, Miceli, Near, and Van Scotter, 2008).

Even if the wrongdoing is brought to an end because of the efforts of the whistleblower, which may happen in some cases (relation 4 in scenario C), leaders or others closely associated with the whistleblower start to behave differently against the individual who blew the whistle (relation 5). The whistleblower is confronted with open hostility of some kind, such as ridiculing or threatening messages. Often the reprisals may be excessive, such as when tasks or responsibilities are instantly withdrawn from the whistleblower or when the social sanctions are presented publicly in a humiliating manner (Bjørkelo, Ryberg, Matthiesen, and Einarsen, 2008). "Silent treatment" (ostracism; cf. Williams, 2007) may take place for a period of time, causing retaliation (short-term harassment). In some cases, the whistleblower also faces being exposed to frequent negative acts. The whistleblower is unable to stop sanctioning acts that are being repeated regularly. If the whistleblower is subjected to long-term workplace bullying (relation 6), this situation should be labelled whistleblower bullying (see later).

Retaliation against and Victimisation of Whistleblowers

The attribution of the whistleblower's motivation and intent can be crucial for the way he or she is treated. According to Miethe (1999), most whistleblowers themselves are unable to say whether their action is either personally or socially motivated. Furthermore, according to Miethe, very few whistleblowers understand or anticipate the consequences of their actions; for instance, they do not consider the risk of being subjected to severe punishment or reprisals, whether some type of short-term whistleblower retaliation or more long-term punishment. Thus, whistleblowers can be met with retaliation in the form of workplace bullying, as outlined in Figure 13.1.

Retaliation against those who blow the whistle may narrowly be defined as "taking adverse action against an employee for opposing an unlawful employment practice or participating in any investigation, proceeding, or hearing related to such a practice" (Cortina and Magley, 2003, p. 248). More widely, it may be defined as "undesirable action taken against a whistle-blower—and in direct response to the whistleblowing" (Miceli et al., 2008, p. 11). This kind of conduct has also been labelled "workplace reprisals" (De Maria and Jan, 1997). Some argue that the purpose of potentially sacrificing the whistleblower is to prevent an "outbreak of an epidemic of ethical and moral responsibility" from other organisation members (Alford, 2008, p. 244). A more detailed definition has been given by Rehg, who describes retalia-tion as the "outcome of a conflict between an organisation and its employee, in which members of the organisation attempt to control the employee by threatening...or actually taking, an action that is detrimental to the well-being of the employee" (1998, p. 17).

Thus, whistleblowers who face negative acts such as aggravated job con-ditions or social rejection because of speaking out are exposed to retali-ation from their leaders, colleagues, or someone else in the organisation. The aggressive conduct directed against the whistleblower can also be seen as a kind of revenge (Rehg et al., 2008). Retaliation may sometimes be motivated by a wish for revenge, but it may also be motivated by other factors. According to Rehg, one may distinguish between the construct of retaliation against whistleblowers from the construct of revenge against an employee for committing some perceived harm. Revenge encapsu-lates a full range of aggressive behaviours, from verbal to physical, from covert to overt, from indirect to direct, and from interpersonally directed to organisationally directed (Aquino, Tripp, and Bies, 2001).

According to Miethe (1999), one should distinguish between individual wrongdoing and organisational wrongdoing (corporate crime). Individual wrongdoing is conducted by one or a few employees, usually not in high-ranking positions. Organisational wrongdoing or corporate crime is when the illegal or unethical acts are carried out or supported by the organisa-tion itself, by the CEO, or by other top leaders (as in the case of Enron). The probability of being punished after whistleblowing is much higher when the voices of concern are made after organisational wrongdoing, as com-pared with action taken after individual-oriented wrongdoing (cf. Near and Miceli, 1996). The two first cases presented in Box 13.1 (the cases of the financial director and the lawyers) should be seen as good examples of whistleblowing after organisational wrongdoing or corporate crime, with extensive media coverage afterwards. The third story, however (the physician's story), should be seen as an example of whistleblowing after individual wrongdoing.

Retaliation: Review of Empirical Findings

Regardless of whether one is an internal or external whistleblower, there are costs as well as benefits for speaking out against organisational misconduct. Some whistleblowers reach personal fame and wide recognition. Such was the case with the whistleblowers in Enron, Worldcom, and the FBI—three employees, all of whom were declared "persons of the year" by *Time* magazine in 2002 (Lacayo and Ripley, 2002). The opposite may also happen: the whistleblower is met with harsh punishment or retaliation, often of a persisting or long-term nature. One example is the case of the director of finance (cf. Box 13.1), who reported organisational wrongdoing and was met with harassment and loss of his job. The fate of the financial director corresponds with the literature, revealing that frequently reported retaliatory acts are ostracism, demotion, denial of opportunity to apply for promotion, removal of benefits, and unjustified use of disciplinary proceedings (e.g., Elliston, Keenan, Lockhart, and van Schaick, 1985).

Other studies have reported that whistleblowers risk losing their jobs and homes. Some whistleblowers have had to file for bankruptcy; others face marriage breakdowns or get divorced. Many also suffer from emotional strain, as well as additional personal problems. They may face career disruption or work-related reprisals, such as being demoted, as well as more personally directed types of retaliation, such as ostracism (Cortina and Magley, 2003; De Maria and Jan, 1997). This division is similar to the distinction reported in workplace bullying literature between work-related bullying, with deterioration or removal of work tasks or responsibility, and person-related bullying in the form of direct social attacks such as verbal threats or ridicule (cf. Tehrani, 1996).

Retaliation was reported by 17%–38% of identified whistleblowers in stratified random samples of federal employees (Miceli, Rehg, Near, and Ryan, 1999). Nearly 1,500 federal survey respondents in the United States who observed and reported misconduct within the preceding 12 months reported the proportion of different types of retaliation (Miethe, 1999). The seven most widespread retaliatory acts were found to be verbal harassment or intimidation (23%), poor performance appraisal (22%), being shunned by coworkers or managers (20%), assignments to less desirable or less important duties (18%), denial of award (13%), denial of promotion (11%), and transfer or reassignment to a different job with less desirable duties (10%).

Some studies based on selected samples of whistleblowers (Jos, Tompkins, and Hays, 1989; Lennane, 1993) and a nationwide interview study (Rothschild and Miethe, 1999) have revealed that the majority of whistleblowers in such investigations suffer from retaliation. Data from random samples of employees have however found that only about one in four whistleblowers experience retaliation (Miceli and Near, 1992). Similarly, in an Australian study, 39% (n = 72) of the whistleblowers reported having been exposed to formal reprimand after they reported wrongdoing at work (De Maria and Jan, 1997).

According to De Maria and Jan, over 30% were punitively transferred and over 20% were referred to psychiatric or psychological examination. In their study, the entire sample of whistleblowers was subjected to ostracism, and over 70% had experienced that their motives for reporting were questioned, or they had been personally attacked in some way.

In a representative sample of Norwegian employees comprising 2,539 individuals and 288 whistleblowers, less than 10% of the respondents faced retaliation after they had blown the whistle (Einarsen et al., 2007). By contrast, the prevalence of whistleblowers who experienced retaliation was almost doubled in a study of the general Norwegian population, conducted by Matthiesen, Bjørkelo, and Nielsen (2008). In this sample of 1,604 individuals, it was found that 18% among 640 whistleblowers had experienced retaliation. Among the whistleblowers, verbal reprimands from immediate superiors and others were reported to be the most widespread retaliatory acts. It should be mentioned that the Matthiesen and associates' study also revealed that the probability of being exposed to workplace bullying was nearly three times higher for the whistleblowers, as compared with the nonwhistleblowers (9.4% versus 3.2%, respectively).

According to Miceli et al. (2008), for most whistleblowers exposed to retaliation, there seems to be a pattern of repeated incidences of such reprisals, often from different sources. The severity of retaliation is therefore probably better assessed on the basis of how many types of reprisals the whistleblowers have experienced or been threatened with. Cortina and Magley (2003) separated work-related retaliation (formal, adverse actions documented in employment records) from social retaliation (antisocial behaviour) in their study. They found that 30% of the whistleblowers experienced social retaliation, whereas 36% experienced both social and work-related retaliation.

Retaliation against whistleblowers in North America seems to have increased in recent years, despite of 25 years of legal protection (Miceli and Near, 2005). As previously mentioned, Miethe distinguishes between two types of whistleblowing as a response to individual misconduct (conducted by single employees, not senior managers) and whistleblowing as a response to organisational wrongdoing (in which the organisation itself, or its senior managers, are the perpetrators). In relation to retaliation, it is reported that the probability of being met with punishment after having blown the whistle is much more likely for whistleblowers of organisational wrongdoing compared with whistleblowers of individual wrongdoing (cf. Miethe, 1999). Moreover, more severe forms of retaliation are found when employees have used external channels as opposed to internal channels to report inappropriate conduct (Near and Miceli, 1996).

A meta-analytic study, in which 26 samples were analysed and common trends identified (n = 18,781), found that retaliation was unrelated to education, job level, or occupying or having a role where it is explicitly prescribed to report wrongdoing at work (Mesmer-Magnus and Viswesvaran, 2005). This study observed that older employees were at greater risk of being on the receiving end of retaliation than were younger ones. To account for this find-ing, one may suggest that younger employees are probably less likely to be in positions to witness illegal or unethical acts of this nature. They are often less experienced, which may cause them to remain passive when illegal or unethical conducts occurs. A more surprising finding is that employees with values congruent with the organisation seem to be more prone to retaliation. A possible explanation may be that these employees to a greater extent than others voice their concern when they see violations of what they consider as organisational norms and values. In other words, they take greater risks and refuse to remain silent. It should come as no surprise that external whistle-blowers—those who choose to go public with their concern about organi-sational misbehaviour, for instance, by speaking to the media—are more exposed to retaliation than those who remain as internal whistleblowers.

The Mesmer-Magnus and Viswesvaran (2005) meta-analytic study also showed that employees who were successful in bringing the misconduct to the fore or were able to provide evidence of wrongdoing were less vulner-able when compared with the employees who were unsuccessful in stopping the wrongdoing or could not provide evidence of it. Furthermore, employees with strong support from their superiors are considerably less prone to face retaliation within their organisation than are their counterparts with little support. The risk of being exposed to retaliation after whistleblowing was also more prevalent when the reported wrongdoing was serious. It should also be mentioned that a recent study revealed gender differences in ante-cedents and consequences of retaliation after whistleblowing (Rehg et al., 2008). According to the findings by Rehg and his associates, lack of support from others was significantly related to retaliation for both men and women. This study found that a positive relationship with one's supervisor was nega-tively related to retaliation for both women and men. A positive leader con-nection was also positively associated with the decision of female employees to report misconduct to external sources.

According to Miethe (1999), prevalence data on retaliation may be biased. Estimating the frequency of retaliation through survey studies may underes-timate the occurrence of the phenomenon because employees who have been fired or expelled as a result of retaliatory acts are not present in the studies. Methodological differences may contribute to the varying reports of retalia-tory acts, although the choice of scientific method hardly accounts for the prevalence variability across studies, according to Miceli and Near (1992).

Victimisation of Whistleblowers

As noted in many studies, whistleblowing may be associated with symp-toms of physical and psychological distress (McDonald and Ahern, 2002), including anxiety, depression, and symptoms analogue to post-traumatic stress (Bjørkelo et al., 2008; Rothschild, 2008). In a survey study of 84 self-selected whistleblowers, 82% reported exposure to harassment, with negative subsequent effects such as troubled emotional state and psychical health, as well as negative influence on social activities (Soeken and Soeken, 1987). In this survey, the whistleblowers (n = 84) as well as their spouses (n = 59) reported increased levels of anxiety, in addition to symptoms of depression.

Furthermore, others have found severe depression or anxiety (84%) to be the most common mental health consequence related to whistleblowing (Rothschild and Miethe, 1999). In this respect, anxiety is generally described as a condition characterised by worry about everyday events and problems, as well as tension of some duration that the person finds hard to control (Andrews and Slade, 2002), whilst depression is described by low mood and lack of energy as well as interests (Olsen, Jensen, Noerholm, Martiny, and Bech, 2003). Seemingly, as a result of their experiences, some whistleblowers get trapped in a "dead end" situation (Alford, 2000). Some whistleblowers also seem unable to find resolution, which implies coming to terms with what has happened to them and being able to move on with their lives (Soeken, 1986). Soeken emphasised that whether the whistleblowers reach this stage of resolution seems to depend upon having a solid job and economic support.

Some whistleblowers develop a "master status" or identity that overshadows all other aspects of life (Bjørkelo et al., 2008; Rothschild and Miethe, 1999). According to Rothschild and Miethe, master status, which is originally a concept applied in sociology, in general describes how we label ourselves and are recognized by others. In relation to whistleblowing, master status signifies the process whereby some employees may have created a new identity based on their having blown the whistle, such that all new experiences following the whistleblowing are interpreted in the light of the whistleblowing (Rothschild and Miethe, 1999). According to a social interactionistic perspective, which assumes that behaviour "is a result of a complex interplay between personality and situational factors" (Greenberg and Baron, 2008, p. 135), employees not only are recipients but also may influence their surroundings by how they behave. Employees who interpret all new events through their master identity as previous whistleblowers may thus elicit responses from leaders and colleagues that may reinforce this identity.

However, some whistleblowers also describe experiences analogous to the concept of post-traumatic growth, which is "the experience of positive change that occurs as a result of the struggle with highly challenging life crises" (Tedeschi and Calhoun, 2004, p. 1). So, whereas some individuals regret having blown the whistle (Alford, 1999), others may not (Branch, 1979). It may be that the division between whistleblowers who claim that they would have done it again (Fitzgerald, 1990) and those who claim that it was not worth it (Alford, 1999) may be associated with whether employees had experienced posttraumatic growth or not. Boatright (1997) refers to the case of Joseph Rose, the U.S. attorney who reported illegal political contribution relating to the Nixon presidential campaign, as an example of how some employees may become more cynical as a result of their experiences. Rose pinpoints that with hindsight, he tends to see all ethical guidelines and focuses on integrity policies as just speeches about principles; for based on his experience, their content seems to disappear "very rapidly when it comes to the practical application of these concepts by strict definition" (Boatright, 1997, p. 111).

Workplace Bullying

The earlier presented behaviours and health consequences experienced by employees who have reported wrongdoing are well known to researchers on workplace bullying. Indeed, several other chapters in this book define and thoroughly reflect on the phenomenon of workplace bullying (e.g., Einarsen et al., this volume). Thus, here we will only briefly outline a few aspects concerning the bullying concept relevant to our investigation. Bullying occurs when an individual is met with humiliating negative behaviour for some period of time and is unable to stop the misconduct to which he or she is exposed, thereby reflecting a type of power imbalance. Exposure to such long-term, systematic negative social acts can be seen as a type of psychological siege, what Leymann (1990) denoted as "psychological terror," which may, in turn, lead to severe physical, psychological, and social problems (see, e.g., Bechtoldt and Schmitt, in press; Matthiesen and Einarsen, 2001, 2004; Mikkelsen and Einarsen, 2002; Tehrani, 1996; Tracy, Lutgen-Sandvik, and Alerts, 2006). Matthiesen (2006) differentiated bullying following whistleblowing as one of 10 bullying types, together with, for example, predatory bullying and conflict-driven bullying suggested by Einarsen (1999). Among a group of 221 victims of bullying (Nielsen, 2003), whistleblowing turned out to be the third most frequent self-reported antecedent for workplace bullying.

Workplace bullying is quite often considered as synonymous with workplace harassment (cf. Brodsky, 1976), indicating an overlap between the two concepts. One or a few extreme negative incidents can, however, also be regarded as harassment. In this chapter we define workplace harassment as one or a few negative and strongly stress-inducing work episodes, whereas workplace bullying is seen as a more long-term negative phenomenon. Retaliation after whistleblowing is here synonymous with work harassment (even though there may exist antecedent differences when compared with other types of harassment). When the retaliation after whistleblowing is of long-term duration, the term *whistleblower bullying* will be applied. We do recognize, however, that there could be an overlap between the two constructs; that is, there are situations in which it will be difficult to categorically label them either as whistleblower retaliation or as whistleblower bullying.

Theoretical Elaborations Regarding the Whistleblowing–Bullying Distinction

Retaliation should not be seen as a kind of dichotomy (that is, whistleblowers are either punished or not punished). Miceli and her associates (2008) claim that the situation of a whistleblower who, for example, experiences being given the cold shoulder from one coworker for a limited period of time and that of a whistleblower who believes that nothing happened may actually be far more similar regarding the strength of psychological consequences than when the short-term reprisal examples are compared with some long-lasting and widespread or degrading punishment. This may strengthen our view that one should try to distinguish between more short-term whistleblower retaliation (harassment), on the one hand, and whistleblower bullying, on the other. Qualitative studies (Bjørkelo et al., 2008) and quantitative studies (Nielsen and Einarsen, 2008) have documented that negative consequences in the form of long-term workplace bullying may follow after whistleblowing. Retaliatory acts may thus develop into workplace bullying when the employee is met with acts that are repeated, are continued over a period of time, and from which the whistleblowers are unable to defend themselves (cf. scenario C, relation 6, in Figure 13.1).

Social interactionistic perspectives view reports and grievances as "attempts of social control that may develop into incidents" (Tedeschi and Felson, 1994, p. 243). Thus, retaliation, and probably also more long-term whistleblower bullying, may be seen as an attempt to gain social control. Miceli et al. (2008) suggest three theoretical explanations why whistleblowing proceeds to retaliation against the person who voiced concern. The first perspective is the *minority influence theory* (cf. Moscovici, 1976). Group members who take a position not held by the majority—as, for instance, do whistleblowers who voice their concern—are more influential if they appear credible, confident, competent, and objective. This perception was the case with the Enron, Worldcom, and FBI whistleblowers, all of whom were actually subjected to a modest level of retaliation. Two of them were also somewhat extra protected because they were leaders of internal audit groups and thus did not have to voice their concerns completely on their own. If you speak out collectively or if it is a natural part of your job to take action when irregularities occur, it is less likely that your personal outcome of the whistleblowing is retaliation or workplace bullying (Miceli and Near, 1992).

The second theoretical perspective concerning the link between the whistleblowing and retaliation hinges on *social power* (French and Raven, 1959). In line with theories of social power, a whistleblower's attempt to influence and terminate the wrongdoing may be seen as a power struggle in which the dominant coalition may either accept or, alternatively, refuse this initiative by bringing the wrongdoing to an end, or may balance the power struggle by retaliating against the whistleblower (Near, Dworkin, and Miceli, 1993).

Powerful whistleblowers, especially those with expert or informal power (i.e., those who possess important expert knowledge or referent power in the form of having high social standing or reputation), are more protected against reprisals than those lacking such bases of power (Miceli et al., 2008). That finding may explain why relatively fewer workers blow the whistle as compared with their colleagues in middle-management positions. When confronted with wrongdoing, workers stay more passive because of their lack of power. By contrast, middle managers have a power base on which they may lean.

The third perspective elaborated by Miceli and her colleagues concerning whistleblowing is based on *resource dependence theory* (Pfeffer and Salancik, 1978). According to this orientation, organisation members are powerful when the organisation depends on them for their resources, or potential contributions. Thus, some whistleblowers may be relatively powerful and to some extent protected against retaliation and subsequent workplace bullying. This may be the case if the organisation depends on those individuals who blew the whistle and if the organisation does not depend on continuation of the wrongdoing or on the wrongdoer (Miceli et al., 2008).

There seem to be few theoretical explanations as to why some whistleblowers are punished by means of retaliation of limited short-term duration (e.g., harassment), whereas others are exposed to more severe long-term harm (e.g., workplace bullying or whistleblower bullying). A group of whistleblowers may be subjected to harassment, but it ceases after some time. As pointed out, whistleblower bullying is more long term and comprises negative social actions for months and possibly years to come. One simple explanation may be that many whistleblowers resign from their job before the reprisals develop into whistleblower bullying. The Norwegian whistleblower survey revealed that members of the whistleblower group, especially those who had blown the whistle several times, had stronger intentions to leave compared with the group of nonwhistleblowers (Matthiesen et al., 2008).

As pointed out earlier, there may be few differences between whistleblower bullying and other types of workplace bullying in terms of the conduct the targets are met with. The antecedent factors may be different, of course. Whistleblowers exposed to bullying may, however, experience the same negative social situations as other employees bullied at work (these are the outcome result). Concerning more short-term retaliation after whistleblowing, misconduct not yet escalated into bullying, it may be easier to outline possible distinctions (Table 13.1).

Table 13.1 lists 15 qualities that we may use to compare "ordinary" workplace bullying with more short-term whistleblower retaliation. We will comment on some of the points in this table. It should be emphasised that no empirical evidence currently exists regarding potential differences between "ordinary" workplace bullying and "ordinary" whistleblower retaliation.

TABLE 13.1

Fifteen Possible Differences between Typical Workplace Bullying and
Whistleblower Retaliation

	Workplace Bullying	Whistleblower Retaliation
Repeated negative acts	Yes	Yes or no
Acts of revenge	Usually no	Yes
Ostracism (silent treatment)	Yes or no	Yes
Social attacks or ridicule	Yes	Yes
Serious threats about physical violence	Usually no	Yes or no
Job content deterioration or demotion to another position within the company	Yes or no	Yes (after organisational wrongdoing and corporate crime whistleblowing)
Bystander learning lesson essential	No	Yes
Learning lesson not to repeat the conducted act	No	Yes
Onset because of general dislike of person	Usually yes	No
Onset because of crucial triggering event	Usually no	Yes or no
Dismissal or fired	Usually no	Yes or no
Reaction to weak social skills	Yes or no	No
Informal norms violated	Yes or no	Usually yes
Formal norms violated	Yes or no	No
Negative organisational learning signal	Yes	Yes

Still, the list in Table 13.1 may prove helpful when it comes to understanding the two phenomena and in inspiring future research. As we propose in the table, workplace bullying consists of repeated negative social acts. Such is not necessarily the case with whistleblower retaliation. Whistleblower retaliation can also be seen as acts of revenge. Revenge is less common in workplace bullying, however. Ostracism is more typical in whistleblower retaliation, especially when accusing the whistleblower for having violated formal norms or laws is difficult.

Threats of physical violence may also be triggered by revealing organisational misconduct and is not common in workplace bullying, at least not in Norway and other Nordic countries (Nielsen et al., 2008). Targets of workplace bullying can be socially attacked; they also find that meaningful job content may be taken from them or removed, or they may be met with both types of sanctions (Hallberg and Strandmark, 2006; Rayner and Cooper, 2006). Whistleblowers, especially those who have voiced their concerns externally and have in this way left superiors or the company itself with an image problem, will more likely face risks to their jobs or are more likely to encounter hostile acts, as described earlier. They may be stripped of a meaningful job or left without a future career opportunity should they not take the hint to quit the organisation. Retaliation or punishment against whistleblowers may also be viewed from the organisational perspective as a learning opportunity, one provoked by a desire to "teach" bystanders, or the whistleblower, to refrain from whistleblowing. We assume, however, that workplace bullying is less commonly undertaken as a way to teach others a lesson.

Personal animosity or dislike may also trigger workplace bullying because the target lacks social skills and consequently may break social norms. The whistleblowers, however, are disliked because of what they have done by way of specific acts (that is, by their voicing concern to stop wrongdoing), not because of general social skills or personality characteristics. Furthermore, it may also sometimes be difficult to trace or identify the triggering event or events of workplace bullying, even though Leymann specified this option in his typical model of the bullying process (Leymann, 1996). What triggers whistleblowing appears to be more concrete—a wrongdoing has occurred. The wrongdoing is observed by someone there who takes action by blowing the whistle; and in response, the whistleblower is later met by reprisal(s) from the wrongdoer(s). What workplace bullying and whistleblower retaliation do have in common is that they can both be seen as signals of negative organisational learning, as an invitation to bystanders and employees in general to behave in a more defensive and less open and creative way (cf. the single-loop learning construct of Argyris and Schön, 1989).

Antecedents of Successful Whistleblowing and Bullying after Whistleblowing

To improve the situation for whistleblowers, it is essential that organisations and their members are aware of the conditions under which an employee is likely to be effective when an individual or group reports wrongdoing. As mentioned previously, employees were more successful in bringing the misconduct to an end when it was possible to provide evidence of wrongdoing and when the employee in question was less vulnerable (Mesmer-Magnus and Viswesvaran, 2005). According to a range of scholars (Fletcher, Sorrell, and Silva, 1998; Greene and Latting, 2004; Johnson, Sellnow, Seeger, Barrett, and Hasbargen, 2004; Martin, 1999; Miceli and Near, 1988, 2002; Miceli, Near, and Schwenk, 1991), it is argued that employees need to take the following necessary steps in order to be successful: (1) ensure that the nature of the wrongdoing is serious enough; (2) attempt to use internal channels of reporting before reporting to external sources such as, for example, the media; (3) provide proof; (4) receive support; and (5) ensure that they report in what is perceived as a "valid" whistleblowing manner.

Zapf (2007), on the other hand, has characterised whistleblower bullying as a token of unsuccessful whistleblowing. Previous studies have shown how many whistleblowers truly believed in advance that if their superiors only knew, the report would be effective and the wrongdoing would be stopped (Bjørkelo et al., 2008; Glazer and Glazer, 1987; Gobert and Punch,

2000; Rothschild and Miethe, 1994). Martin (2003) argued that whistleblowers may initially have been naïve in trusting the system and that they may not have understood the organisational dynamics involved. As successful whistleblowing depends on common perception of the wrongdoing, it may thus be that whistleblowing will more easily elicit workplace bullying if the perceived wrongdoing is about a matter that may be difficult to prove, for example, harassment as opposed to theft. Some working life studies have found that one of the most prevalent types of wrongdoing reported is such unethical social interaction as abusive and intimidating behaviour (Ethics Resource Center, 2005). Other studies have also shown support for this assumption in that in a study that mainly reported about bullying of others, whistleblowers were relatively successful (Bjørkelo, Einarsen, Nielsen, and Matthiesen, in press). Whistleblowing may also lead to bullying because success does not come easily to "solo dissenters" (De Maria, 2006, p. 219). If an employee is isolated in having observed or gained knowledge about the perceived wrongdoing and then reports about the issue, which is not usually considered wrongdoing, whistleblowing may be a risk factor for later workplace bullying.

Prevention and Intervention

Whistleblowing is beginning to gain recognition as a powerful internal instrument for potential positive organisational change (Vandekerckhove, 2008). Research (Baucus and Dworkin, 1994) has documented the potential cost in terms of the risk facing employees who report wrongdoing at work, such costs corresponding with labels such as "retaliation" (Near and Miceli, 1986; Vinten and Gavin, 2005), "victimisation" (Lennane, 1993), "reprisals" (De Maria and Jan, 1997), and "workplace bullying" (Bjørkelo et al., 2008). Such knowledge is needed in order to provide a broader picture of the dangers associated with whistleblowing.

Policies represent one way to increase organisational awareness and to tackle and deal with whistleblowing and its potential consequences (Hassink, de Vries, and Bollen, 2007). Boatright (1997) suggests a five-component policy model consisting of (1) an effectively communicated statement of responsibility, (2) a clearly defined procedure for reporting, (3) well-trained personnel to receive and investigate reports, (4) a commitment to take appropriate action, and (5) a statement of intent or guarantee against retaliation and recrimination. Depending on the national legal context, the last component will, of course, vary (Calland and Dehn, 2004; Lewis and Uys, 2007; Secunda, 2009). In order to be effective, organisational policies also need systems for reinforcing breaches as well as wanted and proactive organisational behaviours (Solano and Kleiner, 2003).

Ways of tackling workplace bullying, similar to those concerning whistleblowing, often address differences regarding awareness and differences in legal regulations across nations (Sheehan, Barker, and Rayner, 1999). Specific suggestions are made, among them establishing bully-free organisational cultures; introducing effective, safe, and fair policies; and providing ways to confront and challenge abusive and bullying management styles (Hoel and Cooper, 2000). Hoel and Cooper (2000, p. 18) also suggest that bullying can be reduced by lowering the general stress levels within organisations and by focusing on situational "issues or antecedents which are under their own control and where intervention can be achieved." Many cases of whistleblowing may evolve into focusing more on the person who has reported a perceived wrongdoing than on the content of the report itself (Paul and Townsend, 1996). Any countermeasures should therefore take into account ways to enhance the essence of the reported case in question rather than focus on the individual reporting them.

Conclusion

Furnham and Taylor (2004, p. 119) write that where "people have faith in the management who will respond to reports not by shooting the messenger but [by] investigating and confronting the problem, the need for whistleblowing is significantly reduced." The problem with this argument is that many whistleblowers blow the whistle thinking that management would like to hear about aircraft brakes that do not work (Ermann and Lundman, 2002), about loss of children's lives as a result of inadequate safety measures (Bolsin, 1998), about unsafe cars (Glazer, 1983) and trouble with operatorless trains (Davis, 2005), and about unwarranted use of force in the treatment of patients (Bjørkelo et al., 2008). The challenge is what to do when management does not investigate and confront the problem because of budget, status, or reputation. Lack of organisational routines or management skills regarding how to handle whistleblowing in a proper manner may lead to a retaliation or harassment of the whistleblower or, even more worrying, to the long-term workplace bullying of those who took action and voiced their concern about organisational misconduct.

References

Alford, C. F. (1999) Whistle-blowers: How much we can learn from them depends on how much we can give up. *American Behavioral Scientist, 43* (2), 264–277.
———— (2000) Whistleblower narratives: Stuck in static time. *Narrative, 8* (3), 279–293.

———— (2008) Whistleblowing as responsible followership. In I. Chaleff, R. E. Riggio, and J. Lipman-Blumen (eds.), *The art of followership: How great followers create great leaders and organizations* (pp. 237–251). San Francisco: Jossey-Bass.

Andrews, G., and Slade, T. (2002) The classification of anxiety disorders in ICD-10 DSM-IV: A concordance analysis. *Psychopathology, 35* (2/3), 100–106.

Aquino, K., Tripp, T. M., and Bies, R. J. (2001) How employees respond to personal offense: The effects of blame attribution, victim status, and offender status on revenge and reconciliation in the workplace. *Journal of Applied Psychology, 86* (1), 52–59.

Argyris, C., and Schön, D. A. (1989) Participatory action research and action science compared: A commentary. *American Behavioral Scientist, 32* (5), 612–623.

Baucus, M. S., and Dworkin, T. M. (1991) Wrongful firing in violation of public policy: Who gets fired and why. *Employee Responsibilities and Rights Journal, 7* (3), 191–296.

Bechtoldt, M. N., and Schmitt, K. D. (in press) "It's not my fault, it's theirs"—explanatory style of bullying targets with unipolar depression and its susceptibility to short-term therapeutical modification. *Journal of Occupational and Organizational Psychology.*

Bjørkelo, B., Einarsen, S., Nielsen, M. B., and Matthiesen, S. B. (in press) Silence is golden? Characteristics and experiences of self-reported whistleblowers. *European Journal of Work and Organizational Psychology.*

Bjørkelo, B., Ryberg, W., Matthiesen, S. B., and Einarsen, S. (2008) "When you talk and talk and nobody listens": A mixed method case study of whistleblowing and its consequences. *International Journal of Organisational Behaviour, 13* (2), 18–40.

Boatright, J. R. (1997) *Ethics and the conduct of business*, 2nd ed. Upper Saddle River, NJ: Prentice Hall.

Bolsin, S. N. (1998) Professional misconduct: The Bristol case. *Medical Journal of Australia, 169* (7), 369–372.

Branch, T. (1979) Courage without esteem: Profiles in whistle-blowing. In C. Peters and M. Nelson (eds.), *The culture of bureaucracy* (pp. 217–238). New York: Holt, Rinehart and Winston.

Breirem, K. (2007) *På BA-HR bakke* [At rock-bottom]. Oslo: Bazar.

Brodsky, C. M. (1976) *The harassed worker.* Lexington, MA: D. C. Heath.

Calland, R., and Dehn, G. (2004) *Whistleblowing around the world: Law, culture and practice.* London: Public Concern at Work (PCaW).

Cortina, L. M., and Magley, V. J. (2003) Raising voice, risking retaliation: Events following interpersonal mistreatment in the workplace. *Journal of Occupational Health Psychology, 8* (4), 247–265.

Davis, M. (2005) Whistleblowing. In H. LaFollette (ed.), *The Oxford handbook of practical ethics* (pp. 539–563). Oxford, UK: Oxford University Press.

De Maria, W. (2006) Brother secret, sister silence: Sibling conspiracies against managerial integrity. *Journal of Business Ethics, 65* (3), 219–234.

De Maria, W., and Jan, C. (1997) Eating its own: The whistleblower's organization in vendetta mode. *Australian Journal of Social Issues, 32* (1), 37–59.

Directorate of Labour Inspection (2007) *Act relating to working environment, working hours and employment protection, etc. (Work Environment Act).* Trondheim, Norway: Author. http://www.arbeidstilsynet.no/binfil/download2.php?tid=92156 (accessed November 7, 2007).

Einarsen, S. (1999) The nature and causes of bullying at work. *International Journal of Manpower, 20,* 16–27.

Einarsen, S., Tangedal, M., Skogstad, A., Matthiesen, S. B., Aasland, M. S., Nielsen, M. B., et al. (2007) *Et brutalt arbeidsmiljø? En undersøkelse av mobbing, konflikter og destruktiv ledelse i norsk arbeidsliv* [A brutal working environment? An investigation of bullying, conflicts and destructive leadership in Norwegian working life]. Bergen, Norway: University of Bergen.

Elliston, F. A., Keenan, J., Lockhart, P., and van Schaick, J. (1985) *Whistleblowing: Managing dissent in the workplace*. New York: Praeger.

Ermann, M. D., and Lundman, R. J. (2002) *Corporate and governmental deviance: Problems of organizational behavior in contemporary society*, 6th ed. New York: Oxford University Press.

Ethics Resource Center (2005) *National Business Ethics Survey: How employees view ethics in their organizations, 1994–2005*. Washington, DC: Ethics Resource Center.

Fitzgerald, K. (1990) Whistle-blowing: Not always a losing game. *IEE spectrum, 27* (12), 49–52.

Fletcher, J. J., Sorrell, J. M., and Silva, M. C. (1998, December 31) Whistleblowing as a failure of organizational ethics. *Online Journal of Issues in Nursing, 3* (3), 1–16. http://www.nursingworld.org/MainMenuCategories/ANAMarketplace/ANAPeriodicals/OJIN/TableofContents/vol31993Dec31998/Whistleblowing.aspx (accessed June 13, 2008).

French, J. R. P., and Raven, B. (1959) The bases of social power. In D. Cartwright (ed.), *Studies in social power* (pp. 150–167). Ann Arbor: University of Michigan.

Furnham, A., and Taylor, J. (2004) *The dark side of behaviour at work: Understanding and avoiding employees leaving, thieving and deceiving*. Basingstoke, UK: Palgrave Macmillan.

Glazer, M. (1983) Ten whistleblowers and how they fared. *The Hastings Center Report*, December, 33–41.

Glazer, M., and Glazer, P. M. (1987) Pathways to resistance: An ethical odyssey in government and industry. *Research in Social Problems and Public Policy, 4*, 193–217.

———— (1988) Individual ethics and organizational morality. In J. S. Bowman and F. A. Elliston (eds.), *Ethics, government, and public policy: A reference guide* (pp. 55–78). New York: Greenwood Press.

Gobert, J., and Punch, M. (2000) Whistleblowers, the public interest, and the Public Interest Disclosure Act, 1998. *Modern Law Review, 63* (1), 25–54.

Greenberg, J., and Baron, R. A. (2008) *Behavior in organizations*, 9th ed. Upper Saddle River, NJ: Pearson/Prentice Hall.

Greene, A. D., and Latting, J. K. (2004) Whistle-blowing as a form of advocacy: Guidelines for the practitioner and organization. *Social Work, 49* (2), 219–230.

Hallberg, L. R. M., and Strandmark, M. K. (2006) Health consequences of workplace bullying: Experiences from the perspective of employees in the public service sector. *International Journal of Qualitative Studies on Health and Well Being, 1* (2), 109–119.

Hassink, H., de Vries, M., and Bollen, L. (2007) A content analysis of whistleblowing policies of leading European companies. *Journal of Business Ethics, 75* (1), 25–44.

Hersh, M. A. (2002) Whistleblowers—heroes or traitors? Individual and collective responsibility for ethical behaviour. *Annual Reviews in Control, 26*, 243–262.

Hoel, H., and Cooper, C. L. (2000) *Destructive conflict and bullying at work*. Manchester, UK: Manchester School of Management.

Johnson, C. E., Sellnow, T. L., Seeger, M. W., Barrett, M. S., and Hasbargen, K. C. (2004) Blowing the whistle on fen-phen: An exploration of Meritcare's reporting of linkages between fen-phen and valvular heart disease. *Journal of Business Communication, 41* (4), 350–369.

Jos, P. H., Tompkins, M. E., and Hays, S. W. (1989) In praise of difficult people: A portrait of the committed whistleblower. *Public Administration Review, 49* (6), 552–561.

Jubb, P. B. (1999) Whistleblowing: A restrictive definition and interpretation. *Journal of Business Ethics, 21,* 77–94.

King, G. (1999) The implications of an organization's structure on whistleblowing. *Journal of Business Ethics, 20* (4), 315–326.

Lacayo, R., and Ripley, A. (2002, December 22) Persons of the year 2002: Cynthia Cooper, Coleen Rowley and Sherron Watkins. *Time.* http://www.time.com/time/subscriber/personoftheyear/2002/poyintro.html (Retrieved January 2, 2008).

Lennane, J. (1993) Whistleblowing: A health issue. *British Medical Journal, 307* (6905), 667–670.

Lennane, J., and De Maria, W. (1998) The downside of whistleblowing. *Medical Journal of Australia, 169* (7), 351–352.

Lewis, D., and Uys, T. (2007) Protecting whistleblowers at work: A comparison of the impact of British and South African legislation. *Managerial Law, 49* (3), 76–92.

Leymann, H. (1990) Mobbing and psychological terror at workplaces. *Violence and Victims, 5* (2), 119–126.

—— (1996) The content and development of mobbing at work. *European Journal of Work and Organizational Psychology, 5* (2), 165–184.

Lundquist, L. (2001) Tystnadens förvaltning. In Socialstyrelsen (ed.), *Utan fast punkt: Om förvaltning, kunskap, språk och etik i socialt arbete.* Stockholm: Modin-Tryck.

Martin, B. (1999) Whistleblowing and nonviolence. *Peace and Change, 24* (3), 15–28.

—— (2003) Illusions of whistleblower protection. *University of Technology, Sydney Law Review, 5,* 119–130. http://www.austlii.edu.au/au/journals/UTSLRev/2003/index.html (accessed September 12, 2009).

Matthiesen, S. B. (2006) *Bullying at work: Antecedents and outcomes.* Bergen, Norway: Department of Psychosocial Science, Faculty of Psychology, University of Bergen.

Matthiesen, S. B., Bjørkelo, B., and Nielsen, M. B. (2008) *Klanderverdig atferd og varsling i norsk arbeidsliv* [Wrongdoing and whistleblowing in Norwegian working life]. Bergen, Norway: University of Bergen.

Matthiesen, S. B., and Einarsen, S. (2001) MMPI-2 configurations among victims of bullying at work. *European Journal of Work and Organizational Psychology, 10* (4), 467–484.

—— (2004) Psychiatric distress and symptoms of PTSD among victims of bullying at work. *British Journal of Guidance and Counselling, 32* (3), 335–356.

McDonald, S., and Ahern, K. (2002) Physical and emotional effects of whistleblowing. *Journal of Psychosocial Nursing and Mental Health Services, 40* (1), 14–27.

Mesmer-Magnus, J. R., and Viswesvaran, C. (2005) Whistleblowing in organizations: An examination of correlates of whistleblowing intentions, actions, and retaliation. *Journal of Business Ethics, 62* (3), 266–297.

Miceli, M. P., and Near, J. P. (1988) Individual and situational correlates of whistleblowing. *Personnel Psychology, 41,* 267–281.

—— (1992) *Blowing the whistle: The organizational and legal implications for companies and employees.* New York: Lexington Books.

—— (2002) What makes whistle-blowers effective? Three field studies. *Human Relations, 55* (4), 455–479.

—— (2005) Standing up or standing by: What predicts blowing the whistle on organizational wrongdoing? In J. J. Martocchio (ed.), *Research in personnel and human resources management,* vol. 24 (pp. 95–136). Oxford: Elsevier.

Miceli, M. P., Near, J. P., and Dworkin, T. M. (2008) *Whistle-blowing in organizations.* New York: Routledge, Taylor & Francis.

Miceli, M. P., Near, J. P., and Schwenk, C. R. (1991) Who blows the whistle and why. *Industrial and Labor Relations Review, 45* (1), 113–130.

Miceli, M. P., Rehg, M. T., Near, J. P., and Ryan, K. C. (1999) Can laws protect whistle-blowers? Results of a naturally occurring field experiment. *Work and Occupations, 26,* 129–151.

Miethe, T. D. (1999) *Whistleblowing at work: Tough choices in exposing fraud, waste, and abuse on the job.* Boulder, CO: Westview Press.

Mikkelsen, E. G., and Einarsen, S. (2002) Basic assumptions and symptoms of post-traumatic stress among victims of bullying at work. *European Journal of Work and Organizational Psychology, 11,* 87–11.

Monsen, P. Y. (2008) *Muldvarp i Siemens* [Mole in Siemens]. Oslo: Spartacus.

Moscovici, S. (1976) *Social influence and social change.* London: Academic Press.

Near, J. P., Dworkin, T. M., and Miceli, M. P. (1993) Explaining the whistle-blowing process: Suggestions from power theory and justice theory. *Organization Science, 4* (3), 393–411.

Near, J. P., and Miceli, M. P. (1985) Organizational dissidence: The case of whistle-blowing. *Journal of Business Ethics, 4,* 1–16.

—— (1986) Retaliation against whistle-blowers: Predictors and effects. *Journal of Applied Psychology, 71* (1), 137–145.

—— (1996) Whistle-blowing: Myth and reality. *Journal of Management, 22* (3), 507–526.

Nielsen, M. B. (2003) Når mobberen er leder: En studie av sammenhengen mellom led-erstiler og psykiske traumereaksjoner hos et utvalg mobbeofre [When the bully is a leader: The relationship between destructive leaders and symptoms of post-traumatic stress disorder among victims of workplace bullying]. Unpublished master's thesis, NTNU, Trondheim, Norway.

Nielsen, M. B., and Einarsen, S. (2008) Sampling in research on interpersonal aggression. *Aggressive Behavior, 34* (3), 265–272.

Nielsen, M. B., Skogstad, A., Matthiesen, S. B., Glasø, L., Aasland, M. S., Notelaers, G., et al. (2008) Prevalence of workplace bullying in Norway: Comparisons across time and estimation methods. *European Journal of Work and Organizational Psychology, 18* (1), 81–101.

Olsen, L. R., Jensen, D. V., Noerholm, V., Martiny, K., and Bech, P. (2003) The internal and external validity of the Major Depression Inventory in measuring severity of depressive states. *Psychological Medicine, 33* (2), 351–356.

Paul, R. J., and Townsend, J. B. (1996) Don't kill the messenger! Whistle-blowing in America: A review with recommendations. *Employee Responsibilities and Rights Journal, 9* (2), 149–161.

Pfeffer, J., and Salancik, G. R. (1978) *The external control of organizations: A resource dependence perspective.* New York: Harper and Row.

Rayner, C., and Cooper, C. L. (2006) Workplace bullying. In K. E. Kelloway, J. Barling, and J. J. Hurrell Jr. (eds.), *Handbook of workplace violence* (pp. 121–145). Thousand Oaks, CA: Sage.

Rehg, M. T. (1998) An examination of the retaliation process against whistleblowers: A study of federal government employees. Unpublished doctoral dissertation, Indiana University, Bloomington. http://www.dtic.mil/cgi-bin/GetTRDoc?A D=ADA351018andLocation=U2anddoc=GetTRDoc.pdf (accessed September 21, 2009).

Rehg, M. T., Miceli, M. P., Near, J. P., and Van Scotter, J. R. (2008) Antecedents and outcomes of retaliation against whistleblowers: Gender differences and power relations. *Organization Science, 19* (2), 221–240.

Rothschild, J. (2008) Freedom of speech denied, dignity assaulted: What the whistleblowers experience in the U.S. *Current Sociology, 56* (6), 884–903.

Rothschild, J., and Miethe, T. D. (1994) Whistleblowing as resistance in modern work organizations: The politics of revealing organizational deception and abuse. In J. M. Jermier, D. Knights, and W. R. Nord (eds.), *Resistance and power in organizations* (pp. 252–273). London: Routledge.

——— (1999) Whistle-blower disclosures and management retaliation: The battle to control information about organization corruption. *Work and Occupations, 26* (1), 107–128.

Secunda, P. (2009) *Retaliation and whistleblowers: Proceedings of the New York University 60th Annual Conference on Labor*. Austin, TX: Wolters Kluwer.

Sheehan, M., Barker, M., and Rayner, C. (1999) Applying strategies for dealing with workplace bullying. *International Journal of Manpower, 20* (1–2), 50–56.

Soeken, D. R. (1986) J' accuse. *Psychology Today, 20* (8), 44–46.

Soeken, K. L., and Soeken, D. R. (1987) A survey of whistleblowers: Their stressors and coping strategies. In *Whistleblowing Protection Act of 1987*, vol. 1st sess. (pp. 156–166). Washington, DC: Supt. of Docs., Congressional Sales Office, U.S. Government Printing Office.

Solano, F., and Kleiner, B. H. (2003). Understanding and preventing workplace retaliation. *Management Research News, 26* (2–4), 206–211.

Tedeschi, J. T., and Felson, R. B. (1994) *Violence, aggression and coercive actions*. Washington, DC: American Psychological Association.

Tedeschi, R. G., and Calhoun, L. G. (2004) Posttraumatic growth: Conceptual foundations and empirical evidence. *Psychological Inquiry, 15* (1), 1–18.

Tehrani, N. (1996) The psychology of harassment. *Counselling Psychology Quarterly, 9* (2), 101–117.

Tracy, S. J., Lutgen-Sandvik, P., and Alerts, J. K. (2006) Nightmares, demons, and slaves: Exploring the painful metaphors of workplace bullying. *Management Communication Quarterly, 20* (2), 148–185.

Vandekerckhove, W. (2008) Protect your whistleblowers! In D. Crowther and N. Capaldi (eds.), *The Ashgate research companion to corporate social responsibility* (pp. 181–198). Aldershot, UK: Ashgate.

Vinten, G., and Gavin, T. A. (2005) Whistleblowing on health, welfare and safety: The UK experience. *Journal of the Royal Society for the Promotion of Health, 125* (1), 23–29.

Williams, K. D. (2007) Ostracism. *Annual Review of Psychology, 2007* (58), 425–452.

Zapf, D. (2007) Mobbing und whistleblowing in organisationen [Mobbing and whistleblowing in organizations]. In M. Boos, V. Brandstätter, and K. J. Jonas (eds.), *Zivilcourage trainieren* (pp. 59–82). Göttingen, Germany: Hogrefe.

Section IV

Managing the Problem

14

Managing Workplace Bullying: The Role of Policies

Charlotte Rayner and Duncan Lewis

CONTENTS

Introduction

A bullying and harassment policy is the employer's statement of intent and a summary of processes as regards bullying and harassment in their organisation. The role of the policy in the management of workplace bullying is central to all concerned. It is a mistake to think that a policy is used only in situations of formal complaint. All policies outline formal procedures, but an effective policy has a far wider purpose and includes, for example, guiding statements about how the organisation intends to prevent bullying and deal with bullying if it occurs. As such, the policy has two immediate roles—as a statement of intent, and also as a document to guide all stakeholders through the formal and informal processes connected to bullying prevention and intervention (Richards and Daley, 2003).

    Those people who work in this area know that one can have a terrific policy that does not make any difference in the workplace itself (Ferris, 2004). This chapter focuses on policies that work. We have needed to include facets beyond what one might find in a policy to include effective communication and monitoring. The nature of the chapter is, therefore, highly applied, and we will find ourselves in areas rarely touched by academics and traditional academic theory. Research in this area is enormously challenging, as all organisations are unique and policies tend to be written only every few years. Hence, we have no systematic data to compare policies and their effectiveness; consequently, in this chapter readers will find no golden bullet for success. That said, some organisations have made strides in this field, learning as they try different approaches (Rayner and McIvor, 2008). In this chapter, we will bring together our own knowledge from working with employers and from the available evidence to outline routes of better practise and highlight avenues of failure in the development and implementation of policies against bullying and harassment at work.

The chapter first considers the nature of the challenge that policies on bullying need to meet. The rest of the chapter follows the sequence of policy making and delivery. First, we consider who should be involved in policy setting and why. Then, we move on to what should be in a policy, the "skeleton" on which one then builds the detail. Second, we reflect on common errors associated with policies and what can go awry in the communication of the policy. Finally, we stress that if the policy is to be a functional system and a useful tool for the organisation, it will need monitoring and review.

The Nature of the Challenge

The introduction to this chapter outlined how a policy on bullying needs to be a statement concerning all aspects of bullying intervention and prevention. To be able to begin to design a policy, one first needs to consider the nature of the problem at hand. Academic knowledge here is very helpful to inform the challenge and map relevant aspects.

What is bullying? This topic has received considerable attention in the academic literature, and perusal of Chapters 1 and 2 of this book will outline the many ways of considering this issue. There is no definitive list of behaviours that constitute bullying; indeed, we understand that people can feel bullied because of the absence of behaviours such as withholding information or being excluded from groups and decisions (Liefooghe and MacKenzie Davey, 2001). As academics we have the luxury of leaving such ambiguities unresolved, instead exploring their nature. The organisation has to confront such difficulties, and doing this provides a major challenge for those designing policies and their use.

Lutgen-Sandvik, Tracy, and Alberts (2007) have offered a delineation of bullying behaviours using the analogy of degree of burns; thus, "third degree" bullying is greater than "first degree." In their continuum, first-degree abuses are low-level abuses that "can cause damage over time, but are common…and usually quick to heal," second-degree abuses are "intensive, frequent and persistent…[and] more painful," and third-degree ones are "extremely escalated…often result[ing] in deep scarring and permanent damage" (ibid., p. 855). Most organisations have no problem with committing themselves to dealing with third-degree bullying; but as the significance lessens, so ambiguities we find in the academic literature are likely to also impact the organisation. A single dramatic incident may count as bullying, but would the organisation take seriously a series of small events that an employee has found very difficult to deal with? Consider also the notion that what is socially acceptable (and not acceptable) at work changes over time (Caza and Cortina, 2007). Because our understanding of what constitutes bullying is constantly under review (Burnes and Pope, 2007), it soon becomes apparent that we seek to define a moving target. As a result, it is common to find very broad conceptualisations of bullying in policies, partly as a pragmatic response to its complexity.

Broad definitions help defend the policy from becoming outdated, and this is the preference for many organisations. On the other hand, broad definitions are less useful for the employee who seeks to know what behaviour is acceptable or not (Bowen and Blackmon, 2003). Hence there is a need for an operational definition that may change—perhaps as part of an employee code of conduct, as, for example, evident in the policy of Staffordshire University (see http://www.staffs.ac.uk/images/behaviour_tcm68-15921.pdf)—which can be reviewed more frequently than the procedural aspects of the policy. Practitioners may look to academic writing for clarity on what bullying is or is not; however, as academics' questions run in parallel to those of practitioners, easy answers will not be found. This said, the analysis of the nature of bullying and its precursors (e.g., Salin, 2003) can provide helpful frameworks for structuring the description of the construct.

A related area as yet undocumented in the academic literature on bullying is the notion of variability of tolerance of behaviour within the organisation. Does one expect the same levels of behaviour from truck drivers and top accountancy employees who all work for the same organisation? Anecdotal evidence suggests that many UK employees think one should (Rayner and McIvor, 2008), as would evidence from the academic literature on fairness and justice (Trevino, Weaver, and Reynolds, 2006). The few studies that have examined bullying amongst hierarchical levels have found the same reported incidence at all levels (Hoel, Faragher, and Cooper, 2004); and although the nature of behaviours may be different because different social contexts provide different cues and opportunities (Tajfel and Turner, 2001), allowing variations in tolerance is hard to defend to employees. It is important that the "variability" debate is brought into open forum so that assurance of evenness of treatment is publicly noted (Schat and Kelloway, 2003).

In these early conceptual stages, it is also helpful to work through the scope of the policy and how existing policies might be linked. Hence, one might need, not to describe the disciplinary process, but to link into the existing policy of the organisation if such a process is already in existence.

How far should one extend the scope of the policy to informal routes? Arguably, bullying is very much a social phenomenon (Lewis, 2002), and so the informal social context is a prime target for intervention through bystander action (Bowes-Sperry and O'Leary-Kelly, 2005) and also prevention (Rayner, 2005). Once again, thinking through the organisational appetite for extending the scope of the policy into selection, training, induction, and management working practises is helpful to have considered early on, and these linkage issues will be returned to later.

The Process of Policy Setting

One decision that needs to be made when considering a policy on bullying is that of ownership. Where does this policy belong in the organisation and who owns it? Much that is written on bullying is from a health perspective (e.g., Vartia, Korppoo, Fallenius, and Mattila, 2003), and some organisations have placed their policy in the health and safety area. There is logic here, because the main outcomes that cause problems to organisations are sickness absence and stress (Giga, Hoel, and Lewis, 2008). Although a health and safety approach may draw on excellence in helping the targets of bullying, the practises needed to prevent or minimise bullying lie in line management, training, and the disciplinary or investigative routes that are normally the province of personnel or human resource (HR) departments (Richards and Daley, 2003). Integration into other organisational policies as well as new legislation concerning equality and human rights means that policies on bullying are most usually situated in the HR department. Whichever department owns the policy must be able to exert sufficient influence to enable delivery of the policy itself (Salin, 2009).

Who else should be involved in policy making? At the end of this chapter, we will summarise by stating that success is to be gained from buy-in and commitment from all players to undertake whatever is stated in the policy. Conversely, policy failure is related to the organisation's being unwilling or unable to deliver what has been committed to in the policy statement. The ability to avoid failure lies in starting at the right place—in the policy design process. A large study in the United Kingdom (Rayner and McIvor, 2008) found that organisations reported great benefits when enabling high levels of involvement in the design process. The same study discovered that the best organisations approached bullying as "our" problem, sharing responsibility throughout the organisation (ibid.). Hence while a policy may be owned by the personnel or HR department, the design process needs to include those people who will be delivering the various aspects of the policy. Thus the policy will include HR and trade unions or worker representatives (Lewis and Rayner, 2003), as well as managers. One might also include occupational health and safety and/or organisational counselling services, complaints managers, trainers, finance officers (who will need to monitor extra costs), and others such as diversity and equality officers. Organisations often use advice and guidance from external agencies and professional bodies. For example, in a UK context, the Chartered Institute for Personnel Development (CIPD) and the Advisory, Conciliation and Arbitration Service (Acas) offer sensible advice and guidance to ensure legal compliance and equitability. Acas, for example, regularly produces policy discussion papers on a range of matters, including bullying (see, e.g., Lewis, 2006). Support from the chief executive or senior management team is needed to add validity to the whole exercise and host the passage of the policy through to organisational approval. Policies that are not supported by the company's board will fail.

The final aspect of this first step is to consider what type of policy one wants. Unfortunately, no current data exist on organisational practise worldwide. Local law may be dominant in this choice (Leka and Kortum, 2008). There appear to be three main approaches in existence. The first is to have a policy specifically for bullying and is most appropriate where the law might demand specific conditions be met within a bullying policy as distinct from other policies. Another approach, which has become very popular in Europe in the last few years (where one sees new legislation in discrimination, harassment, and equality), is to combine the areas of harassment and bullying into a single inclusive policy. Finally, instead of an antibullying policy, one might choose a positive "dignity at work" policy. Those people who prefer the last suggest it is easier and more functional to encourage employees to go towards a positive statement and embed "good behaviour" into the organisation, effectively using all aspects of antibullying policies when they examine situations where dignity and respect have not been adhered to. Different styles will appeal to different organisations, but in our experience organisations that are more mature in their approach generally take a dignity approach. We have seen good organisations using any of these three approaches, as in the end it is implementation that matters the most.

Sequencing Policy Setting

If the route of participative policy setting is adopted, then it is often helpful to bring together a group very early on in an open discussion forum to flush out early concerns with the existing system (or lack of systems). Not presenting a draft of the new policy will enable participants to see that they are genuinely being involved from the beginning (Costantino and Merchant, 1996).

Often an organisation changes an existing policy rather than starting from new. If the current policy is not respected by staff members, then it may be a positive move to begin with a very new format to counter perceptions of a laissez-faire attitude (Skogstad, 2005). Drafts and revisions need to be taken to the group before being signed off at a high level, thereby creating a genuine consultation mechanism that can avoid dislocation from the employee group the policy should be serving (Ferris, 2004). The group can be extremely useful as a mechanism for wider engagement and information gathering, as the whole process of policy development and revision can be a positive engagement experience in itself.

The Contents of a Policy

Setting the Scene

The initial section of a policy should make clear the scope and purpose of the document: the style needs to be welcoming based on a clear statement of intent that is unambiguous in the organisation's opposition to bullying.

The style of writing communicates how much the owners of the policy want it to be used (Pilbeam and Corbridge, 2006). The best organisational policies are short, clear, and simple to read, allowing easy access by all staff members (Richards and Daley, 2003). Policies that use inaccessible language or formats risk that staff will interpret that the organisation does not want to use the policy (Harlos and Pinder 1999).

After introductory antibullying statements, a policy will normally move to a definition of bullying. One should not underestimate the amount of reading any definition will receive. The definition may, in the worst case, become a legal issue, and it should always be checked by a legal representative (Spurgeon, 2003). Most organisations use a very broad definition of workplace bullying that enables the policy not to need change over several years.

As well as a textual definition of bullying, including a set of examples is important in order to "bring alive" the issue in context. What is important here is that a full range of behaviours be used. One might consider cross-checking against academic questionnaires, but care should be taken because such questionnaires can be overly prescriptive (Fevre, Robinson, Jones, and Lewis, 2010; Lewis, Sheehan, and Davies, 2009). Generating the set of behaviours can cause debate and may be where one begins to unpick a key aspect for policy writers: how is bullying interpreted within this organisation? The question is one we have already considered earlier.

These opening paragraphs of an organisation's antibullying policy may need reference to other documents (e.g., a code of conduct) and should provide sources of informal help should someone need clarification regarding the bullying concept. Some organisations have networks of regular employees who volunteer to be signposters and confidential listeners for their colleagues (in Britain, such persons are commonly called dignity advisers or first contact officers). They, or other sources of assistance, need to be highlighted in the organisation's document so employees can consult with them. The statements, definitions, examples, and sources of help for clarifying bullying can be replicated in recruitment literature and websites, induction, posters, literature used by trade unions, and all other information sources, including workplace posters, e-mails, and staff surveys. By providing such a documented plan, a coherent approach is created for employees and is likely to heighten the perception of organisational intent from employees. Inconsistency is likely to reduce faith in any system (Yamada, 2000).

Why bother with these extensive techniques and care for consistency? Early intervention can stop situations from escalating (Andersson and Pearson, 1999; Hoel and Cooper, 2000) and avoid an expensive formal process. Add to these benefits the avoidance of prolonged stress on the target, the accused bully, and work colleagues, and the related sickness absence or risk of personnel leaving their jobs, then the argument for early informal resolution is both financially and morally sound. To use such informal systems, one has to be alerted to situations of low-level conflict, and clear examples can enable employees to know a situation is not right and prompt informal and speedy resolution. It is to these systems that we now turn.

Dealing with Issues Informally

If one accepts the position that many situations of bullying can be described as escalated conflict (Leymann, 1996), the argument that in many cases bullying is a failure of ordinary line management to detect and deal with negative experiences quickly becomes persuasive. To limit trauma to those involved, the speed, skills, and confidence of those who may potentially defuse and resolve the conflict situation, or at least intervene, are salient. Line managers are likely to be closest and best able to intervene. Ideally, managers can seek out potentially damaging behaviours in a proactive, preventative manner. But do they have the skills and confidence to do so? If the policy suggests, as it should, that line managers are the "first stop," then there are training implications to ensure such skills and confidence exist. For example, direct intervention by a manager in a colleague-on-colleague situation needs to be very skilful to avoid escalating the situation. In the United Kingdom, we are currently seeing informal approaches being developed. For example, such organisations as BT, the UK-based telecommunication company, are bringing mediation skills into everyday management training to enable conflict resolution to be undertaken quickly by line management as part of their regular activities. This early first stage is endorsed by most writers in the field (e.g., Einarsen, 1999; Keashly and Jagatic, 2000; Zapf and Gross, 2001).

Larger organisations often have a network of staff who can also be reached. Employees may be reluctant to see their line manager (or any superior, for that matter) if they perceive that person to be the problem or a part of their problem, and so the network of volunteers can be used at this stage. How much the networks actually advise varies; many HR departments prohibit such staff from giving advice because of concerns about the legal correctness of any advice and implications in the event a case reaches a courtroom. Naturally, confusion can arise for a consumer if the name of the network suggests that the network gives advice. The military in the United Kingdom has recently extended its training of advisers to lower ranks so that all service personnel can talk to someone close to their own level. Encouraging is the example of Glasgow Caledonian University, where the university's in-house volunteers advise all parties and are trained to act as mediators at an early stage. Again, the university group reflects a range of seniority in the hierarchy so that senior staff can mediate for their own level and so on. We are unaware of any systematic research that has evaluated exactly who is seen by whom; but because data could be collected straightforwardly and anonymously, this would be a fruitful area for research.

Although it is ideal to leave informal resolution to line management, no organisation is perfect and staff may choose to use informal resolution routes (e.g., advisers) outside the normal hierarchy. What is crucial is that informal routes work.

Most written policies progress from definition to the first level of resolution, that of the informal process. Unfortunately, many written policy statements have a short paragraph stating, "Wherever possible, informal methods should be used to resolve situations informally," which is then followed by many pages of text concerning the formal policy. The message that can be conveyed is that the organisation is focused on formal systems. In countries with a tradition of informal solutions (e.g., in Scandinavia), the situation may be the opposite; thus more weight should be placed on developing sound formal systems to complement the existing informal ones. Very often UK HR and trade unions have a focus on the formal, which is understandable (Hoel and Beale, 2006; Lewis and Rayner, 2003), but the reality is that the informal process is completely different from the formal, requiring different skills and approaches (Rayner and McIvor, 2008), which leverage trust and positive relationships between both parties and line management (Hubert, 2003). In some countries, developing trust will be a marked change in role for both groups.

How the bullying policy sits in the organisation is important for HR personnel and their understanding of the process. If one adopts a health and safety perspective, then bullying is a risk that must be dealt with as soon as it is known about, for it may affect someone's psychological health (i.e., it is an organisation-owned risk to be dealt with). However, most HR departments take a "complaints policy" route, where one can act only if a written complaint is received and the complainant is willing for the accused to know details of the problem (i.e., this is an interpersonal issue between two or more employees). This written requirement can be the source of HR's effectively negating the informal stage. Thus it is essential that this "competing approaches" dilemma be resolved, and we suggest the health and safety angle should predominate. The same issues can affect trade union representatives' perceptions of a bullying situation. The policy-drafting process is an ideal place for both sides—management and workers—to work through their protocols to facilitate a balance and compatibility between formal and informal solutions.

The Formal Process

Some situations cannot be resolved informally. Very often the reality is that mechanisms do not exist or do not work at the informal level as they should (Salin, 2009). Other times, even effective systems are unable to resolve the situation (Kim, Ferrin, Cooper, and Dirks, 2004). Sometimes formal action is seen as desirable from the complainant, HR, and trade unions—such as when very serious behaviour is reported (e.g., overt abuse in initiation rites) and the organisation management wants to be seen to take action.

At this stage, one should then be able to link into existing policies of complaints and the disciplinary and grievance procedure if these exist (Richards et al., 2003). As these topics are the purview of other academic and practitioner areas, we will not dwell on them (see also Hoel and Einarsen, this volume). Instead, we will highlight aspects of formal policies that can fail in situations of bullying.

Early resolution is important for the formal stage, just as it is for the informal. Prolonged processes can cause trauma, and time becomes a key factor. Policies that contain service-level agreements (e.g., "the investigation will commence within 10 working days of receiving the complaint") can be useful. All parts of the process need to be done well, and sometimes HR and trade unions balk at time-bound systems because they worry that complex cases will not be dealt with properly; consequently, all parties become "slaves to the policy," rather than the policy working for effective outcomes. This dilemma can be resolved through having working practises that allow for exceptions, but where the nature of such exceptions is very well defined (see Hoel and Einarsen, this volume).

Investigators in bullying cases, especially bullying composed of many small occurrences, need a higher level of sensitivity than they might require in other complaint situations (e.g., persistent lateness or unfair pay). The need for training for investigators and decision-making panels has emerged as bullying has been brought onto organisations' agendas. Royal Mail (the United Kingdom's postal service) has developed a specialist team of investigators who can be called on if line managers choose to do so. In response to this trend, specialist training for line managers to act as investigators is now more widely available in the United Kingdom (for high-quality examples of training, see http://www.nweo.org.uk). Our anecdotal evidence suggests that investigators can often spend much time working on bullying cases, and resourcing investigators can be an area of difficulty for those managing the process (Tehrani and Monkswell, 2001). Having a team of specialists takes such pressures away from line management. However, we suggest that undertaking investigations is an excellent way of exposing line managers to the nuances and "differing realities" one finds in bullying cases (Rayner and McIvor, 2008). Our experience is that the more mature organisations do use specialist external investigators only in highly complex (hence excessively time consuming) or very high-level cases (e.g., at board level), with all remaining cases being investigated internally. There are cost and other resourcing implications for undertaking investigations that need clarity at the time of policy making and acceptance of budgetary implications by senior management.

In making judgement on cases after a report has been submitted, HR can find a conflict of interest (Lewis and Rayner, 2003), and an inconsistency in outcomes can be perceived (Rayner, 1998; Salin, 2009). If targets are moved after they have won a case, other employees can become confused, for why should a target have to have the upheaval of changing jobs when his or her case has been proven?

A negative aspect of traditional UK disciplinary procedures is that the outcome is confidential. In bullying, this secrecy can leave us with no feedback loop to the rest of the organisation; furthermore, the organisation's officers remain unable to let employees know they are taking action. Royal Mail now takes a small space in the internal newspaper to summarise the nature of disciplinary outcomes on bullying and harassment to show that there is a penalty for those found to have been bullying. This is a simple and innovative way to close the learning loop and get the unacceptability message across without using any names. Commitment to feedback can be part of a written policy. Other organisations allow the nature of the discipline to "get out into the grapevine" and hence circulate the message to staff. We would prefer to see the former system because it means the organisation is dealing with cases as normal routine behaviour and not allowing the grapevine to possibly get it wrong!

Common Mistakes in Policy Making and Delivery

Bullying is an area about which most employees will have emotion (Richards and Daley, 2003). Having written the policy, it can be a common mistake to then not promote it effectively and thereby fail to confirm seriousness over the issue—and employees notice this (ibid.). An organisation that wants a policy to work must promote the policy by embedding it in training, induction, websites, and publications for those who do not access computers. The trade union can be a useful promoter and will, of course, do a better job if union representatives have been involved in the process of policy making.

Sequencing is critical such that to write a policy and then launch it without giving sufficient training to managers may actually be counterproductive. Hence, the implementation of the policy needs to work through ensuring provisions for those using the policy are in place before any launch. Aspects of our final section might be useful to map prelaunch checklists.

Communication is a common feature to many problems we have come across. Failure to communicate the policy or have it available to hand has already been mentioned. Within the antibullying process, much communication is needed. Protocols regarding confidentiality need to be made and carefully adhered to. Communicating during an investigation, especially if it runs late, is essential and a mitigating factor for stress to those concerned (Dollard, Siunner, Tuckey, and Bailey, 2007). Communicating the outcome of any formal complaint and discipline to any party needs to be extended to the wider community to enhance the seriousness with which a policy is taken by the organisation. Finally, communicating that the policy still exists after launch is essential. This communication can occur during the monitoring and review process, to which we now turn.

Monitoring and Review

A bullying policy is a whole system for prevention and intervention in cases of bullying. Policies do become out-of-date, and sometimes external factors such as new law mean that review is forced on organisations. We argue that any antibullying system needs to have effective monitoring built in. How does one know if a policy is successful or not?

Staff perception surveys are now being run in many organisations on an annual basis. This is a good place to pick up how many people consider themselves bullied (to be compared to the number of complaints received), and over time this figure should reduce. Surveys can identify users' attitudes to and use of the various informal and formal mechanisms as well as show barriers to use (e.g., not trusting confidentiality) or awareness of systems. Naturally, such figures are only indicators, but over a number of years one would expect to see improvement. If none were found, then local focus groups, exit interviews, and other investigative techniques might be used to redesign the system and the policy.

Managers are central to successful policy by not applying bullying tactics themselves and by demonstrating clear leadership in tackling bullying events immediately. Hence, monitoring managers' feelings and attitudes about the support in place for them to act effectively is essential. When involved with systems that deal with bullying, the interconnectedness of many aspects of the organisation becomes swiftly apparent. Are we hiring and promoting the right people? Do we train in conflict management and mediation? Are we giving our staff an accurate understanding of what is acceptable behaviour? Are we managing the information flows in cases that are known and being watched by employees? Are our decisions consistent, or do we avoid accepting cases where managers are put in the spotlight? Do we have any "untouchable" employees who appear to escape systems or sanction? We have known many organisations and can only advise policy designers that you will be judged by employees on your worst cases, not the best ones. As such, close monitoring allows the designers to be aware of seismic shifts (small and large) as employee attitudes make or break the use of such systems. It is essential that formal reviews of the policy and the data connected to it are undertaken at least once a year (Rayner and McIvor, 2008).

Conclusion

The good news is that a well-designed and coordinated antibullying policy can work, but conversely a policy that is designed by one department in isolation from users and other service deliverers can have no impact at all. A bullying policy weaves together many strands of the organisation—personnel/HR functions such as recruitment, selection, training, discipline, and complaint systems; trade union activity in prevention, policy promotion, and intervention; line management in role modelling and dispute handling; as well as helping services such as occupational health, counselling, advice networks, and external providers of assistance. The answer to maximising a policy is to ensure two things: first, that employees and service deliverers are all involved in its design; and, second, that dilemmas and discrepancies are actually resolved rather than ignored. The policy can then move with the issues and the organisation as it matures away from destructive conflict.

References

Andersson, L. M., and Pearson, C. M. (1999) Tit for tat? The spiraling effect of incivility in the workplace. *Academy of Management Review*, 24 (3), 452–471.

Bowen, F., and Blackmon, K. (2003) Spirals of silence: The dynamic effects of diversity on organizational voice. *Journal of Management Studies*, 40 (6), 1393–1417.

Bowes-Sperry, L., and O'Leary-Kelly, A. M. (2005) To act or not to act: The dilemma faced by sexual harassment observers. *Academy of Management Review*, 30 (2), 288–306.

Burnes, B., and Pope, R. (2007) Negative behaviours in the workplace: A study of two Primary Care Trusts in the NHS. *International Journal of Public Sector Management*, 20 (4), 285–303.

Caza, B. B., and Cortina, L. M. (2007) From insult to injury: Explaining the impact of incivility. *Basic and Applied Social Psychology*, 29 (4), 335–350.

Costantino, C. A., and Merchant, C. S. (1996) *Designing conflict management systems.* San Francisco: Jossey-Bass.

Dollard, M., Siunner, N., Tuckey, M. R., and Bailey, T. (2007) National surveillance of psychosocial risk factors in the workplace: An international overview. *Work and Stress, 21* (1), 1–29.

Einarsen, S. (1999) The nature and causes of bullying at work. *International Journal of Manpower, 20* (1–2), 16–27.

Ferris, P. (2004) A preliminary typology of organisational response to allegations of workplace bullying: See no evil, hear no evil, speak no evil. *British Journal of Guidance and Counselling, 32* (3), 389–395.

Fevre, R., Robinson, A., Jones, T., and Lewis, D. (2010) Researching workplace bullying: The benefits of taking an integrated approach. *International Journal of Social Research Methodology, 13* (1), 71–85.

Giga, S. I., Hoel, H., and Lewis, D. (2008) *The costs of workplace bullying.* London: Unite the Union/Department for Business, Enterprise and Regulatory Reform.

Harlos, K., and Pinder, C. C. (1999) Patterns of organizational injustice: A taxonomy of what employees regard as unjust. *Qualitative Organizational Research, 2,* 97–125.

Hoel, H., and Beale, D. (2006) Workplace bullying, psychological perspectives and industrial relations: Towards a contextualized and interdisciplinary approach. *British Journal of Industrial Relations, 44* (2), 239–262.

Hoel, H., and Cooper, C. L. (2000) *Destructive conflict and bullying at work.* Manchester, UK: University of Manchester Institute of Science and Technology.

Hoel, H., Faragher, B., and Cooper, C. L. (2004) Bullying is detrimental to health, but all bullying behaviours are not necessarily equally damaging. *British Journal of Guidance and Counselling, 32* (3), 367–387.

Hubert, A. B. (2003) To prevent and overcome undesirable interaction: A systematic approach model. In S. Einarsen, H. Hoel, D. Zapf, and C. Cooper (eds.), *Bullying and emotional abuse in the workplace: International perspectives in research and practice* (pp. 299–311). London: Taylor & Francis.

Keashly, L., and Jagatic, K. (2000) The nature, extent, and impact of emotional abuse in the workplace: Results of a statewide survey. Paper presented at the Academy of Management Conference, Toronto, Canada.

Kim, P. H., Ferrin, D. L., Cooper, C. D., and Dirks, K. T. (2004) Removing the shadow of suspicion: The effects of apology versus denial for repairing competence- versus integrity-based trust violations. *Journal of Applied Psychology, 89* (1), 104–118.

Leka, S., and Kortum, E. (2008) A European framework to address psychosocial hazards. *Journal of Occupational Health, 50* (3), 294–296.

Lewis, D. (2002) The social construction of workplace bullying. Unpublished doctoral thesis, University of Wales, Cardiff.

——— (2006) *Fourth policy discussion paper on workplace bullying and harassment.* Report for Acas. London: Advisory, Conciliation and Arbitration Service.

Lewis, D., and Rayner, C. (2003) Bullying and human resource management: A wolf in sheep's clothing? In S. Einarsen, H. Hoel, D. Zapf, and C. L. Cooper (eds.), *Bullying and emotional abuse in the workplace: International perspectives in research and practice* (pp. 370–383). London: Taylor & Francis.

Lewis, D., Sheehan, M., and Davies, C. (2009). Uncovering workplace bullying. *Journal of Workplace Rights, 13* (3), 279–299.

Leymann, H. (1996) The content and development of mobbing at work. *European Journal of Work and Organizational Psychology, 5* (2), 165–184.

Liefooghe, A. P. D., and MacKenzie Davey, K. (2001) Accounts of workplace bully-ing: The role of the organization. *European Journal of Work and Organizational Psychology*, 10 (4), 375–392.

Lutgen-Sandvik, P., Tracy, S. J., and Alberts, J. K. (2007) Burned by bullying in the American workplace: Prevalence, perception, degree and impact. *Journal of Management Studies*, 44 (6), 837–862.

Pilbeam, S., and Corbridge, M. (2006) *People resourcing: Contemporary HRM in practice.* Harlow, UK: Financial Times.

Rayner, C. (1998) Workplace bullying: Do something! *Journal of Occupational Health and Safety–Australia and New Zealand*, 14 (6), 581–585.

———— (2005) Reforming abusive organizations. In V. Bowie, B. S. Fisher, and C. L. Cooper (eds.), *Workplace violence: Issues, trends, strategies* (pp. 60–74). Cullompton, Devon, UK: Willan.

Rayner, C., and McIvor, K. M. (2008) *Tackling bullying and harassment: Final report to the Dignity at Work Steering Committee.* http://www.port.ac.uk/workplacebullying (accessed May 1, 2009).

Richards, J., and Daley, H. (2003) Bullying policy: Development, implementation and monitoring. In S. Einarsen, H. Hoel, D. Zapf, and C. L. Cooper (eds.), *Bullying and emotional abuse in the workplace: International perspectives in research and practice* (pp. 127–144). London: Taylor & Francis.

Salin, D. (2003) Ways of explaining workplace bullying: A review of enabling, moti-vating and precipitating structures and processes in the work environment. *Human Relations*, 56 (10), 1213–1232.

Salin, D. (2009) Organisational responses to workplace harassment: An exploratory study. *Personnel Review*, 38 (1–2), 26–44.

Schat, A. C. H., and Kelloway, E. K. (2003) Reducing the adverse consequences of workplace aggression and violence: The buffering effects of organizational sup-port. *Journal of Occupational Health Psychology*, 8 (2), 110–122.

Skogstad, A. (2005) The destructiveness of laissez-faire leadership behaviour. Paper pre-sented at the Workplace Bullying Conference, Portsmouth University, England.

Spurgeon, A. (2003) Bullying form a risk management perspective. In S. Einarsen, H. Hoel, D. Zapf, and C. L. Cooper (eds.), *Bullying and emotional abuse in the workplace: International perspectives in research and practice* (pp. 327–338). London: Taylor & Francis.

Tajfel, H., and Turner, J. (2001) An integrative theory of intergroup conflict. In M. A. H. D. Abrahams (ed.), *Intergroup relations: Essential readings in social psychology* (pp. 94–109). Philadelphia: Psychology Press/Taylor & Francis.

Tehrani, N., and Monkswell, J. (2001) *Building a culture of respect.* London: Taylor & Francis.

Trevino, L. K., Weaver, G. R., and Reynolds, S. J. (2006) Behavioral ethics in organiza-tions: A review. *Journal of Management*, 32 (6), 951–990.

Vartia, M., Korppoo, L., Fallenius, S., and Mattila, M. (2003) Workplace bullying: The role of occupational health services. In S. Einarsen, H. Hoel, D. Zapf, and C. L. Cooper (eds.), *Bullying and emotional abuse in the workplace: International perspec-tives in research and practice* (pp. 285–298). London: Taylor & Francis.

Yamada, D. C. (2000) The phenomenon of "workplace bullying" and the need for status-blind hostile work environment protection. *Georgetown Law Journal*, 88 (3), 475–536.

Zapf, D., and Gross, C. (2001) Conflict escalation and coping with workplace bul-lying: A replication and extension. *European Journal of Work and Organizational Psychology*, 10 (4), 497–522.

15

Investigating Complaints of Bullying and Harassment

Helge Hoel and Ståle Einarsen

CONTENTS

Introduction

Organizations are increasingly acknowledging the need to have in place proper policies and procedures to deal with cases of bullying and harassment should they arise (see also Rayner and Lewis, this volume). Reflecting such a need and to ensure an equal playing field, the social partners in Europe, represented by various employer associations (e.g., Business Europe) and the European Trade Union Confederation (ETUC) signed a framework agreement in 2007 on the prevention of violence and harassment. The agreement ensures the right of employees to file complaints against alleged perpetrators with the aim of having their case heard by means of an impartial investigation and, given the outcome, the appropriate actions taken (European Social Dialogue, 2007).

In an increasingly diverse workplace (see Lewis et al., this volume), outlets or institutional mechanisms for dealing promptly and fairly with complaints are essential. Hence, whilst increased diversity represents potential for organizational opportunities and strength, it also gives rise to situations where misunderstandings, disagreements, and resentment may be rife (Baron and Neuman, 1998). Whilst such situations can be minimized if they are properly managed, all organizations should be prepared for the need to investigate allegations of behavioural misconduct. The presence of well-planned procedures for investigation and their proper implementation provides the organization with an opportunity to make correct decisions, to reestablish fairness, and to bring the matter to a conclusion. Where appropriately applied, they also provide security for the individual and send a strong signal to employees that these issues are taken seriously and not tolerated by the organization (Stockdale and Sagrestano, in press). By contrast, where such response mechanisms are not in place, cases of this nature often remain unresolved for a long time, sometimes even for years, causing frustration and resentment to many or even all of those involved (Einarsen et al., this volume). In addition to the negative organizational effects such ongoing cases or scenarios are likely to generate, which also increasingly show up on the organization's balance sheet (see Hoel et al., this volume), organizations may face the prospect of litigation. In this respect it is important to point out that whilst organizations vary with respect to the risk of bullying, no organization can be considered "bully-proof" (Rayner et al., 2002), thus the need to have in place policies and procedures on bullying applies to all organizations. The case for having a complaint of bullying or harassment heard and impartially investigated can equally be made on the grounds of justice and fairness, as reflected in the growing attention being paid to issues associated with respect to the dignity of individuals in terms of their work experience (Di Martino et al., 2003).

Where proper procedures exist and these are applied correctly in terms of a fair investigation process, the conclusions reached and the appropriate sanctions towards the perpetrators taken, their presence may also have an impact on behaviour within the organization (Hulin et al., 1996). Although it is vital that organizations put in place processes and mechanisms to prevent bullying and harassment in the first place, organizations must equally be ready to respond to those cases that slip through the safety net and where a formal investigation is warranted. Having procedures in place for such situations also ensures predictability for everyone involved and their perceptions of procedural justice (Neuman and Baron, this volume), a key element in perceptions of fairness. In addition, a written procedure that allocates clear roles and responsibilities also acts as a guarantee of a planned and systematic process in which the impact of potentially interfering factors such as organizational politics (Salin, 2003) and heightened emotions can be reduced to a minimum.

In the following section, we describe how to conduct a fair and correct internal investigation in response to complaints about bullying and harassment at work. Although the previous edition of this book included a chapter on investigation, the chapter focused on investigations carried out by external investigators (Merchant and Hoel, 2003). In principle, every investigation into complaints of bullying instigated by the employer is an internal investigation (as opposed to any such investigation conducted by national or local bodies or other official inspectorates), though the employer may, of course, use both internal staff and externally hired consultants. The focus of the present chapter is, however, on the former because this should be the rule. In cases where the complaint involves very senior members of staff, it may, however, be advisable to employ external expertise, not least to ensure that the investigation is taken seriously by those involved and that the result will be respected.

Initially, the chapter explains the basic principles that must govern the investigative process. In this, we stress the importance of anchoring the investigation process within the organization's bullying policy. This is followed by a systematic exploration of the investigation process and its various stages. We close by discussing some of the obstacles often encountered in providing a fair hearing in such cases.

Principles Governing the Investigation Process

Predictability of process and how it is progressed and concluded is a key issue for trust in the process to be established. This can be best achieved by embedding the investigation process firmly in existing organizational policies and procedures. This section examines some core principles that need to be observed for the investigation process to be successful.

Investigation Informed by Local Policy

It is left to the employer to ensure that any complaints of bullying and harassment are dealt with in a fair as well as ethically and legally correct manner, ensuring the rights of both targets and alleged perpetrators (Einarsen and Hoel, 2008). To avoid having to respond to an aggravated situation unprepared or in haste, when emotions are running high, it is essential to have thought through the process of how a complaint should be handled independently of, or divorced from, any ongoing bullying scenario. Although such local policies and other organizational antibullying measures are often introduced in direct response to awareness of local problems (Salin, 2009), we suggest that to avoid suspicion and defensiveness, polices and procedures should be generated in "peacetime," and not in connection with any particular dispute.

The case for policy is made elsewhere (Rayner and Lewis, this volume). In terms of investigation of complaints of bullying, the policy has specific functions and fulfils certain needs. For the individual, it should provide the opportunity to have one's case heard and thus offers the prospect of personal vindication or redress (Meglich-Sespico et al., 2007). Furthermore, the presence of a policy and accompanying procedures for dealing with formal complaints represents predictability and security for the individual in terms of how his or her case is handled as well as the implications for any perpetrators found guilty of an offence. In this respect, the policy can be seen to represent an expression of the balance between the employers' duty of care to their employees, on the one hand, and the "managerial prerogative," or managers' right to manage, on the other. For the organization, it is a way of resolving or bringing an ongoing case to conclusion and reduces the chance of litigation.

It is suggested that effective policies on bullying and harassment should emphasize the following principles (e.g., Einarsen and Hoel, 2008, p. 161; Stockdale and Sagrestano, in press):

- The right of every employee to work in an environment free of harassment, bullying, and intimidation
- Nontolerance of bullying and harassment, emphasizing the seriousness with which any breach of the policy will be considered, highlighting that disciplinary action may be taken depending upon the severity of the offence
- The demand for compliance applied to all employees, workers, and managers, as well as any individuals subcontracted or seconded to work for the organization
- Nontolerance of any attempts at recriminations against or further victimisation of anyone using the policy to complain, including a nontolerance of malicious complaints

In addition to highlighting the foregoing principles, the policy should also contain the following elements:

- Standards for behaviour or conduct (against which the investigators judge the complaint and the evidence in its support). Because many people will not label their experience as bullying or harassment, it could be even more appropriate to include a range of examples of unacceptable behaviour rather than to provide a definition (Eberhardt et al., 1999).
- Designated, reasonable timeframes for the various stages of the complaint process.
- A description of the complaints procedures and the nature of the investigative process.
- Systems for monitoring, recording, and internally publicising complaints and their outcomes (Stockdale and Sagrestano, in press).
- Due process and natural justice.

To treat everyone involved fairly and in order to arrive at a just conclusion that is also perceived as such by those directly involved and by the wider organizational community, certain principles need to be taken on board. Of particular relevance here is the principle of "natural justice," or what is also referred to as due process, "typically construed as the right to know the charges and respond to evidence against oneself" (Stockdale and Sagrestano, in press, p. 26). Fair treatment means that someone accused of bullying should know what he or she is being accused of and the exact nature of the complaint. Furthermore, alleged perpetrators must be given the opportunity to defend themselves against the complaint by being given access to any evidence relevant to the outcome of the case as well as having the opportunity to call witnesses and present other evidence that may support their side of the case. Moreover, in order for fairness and justice to prevail, the parties must be treated equally throughout the process. This treatment implies that no one should be judged before the investigation is concluded, which is, according to Cropanzano et al. (2007), essential in order to preserve workplace harmony. This also suggests that investigators must refrain from making moral judgments or obtaining character statements and that any conclusion must be based entirely on the facts of the case and not on sympathies or personal feelings (Ishmael, 1999; NWEO, 2005).

Subjective versus Objective Experiences

Bullying and harassment as psychological phenomena are, of course, subjective experiences, to some extent being "in the eyes of the beholder" (see Einarsen et al., this volume). By contrast, the aim of an investigative process is to look at bullying from an objective perspective to see if experiences can be verified or confirmed by other third parties or by other objective evidence (Brodsky, 1976). Although individuals may have different thresholds with respect to their level of tolerance and, therefore, assess similar experiences differently, organizations need clear criteria for what is tolerated and what is not. Moreover, in order to take action against someone accused of bullying, the organization needs to establish objectively whether the alleged behaviour has taken place, the nature of such behaviour, and whether the actual events must be considered as a breach of either internal codes of conduct or possibly the relevant national legal code.

Confidentiality versus Anonymity

Confidentiality and anonymity are two other principles of great relevance to investigations. The issue of confidentiality in connection with complaints about bullying and harassment has been the subject of considerable debate (Rayner et al., 2002). It is essential to emphasize that the confidentiality that can be offered to the complainant and others implicated is an assurance that the case and the information received will be treated in confidence as far as it is possible to progress the investigation (Merchant and Hoel, 2003). Thus, complete confidentiality cannot be promised because it would jeopardize the organization's duty of care, not only to the complainant but also to other organizational members who may be at risk of similar treatment (Ishmael, 1999). By contrast, because doing so would breach the principles of natural justice, organizations should not offer anonymity. The potential downside of this is that it could make witnesses reluctant to come forward. But here it is relevant to point out that we are talking about an organizational process and not a criminal investigation.

Vexatious, False, and Malicious Complaints

The complaint procedure applies to genuine complaints about bullying and harassment. In this sense, it is important to stress that most complaints that are not upheld or proven may still be genuine and made in good faith. By contrast, vexatious complaints, for their part, refer to grievances that are considered not to be made in good faith, as, for example, when made for personal gain (see http://thebullyinghelpline.blogspot.com/2009/09/unfounded-or-vexatious.html). False or malicious complaints signify complaints that are fabricated or not rooted in reality and deliberately set out to harm one or more individuals. Vexatious, false, and malicious complaints must not be tolerated by the organization, a principle also laid down in the European Framework Agreement on Violence and Harassment developed by the European Social Dialogue (2007) and mentioned in the introduction.

Investigation Based in Law and Statutory Requirement

An investigation must be rooted in relevant organizational policies and procedures. It must also, in principle, be governed by statutory regulation relevant to the case (e.g., health and safety regulations) and specific regulations on bullying and harassment where such legislation exists. It must therefore be expected that the investigators will have a clear understanding of these issues and their implications for the process and the role of the investigation. Understanding of the case is particularly evident when the investigators draw their conclusions, as it is against standards laid down in statutory regulations as well as internal policies that the actions and behaviour revealed by the investigation will be assessed and judged. Where such standards might not be specific or descriptive but, rather, refer to general principles of fair human interaction and management, it is to some extent left to the discretion of the investigators to judge whether the conduct in question represents a breach of such standards.

Investigation Supported by Collective Agreements

As reported initially, the European Framework Agreement on Violence and Harassment at Work (European Social Dialogue, 2007) relies on existing legislative instruments (see, e.g., European Economic Community, 1989, concerning measures for improvements in the safety and health of workers at work; European Trade Union Confederation, 2009) and commits the members of the signatory parties or of organizations on the employer and trade union sides to implement it within a time scale of three years. The agreement suggests that "a suitable procedure will be underpinned but not confined to the following" (European Social Dialogue, 2007):

- It is in the interest of all parties to proceed with the necessary discretion to protect the dignity and privacy of all.
- No information should be disclosed to parties not involved in the case.
- Complaints should be investigated and dealt with without undue delay.
- All parties involved should get an impartial hearing and fair treatment.
- Complaints should be backed up by detailed information.
- False accusations should not be tolerated and may result in disciplinary action.
- External assistance may help.

In is important to note that establishment, revision, and monitoring of procedures should be done in partnerships, across Europe, and independent of custom.

Investigative Responsibilities

To ensure that the investigation is carried out in line with principles for fairness and reasonableness, clear allocation of roles and responsibilities is essential. In terms of the organization's role, it must ensure that the investigation is carried out in line with the requirements of any relevant policy and any written accompanying internal procedures. Furthermore, the employer, or the employer's representative as set out in internal policies, gives the mandate for the investigation and in this way agrees that it can go ahead. In practice, this role is often allocated to the head of human resources or head of personnel. As has already been emphasized, the organization must ensure that the rights of complainants as well as alleged perpetrators or those complained against are respected as previously argued. Any support system available to the parties involved must also be established and explained by the employer. To ensure that the remainder of the process is carried out in a way that is fair and defensible should the case still at one stage go to court, those entrusted with the role of investigator must have received training in the principles guiding such investigations and in their practical application.

As far as the investigators are concerned, their overall responsibility is to conduct an assessment of the extent to which the complaint is upheld or supported. To this end, they will gather the necessary evidence from interviews with those directly involved and from witnesses, and if necessary, by obtaining other forms of relevant proof or evidence. Furthermore, their role is to assess and weigh up the evidence, drawing a conclusion in terms of fully or partially supporting the complaint or of dismissing it where no, or insufficient, evidence has come to light. As stated earlier, the evidence must be assessed against any relevant statutory regulation as well as standards laid down in internal policies.

The Investigation Process

There is, of course, a requirement that successful outcomes rely on investigators being comprehensively trained for their task. Furthermore, it is advantageous that the investigators come from different organizational units, although optimal solutions might sometimes be difficult to achieve. These requirements also apply to the number of internal investigators employed for the task. Here, a team of investigators (two or more) has clear advantage in terms of speed of investigation and the opportunity to reflect on the case and the evidence gathered as the investigation progresses.

Although a dose of realism is healthy when considering schedules for progressing a case, time is at a premium here; and the longer the case drags on, the greater the negative impact on individuals involved as well as on the organization. Therefore, the organization should state within their policy a reasonable, specified timeframe within which the investigation should be completed (e.g., within 20 working days after the submission of the complaint). Organizations aiming higher, say, 10 days, may find themselves fighting a losing battle. However, as circumstances may vary in terms of the nature of the complaints (e.g., the seriousness of the case and the number of people involved), the context (e.g., to what extent the nature of their work allows people to leave their other duties), and the particular situation within which they emerge (e.g., unanticipated work pressures, or vacation time), there must be room for flexibility because such targets cannot always be achieved. All in all, considerations relating to organization of the process at the outset should not be underestimated, and as was argued by Merchant and Hoel (2003), much can go wrong that cannot easily be repaired if attention to detail slips, particularly at the planning stage.

The investigation process can therefore be seen to consist of four separate stages: (1) preparation or planning, (2) gathering evidence, (3) reaching a conclusion, and (4) writing a report.

1. Preparation

It is the employer, typically the head of personnel or HR, who identifies investigators and gives the mandate for the investigation. Such a mandate is essential because it provides investigators with the necessary authority, acting on behalf of the employer, to call individuals to interviews; to access relevant information, including personnel files (within the confines of national data protection legislation); and to gain admittance to work areas where necessary. Implicitly, it also grants the investigators the time to carry out the investigation. As far as timeframes are concerned, the investigators need to consider the entire process, taking into consideration possible time constraints of an operative nature as well as constraints related to demands from their normal job, ensuring that the investigation can proceed in line with requirements laid down in internal procedures. Where two or more investigators are identified for the task, allocation of responsibilities and division of labour between the team members must be clarified. Because it is essential that information obtained is effectively shared between the team members, channels for communication need to be agreed upon. Similarly, the information that needs to be communicated to the organization during the investigation needs to be considered.

In order to devise a strategy for the investigation, investigators need to study the complaint in detail. Does the complaint suggest that any particular regulation or policy has been breached? In particularly serious cases—for example, in cases of assault or physical violence or where there is fear of repetition or retaliation—suspension of the alleged perpetrator (on full pay) may be considered. Alternatively, it may be necessary to physically separate the parties for the period of the investigation. To ensure that the process is fair and to limit the influence of potential biases, the investigators ought, at the outset, to reflect on their own attitudes and possible prejudices to the parties involved.

As far as letters or other written communication regarding interviews is concerned, Merchant and Hoel (2003) argue, their wording is of vital importance and needs particular consideration. Considerations of fairness suggest that the same message should be communicated to complainants and to the alleged perpetrator in terms of the process, including any statements about confidentiality. To avoid interference with the process, it is useful to include a passage emphasizing that potential witnesses should not be approached and that any intimidation of witnesses or the opposing party will not be tolerated (Merchant and Hoel, 2003). Similarly, to prevent distortion of facts, witnesses should be advised to maintain confidentiality (North Western Local Authorities' Employers' Organization, 2005).

In order to create a nonconfrontational, albeit formal, atmosphere in what may be a highly charged situation, attention needs to be paid to the physical layout of the interview room in terms of how furniture is placed. Where a case is highly visible—for example, where the organization is small and the nature of the case may be associated with personal embarrassment on the part of those involved—it may be necessary to locate the interviews on neutral ground (e.g., off the premises). Interviewees should be informed about their right to be accompanied during interviews (e.g., by a trade union representative or a colleague), although these persons should not represent the interviewees by speaking on their behalf. By providing interviewees with support, such arrangements could help calm down an emotionally heightened situation and may, therefore, be beneficial to the progress of the investigation.

The order of interviews is of importance for the effectiveness of the investigation and to uphold principles of natural justice. Thus, it is recommended to start by interviewing the complainant, followed by the alleged perpetrator (North Western Local Authorities' Employers' Organization, 2005). Only when these interviews have been carried out will it be possible to have a clear picture of which witnesses need to be interviewed.

2. Gathering of Evidence

Because the investigators should establish whether and to what extent the complaint is substantiated (or proven), their objective is to identify and uncover any evidence that may help them arrive at a conclusion about the complaint one way or the other. Typically, the evidence is collected by means of interviews of those directly involved as well as possible witnesses or by other evidence, an issue we will return to later. Throughout this process, the focus must be on facts, disregarding any statement based on beliefs, opinions, or personal characteristics.

Because the outcomes of the interviews are the basis for the progress of the investigation and the ability to draw fair conclusions, due attention must be given to the preparation for the interviews. However, our experience of training investigators tells us that many would-be investigators severely underestimate the difficulty of getting the optimal outcome from an interview. Thus, we argue that even extensive experience of interviewing employees for selection, appraisal, and even disciplinary hearings does not automatically translate into the skills needed for the investigation interview. It is a question of not only dealing with emotionally difficult situations in a constructive manner but also not allowing personal biases to prejudice or interfere with the process. Investigation interviews should neither be open for dialogue, as in appraisals, nor be an opportunity for interviewees to take the initiative or respond entirely on their own terms, steering the interview in their own preferred direction. Nevertheless, it is important that the interviewee initially be given the opportunity to tell the story the way he or she sees it, elaborating on issues the interviewee considers relevant to the case. This scope is particularly important for the complainant and the alleged perpetrator, whereas witnesses are to be seen more as a source of specific information that the investigators may need in order to draw conclusions about the case. Giving interviewees such scope also tends to calm down the situation and help the interviewees relax. Doing this is paramount, for where there is a lot at stake, interviewees, particularly those accused of harassment or bullying, but also complainants in their defense or in their allegations, will attempt to portray themselves in a positive light whilst often trying to weaken the case of their opponent.

Thus, investigators must study the complaint and think through the interviewing process in advance, identifying issues in need of clarification. The way questions are best phrased in order to discover and obtain the facts of the case must be given consideration. For example, open questions should be used for interviewees to elaborate or give their account of issues or events, whereas prompts should be provided for confirmation of facts (North Western Local Authorities' Employers' Organisation, 2005). Although the ability to phrase questions in a way that unlocks and provides access to facts, access to chains of events and their underlying motives, and the capacity to capture and follow up leads during the course of interviews will greatly improve with practice, experience should never replace the need for preparation. Being part of a team of investigators will also ease these challenges. Given the importance of being proactive throughout the interview, not letting slip opportunities to follow up a lead requires concentration. As has been argued by Merchant and Hoel (2003), a failure to pick up threads of evidence could actually lead to injustice. Nevertheless, whilst being fully in control of the process, interviewers should let the interviewees do most of the speaking and strictly avoid prompting particular responses or putting words or arguments in their mouths. Furthermore, any request from interviewees to speak off the record should always be denied.

Having interviewed the complainant(s) and then the alleged perpetrator(s), witnesses who could throw light on the case should then be called. Again, it is important to highlight that personality characteristics, beliefs, or opinions should have no role to play. Witnesses are sources of fact only and do not have a say regarding the conclusion. The number of witnesses to be interviewed will be determined by the needs of the case in terms of the ability of the investigators to gather sufficient evidence to reach a defensible conclusion. Attention to fairness, in terms of a balance of witnesses interviewed for each side, will also have an impact on the total number of witnesses interviewed, as will the organizational resources available. One should note that it may be necessary to interview an informant more than once to clarify matters or when the emergence of new information requires this. A cautionary note needs to be sounded with respect to other issues giving rise to concern that may surface during the course of the investigation. In this respect, it is advised that these issues are kept separate from the ongoing investigation and are, if necessary, reported to the employer. In this respect, it is important to stay on track and not be diverted into investigating other issues that may surface during the investigation process.

How best to take records of interviews has also been a matter for discussion. Whilst Merchant and Hoel (2003) argue in favour of tape-recording, emphasizing that this removes any questions about the accuracy of the account and leaves the interviewer to focus on the interactions, traditional note-taking on paper or computer may be a more appropriate approach depending on the nature of the situation and as a way of reducing paper-work by focusing on key issues. A useful compromise may be to rely on note-taking for the summary or minute and use the recordings only as a backup where there may be disagreement about what was actually said. However, to maintain trust in the process, it is essential that the intervie-wee agrees that the minutes represent a true account of the interview by signing it.

Where there are no witnesses or none is willing to come forward, or where it is one person's word against another's, investigators should seek out other evidence which could corroborate any claim made. For example, physical marks could be presented as evidence of an assault. Similarly, unusual behaviour and emotional reactions exhibited by the target could equally back up a claim of bullying. As part of the investigation, it can sometimes also be necessary to inspect the physical environment in which an incident is claimed to have taken place to get a clear picture of events or, where there is doubt, to verify the feasibility of the said events (Merchant and Hoel, 2003). For the same purpose, it may be necessary to check information available on personnel files or on personal computers (e.g., job-descriptions, sickness records, or e-mails), although national legal provisions may restrict access.

3. Reaching a Conclusion

Investigators should work on the assumption that their task is to involve a sufficient number of witnesses to be able to draw a defensible conclusion and not to carry out a complete or exhaustive process (Merchant and Hoel, 2003). In reaching a conclusion, they should base their judgment on the principle of probability, rather than providing evidence to prove the case beyond reason-able doubt as would happen in court. However, the more serious the case is in terms of the potential implications it could have for alleged perpetrators should the investigation support the complaint, the higher the level of proof that will be required. Thus, in serious cases where dismissal is a likely out-come should the complaint be upheld, the investigators must have strong evidence on their side. Such cases may well end up in court at a later stage. Therefore, whilst less serious cases would require only a probability greater than 50% to be supported, very serious complaints may require a probability in the order of 70–80%, or higher. However, where it is a case of one person's word against another's and insufficient evidence emerges to enable investi-gators to reach a particular conclusion, the complaint cannot be supported. Yet, as has been argued by Merchant and Hoel (2003), this does not mean that the complaint is false or malicious, simply that insufficient evidence has emerged to substantiate it.

Based on the facts and an overall assessment of the case, it may be necessary to conclude that the complaint is vexatious or, in some rare cases, even malicious or false, as described in a previous section.

4. Writing a Report

To complete the process, investigators are required to write an investigation report summarizing the process, the facts, and the conclusions. The role of the report is to document the investigation process, provide an account of the evidence, and present the investigators' conclusions. Because employers will rely on the report and its conclusion in their final judgment with respect to the nature and severity of sanctions, if any, and as a basis for organizational intervention where important shortcomings are revealed, the report is of considerable importance. As a record of the investigation process, it is of great importance should the case in the end go to court. Whilst format and order will vary, the report will normally contain the following elements: (a) the complaint itself and the response from the alleged offender, (b) external (e.g., laws and statutory regulations) and internal (e.g., policies and procedures) provisions applicable to the case, (c) documentation of the investigation process and the evidence revealed, (d) assessment of evidence and conclusions, and (e) reference to other potential organizational problems that become evident during the process (North Western Local Authorities' Employers' Organization, 2005).

Following the submission of the report, it is left to the employer to ensure that the investigation conclusion is translated into appropriate action depending on the seriousness of the offence and in line with policy and organizational custom and practices. Appropriate actions must first of all provide a restoration of the working environment of the complainant, securing a bullying-free work situation. Second, in line with the organizational policy, the employer must decide on any disciplinary action to be taken against the perpetrator.

Barriers to a Fair Hearing

As outlined in the foregoing, the primary objective of investigators is to establish whether there is sufficient evidence to support or uphold a complaint. However, to meet such an objective investigators need to have the skills to obtain the information necessary to reach such a decision. We argue that this process is far from straightforward and that success relies largely on investigators' knowledge about bullying phenomena as well as on their acquired investigative skills. This is evident as a number of factors associated with the bullying process may militate against the opportunity for targets to receive a fair hearing (Hoel, 2009). First, because much bullying may take place in private, out of sight of observers, it may come down to one person's word against another's. Second, even where alleged perpetrators agree about the nature of individual events, they may argue that these represent separate, isolated events, each with its own logical explanation, thus contradicting targets' perceptions that a pattern of abuse exists (Einarsen and Pedersen, 2007). Yet, the investigators have to reach a conclusion based on the facts of the case and on the principle of probability, and not on any admission of guilt by the alleged perpetrator. Third, as the parties are likely to have a past (and a future) together, individual acts and their meaning for the protagonists cannot be understood in isolation (Hoel et al., 1999). For example, acts that on the face of it appear harmless or even friendly to observers may be seen as ironic and directly unfriendly when seen in connection with previous negative interaction and disagreement (Hoel et al., in press). In addition, such factors as the alleged perpetrator acting differently in different situations and with different people (Rayner, 1999), the tendency to believe those with greater power (Vickers, 2006) or to side with an alleged perpetrator who fulfils a key organizational role (Hoel et al., in press), as well as victims' failure to attract sympathy for their cases as a result of their mental instability (Einarsen and Mikkelsen, 2003), all could have implications for how their cases are treated. Given this, we argue that the ability to uncover the facts of the case and act impartially throughout the process hinges on a combination of knowledge of the phenomena and their effects on those involved, the ability to organize and carry out a complex investigation, interviewing and listening skills, and the ability to consider the facts and draw a balanced conclusion. Finally, investigators need to have a sound ethical stance towards the problem and their role.

In addition to the points raised here, there may be further barriers to a fair process:

- Where employees question the fairness of the procedures or consider complaining to be unsafe in terms of possible retaliation (Hoel and Cooper, 2000), their use will gradually cease. It is, therefore, important to reiterate that even where a complaint filed in good faith is not substantiated, it must never be used as a justification for punishment or retaliation.

- Because investigations can be costly in terms of the resources they absorb, there can be pressure within the organization to limit the number of complaints by introducing measures to assess their seriousness and likely outcome. Such moves could lead to unfairness and could involve some kind of character judgment, particularly where such prescreening processes are not transparent. Such processes could also be seen to protect particular groups or members of the organization, making the powerful or seemingly irreplaceable "untouchable" or putting them beyond complaint.

- As part of drives to get organizations to adopt mediation more widely as a conflict resolution tool, complainants of bullying and harassment could find their cases "forced down" the mediation route, possibly putting pressures on individuals to agree to such an approach where this is not, strictly speaking, in the interest of the target—for example, where the case has progressed to a point where mediation is seen as ineffective or even directly harmful to the target (see Keashly and Nowell, this volume).

Where investigations are taken seriously and dealt with in a professional and fair manner, and their conclusions are followed up by means of appropriate sanction that makes it clear that the organization does not tolerate bullying, they may not only represent a powerful tool to bring a complaint to a close but may also contribute to prevention by leading to a behavioural change in the work environment. Thus, bullying and harassment are less likely to occur in an environment intolerant of bullying and harassment where complaints are seen to be taken seriously and where there are serious consequences for perpetrators found guilty of such behaviour (Stockdale and Sagrestano, in press).

An argument against introducing a policy against bullying, including complaints procedure and investigation, may be that the organization risks being flooded with complaints. However, the reality is that few people decide to label their experience of bullying or harassment for what it is (see also Nielsen et al., this volume), and among those who do, only a minority proceed to file a complaint within their organization or through any other institutional arrangement available to them (Stockdale and Sagrestano, in press). The arguments for not proceeding to a complaint could vary but may include being less concerned with punishing any perpetrator than ensuring that any harassment or bullying stops, the belief that the complaint would not be taken seriously, reluctance to have a complaint filed on one's personal employment record, and fear of repercussions and "secondary victimization" (Berdahl and Raver, in press). A fair policy and fair investigation processes are vital guarantees against such obstacles.

References

Baron, R. A., and Neuman, J. H. (1998) Workplace aggression—the iceberg beneath the tip of workplace violence: Evidence of its forms, frequency, and targets. *Public Administration Quarterly*, *21*, 446–464.

Berdahl, J. L., and Raver, J. L. (in press) Sexual harassment. In S. Zedeck (ed.), *Handbook of industrial/organizational psychology*. Washington, DC: American Psychological Association.

Brodsky, C. M. (1976) *The harassed worker*. Lexington, MA: D. C. Heath.

Cropanzano, R., Bowen, D. E., and Gilliland, S. W. (2007) The management of organizational justice. *Academy of Management Review, 21*, 34–48.

Di Martino, V., Hoel, H., and Cooper, C. L. (2003) *Preventing violence and harassment in the workplace*. European Foundation for the Improvement of Living and Working Conditions. Luxembourg: Office for Official Publications or the European Communities.

Eberhardt, B. J., Moser, S. B., and McFadden, D. (1999) Sexual harassment in small government units: An investigation of policies and attitudes. *Public Personnel Management, 28*, 351–364.

Einarsen, S., and Hoel, H. (2008) Bullying and mistreatment at work: How managers may prevent and manage such problems. In A. Kinder, R. Hughes, and C. L. Cooper (eds.), *Employee well-being and support: A workplace resource* (pp. 161–173). Chichester, UK: John Wiley.

Einarsen, S., and Mikkelsen, E. G. (2003) Individual effects of exposure to bullying at work. In S. Einarsen, H. Hoel, D. Zapfand, and C. L. Cooper (eds.), *Bullying and emotional abuse in the workplace: International perspectives in research and practice* (pp. 127–144). London: Taylor & Francis.

Einarsen, S., and Pedersen, H. (2007) Håndtering av konflikter og trakassering i arbeidslivet: Jus og Psykologi (2. edition). Oslo: Gyldendal Akademisk.

European Economic Community (1989) Directive 89/391/EEC, June 12. http://eur-lex.europa.eu/LexUriServ/LexUriServ.do?uri=CELEX:31989L0391:EN:HTML (accessed May 14, 2010).

European Social Dialogue (2007) *Framework agreement on harassment and violence at work*. http://www.etuc.org/a/3574

European Trade Union Confederation (ETUC). (2009) Autonomous framework agreement on harassment and violence at work: An ETUC interpretation guide. http://www.etuc.org/IMG/pdf_pdf_CES-Harcelement-Uk-2-2.pdf (accessed November 14, 2009).

Hoel, H. (2009) Regulating for negative human interaction: An evaluation of the effectiveness of anti-bullying regulations. Paper presented at the Conference of the Regulating for Decent Work Network, International Labour Office, July 8–10, Geneva, Switzerland.

Hoel, H., and Cooper, C. L. (2000) Destructive conflicts and bullying at work. Unpublished report. Manchester, UK: University of Manchester Institute of Science and Technology.

Hoel, H., Glaso, L., Hetland, J., Cooper, C. L., and Einarsen, S. (in press) Leadership as predictor of self-reported and observed workplace bullying. *British Journal of Management*.

Hoel, H., Rayner, C., and Cooper, C. L. (1999) Workplace bullying. In C. L. Cooper and I. T. Robertson (eds.), *International review of industrial and organizational psychology*, vol. 14 (pp. 195–230). Chichester, UK: John Wiley.

Hulin, C. L., Fitzgerald, L. F., and Drasgov, F. (1996) Organizational influences on sexual harassment. In M. S. Stockdale (ed.), *Sexual harassment in the workplace: Perspectives, frontiers, and response strategies* (pp. 127–150). Thousand Oaks, CA: Sage.

Ishmael, A. (1999) *Harassment, bullying and violence at work. A practical guide to combating employee abuse*. London: The Industrial Society.

Meglich-Sespico, P., Faley, R. H., and Knapp, D. E. (2007) Relief and redress for targets of workplace bullying. *Employee Response Right Journal, 19*, 31–43.

Merchant, V., and Hoel, H. (2003) Investigating complaints of bullying. In S. Einarsen, H. Hoel, D. Zapf, and C. L. Cooper (eds.), *Bullying and emotional abuse in the workplace: International perspectives in research and practice* (pp. 259–269). London: Taylor & Francis.

North Western Local Authorities' Employers' Organization (2005) *Investigating bullying and harassment complaints: A practical guide for investigators.* Manchester, UK: North Western Local Authorities' Employers' Organization.

Rayner, C. (1999) From research to implementation: Finding leverage for prevention. *International Journal of Manpower, 20* (1/2), 28–38.

Rayner, C., Hoel, H., and Cooper, C. L. (2002) *Workplace bullying: What we know, who is to blame, and what can we do?* London: Taylor & Francis.

Salin, D. (2003) Bullying and organisational politics in competitive and rapidly changing work environments. *International Journal of Management and Decision Making, 4,* 35–46.

——— (2009) Organisational responses to workplace harassment: An exploratory study. *Personnel Review, 38* (1), 26–44.

Stockdale, M. S., and Sagrestano, L. M. (in press) *Strategies and resources for institutions and targets of sexual harassment in employment and education.*

Vickers, M. H. (2006) Towards employee wellness: Rethinking bullying paradoxes and masks. *Employee Response Rights Journal, 18,* 267–281.

16

Interventions for the Prevention and Management of Bullying at Work

Maarit Vartia and Stavroula Leka

CONTENTS

Introduction

Research and practice from the past 15 years have provided us with plenty of knowledge about bullying and negative treatment at work. Bullying is widely recognized and acknowledged by international and national organizations as a serious psychosocial hazard that leads to negative consequences for the targets as well as for the organization and society. For that reason, interventions for the prevention and management of bullying are seen to be important (Cassitto et al., 2003; Di Martino et al., 2003; Leka and Cox, 2008). The work environment hypothesis, which suggests that a number of factors associated with the work environment and organization act as antecedents of bullying, is nowadays the most favoured model to explain the onset of bullying at work. Based on practical and clinical work, and on experience from situations where bullying has taken place, it is argued that bullying is an escalating process that needs to be stopped as early as possible. A question often asked by different organizational actors is, what actually works? That is, what kind of activities should organizations undertake to prevent, reduce, and manage bullying and harassment and their detrimental effects?

In this chapter, the different types of interventions and strategies used in the prevention and management of bullying are presented and their effectiveness is discussed. Particular attention is paid to the strategies and methods used in organizational interventions. Furthermore, key principles of planning and implementation of interventions for the prevention and management of bullying are outlined and examined.

Levels of Interventions

According to Oeij et al. (2006) the term *intervention* has been applied to indicate a process of change set in motion within and with regard to work organizations. Strategies used in the prevention and management of psychosocial risks at work, including bullying and harassment, differ in many ways. A distinction is commonly made between primary, secondary, and tertiary prevention, on the one hand, and organizational, job- or task-level, and individual-oriented interventions, on the other. In addition, a further fourth level of prevention can be identified, that of policy-level interventions, aimed at bringing about change through their influence on the macro level nationally and internationally.

Primary stage interventions are proactive by nature and aim to prevent the harmful phenomena or effects emerging in the first place by reducing their risks. *Secondary stage* interventions aim to reverse, reduce, or slow the progression of the situation or of ill-health, to stop the event from recurring, and/or to increase the resources of individuals to cope with the situation. *Tertiary stage* interventions, for their part, are rehabilitative by nature, aiming at reducing the negative impacts caused by different occupational hazards and restoring the health and well-being of employees, as well as restoring a safe and healthy workplace.

In relation to bullying, organizational or employer-level interventions aim to influence the attitude towards bullying and inappropriate behaviour, to develop organizational culture where there is no room for bullying, and to introduce policies and procedures for prevention, as well as staging interventions when a problem occurs. The job-level strategies aim to prevent and tackle the problem by influencing the work environment, the job descriptions, the work organization, and the functioning of the work unit. Finally, individual-level interventions aim to change characteristics of the ways individuals interface with the job, such as perceptions, attitudes, or behaviour or the individual's health and ability to do his or her job.

According to many experts, intervention programmes are most effective and mutually reinforcing when they include both individual-directed and organizational or environment-directed strategies on primary, secondary, and tertiary levels (e.g., LaMontagne et al., 2007). Moreover, intervention programmes should consider changing not only the attitudes and behaviours of individuals but also the social context and the work environment.

In the next sections, different level interventions for the prevention and management of bullying will be presented.

Policy-Level Interventions

Policy-level interventions are most often targeted at preventing bullying and forcing employers to manage cases firmly and ethically. Such interventions encourage and oblige actors to take action for the prevention and reduction of bullying. They include development of statutory regulation and legislation and national policy, specification of best practice standards at national or stakeholder levels, the signing of stakeholder agreements for a joint strategy for the problem, and the signing of declarations at the European or international levels (Leka et al., 2008a; Zwetsloot and Starren, 2004). However, according to the intervention literature, interventions at the policy level are often ignored (Leka et al., 2008a).

Over the last few years, antibullying agreements and regulations have been passed for the management of bullying and harassment at work at national levels as well as at the European level. In this respect, an important recent initiative is the autonomous Framework Agreement on Harassment and Violence at Work, signed in 2007 by the European trade union, business, and economic interest groups ETUC/CES, BUSINESSEUROPE, UEAPME, and CEEP* (http://www.tradeunionpress.eu/Agreement%20violence/ Framework%20Agreement%20Harassment%20and%20Violence%20at%20 Work2.pdf).

---

* European Trade Union Confederation (ETUC), the Confederation of European Business (BUSINESSEUROPE), European Association of Craft Small- and Medium-sized Enterprises (UEAPME), the European Centre of Enterprises with Public Participation and of Enterprises of General Economic Interest (CEEP).

The agreement aims to increase among employees, workers, and their representatives awareness and understanding of workplace harassment and violence and to provide employers, workers, and their representatives at all levels with an action-oriented framework to identify, manage, and prevent problems of harassment and violence at work. According to the agreement, enterprises need to have a clear statement emphasizing that harassment and violence will not be tolerated. Procedures outlining how to deal with individual cases, should they arise, should also be included.

The agreement should be implemented and monitored within three years of the signing at the national level. So far, it has been translated into many languages, and the signatory partners have started discussions with the aim of finding appropriate measures to implement the agreement in their respective countries. Evidence from different countries shows that the agreement has evoked activities at national levels and contributed to raising public awareness of the issue. In some countries, employee and employer organizations have concluded further agreements for the implementation of the agreement at a national level, information about the agreement has been distributed to the members of the workers' and employers' organizations, and working groups have been established to plan the implementation of the agreement and to develop material to support work against harassment. Obviously, the European 2007 framework agreement has raised public awareness of the issue of workplace bullying and harassment (ETUC/BUSINESSEUROPE-UEAPME/CEEP, 2009).

At national levels, specific legislation against bullying and harassment has been passed in some countries, and in others the general safety and health regulation also cover bullying (see also Yamada, this volume). In Sweden, the Victimisation at Work Ordinance had been enacted earlier, in 1993. Thereafter, specific legislation to combat bullying and harassment at work has been enacted in such countries as the Netherlands, France, and Belgium; and in Finland and in Norway the Occupational Safety and Health legislation includes sections on bullying and harassment. The Finnish Health and Safety Act is an example of legislation that includes several sections that tackle bullying preventively and also respond to individual incidents of bullying (http://www.finlex.fi/en/laki/kaannokset/2002/en20020738). (See Figure 16.1.) Similarly, the first antibullying law in North America came into effect in 2004 in Quebec, Canada (see Yamada, this volume).

The work by the Health and Safety Authority (HSA), a cooperative effort between the Irish government and social partners in Ireland, is an example of a government-initiated countrywide strategy for tackling bullying at work (Murray, 2002). The HSA implemented its initiative at the dawn of the 21st century, with the latest code of practice for employers and employees on the prevention and resolution of bullying and harassment at work introduced by HSA in 2007 (http://www.hsa.ie/eng/Publications_and_Forms/Publications/Health/CoP%20Bullying.pdf).

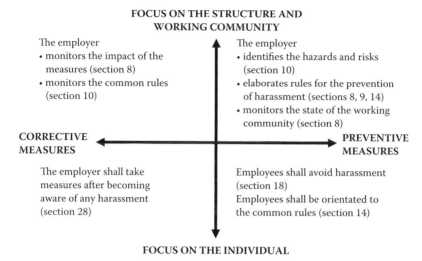

**FOCUS ON THE STRUCTURE AND
WORKING COMMUNITY**

The employer
- monitors the impact of the measures (section 8)
- monitors the common rules (section 10)

The employer
- identifies the hazards and risks (section 10)
- elaborates rules for the prevention of harassment (sections 8, 9, 14)
- monitors the state of the working community (section 8)

**CORRECTIVE
MEASURES**

**PREVENTIVE
MEASURES**

The employer shall take measures after becoming aware of any harassment (section 28)

Employees shall avoid harassment (section 18)
Employees shall be orientated to the common rules (section 14)

FOCUS ON THE INDIVIDUAL

FIGURE 16.1

Finnish Occupational Safety and Health Act: Dimensions in preventing harassment and inappropriate behavior at work.

The role of the HSA is to promote best practice in this area and to ensure employer compliance with the code. Unfortunately, so far no evaluation of the effects of the code is available.

Although a number of initiatives in the form of legislation and policies have been implemented, analysis and overall evaluation of these is relatively rare (Leka et al., 2008a). What exists by way of evaluation has, however, revealed that although enactment of legislation on bullying appears to have positive effects, there are also shortcomings and problems associated with such regulation. For example, the evaluation of the Swedish Ordinance for Victimization at Work (Hoel and Einarsen, 2010) showed that the ordinance has not been as effective as had been hoped. Shortcomings associated with the lack of involvement of key actors and with the antibullying regulatory framework itself were suggested to be the main reasons for the limited success of the ordinance. It also appears that the regulation has been introduced too early, at a time when the levels of awareness, recognition, and knowledge of the issue among employers were not sufficiently developed. Hoel and Einarsen, however, argue that the ordinance still gave the issue credibility. They conclude that statutory intervention should be supplemented with "well-informed, trained and motivated employers and trade unions who, in collaboration, are willing to deal with the problem proactively on an organisational level as well as responding to individual cases when they occur, supported by an enforcement agency or inspectorate which is equipped and geared up for its role" (Hoel and Einarsen, 2010, pp. 47–48). Similarly, in her assessment of the French law against moral harassment, Bukspan (2004, pp. 405–406) also concludes that "one must be vigilant, so that the legislation passed does not create hope, which, in the absence of workable and precise provisions, might be seen as a betrayal causing further suffering and injustice by the victims and their families."

A study on the implementation of the Finnish Occupational Health and Safety Act (Salminen et al., 2007) found that the new section on harassment and other inappropriate treatment from 2003 was recognized in most workplaces. The regulation was seen to have motivated initiatives against bullying in organizations, leading to the introduction of organizational antibullying policies and guidelines.

In addition, other positive experiences of the effects of legislation have been reported. Experts from different European countries (i.e., researchers, practitioners, clinicians, health and safety authorities, and union officials) interviewed for the Psychosocial Risk Management–European Framework (PRIMA-EF) policy project funded by the EU Sixth Framework programme (Leka et al., 2008b) reported that regulations and mutual agreements have been seen to increase awareness and knowledge about bullying and to push different employers into taking action. However, the panel of experts suggested that laws and regulations would never be sufficient in themselves and that any effective strategy needs to include education for all stakeholders as well as mechanisms for conflict resolution.

Organizational and Workplace-Level Interventions

At the organizational level, activities and interventions primarily focus on primary and secondary preventions. In most cases, the main aim of the activities is to prevent and reduce the occurrence of bullying by increasing awareness and recognition of and knowledge about bullying and by reducing the potential risks of bullying in the psychosocial work environment and by developing the functioning of the work unit. Preventive strategies also focus on organizational culture and management practices with the aim of increasing the capability of management and supervisors as well as other actors to investigate complaints and resolve conflicts and bullying in the workplace.

Organizational and workplace-level activities are focused on all employees, whether they be targets of bullying, perpetrators, or observers. Activities are also targeted at different actors in the organization—that is, at those who have the responsibility for securing a safe work environment for all employees (employers, managers, supervisors) and those who may help and support them in this work (health and safety representatives, union representatives). Different measures at all levels of prevention are used in organizations to prevent and reduce bullying at work, but assessment and evaluation of different strategies in relation to their effectiveness have so far been scarce.

Antibullying Policies

Researchers and practitioners (e.g., Hoel and Cooper, 2000; Hubert, 2003; Richards and Daley, 2003) have recommended the introduction of organizational antibullying policies and specific guidelines for prevention of bullying and for handling complaints and incidents of bullying (see also Rayner and Lewis, this volume). Written policies typically communicate a no-tolerance policy and require managers to take action (Richards and Daley, 2003). The effectiveness of written antibullying policies has, however, also been questioned (Salin, 2009).

It is suggested that an antibullying policy should include, for example, a clear statement from management that any kind of bullying and harassment is unacceptable, description of bullying and harassment, reference to legislation and other relevant regulations, and clarification of responsibilities as well as allocation of roles and responsibilities of management and other players. In addition, the policy should include clear guidance for the persons experiencing bullying, for witnesses, and for the persons accused of bullying. It must contain complaint procedures, information on support mechanisms, and measures to prevent bullying in the organization, as well as measures to monitor and evaluate the policy (Einarsen and Hoel, 2008; Leka and Cox, 2008).

The obligation to draw up organizational policies in this area differs between countries. For example, in Belgium, the law obliges employers to include appropriate measures against abusive behaviour in their policies to prevent psychosocial risks at work. In Finland, health and safety inspectors have played an active role in increasing the number of policies in organizations. Consequently, when carrying out inspections in organizations, inspectors now always ask if the organization has a written policy and advise organizations on how to draw up a policy where there is none. A study in the municipal sector in Finland (Salin, 2008) also found that the introduction of written antibullying policies and the provision of information were the most common measures adopted by organizations to counteract workplace bullying.

It has also been emphasized that developing and implementing a policy is as important as its contents (Einarsen and Hoel, 2008). In this respect, it is recommended that organizational antibullying policies should be drawn up jointly among the employer or HR, employee representatives, safety representatives, and occupational health care personnel, as well as union representatives. In connection with the implementation of the policy, all members of staff should also be given basic training on the policy. The use and effectiveness of the policy should be monitored and evaluated on a regular basis.

Although many organizations have policies in place, it seems that the awareness and use of the policies remain insufficient. In a large organization (i.e., one employing more than 35,000 people), awareness and use of the policy over time were assessed among a sample of supervisors, safety delegates, and employees: first when the policy had been in place for two years (N = 185) and again after eight years (N = 575) (Vartia, 2002, personal communication). In 2002, 57% of the supervisors and 11% of the employees were well acquainted with the policy; in 2008, only 12% of the respondents were well acquainted with the policy, a total of 33% had acquainted themselves superficially with the policy, and 15% were unaware of its existence. The number of people who had attended training sessions or information meetings dealing with the issue was quite low in both time points. (For more information on antibullying policies, see also Rayner and Lewis, this volume.)

Interventions for the Prevention and Reduction of Bullying

Assessment of Risks and Antecedents of Bullying

A core component of any workplace bullying prevention strategy is designing out the risks (i.e., the causes or antecedents) of bullying in an organization (see Salin and Hoel, this volume). An analysis of risks or of potential organizational antecedents of bullying can be conducted in various ways. One example is the application of particular instruments such as the Bullying Risk Assessment Tool (BRAT), which was developed for measurement of potential risk factors of bullying (Hoel and Giga, 2006). The tool includes items assessing the following factors: organizational fairness, team conflict, role conflict, workload, and leadership. Although the full applicability of the tool has yet to be tested, a validation study has confirmed the validity of the instrument with all five factors emerging as predictors of negative behaviour and self-labelled bullying (Hoel and Giga, 2006). Another example is the Val.Mob. scale, an Italian instrument developed for assessing both the aspects and the characteristics of mobbing as well as factors contributing to mobbing (Aiello et al., 2008). The instrument consists of three sections: the Mobbing scale, the Symptomatological scale, and personal and social data.

Preintervention surveys, interviews, focus groups, introductory meetings, and joint discussions are methods used to map the risks and antecedents of bullying and harassment at the organizational and workplace level and to establish preintervention baseline measures (Hoel and Giga, 2006; Mikkelsen et al., 2008). In addition to mapping the need for change, the aim of the preintervention measurements is to reach a common understanding of the situation. In line with this, a Danish intervention study was designed to develop, test, and evaluate interventions aimed at preventing severe interpersonal conflicts and bullying at work, using survey data, introductory meetings, interviews, and focus groups to increase researchers' and external consultants' knowledge of the workplaces, enabling a good workplace-intervention fit. Among the themes discussed were organizational culture, quality of the psychosocial work environment, conflicts, bullying, and management skills in relation to handling conflicts and bullying (Mikkelsen et al., 2008). Similarly, for a training-based intervention programme in the United Kingdom, a preintervention questionnaire comprising a variety of instruments to measure negative behaviour and experiences and consequences of bullying (Hoel and Giga, 2006) was devised and informed by a large-scale focus group study.

Among the various intervention measures employed, carrying out training of all employees as well as targeted training of supervisors and management is the method most often used. Other measures include, for example, meetings with key persons or meetings of all employees, steering group meetings, dialogue meetings, and group work (Mikkelsen et al., 2008; Vartia, 2009). In some cases, knowledge of bullying has been disseminated by newsletters and information folders (Mikkelsen et al., 2008) as well as through public education campaigns (Roscher, 2007). In an intervention program in health care, establishment of internal conflict counselling services and accreditation of advisors also was applied (Roscher, 2007).

One example of a training intervention is the management intervention programme carried out in five British public-sector organizations (Hoel and Giga, 2006), the aim of which was to develop, implement, and evaluate the effectiveness of three different intervention programmes. The programme focused on training in three areas: policy acknowledgement and communication (30 minutes), stress management (three hours), and awareness of negative behaviour (three hours), which were used in different combinations in six intervention groups. Although the programme was rated favourably by participants in terms of participant interest and applicability, it did not show any statistically different results for any of the key variables (e.g., being bullied, witnessing bullying, or negative behaviours) for the various interventions or combination of interventions. In explaining their findings, the researchers suggest that perhaps the managers and supervisors who needed the training most did not take part and that altogether too few people were perhaps trained. They also remark that the training sessions were quite short, thus not necessarily allowing for experiential learning to take place. Furthermore, the time between the interventions and the postintervention measurement might have been too short (six months).

In the intervention project conducted in three Danish organizations (Mikkelsen et al., 2008), different methods were used, each tailored to the needs and situation of the organization. On the basis of the data collected, the interventions in relation to bullying were targeted—for example, to increase awareness of conflicts and conflict management skills, to increase knowledge of and awareness of bullying, and to increase organizational competence in preventing and managing bullying. In two organizations, interventions included six steering group meetings, 60–90-minute lectures on bullying for all employees, a two-day course on conflict prevention and management for all managers and key employees, dialogue meetings focusing on psychosocial work environment issues, and four newsletters and a folder on bullying. In the third organizational intervention, the main target of the intervention was the work environment organization (WEO) and the interventions included six steering group meetings, six hours' coaching of the principal security managers, 90-minute lectures on bullying for the main WEO, 90-minute lectures on bullying for all employees, and three newsletters and a folder on bullying. The process evaluation showed positive changes for all three workplaces.

An intervention project by Keashly and Neuman (2009) carried out within the U.S. Department of Veterans Affairs and aimed at changing the nature of the conversation and interactions among employees focused on abusive behaviour and the way people treat each other. The project adhered to the idea that a sound workplace aggression intervention requires changes in the very nature of the conversation that organizational members are having with each other, as well as changes in the context in which these conversations occur (Kowalski et al., 2003). The approach taken involved the Collaborative Action Inquiry (CAI) process grounded in action research. The process included such various methods as having members of the project team modelling the respectful and collaborative behaviour that they expected from others. Situations where this ideal was not met were used as opportunities for learning. Individuals were also educated about the nature of workplace aggression and bullying in an effort to sensitize them to this issue and make them aware of their own assumptions and behaviours. This work was proved to be effective, and the process changed the nature and the character of conversations within the organization and created an atmosphere of trust.

Some positive results were also received in a one-year intervention programme where the entire personnel in eight primary schools participated in a one-year intervention project with the aim of reducing inappropriate treatment among members of staff (Vartia, 2009). The intervention included meetings and training sessions in all schools for all members of staff focusing on the bullying phenomenon, antecedents for and consequences of bullying, discussions and group work, and a joint half-day event where bullying was dealt with, for example, by means of improvisation theatre. A slight decrease in some forms of inappropriate behaviour was found, and one out of five employees reported that the level of inappropriate behaviour had decreased. The project also seemed to have changed employees' own behaviour, with one out of four employees reporting that if they were to see someone being treated inappropriately, they would intervene in the situation more readily than before. Moreover, almost half of the employees reported that they take more notice of their own behaviour towards their coworkers than before the intervention programme.

When Bullying Has Taken Place

Bullying has been described as an escalating process (Björkqvist, 1994; Leymann, 1996). In their response, researchers and practitioners have frequently discussed what would be the appropriate means to intervene and stop the escalating process as well as how to provide support to the targets. Various models based on stages of conflict escalation models have been presented (Fisher and Keashly, 1990; Glasl, 1994; see also Keashly and Nowell, 2003), each with its own set of proposed interventions. For example, the contingency model of third-party interventions in conflicts by Fisher and Keashly (1990) differentiates between four stages of interventions: discussion, polarization, segregation, and destruction. Arbitration, mediation, consultation, and peacekeeping are methods suggested in stages of segregation and destruction (see Keashly and Nowell, this volume).

An interview study involving a sample of German consultants specializing in bullying consultation (Saam, 2010) found that consultants differentiated between a conflict view and a multilevel view (meaning that there are problems on the dyadic, group, and organizational levels). The main measures applied in bullying cases by the consultants were found to be conflict moderation or mediation, coaching, and organizational development. Saam (2010) commented that this was interesting because dominant contingency models of conflict intervention (see also Keashly and Nowell, this volume) do not recommend these approaches on particularly escalated conflicts. Furthermore, consultants favouring conflict moderation or mediation conceived bullying as a particularly escalated form of conflict. On the other hand, consultants who favoured coaching or organizational development conceived bullying as a multilevel phenomenon. The latter approach saw bullying as an interpersonal conflict on a dyadic level (between the bully and the target), yet embedded in a particular group that was itself embedded in a particular organization. Hence, they saw the intervention strategy as a combination of interventions on all three levels. In addition, a Finnish study on the measures that personnel managers in the municipal sector have taken to intervene in workplace bullying showed that the organizations relied heavily on reconciliatory measures and that punitive measures were seldom used (Salin, 2009).

Although it seems that many practitioners use mediation in bullying situations, several experts on bullying see it as an unsuitable measure in escalated bullying cases (e.g., Keashly and Nowell, 2003; Rayner, 1999). Evaluation studies on the effectiveness of mediation in conflict and bullying cases are, however, so far few in number. Studies on mediation in hierarchical conflicts between supervisors and subordinates found, though, that hierarchical position plays a crucial role in mediation and that the power imbalance remains to some extent in the mediation process. Also, the experience of procedural justice and situational uncertainty are found to be important determinants in the mediation process (Bollen et al., 2009; Ittner et al., 2009). In this respect, supervisors perceived greater procedural justice in the mediation process than did subordinates, and there was a strong relationship between perceived procedural justice and perceived effectiveness of mediation (Ittner et al., 2009). The authors conclude that procedural justice is especially impor-tant for the party with the lower hierarchical position. It should be noted that during mediation, subordinates experienced more situational uncertainty than did supervisors, and this uncertainty was negatively related to satisfac-tion with mediation.

Supporting the Victims of Bullying

Individual- and group-oriented tertiary interventions have been used to treat the negative consequences of bullying in individuals, and some positive results have emerged from programmes where different measures have been used. In a Danish two-year rehabilitation programme among victims of bul-lying who had been unemployed or who had been on long-term sick leave because of bullying, several measures were used: psychological counselling, physiotherapy and physical exercise, and counselling or on-the-job training (Mikkelsen and Einarsen, 2006). The aim of the programme was to explore whether such a rehabilitation regimen has beneficial effects on the health and well-being of victims of bullying and whether it increases the likelihood of their returning to work. A significant decrease was found in psychological and psychosomatic symptoms among the victims during the course of the programme. Encouraging results have also been found, for example, with behavioural therapy (see Schwickerath and Zapf, this volume, and Tehrani, this volume).

Table 16.1 summarizes the different levels of intervention strategies for the prevention and management of bullying at work.

Key Principles in Planning and Implementing Interventions for Bullying at Work

Planning and implementing a successful and effective intervention for bullying and harassment in an organization is challenging. The general key components for successful interventions for all psychosocial risks at work, including bullying, relate to the content of the intervention, the implementation process, and the evaluation of the given interventions. In addition, some special issues need to be taken into consideration in plan-ning and implementing interventions for the prevention and management of bullying. (Leka et al., 2008a; Leka et al., 2008b; also see http://prima-ef.org/Documents/10.pdf and http://prima-ef.org/Documents/PRIMA-EF%20BROCHURE_English.pdf).

The key components for the successful planning and implementa-tion of interventions for bullying at work are explained in the following sections.

TABLE 16.1

Different Levels and Some Examples of Bullying Interventions

Level Of Work Organization Interventions	Stage Of Prevention		
	Primary interventions for the prevention and reduction of bullying	Secondary interventions to reverse and stop the process or increase the resources of individuals	Tertiary interventions for reducing the negative impacts of bullying and restoring the health of the victims of bullying
Society / Policy	Law/regulation Collective agreements Policy/code of conduct	Court case Industrial tribunal	Provision of rehabilitation opportunities
Organization / Employer	Antibullying policy Code of conduct Development of organizational culture Management training Organizational surveys	Handling procedures Mediation Investigation	Corporate agreements, programmes, and contracts of professional aftercare
Job / Task	Psychosocial work environment redesign Risk analysis Training (awareness, recognition of bullying)	Staff surveys Case analysis Training (e.g., conflict management, investigative skills) Conflict resolution, mediation	Group recovery programmes
Individual / Job Interface	Training (e.g., assertiveness training)	Social support, counselling	Therapy counselling, physical activities, redress

Source: Adapted from Murphy and Sauter, 2004; Leka et al., 2008b; see also Hoel, 2008. With permission.

Responsiveness to the Needs of the Organization and the Situation

Interventions should respond to the respective needs of the organization and the situation. In order to tailor the intervention to the specific problems and needs of the respective organization and individuals, interventions should be based on a proper risk assessment or analysis of the situation. Depending on the situation and size of the workplace, different measures—for example, questionnaires and individual and (focus) group interviews—can be used to assess the situation and needs for change. Analysis of the antecedents of bullying should include the features of the work environment found to be risk factors for bullying.

Theory-Based Interventions

Interventions for bullying should be founded on scientific research–based knowledge and evidence-based practice both as far as the contents of the interventions (e.g., knowledge about the antecedents of bullying, escalating nature of the bullying process) and the implementation methods (e.g., knowledge about effects of training) are concerned.

Commitment of Management and Ownership by Those Subjected to Activities

The overall commitment, support, and active participation of management and supervisors throughout the intervention—for the aims, measures, implementation, and evaluation—are crucial. Those individuals who are the subject of the activities should have ownership of the process, and they or their representatives should be given the opportunity to participate actively and be consulted in the planning of the interventions. Active participation and an opportunity to influence the planning of the intervention increase one's motivation and commitment to the activities. It is also recognized that occupational health and safety staff as well as trade unions may be useful partners.

It is of utmost importance that the aims of the interventions and the overall importance of the activities are understood and agreed on by both management and employees. Furthermore, the social partners (i.e., employers, employees, and trade unions) need to share a mutual understanding about the phenomenon of bullying. For example, major disagreement on what may cause bullying at work among the social partners may make the social dialogue process, and consequently the implementation of interventions, difficult. Input from external practitioners and researchers can, in such situations, help employers and employees reach a common way of thinking about bullying and the interventions needed.

Outside experts need to work in close contact with all those individuals from the organization—management, key persons, supervisors, and employees—who are involved in the intervention process. Adequate and continuous communication between the outside consultants and all those involved in the interventions is fundamental in all stages of the process.

Systematic and Stepwise Implementation

The intervention should be designed to be implemented in a system of steps that needs to be carefully and thoughtfully managed. The aims, objectives, and implementation strategy of the intervention need to be clearly defined and outlined. Also, the roles and tasks of different players need to be discussed and agreed to.

Evaluation of Interventions

An evaluation process that is clearly linked to the aims of the interventions and the identified problems in the organization should be applied as part of any intervention project. Such evaluations need to combine a focus on the aims and objectives of the interventions (evaluation in relation to the goals), as well as on the quality and effectiveness of the implementation process (process evaluation). However, in order to be able to evaluate the results and effectiveness of the interventions, the aims of the interventions should be explicitly formulated.

A variety of methods should be used to evaluate the success of interventions (e.g., survey, interviews, and group discussions), and short-, medium-, and long-term outcomes should be measured. More than one evaluation point is also preferable. Suitable methods are dependent on the aims of the intervention and also on the size and the available resources of the organization. In assessing the effects of the interventions, pre- and postintervention measurements are recommended. In large organizations, preintervention surveys are practicable, but in very small workplaces other methods may be more suitable.

Readiness to Take Action

In planning interventions for bullying, one needs to take account of the readiness of organizations and individuals to take action against bullying. The awareness and recognition as well as knowledge of bullying vary widely between organizations and countries, as shown in several recent studies (European Agency for Safety and Health at Work, 2010; Leka et al., 2008b). Low levels of awareness and knowledge influence the readiness and willingness of organizations, as well as managers and employees, to take action.

A Nonblame Culture

The sensitivity of the issue needs to be taken into account when discussing and planning interventions for bullying. Bullying at work creates strong emotions among all those involved, whether one refers to the target, witnesses, those accused of bullying, or even those providing support to resolve the situation. Bullying at work also arouses shame and guilt, which makes implementing an organizational level intervention on bullying even more difficult to do than putting in place one on work-related stress, for example, for the sensitivity of the issue requires that discussions and negotiations in all phases of any intervention are dealt with professionally in a nonblame culture.

Lessons Learned

In this chapter, the intervention strategies and methods for the prevention, reduction, and management of bullying at work have been described. In addition, key principles to take into account in planning and implementing interventions for bullying to ensure the effectiveness of the actions have been addressed.

In recent years, many kinds of measures, at the policy level internationally and nationally as well as at the organizational and individual levels, have been instituted for the prevention and reduction of bullying. Although evaluated interventions have so far been limited in number and have not always produced considerable results, some have shown positive results. Above all, they have taught us a great deal and given guidance for implementing interventions for bullying at work at different levels of organizations as well as at the societal level.

The few existing evaluations of the policy-level interventions have not been very reassuring. In the organizational and group level, the use of training and different kinds of group measures have brought about positive developments. Written antibullying policies are common in organizations, but their effectiveness has not been evaluated properly. Although prevention and reduction of potential antecedents and risks for bullying in the work environment have been emphasized, no intervention has so far very clearly shown the effectiveness of such activities. The use of mediation in severe interpersonal conflicts and bullying is a controversial issue, and therefore it needs to be studied and evaluated more.

The question that needs to be asked in connection with the effectiveness of interventions is, are the measures wrong, or have we failed in the implementation of the interventions? One of the important issues to get right is the right timing of the interventions. Very low awareness and insufficient knowledge of bullying in organizations, among both employers and employees, may bring about many problems and undermine the success of the interventions. Experience has also shown that so far, it has often been difficult to get organizations to take part in intervention studies for bullying (Mikkelsen et al., 2008). There might be many reasons for this.

One of the reasons is probably low awareness and recognition, although this understanding varies widely between organizations and counties, as does having insufficient "correct" knowledge about the problem. One reason may be the fear of acquiring a negative image that may attach itself to an organization carrying out an intervention to tackle and reduce bullying at work.

For the interventions to be successful, there has to be mutual understanding about the phenomenon and the antecedents or risks for bullying, as well as a common language to talk about bullying across the organization, involving employers, employees, union representatives, and health and safety professionals. Bullying needs also to been seen as a work environment problem, with interventions aimed at the organizational level focusing on the environment, the culture, and organizational structures.

Before an organization makes a decision to implement an intervention, management needs to have sufficient information about the bullying phenomenon as well as knowledge about how to tackle the problem. Then, before any interventions are implemented, enough information must be given to all those who may be targeted by the interventions. Where there is insufficient knowledge about the problem within the organization for the company to implement a successful intervention based on its own resources, it might be necessary to seek help by means of external expertise. Commitment of management is an absolute prerequisite for an intervention to succeed. If line managers, for example, do not take part in training, no change can be expected.

The aims of the interventions are always to change the situation for the better. It is, however, important to acknowledge that by increasing the awareness of bullying, expectations may arise that could affect results so that real improvement will go unnoticed. In a follow-up study to the Finnish intervention programme in primary schools (Vartia, 2009), employees were also asked about how successful the programme had been in removing bullying from their workplace. Over 40% of respondents reported that more could have been done. The difficulties in discussing the issue and the lack of time available to do so were the factors most often mentioned as reasons why not enough had been done. Interventions may also raise false expectations—that "the bully will be fired," "the manager will be removed," and "everything will be as before"—which can manifest themselves in dissatisfaction.

Intervention measures applied must correspond with the needs and resources of the respective organization, group and/or individuals. Furthermore, different strategies apply to different levels. Bullying is a complex phenomenon in terms of both its antecedents and its consequences. Therefore, to ensure success, simultaneous interventions at various levels might have to be considered.

Every intervention programme should include proper evaluation. Attention should be paid to the length of the follow-up period. Some positive effects may be achieved during a one-year process, as some interventions have shown. However, positive results in the form of reduction of bullying within a shorter timeframe are unlikely to be achieved.

References

Aiello, A., Deitinger, P., Nardella, C., and Bonafede, M. (2008) A tool for assessing the risk of mobbing in organizational environments: The "Val.Mob." scale. *Prevention Today. Quarterly Journal of Multidisciplinary Research on Occupational Safety and Health, 4* (3), 9–24.

Björkqvist, K. (1994) Trakassering förekommer bland anstälda vid Åb Akademi. [Bullying among staff in Åbo Akademi University]. *Meddelanden från Åbo Akademi, 9,* 14–17.

Bollen, K., Euwema, M., and Müller, P. (2009) Mediation in hierarchical conflicts at work: Managing power differences in a constructive manner. Paper presented at the 13th European Congress of Work and Organizational Psychology, May 13–16, Santiago de Compostela, Spain.

Bukspan, E. (2004) Bullying at work in France: A personal view. *British Journal of Guidance and Counselling, 32,* 397–406.

Cassitto, M. G., Fattorini, E., Gilioli, R., Rengo, C., and Gonik, V. (2003) *Raising awareness of psychological harassment at work.* Geneva: WHO. http://www.who.int/occupational_health/publications/en/pwh4e.pdf

Di Martino, V., Hoel, H., and Cooper, C. L. (2003) *Preventing violence and harassment in the workplace.* European Foundation for the Improvement of Living and Working Conditions. Luxembourg: Office for Official Publications of the European Communities.

Einarsen, S., and Hoel, H. (2008) Bullying and mistreatment at work: How managers may prevent and manage such problems. In A. Kinder, R. Hughes, and C. L. Cooper (eds.), *Employee well-being and support: A workplace resource* (pp. 161–173). Chichester, UK: John Wiley.

ETUC/BUSINESSEUROPE-UEAPME/CEEP (2009) Implementation of the ETUC/BUSINESSEUROPE-UEAPME/CEEP Framework Agreement on Harassment and Violence at Work. http://resourcecentre.etuc.org/linked_files/documents/Final_joint_table_2009%20harassment_violence_EN.pdf

European Agency for Safety and Health at Work (2010) *Violence and harassment at work.* Bilbao: European Agency for Safety and Health at Work.

Finnish Occupational Safety and Health Act (2002) http://www.finlex.fi/en/laki/kaannokset/2002/en20020738

Fisher, R. J., and Keashly, L. (1990) Third party consultation as a method of intergroup and international conflicts resolution. In R. J. Fisher (ed.), *The social psychology of intergroup and international conflict resolution* (pp. 211–238). New York: Springer Verlag.

Framework Agreement on Harassment and Violence at Work (N.d.) http://www.etuc.org/a/3574

Glasl, F. (1994) *Konfliktmanagement: Ein handbuch fur führungskräfte und berater* [Conflict management: A handbook for managers and consultants]. Bern, Switzerland: Haupt.

Health and Safety Authority (HSA) (2007) Code of practice for employers and employees on the prevention and resolution of bullying at work. http://www.deti.ie/publications/employment/2007/hsabullyingcop2007.pdf

Hoel, H. (2008) Intervening against workplace bullying: Exploring key issues. Paper presented in the PRIMA-EF Symposium, May 20–21, Helsinki, Finland.

Hoel, H., and Cooper, C. L. (2000) Destructive conflict and bullying at work. Unpublished report. Manchester: University of Manchester Institute of Science and Technology.

Hoel, H., and Einarsen S. (2010) Shortcomings of anti-bullying regulations: The case of Sweden. *European Journal of Work and Organizational Psychology*, 19 (1), 30–50.

Hoel, H., and Giga, S. (2006) *Destructive interpersonal conflict in the workplace: The effectiveness of management interventions.* Manchester, UK: University of Manchester.

Hubert, A. (2003) To prevent and overcome undesirable interaction: A systematic approach model. In S. Einarsen, H. Hoel., D. Zapf, and C. L. Cooper (eds.), *Bullying and emotional abuse in the workplace: International perspectives in research and practice* (pp. 299–311). London: Taylor & Francis.

Ittner, H., Bollen, K., and Euwema, M. (2009) Mediation in hierarchical conflict—no place for perceived justice. Paper presented at the 13th European Congress of Work and Organizational Psychology, May 13–16, Santiago de Compostela, Spain.

Keashly, L., and Neuman, J. H. (2009) Building a constructive communication climate. In P. Lutgen-Sandvik and B. D. Sypher (eds.), *Destructive organizational communication: Processes, consequences, constructive ways of organizing* (pp. 339–362). New York: Routledge.

Keashly, L., and Nowell, B. L. (2003) Conflict, conflict resolution and bullying. In S. Einarsen, H. Hoel., D. Zapf, and C. L. Cooper (eds.), *Bullying and emotional abuse in the workplace: International perspectives in research and practice* (pp. 339–369). London: Taylor & Francis.

Kowalski, R., Harmon J., Yourks, L., and Kowalski, D. (2003) Reducing workplace stress and aggression: An action research project at the U.S. Department of Veterans Affairs. *Human Resource Planning*, 26 (2), 39–53.

LaMontagne, A. D., Keegel, T., Louie, A. M. L., Ostry, A., and Landsbergis, P. A. (2007) A systematic review of the job-stress intervention evaluation literature, 1995–2005. *International Journal of Occupational Environmental Health*, 13, 268–280.

Leka, S., Aditya, J., Zwetsloot, G., Vartia, M., and Pahkin, K. (2008a) Psychosocial risk management: The importance and impact of policy level interventions. In S. Leka and T. Cox (eds.), *The European Framework for Psychosocial Risk Management: PRIMA-EF* (pp. 115–135). Nottingham, UK: Institute of Work, Health and Organizations (I-WHO).

Leka, S., and Cox, T. (2008) *PRIMA-EF: Guidance on the European Framework for Psychosocial Risk Management: A recourse for employers and worker representatives.* Protecting Workers' Health Series No. 9. Geneva: WHO.

Leka, S., Vartia, M., Hassard, J., Pahkin, K., Sutela, S., Cox, T., and Lindström, K. (2008b) Best practice in interventions for the prevention and management of work-related stress and workplace violence and bullying. In S. Leka and T. Cox (eds.), *The European Framework for Psychosocial Risk Management: PRIMA-EF* (pp. 136–173). Nottingham, UK: Institute of Work, Health and Organizations (I-WHO).

Leymann, H. (1996) The content and development of mobbing at work. *European Journal of Work and Organizational Psychology, 5*, 165–184.

Mikkelsen, E. G., and Einarsen, S. (2006) Psychosocial rehabilitation positively affects victims' health and well-being. Paper presented at the Fifth International Conference on Bullying and Harassment in the Workplace, June 15–17, Dublin, Ireland.

Mikkelsen, E. G., Høgh, A., and Olesen, L. B. (2008) Prevention of bullying and conflicts at work—an intervention study. Paper presented at the Sixth International Conference on Workplace Bullying, June 4–6, Montreal, Canada.

Murphy, L. R., and Sauter, S. L. (2004) Work organization interventions: Stage of knowledge and future directions. *Social and Preventive Medicine, 49* (2), 79–86.

Murray, P. (2002) Intermediary strategy for prevention of workplace bullying. In European Week for Safety and Health at Work, European Agency for Safety and Health at Work, *Prevention of psychosocial risks and stress at work in practice* (pp. 70–72). Luxembourg: Official Publications of the European Communities.

Oeij, P. R. A., Wiezer, N. M., Elo, A. L., Nielsen, K., Vega, A., Wetzstein, A., and Zolnierczyk, D. (2006) Combatting psychosocial risks in work organizations. In S. McIntyre and J. Houdmont (eds.), *Occupational health psychology: European perspectives on research, education and practice*, vol. 1. Nottingham, UK: University of Nottingham Press.

Rayner, C. (1999) From research to implementation: Finding leverage for prevention. *International Journal of Manpower, 20* (1/2), 28–38.

Richards, J., and Daley, H. (2003) Bullying, policy: Development, implementation and monitoring. In S. Einarsen, H. Hoel., D. Zapf, and C. L. Cooper (eds.), *Bullying and emotional abuse in the workplace: International perspectives in research and practice* (pp. 247–258). London: Taylor & Francis.

Roscher, S. (2007) Preventing bullying in hospitals: Results of a field study. Paper presented at the 13th European Congress of Work and Organizational Psychology, May 9–12, Stockholm, Sweden.

Roscher, S., Nienhaus, A., and Harms, P. (2006) Prevention von mobbing in krankenhäusern [Prevention of bullying in hospitals]. *Arbeitsmedizin, Sozialmedizin, Umwelsmedizin, 41* (3), 118–119.

Saam, N. J. (2010) Interventions in workplace bullying: A multilevel approach. *European Journal of Work and Organizational Psychology, 19* (1), 51–75.

Salin, D. (2008) The prevention of workplace bullying as a question of human resource management: Measures adopted and underlying organizational factors. *Scandinavian Journal of Management, 24*, 221–231.

———— (2009) Organizational responses to workplace harassment: An exploratory study. *Personnel Review, 38* (19), 26–44.

Salminen, S., Ruotsala, R., Vorne, J., and Saari, J. (2007) *Työturvallisuuslain toimeenpano työpaikoilla: Selvitys uudistetun työturvallisuuslain vaikutuksista työpaikkojen turvallisuustoimintaan.* [Implementation of the Finnish Occupational Health and Safety Act at workplaces: A study on the effects of the revised act on the occupational safety and health actions of workplaces] (in Finnish, English summary). Helsinki: Sosiaali- ja terveysministeriö [Ministry of Social Affairs and Health].

Vartia, M. (2002) Implementation of a non-bullying policy: A follow-up. Paper presented at the closing event of the European Week for Safety and Health at Work, November 25, Bilbao, Spain.

——— (2009) Reduction of inappropriate behaviour and bullying at work—a workplace intervention among school staff. Paper presented at the 13th European Congress of Work and Organizational Psychology, May 13–16, Santiago de Compostela, Spain.

Zwetsloot, G., and Starren A. (2004) *Corporate social responsibility and safety and health at work*. Luxembourg: European Agency for Safety and Health at Work, Office for Official Publications of the European Communities.

17

Workplace Bullying: The Role for Counselling

Noreen Tehrani

CONTENTS

Introduction

Whilst most internal bullying policies recommend that counselling should be offered for targets of bullying and occasionally for the bullies, there is very little in the literature to describe what an appropriate approach or model of counselling might be or, indeed, whether counselling is an effective method for reducing the distress experienced by those involved. This chapter looks at the nature of counselling and what an employee might expect when attending an assessment and counselling session. Whilst there is little objective evidence to show any significant difference in the effectiveness among different counselling models, the important success factors are described. Furthermore, counselling employees distressed by workplace bullying requires the counsellor to understand the nature of the "bullying drama" and the different players involved.

Achieving such understanding is easier where the organisation's counselling is provided by a single counsellor or counselling organisation and data are collected on the number, nature, and location of bullying cases, allowing for trends or bullying hotspots to be identified. Most counselling training, however, relates to one-to-one counselling relationships. This traditional form of counselling can be counterproductive when dealing with bullying, particularly where the problem involves a systemic failure or flaw in the organisational culture. Systemically and organisationally trained counsellors will deal with the issues raised by the individual employee, but their focus will go beyond the employee to include subordinates, colleagues, managers, and the organisation or system that may have caused, facilitated, or supported the bullying behaviours. Organisationally aware counsellors are in a good position to identify potential flaws within a team, a section, or the whole organisation. These flaws can be addressed through team training, mentoring, or counselling employees or working with the organisation to create a culture that is much more respectful and able to deal creatively with the normal clashes and conflicts that arise in relationships.

Therefore, this chapter presents an integrated approach to counselling in cases of bullying, together with a specific model of integrated counselling. Reference is also made to the need to counsel and support the organisation in dealing with bullying in a positive and creative way.

What Is Counselling?

At its most simple level, counselling is a helping relationship that enables the counsellor to work with one or more people to explore a difficulty that he or she is experiencing. It is provided at the request of the people involved and is entered into voluntarily. Organisations may also send employees for counselling. However, for the counselling relationship to work, it needs to be entered into freely. Whilst there are hundreds of models or types of counselling, there are some basic principles that underpin the counselling relationship: "a warm concern for and acceptance of the other; openness and attunement to the other's experiential reality; a grasp of what the other needs; an ability to facilitate the realisation of such needs and an authentic presence" (Heron, 2001, p. 11). In counselling sessions, the person seeking help should be treated as a client and not as a patient or customer. This relationship is therefore more equal than is commonly the case with patients or customers. During the counselling process, clients are encouraged to explore aspects of their lives and feelings. The counselling relationship is designed to make it easier to talk freely and openly in a way that is rarely possible with family members or friends.

The impact of bullying behaviours on its targets has been shown to cause intense feelings of anger, anxiety, grief, and embarrassment (see also Høgh, Mikkelsen, and Hansen, this volume). Counselling offers the client an opportunity to explore his or her responses to the bullying within a safe relationship in order to gain an understanding of why the bullying had occurred, what can be learnt from the experience, and what needs to be done to regain a sense of autonomy and self-worth. During the counselling, the counsellor will be alert to the presence of hidden or repressed thoughts, feelings, and responses that may have gone unrecognised or unacknowledged by the client. Depending on the counsellors' orientation and method of working, appropriate training and supervision should help them to accept and reflect upon their client's problems without becoming burdened or injured by them (BACP, 2009).

One of the questions many people want answered when seeking professional help is, "Which form of counselling is most effective?" Whilst there is a current vogue for promoting counselling techniques that easily lend themselves to random-controlled trials, where studies have been carried out to compare the effectiveness of the different counselling models, the evidence has shown that broadly similar results are achieved by all the therapeutic models (Cooper, 2008). It has been found that the major differences in counselling outcome is related to three factors:

1. The qualities and capabilities of the individual therapist (Okiishi et al., 2003)
2. The strength of the relationship between the counsellor and client (Cooper, 2008)
3. The extent to which the client has been willing to engage in the counselling process (Orlinski et al., 1994)

In 2001, Wampold undertook a meta-analysis of counselling outcomes and estimated that only 1% of the variance in counselling effectiveness could be accounted for by the counselling model or approach.

Emotional Competence and Bullying

Whilst many people, particularly those who regard themselves as victims of bullying, hold the view that within the workplace there is a clearly defined group of antisocial workers and that these workers engage in bullying wholly innocent and unsuspecting colleagues, the reality is much less clear. There are without doubt some personality-disordered psychopaths within organisations who, through their use of charm, manipulation, and ruthlessness and their lack of any feelings or conscience, selfishly take what they want and do as they please without a thought for their victims (Hare, 1993). However, this is not the situation found by the author, who has been called into a wide range of organisations to help in the formal and informal investigations of bullying as well as undertaking counselling with large numbers of employees. The experience of dealing with organisations and employees caught up in the warp and weft of bullying indicates that it is rare to find a totally innocent victim or a wholly guilty oppressor (Page, 1999). Frequently, players will describe themselves as the victim of the other party whilst attributing their own behaviour to the circumstances in which the behaviour occurred, each side providing evidence to support its own version of events. Then comes the question, how is it possible for people to have such disparate views of what are, on the surface, the same events? The answer appears to be attributable to the way people view the world and the actions of others (Kelley, 1972). There is a bias to view one's own negative actions as a consequence of situational factors ("I was very cross that day because the baby was teething and I did not get any sleep") whilst, on the other hand, viewing the negative actions of others as indicative of a persistent trait or intent ("That is typical of Jenny—she always gets angry when she does not get her own way"). Fortunately, the principles underpinning most of the counselling models provide counsellors with some insight into how individuals become engaged in, and maintain, negative behaviours. It has been suggested that the reason for this displacement of personal responsibility onto others lies within society, particularly in those cultures where there is little or no emphasis on enabling children to learn how to recognise, accept, and deal with their negative emotional responses (Heron, 2001). Child-raising patterns frequently place on children the requirement to control their emotional responses in order to be socially acceptable. This emotional control is achieved without the child or its parent recognising or understanding the important purpose of the emotional expression as a mechanism for responding to and meeting basic human needs. As a result of this suppression of emotional process, children are left with the distress of their unmet needs. Instead of children achieving a healthy understanding of their deeper needs, they are taught to control their emotions through a toxic mix of suppression, repression, and denial. Children begin to release their emotional distress resulting from displacing their emotions into a range of behaviours including aggression, demanding, blaming, clinging, and whining. This pattern of learnt behaviour continues into adult life, with the earlier life traumas being added to and triggered by new traumatic events. Unless as adults people come to recognise and meet their hidden needs, this process continues throughout life and emerges in automatic, compulsive, unwarranted, unacceptable, and unexpected forms of behaviour.

Most adults are unaware of the influence of these intrapsychic processes, and as a result their rational and sensitive responses to situations become mixed with compulsive, distress-driven reactions powered by a hidden inner distress. These displacement behaviours are often reinforced within the workplace, where they may become valued and legitimised, with the result that the rational and compulsive become intertwined. Employees become skilled at dressing up irrationally driven behaviours as being reasonable; for example, being driven to succeed in achieving career goals regardless of the cost to themselves or their family, they justify their behaviour on the grounds of business need rather than recognising that their hyperactivity is a displacement of the distress or anger they experienced but failed to express when they did not achieve the recognition or validation they craved as a child. On the other hand, where employees have been found to have reported being victims of bullying behaviours by colleagues or managers in every team or organisation they join, the employees will blame others for causing their distress whilst failing to recognise that their difficulty in establishing positive working relationships originates from a repressed destructive anger that has followed their unexpressed feelings of the loss of a loved parent or sibling, a loss that manifests itself in an insatiable need for reassurance and support and feelings of being rejected.

People develop emotional competence through understanding and accepting their drives, responses, and reactions to their environment and their relationships. Emotionally competent individuals are able to accept their positive and negative emotions and willingly recognise that their experience of distress is a mechanism of self-healing rather than provide evidence of their imminent psychological breakdown. Emotional competence requires people to identify when old emotional pain and distress are being reenacted in displacement or other forms of distorted behaviour and then to attempt to recognise and understand and deal with the original hurt. To achieve emotional competence, therefore, people need to

1. Gain insight into their early traumatic experience and the influence it is having on their adult behaviour.
2. Notice when the conditioned or obsessional forms of displacement prevent more rational and flexible behaviours.
3. Learn how to appropriately challenge others who are unconsciously acting out their denied distress in negative and other forms of disrupted behaviour (Heron, 2001; Steiner, 1999).

This is not to say that employees may not be severely mistreated at work by others who abuse their power or who accept or condone bullying behaviours.

The Bullying Drama

There are many forms of bullying, including the simple forms such as predatory, dispute-related, and escalating bullying (Einarsen, 1999). However, the more complex forms are more difficult to recognise and resolve, such as delegated bullying, in which the bully persuades or coerces someone else to undertake the bullying on his or her behalf; subordinate bullying, which is the bullying of a more senior person by a subordinate person; passive-aggressive bullying, which involves nonassertive, undermining behaviours; or personality disordered bullying, which is related to a personality disorder in the perpetrator (Tehrani, 2005). When dealing with bullying, counsellors provide therapeutic support to those who have become distressed by the negative behaviours of others. The distress that clients experience when faced with perceived or actual negative experiences can be extremely distressing. The stories surrounding bullying are dramatic, involving high emotions, with references to acts of deceit, favouritism, humiliation, intrigue, and undeserved punishments. However, it is important for the counsellor to recognise that what is described by each of the players, particularly in complex bullying dramas, may become distorted and changed as it passes through the lens of unmet needs and unresolved life experiences. Identifying the true nature of the drama is a skilled process requiring a detailed understanding of the life history of the client, an identification of the signs of emotional displacement observed during the interactions that take part within the counselling relationship, and an awareness of the signals that provide an insight into what may be influencing the current perceptions and behaviours of the client, for abusive relationships involve interactions in which all the players play their part. When a client tells his or her counsellor, "I did nothing to cause this to happen to me," the counsellors should be aware that this is unlikely to be the real situation and that their client may have failed to recognise, unwittingly or unknowingly, his or her own part in the bullying drama. This does not mean that the target of bad behaviour has done anything intentional or wrong; rather, the bullying target's behaviour or presence was perceived by the perpetrator as a threat to his or her own power, position, or self-esteem. Therefore, the true extent and complexity of the bullying drama may remain hidden unless the counsellor has the opportunity to observe and work with all the players. Only by understanding the real causes of the conflict and by resisting the temptation to apportion blame and responsibility will it become possible for the counsellor to address the underlying issues. This position of unconditional positive regard (Rogers, 1957) is an essential prerequisite of all counselling relationships. Whilst counsellors may find the behaviours of their clients unwise, irresponsible, or cruel, in order to bring about any change, counsellors must form a working alliance (Satterfield and Lyddon, 1998) with their clients based on respect for them as human beings. The drama triangle (Karpman, 1968), which describes a dynamic relationship among the persecutor, victim, and rescuer, is a useful way of looking at abusive relationships (Proctor and Tehrani, 2001). However, the drama triangle fails to recognise the importance of a fourth role common in bullying cases—that of the "avenger." The avenger is typically someone who has experienced bullying in the past and attempts to deal with unresolved distress by taking action on behalf of others. In this new model, it is proposed, the four roles interact with

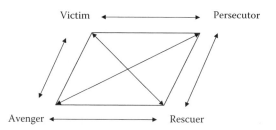

FIGURE 17.1
Roles within the bullying drama.

each other (Figure 17.1). Each of the roles involves a compulsive maladaptive pattern of behaviour that is triggered by particular situations connected to an unresolved trauma from the past. Whilst occasionally a single individual will fulfil the characteristics of a single role, never moving into any of the other roles, this is a rare situation. More typically, an individual will have a preference or feel "comfortable" in one or two of the roles and will alternate between these roles as circumstances change. For the bullying drama to be played out, individuals will occasionally be forced to play roles with which they are less familiar; this situation may occur when there are changes in the balance of power within the system.

For example, a victim may seek the support of a colleague regarding what he or she believes to be bullying. The victim may feel that the oppressive behaviour of his or her manager could be unintentional and discusses this with a colleague with the aim of seeking support to raise the issue sensitively in a team meeting. However, the colleague has had difficulties with the same manager and begins to push the victim to make a formal complaint against the manager. In this situation, the avenger uses his or her own "victim status" coercively and as a result becomes an oppressor of the victim, who has wished to take a more conciliatory action. The victim is now in a situation where he or she is faced with deciding whether to collude with the avenger and take out a formal complaint against the manager or to pursue the original decision, which would involve rescuing the manager from the aggression of the avenger. The bullying drama will continue for as long as the players remain unaware of the true nature of the games being played and begin to recognise their part in sustaining the drama interactions. Self-awareness is essential, as is the ability to recognise that by moving beyond the game, it becomes possible to recognise all the players' needs and frailties, which are so often obscured within these archetypal roles.

The restored self model (see Figure 17.2) illustrates how the bullying drama, shown at the lower level, is connected to a higher level of understanding and meaning. It is by engaging with this higher level that individuals can become aware of their sense of vulnerability, power, responsiveness, and wish for justice. Counsellors need to encourage their clients to recognise that remaining within the bullying drama will lead not to justice and peace but, rather,

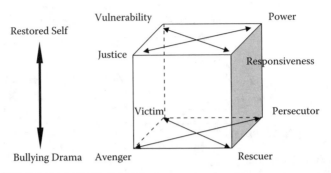

FIGURE 17.2
The restored self model.

to a continuation of the situation. Breaking the bullying drama requires the players to understand and operate at this higher level of functioning freed from the negative feelings and fears found within the dynamics of the bullying drama. However, embracing vulnerability, power, responsiveness, and justice requires determination and courage and considerable trust between client and counsellor. The role of the counsellor is to demonstrate that whilst being vulnerable may make it easier for an individual to become a victim, true vulnerability is a strength when it enables the vulnerable person to recognise the reality of a difficult situation and encourage the seeking of support and understanding of others, rather than trying to take control of everything alone. Power can corrupt and lead to abuse, but it can also be used in a positive way when it enables the powerful person to act with justice or compassion for others. Being overly responsive can turn a counsellor or other supporter into a compulsive carer, whereas the ability to listen and reflect without taking over can have a positive influence. The application of justice without compassion and understanding can be experienced as punishing and damaging even when the just solution is recognized by all concerned as appropriate, fair, and reasonable. In order to work with the bullying drama, counsellors must work at this higher level if they are to achieve an understanding of the patterns of behaviour being acted out and recognise their own vulnerability to becoming caught up in the drama (Page, 1999).

The Bullying Drama and Organisations

The restored self model can also be seen operating at an organisational level, with organisations becoming trapped in the drama of bullying instead of recognising their own vulnerabilities, power, and need for responsiveness and justice. Organisational counsellors can use their counselling skills to help organisations deal with bullying in a way that moves beyond inaction, sentimentality, vengeance, and punishment to a place where the organisation demonstrates a positive regard for people with bad behaviours, including bullying as well as being handled firmly and appropriately by organisations through fair investigation and disciplinary procedures (Hoel and Einarsen, this volume; Rayner and Lewis, this volume). Whilst organisations have the responsibility for dealing with bullying, they should also recognise that they are not immune from becoming trapped in the bullying drama through a parallel process (Talbot, 1990), one in which the processes and behaviours at play in the relationship between bully and victim are reflected in the behaviour of the organisation. To become an agent of the restorative self, organisations need to balance their use of power, with justice, vulnerability, and responsiveness, to establish solutions that not only solve the problem of bullying but also address the underlying causes of bullying.

The Wounded Helper

When working with bullying, a counsellor needs to be able to work within a number of different models of intervention, continually monitoring the value and benefit of the particular intervention in use. Knowing how and when to move from one style of working to another and understanding how to maintain the creative balance in the use of their own personal power and vulnerability whilst simultaneously capturing and responding to the power and vulnerability of their client is essential for counsellors. However, counsellors and other helpers are not immune to experiencing the anxiety and distress of their own unresolved traumas; indeed, some practitioners unconsciously choose to work with victims of bullying or their oppressors as a means of dealing with their own unresolved and hidden distress. Fundamental to counselling training and ongoing personal development is the professional requirement for all counsellors to present their client work in supervision. In this, the supervisor will not only deal with the nature and appropriateness of the counselling process and techniques used but also spend time looking at the impact of the content of the counselling on the counsellor, including any of the hidden drives or motivations that may be influencing the counsellor's behaviour within the therapeutic relationship. The professional requirement for all counsellors is to attend regular supervision sessions and to maintain their continuing professional development through attending conferences, undergoing training, and engaging in other learning opportunities. Fulfilling this requirement is essential in order to protect clients from the abuse of power that could so easily occur within the counselling room, particularly so when the helper turns into the persecutor or the avenger.

An Integrated Counselling Approach

There are many hundreds of counselling theories available to counsellors, let alone the variations that counsellors introduce to adapt a particular model to meet their own preferences or needs. For many counsellors, the ability to adopt an integrated approach allows a certain freedom of practice to suit their style of work, an approach appearing to be one of the most popular approaches in contemporary counselling practice (Dryden, 1992). Whilst to the outsider the choice of technique or approach may appear random, to the experienced integrative counsellor this could not be further from the truth. Integrative counselling approaches have developed out of the lack of any strong evidence to indicate that any one approach is significantly more effective than others (Norcross and Arkowitz, 1992).

In simple terms, counselling falls into three main groups. The first group includes the psychodynamic schools, which have their origins in the work of Sigmund Freud. These approaches recognise and deal with the unconscious conflicts relating to instinctual drives and repression. The second group, the behavioural school, has its origins in the work of Pavlov and Skinner and is characterised by conditioned learning. Finally, there are the humanistic or existential schools, with their roots in the work of Aristotle, Maslow, Rogers, and Berne, with a belief in self-actualisation. Whilst counselling training tends to teach each of the counselling models in isolation, many counsellors will pick and choose between models or approach to address an identified need or a client preference, or perhaps they will blend several theories together to form a hybrid approach (Lapworth et al., 2004). This integrative movement can be observed in recent developments within cognitive behavioural therapy (CBT), with its origin within the behavioural school. Over the past few years, CBT has begun to incorporate aspects of attachment theory, which is more closely associated with the psychodynamic school and mindfulness, with its origin in spirituality and self-actualisation.

An Integrated Model of Counselling

Whilst there is no evidence to support a particular model of counselling, some models lend themselves more easily to the organisational need for a clearly defined process, ease of access, and speed of recovery. The following outline has been used by the author for many years with positive outcomes. The model involves five main elements that are utilised throughout the counselling process. As the emphasis or importance of a particular element may change during the counselling progress, the counsellor will need to be continually aware of the process, the elements, and the ultimate ending of the counselling relationship and contract. The five elements involved in this approach are (1) assessment, (2) education, (3) symptom reduction and other interventions, (4) integration and understanding, and (5) rehabilitation and return to work.

Assessment

Employees attending a counselling assessment session tend to fall into one of three groups. First are those employees who come to the assessment unaware that their symptoms of distress are related to an oppressive or negative relationship. For these people, it may come as a surprise to recognise that the behaviour they are experiencing is unacceptable and that they have the right to be able to ask for it to be stopped. The second group comes because they have been referred by management following a disciplinary hearing in which the employee has been identified as the perpetrator or victim of bullying. Finally, there are employees who recognise that they are involved in a conflict with one or more people and want to find out what they can do to resolve it. Regardless of the reason for attending the assessment, each employee is made aware of the boundaries within which the counselling will take place, including that within certain limits the counselling will be confidential,* that the employee needs to attend voluntarily, and that the counselling is not an easy option and will require the employee to fully engage in the process. The assessment begins with an outline of what will happen during this first session and the agreement that depending on the outcome of the assessment, a decision will be made on whether there is a basis for the client and counsellor to work together. In addition to the initial contracting assessments, there are three further sections. First, there is a structured interview, which involves the counsellor taking a life history of the employee, providing an opportunity for the client to describe what has caused him or her to seek counselling. Second, there is a review of results of a questionnaire that the client completed previously. The questionnaire measures clinical symptoms, including anxiety, depression, and post-traumatic stress. In addition there is a set of well-being and occupation questionnaires that assess general health and lifestyle, locus of control, behavioural style, coping skills, and personality. During the assessment, the counsellor provides feedback on the questionnaires. Finally, there is an opportunity for the employee and the counsellor to talk about what the employee would like as an outcome and to decide on the best approach. It is important to ensure that prior to commencement of the counselling, the employee is not exposed to any further bullying or other negative acts. Attempting to undertake counselling whilst the client is in real danger of further traumatic exposures undermines the counselling. Following the assessment, the counsellor

---

* Confidentiality is broken only if there is a serious risk to the life or health of the client or of another person or if a crime has been committed.

provides the employee with a written report of the assessment and, where appropriate, an agreed programme or plan for future sessions. Only when the counsellor and client have agreed on the content of the report and counselling programme does the actual counselling begin.

Education

Education is an important part of this integrated approach. The education process is continual and fully integrated into all the other counselling activities. In the assessment, typically there will be a description of the biological basis of stress and trauma, including the use of drawings and descriptions of how the brain copes with aversive events by locking the sensory memory traces within the amygdala, where they are inaccessible to conscious review or understanding (Damasio, 2006). This education is augmented by other information taken from the psychometric results with, for example, clients being shown where they may be lacking a basic coping skill and what they need to do to develop this ability. For clients, the whole counselling experience is one of action learning in which they are asked to seek answers and meaning from their experiences. Many clients will also take the opportunity to read recommended books or formulate hypotheses on their reactions and responses by drawing on information from their own areas of expertise, family feedback, and reading. The action learning process encourages the client to become intrigued by his or her own responses and reactions rather than a victim of them. Occasionally, the trauma response to the bullying situation involves strong physical reactions, including panic attacks, palpitations, diarrhea, and trembling. Where such reactions occur, the education will include information on the nature of human responses to strain and how these responses are important to survival. A simplified version of Damasio's tree (Damasio, 2003)—which shows the intimate relationship among the immune responses, pain and pleasure behaviours, drives and motivations, and their connection to the expression of emotions and feelings—is often used to illustrate the way that early life traumas may be restimulated without conscious awareness.

Symptom Reduction

The main body of counselling work involves symptom reduction. Each activity will be discussed with the client and introduced flexibly in response to the client's needs. The initial work tends to be a response to the needs identified in the psychometric testing and personal history, particularly where the client has been found to have an issue that may be making him or her vulnerable to distressing symptoms. When the assessment has shown, for example, that the client has a poor lifestyle, this problem will be addressed by talking about the benefits of reducing caffeine intake, taking regular exercise, cutting out excessive alcohol, and eating regular meals. The inadequate use of a particular coping skill will be addressed by organising training or coaching in the required skill. Whilst these practical steps may not be regarded as a normal part of the counsellor's role, particularly by those counsellors who take a nondirective stance to their counselling work, in this pragmatic approach to counselling there are clear benefits for the client engaging in some practical activities that provide them with relief from their symptoms and increases their self-confidence.

Typically, one of the first tools to be offered to a client is an ability to relax and create a "safe place." This sense is achieved by the use of a relaxation script in which the client is encouraged to find an imaginary place where he or she feels at home and at peace. Only when this sense of safety has been achieved is it advisable to go to the next stage, where some of the reexperience symptoms are addressed. Reexperience comes in a number of forms, including flashbacks, nightmares, hallucinations, and emotional or physical reactivation. A reexperience is different from recall or remembering a distressing event; within the reexperience, the client responds as if the incident is actually occurring. The main task involved in reducing the power of the reexperienced responses is to encourage the client to summon up the trigger of this reaction, which may be an upsetting image, sound, smell, body response, or taste. For each of the five senses, the process is similar, involving the client going through the relaxation exercise, then gradually, and at the client's own pace, purposefully bringing the feared stimuli to mind, and then finding ways to manipulate them by changing their size, shape, colour, smell, or intensity. Through this process, the conscious mind comes to recognise that the feared perception is a construction rather than a real event. At the end of each exercise, the client is asked to switch the stimuli off and to check that they remain off until he or she decides to visit the stimuli once more. Through this process of desensitisation, the client becomes aware that the feared flashbacks, hallucinations, and sensory reactivations can be brought under conscious control.

Integration and Understanding

Throughout the counselling, clients need to learn how to reflect upon and become interested in their responses to events. The mindfulness approach to reactions requires them to adopt a concentrated awareness of their present thoughts, actions, and motivations without becoming emotionally attached or distressed by them (Brown and Ryan, 2003). Writing and keeping journals encourages clients to reflect on their changing perceptions through the identification of repeating patterns of behaviours, events, or emotions. This process is helpful in creating deeper meanings and reducing distress (Pennebaker and Chung, 2007). For many clients, it is difficult to accept that they may have played any role in the bullying drama apart from being the innocent victim. It may be difficult to contemplate that one of the more negatively perceived roles is anything but bad. However, despite the view, it is clear that the seemingly negative roles involve qualities that are not in themselves negative but can possess positive elements (Karpiak, 2003). The bullying drama model predicts that where clients are able to accept their shadow side through the expression of their vulnerability, power, responsiveness, and need for justice—all virtues inherent within the victim, oppressor, rescuer, and avenger roles—this acceptance will enable them to gain self-awareness and wisdom.

Returning to the Workplace

Some of the practical aspects of returning to work and rehabilitation of victims of bullying are to be found in the chapter by Schwickerath and Zapf in this volume. However, counsellors also have an important role in preparing their clients for returning to work. Although many people involved in bullying do not return to the same role or even to the same organisation, the workplace is the place where the traumatic bullying experiences took place; therefore, it can hold the triggers to memories that have not been fully processed. The use of visualisation prior to the actual return to work can be extremely helpful to flush out some of these triggers and deal with them. Walking along a corridor, going into a meeting room, or coming face to face with a person associated with the bullying can reactivate high levels of distress. It is helpful to practice the return to work through the medium of visualisation prior to the actual return. This virtual return to work can identify areas of counselling or practical steps that need to be undertaken if the return is to be successful. Even if the initial return is successful, it is important to arrange for the counselling to continue during the first few weeks following the return to work in order to deal with the unexpected responses to the working environment.

Conclusions

Bullying involves complex human interactions, some of which are open and observable, whilst others are hidden and frequently denied. Bullying and the bullying drama are inevitable consequences of a cultural and personal failure to accept and understand the shadow side of our nature. The process of splitting in which people are labelled as victims, targets, bullies, or perpetrators adds to the problem by failing to recognise that within every human being is the potential of both good and evil, kindness and cruelty. At its best, counselling can contain the anxiety and hurt of all those involved in the bullying drama and with wisdom and patience create a place of learning where the shadow can meet the light.

References

BACP (2009) *Ethical framework for good practice.* Lutterworth, UK: British Association of Counselling and Psychotherapy.

Brown, K., and Ryan, R. (2003) The benefits of being present: Mindfulness and its role in psychological well being. *Journal of Personality and Social Psychology, 84,* 822–848.

Cooper, M. (2008) The facts are friendly. *Therapy Today, 19* (7), 8–13.

Damasio, A. (2003) *Looking for Spinoza: Joy, sorrow and the feeling brain.* London: Heinemann.

——— (2006) *Descartes' error.* London: Vintage.

Dryden, W. (1992) *Integrative and eclectic therapy: A handbook.* Buckingham, UK: Open University Press.

Einarsen, S. (1999) The nature and causes of bullying at work. *International Journal of Manpower, 20* (1–2) 16–27.

Hare, R. D. (1993) *Without conscience: The disturbing world of the psychopaths among us.* New York: Guilford.

Heron, J. (2001) *Helping the client: A creative practical guide.* London: Sage.

Karpiak, I. E. (2003) The shadow: Mining its dark treasury for teaching and adult development. *Canadian Journal of University Continuing Education, 29* (2), 13–27.

Karpman, S. B. (1968) Fairy tales and script drama analysis. *Transactional Bulletin, 7* (26), 39–43.

Kelley, H. H. (1972) Attribution in social interaction. In E. E. Jones, D. E. Kanouse, H. H. Kelley, R. E. Nisbitt, S. Valins, and B. Weiner (eds.), *Attribution: Perceiving the causes of behavior.* Hillsdale, NJ: Lawrence Erlbaum.

Lapworth, P., Sills, C., and Fish, S. (2004) *Integration in counselling and psychotherapy: Developing a personal approach.* London: Sage.

Norcross, J. C., and Arkowitz, H. (1992) The evolution and current status of psychotherapy integration. In W. Dryden (ed.), *Integrative and eclectic therapy.* Birmingham, UK: Open University Press.

Okiishi J., Lambert, M. J., Neilsen, S. L., and Ogles, B. M. (2003) Waiting for the supershrink: An empirical analysis of therapist effects. *Clinical Psychology and Psychotherapy, 10,* 361–373.

Orlinski, D. K., Grawe, K., and Parkes, B. (1994) *Process and outcome in psychotherapy and behavior change.* Chicago: John Wiley.

Page, S. (1999) *The shadow and the counselor: Working with darker aspects of the person, role and profession.* London: Taylor & Francis.

Pennebaker, J. W., and Chung, C. K. (2007) Expressive writing, emotional upheavals, and health. In H. Friedman and R. Silver (eds.), *Handbook of health psychology* (pp. 263–284). New York: Oxford University Press.

Proctor, B., and Tehrani, N. (2001) Issues for counselors and supervisors. In Noreen Tehrani (ed.), *Building a culture of respect: Managing bullying at work.* London: Taylor & Francis.

Rogers, C. R. (1957) The necessary and sufficient conditions of therapeutic personality change. *Journal of Consulting Psychology, 21,* 95–103.

Satterfield, W. A., and Lyddon, W. J. (1998) Client attachment and the working alliance. *Counselling Psychology Quarterly, 11,* 407–416.

Steiner, C. (1999) *Achieving emotional literacy*. London: Bloomsbury.

Talbot, A. (1990) The importance of parallel process in debriefing crisis counselors. *Journal of Traumatic Stress*, 3 (2), 265–277.

Tehrani, N. (2005) *Bullying at work: Beyond policies to a culture of respect*. London: CIPD.

Wampold, B. E. (2001) *The great psychotherapy debate: Models, methods and findings*. Mahwah, NJ: Lawrence Erlbaum.

18

Inpatient Treatment of Bullying Victims

Josef Schwickerath and Dieter Zapf

CONTENTS

Introduction

Because victims of bullying can develop a mental or somatic illness (see Høgh et al., this volume), bullying thus becomes a subject matter in the field of psychotherapy or inpatient treatment (Schwickerath, 2009). In the area of therapy for patients who experienced bullying, there is a big research gap with only a few studies restricted to quite specific questions (cf. Tehrani, 2003; Vartia et al., 2003). The therapy concept described in this chapter, which is based on results of research on bullying, has been introduced within inpatient treatment in Germany. It is one of the first inpatient therapy programmes for patients suffering from workplace bullying (Schwickerath, 2001, 2009). The range of treatment is based on special offers for groups in behaviour therapy, and the overall programme is distinguished by its goal-oriented procedure. It contains educative units as well as problem-solving-oriented and process-oriented units. What makes the treatment unique is that well-established clinical concepts and clinical experience are linked to findings from bullying research.

Negative effects of bullying on a person's health often result in unusual symptoms and normally go beyond the findings of most other stress and health studies in work psychology (Zapf and Semmer, 2004; cf. Zapf et al., 1996). In a study conducted by Meschkutat et al. (2002), for example, 43.9% of targets of bullying fell ill, almost half of these for more than six weeks. Health problems associated with bullying include psychosomatic complaints such as being tense or nervous, having headaches or sleeping problems, experiencing depressive moods, being obsessive, having anxiety disorders, and experiencing symptoms that resemble post-traumatic stress disorder (PTSD) (see Hogh et al., this volume).

This chapter is based on the scientific research and the practical work done in Germany in AHG Klinik Berus, a rehabilitation clinic where the first author is employed as senior psychotherapist. Between 1999 and November 2008, more than 1,400 patients with bullying experiences were treated in the AHG Berus clinic. According to German law, rehabilitation should prevent or postpone the "impairment of earning capacity" and, accordingly, an "early retirement from working life" (Steffanowski et al., 2007). Therefore, one of the main aims of therapeutical work is to obtain or reestablish the patient's ability to work. Every patient is given an initial diagnosis at the beginning of the rehabilitation programme. The main diagnoses for treatment at the clinic (Schwickerath, 2009) show that most of the victims of bullying so far were admitted because of depressive symptoms. The values of depression of "bullied patients"—our term for patients who are being treated at the clinic because of their experiences of being bullied—are comparable to those of other psychosomatic patients (Schiller et al., 2004). However, bullied patients show a considerable correlation between job stressors and depression, which is not true for other patients. This finding is in line with Kivimäki et al. (2003), who were able to demonstrate a relationship between bullying and the occurrence of depression and cardiovascular diseases.

Indication to Inpatient Treatment

In this chapter, we talk about inpatient treatment of bullying victims. From our practical experience, we know that there are many victims or former victims who need therapeutic help but have never received any. However, we know that there are also many victims who are able to solve the bullying problem by other means and, maybe, without any help. Our experiences may reflect the findings of several researchers who differentiated groups or clusters of victims with differing profiles (e.g., Matthiesen and Einarsen, 2001; Notelaers et al., 2006; Zapf, 1999b). It is likely that some of these clusters contain individuals who are likely to seek therapeutic help, whereas individuals belonging to other clusters may likely be able to solve the problem without therapeutic help, by, for example, seeking a job in another organisation. Thus, what we say about victims and therapeutic help in the remainder of this chapter may not apply to every victim, but it may apply to a substantial group of victims, in particular, for all those who have taken part in a therapy programme.

If patients are to receive a psychotherapeutic treatment, the question is raised whether a special kind of therapy, especially an inpatient therapy, is advisable. The question of what indicates a special kind of therapy is considered to be one of the most important problems of psychotherapy, but there is little research carried out on this issue (see Margraf, 2000; Zielke, 2010). Regarding the treatment of victims of bullying, this question is raised in an exceptional manner, because a variety of newspaper reports as well as personal feedback from patients have indicated that the conventional treatment of bullying victims—that is, a therapy programme that has not been adapted to the specific requirements of a bullying case—could often not master the problem.

An inpatient treatment normally starts with an initial diagnosis carried out by the general or family physician (GP) or a specialist (e.g., according to the International Classification of Diseases, ICD-10). The aim is to clarify whether therapy should come into consideration, and if so, which method should be applied. Although classification systems like ICD-10 (Dilling et al., 1993; WHO, 2007) or the *Diagnostic and Statistical Manual of Mental Disorders,* 4th ed. (DSM-IV), of the American Psychiatric Association (2000) provide differentiated descriptions of diseases, including detailed lists of symptoms of every disease as well as references to empirical validation (Chambless et al., 1997), these manuals often fail to provide sufficient information for the indication of a specific treatment, at least not for the treatment of bullying victims. Because disease classifications do not offer sufficient information for a specific treatment, which additional criteria will be used to favour inpatient treatment depends on the consulted doctor. Criteria should include the patient's imminent loss of earning capacity. Does treatment require taking the patient out of the pathogenic (work and nonwork) environment? Has the patient developed distinctive behavioural deficiencies? Furthermore, is the patient suffering from any comorbid psychosomatic or psychiatric diseases, and finally, will outpatient treatment be enough (cf. Zielke 1994, 2010)?

Because bullying victims experiencing high levels of stress tend to show a poor ability to distance themselves from the bullying situation and to cope with problems adequately (Schwickerath, 2009), the first step to gain distance and become emotionally stabilized is to take them out of the work environment. This is done by signing the patients off as sick, which would free them from their work duties. Empirical evaluation data show that signing a patient off as sick is often insufficient, however, because the symptoms of many affected persons do not regress solely as time goes by (Schwickerath, 2009; see also the evaluation study discussed later). Though no longer being exposed to the bullying, many patients cannot stop ruminating about the problem, at the same time making no progress and coming up with no solution. This is so because a variety of both organisational and individual factors contribute to the bullying process. These factors may prevent an easy solution and therefore have to be considered in deciding whether or not therapeutic treatment is advisable. On the part of the organisation, limited opportunities to communicate, lack of socioemotional gratification, and lack of social support (Gross, 2004) have to be considered. On the part of the individual, the great importance of work in the personal lives of bullying victims, which makes it difficult for them to give up their particular jobs, as well as a strong feeling of injustice and increased sensitivity to rejection, makes any "fast and easy" solutions to the bullying problem almost impossible. Other observable behavioural weaknesses or deficits of bullying victims are an increased proneness to resignation in case of failure, low self-confidence, and lack of assertiveness and problem-solving strategies (Schwickerath, 2009). These psychological problems and behavioural deficits suggest an inpatient therapeutic setting, which allows the treatment of these problems and accommodates the development of a reasonable perspective for the following years.

The specific characteristics of patients suffering from bullying can be complemented by behavioural aspects of chronic diseases (Zielke and Sturm, 1994). These behaviours include, among others, passivity and helplessness, loss of self-management strategies, and limited opportunities for passive relaxation. These behaviours mean that an inpatient therapy is indicated if the following criteria apply: (1) most of the problem areas typical of bullying victims apply, (2) behavioural patterns of chronic diseases have been developed, and (3) the patients show a basic motivation and readiness to take on responsibility and deal with the problems related to workplace bullying (Schwickerath, 2010; Zielke, 2010). In such circumstances, an appropriate institution with special treatments for victims of bullying has to be found. Concerning this matter, the consulting doctor needs to take into account the principle of minimal intervention (see Kanfer et al., 2006). This principle postulates that a patient's goal has to be reached using minimum expenditure. In practise, there is often the problem that persons who are consulted first by the victims (e.g., family doctors), have too little knowledge about bullying and its health consequences and are unaware of the criteria mentioned earlier. This lack makes it sometimes difficult for the victims to gain access to a suitable therapy programme.

Concepts of Inpatient Bullying Therapy

The therapy concept of the AHG Berus hospital is based on scientifically founded methods of behaviour therapy, in particular of cognitive-behavioural therapy (e.g., Kanfer and Schefft, 1988; Mahoney, 1989; O'Donohue et al., 2003). The therapeutic work integrates cognitive-behavioural therapy with results from workplace bullying research. It can be characterised by the following elements: (1) a therapy process that includes the phases "distancing," "understanding," "decision making," and "taking action"; (2) a specific approach that involves working on organisational aspects first and, thereafter, on the patient's own contributions to the bullying situation; (3) the formulation of a dysfunctional model (see later) applicable to bullying; (4) the development and examination of a future perspective concerning work as well as other areas of life; (5) specific practical exercises (e.g., in distancing oneself from the problem); and (6) such further therapeutic measures as sports or occupational therapy, which are part of the inpatient setting.

The starting point of the treatment involves patients' individual therapeutic contacts with their therapists, who are physicians or certified psychologists. In these initial therapeutic sessions, the motivational preconditions for active participation in treatments during the hospital stay have to be established. The individual therapy programme has to be developed and coordinated within this setting. Every patient is assigned to a particular therapist who is his or her main contact person throughout the stay in the hospital and who is responsible for all problems that may unfold during the entire programme. The one-on-one contacts with the therapist provide a basis for individual objectives and therapy steps, which can then also be transferred into activities in group therapy. Weekly supervised sessions by a clinically experienced psychological psychotherapist and a medical specialist assure a broad and differentiated approach to the complex set of psychological and medical problems of the patients suffering from bullying.

In the following, the therapy process will be described, followed by a description of a group therapy concept for bullying patients. In doing so, aims, structure, contents, and procedure in the group of bullying patients will be described. Further therapeutic steps will be introduced thereafter. An overview is presented in Figure 18.1.

Inpatient Bullying Therapy: The Therapy Process

As mentioned, the inpatient therapy or medical rehabilitation of bullied patients in the (AHG) Berus hospital takes a characteristic course, consisting of the phases "distancing," "understanding," "decision making," and "taking action," as described in this section (cf. Figure 18.2).

Establishing Distance

Clinical experience has shown that it is important for the bullied patients to distance themselves from the bullying incidents to be able to focus on a constructive solution to the problem. This was confirmed in an empirical study with victims of bullying having a lower ability to distance themselves from problems at work in comparison to nonvictims (Schwickerath, 2009). The stay in the clinic allows patients to be physically and psychologically separated

Phase	Goals	Contents	Selection of methods	Comprehensive aspects Explanations	Reference to the phase model by Kanfer et al. (2006)
D I S T A N C E	Establish distance to the bullying situation; establish work relationship; emotional stabilization	Reception in the clinic; Spatial separation; Development of a working collaboration for the therapy; delivery of information	Relaxation Sports Euthymic offers Exercises to practise establishing distance	The phases describe **key aspects**; contents also extend into other phases. **Transparency** of the therapy is important for the bullied patients.	1. Initial phase: Establishing convenient starting conditions 2. Developing motivation to change and preliminary choice of areas to be modified
U N D E R S T A N D I N G	Elaborate dysfunctional model to understand the bullying situation with regard to the organisation, the bully, and the patient's own contributions	Individual behaviour analysis and complementary elements to properly classify the bullying events	Microanalysis; organisational charts using a flip-chart; analysis and changing of inner forces; image or narrative of the dysfunctional model; approaches of cognitive behaviour analysis; cake model; curriculum vitae; Four-ears-model; coping with anger; visualisation; aspects of professional decision making and responsibility (Sonntag, 2004); triangle of can, should, and want; being able to deal with different personality styles; cognitive restructuring	**Starting point** is the individual behaviour analysis as a basis to elaborate therapy goals. **Procedure**: from external aspects to internal aspects. **Comprehensive and central**: • Motivation to change • Perspective • Goal • Meaning • Motto containing the question "What else do I want to achieve in my life?"	3. Analyzing behaviour and conditional functional model

			Therapeutic components:	
D E C I D I N G	Clarify the patient's future professional direction (return to the workplace, transfer within organisation, turnover, retirement)	Technique of columns – pros and cons Image or narrative as a goal (motto) Decision tree	Individual therapy Bullying group Sports therapy Sociotherapy Occupational therapy with work therapy	4. Agreeing on therapy goals
	Help with making a decision within the elaborated "motto"			
T A K I N G A C T I O N	Learn to develop abilities/capabilities to put decision into practise; develop problem solving competencies; learn strategies of distancing oneself	Role plays, exercises of distancing, yes–no exercise, defined ritual, acceptance and commitment—therapy; mindfulness exercises, perception and acceptance of emotions, assertiveness training, problem-solving strategies, learning abilities of planning and organising at work, time management, elements of wisdom therapy such as "method of unsolvable problems," clarifying social aspects, retirement, dismissal, legal measures	Project group Euthymic offers Testing of behaviours Relaxation training Assertiveness training Groups with a defined topic (e.g., tinnitus), Group cohesion	5. Planning, choosing, and executing specific methods 6. Evaluating therapeutic progress 7. Final phase: optimizing success, termination of the therapy
	Problem-solving process in terms of self management, coping with stress, encouragement of assertive communication and conflict-solving behaviour Strengthening self confidence and assertiveness			

FIGURE 18.1
Overview of bullying therapy.

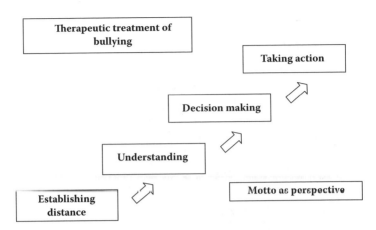

FIGURE 18.2
Therapeutic treatment of bullying.

from their problems at work and at home. This phase is about acknowledging the issues that concern the patients and addressing them, emotionally stabilizing the patients by providing a therapist who is responsible for them, and helping them to establish a therapeutic relationship. It is of primary concern to communicate the therapy rationale and define the working steps of the therapy based on preliminary decisions with regard to the areas to be changed. In doing so, therapists ensure that patients suffering from bullying attend group therapy meetings, referred to as the "bullied patients' group." If patients are not able to participate in group meetings from the beginning, they are prepared specifically through individual therapy sessions or by attending purely educational group meetings. In essence, this phase is consistent with Kanfer's initial phase of role structuring, creating a therapeutic alliance, developing a commitment for change, establishing positive starting conditions, developing a motivation to change, and providing a preliminary selection of areas where change is considered feasible (Kanfer and Schefft, 1988; Kanfer et al., 2006).

Understanding

The key feature of this phase is the development of an individual *dysfunctional model*. This is a model describing the factors contributing to the patients' problems. A dysfunctional model includes the following parts: a situation—in our case, the bullying situation—and the thoughts, emotions, bodily reactions, and actions occurring in this situation. Each of these areas can affect the others. For example, how one thinks about a problem can affect how one feels physically and emotionally. With regard to the bullying situation, the dysfunctional model describes the contribution to the bullying problem of the organisation, the bully, and the patients themselves (see Figure 18.3).

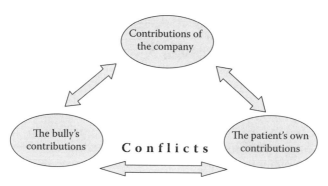

FIGURE 18.3
The conflict model.

The development of the dysfunctional model is based on individual behaviour analysis, which is at the core of behaviour therapy (Kanfer et al., 2006). Additional elements, for instance, from systemic therapy approaches, are used in this phase to classify the bullying incident and to support the development of the dysfunctional model for the patient. The main aim is to impart a conflict or stress model relating to the bullying problem. Here, the different perspectives of victims and perpetrators are considered (Zapf and Einarsen, 2005); for example, the perspective of a group of perpetrators who in their view hassle the victim only every now and then and who do not attach much importance to their actions may be compared with the perspective of the victims, who consider such isolated actions by several perpetrators to be aimed deliberately and systematically against them. This phase largely corresponds to the third phase by Kanfer et al. (2006), namely, behavioural analysis and selecting target behaviours.

Decision Making

This phase aims to resolve the direction of the patients' further occupational activities. Can the patients return to their former places of work? Is relocation to another department or even leaving the organisation to be considered? Is there the prospect of finding a job elsewhere, or is retirement an option? Already being initiated in the previous phases, this decision-making process is dealt with once again, most importantly by clarification of the patients' mottos for the future (see later). Only with such clarification is it possible to make a sustainable decision. This phase partly corresponds to phase 4 by Kanfer and Schefft (1988), namely, negotiating objectives and methods of treatment.

Taking Action

The purpose of this phase is the acquirement of abilities or skills to develop a new perspective, which has been worked out in the previous phase. Thereto it is necessary to learn adequate problem-solving strategies as well as strategies to distance oneself. The practise part involves role plays in group settings and is essential to strengthen self-confidence and to develop more self-assurance. This phase partly corresponds to phases 5–7 by Kanfer and Schefft (1988): implementing treatment and maintaining motivation by establishing coping skills and strategies, monitoring and evaluating progress, maintenance, generalisation, and termination of treatment.

Bullied patients, who are often depressive, frequently experience their situation as hemmed in. They need sustainable motivation to change things to be able to escape their limiting and gridlocked situation. Therefore, therapists must help patients develop short- and long-term goals. From our point of view, changes in behaviour are closely connected to this motivation and to the question, "What is the use of it?" The motivation to change and to develop new goals usually affects general goals and values in a patient's concept of life (Frankl, 1997). To emphasise the importance of developing a new perspective, we coined the term *motto* as a label for the new goal.

Inpatient Bullying Therapy: The Procedure

At the core of the therapy programme are the regular sessions of the bullied patients' group. Usually, there are at least eight group sessions. Admission usually takes place in the first and fifth sessions, because these include educational elements for new patients. The group is led by two experienced therapists, and the attendance is limited to 12 participants at most. Group therapy is an effective way to teach the patients about bullying, imparting knowledge that is relevant to the therapy. Another advantage of group therapy is that patients feel understood and taken seriously by other members of the group who have had similar experiences. In addition, from the other patients they receive social support, which they were lacking so much while they were bullied at their places of work (Zapf and Einarsen, 2005). The group setting acts as a social microcosm where problems are dealt with collectively, where shared personal experiences with bullying play an important part, and where group-dynamic processes among the patients are evoked. This microcosm forms a good precondition for a better understanding of how bullying conflicts develop and what their antecedents and consequences are.

Before attending their first group session, patients receive information on procedures, rules, and goals, which have to be transparent and comprehensible for each patient. Constructive group conditions, as defined by Yalom (1970), are communicated to the patients. Examples for such group conditions are group cohesion, openness, mutual trust, and model learning, as well as hope and confidence.

The procedure of the interactive group sessions is based on Grawe's problem-solving approach (2002), in which principles of psychotherapy such as problem solving, assessment of the patient's motivation, actualisation of problems, and activation of resources are taken into account. Because patients come with different prerequisites and differing knowledge about bullying, they are given information about the main manifestations, causes, and consequences of workplace bullying. Informing patients about bullying facilitates their entry into the group, for patients partly recognize their personal situations in the general descriptions of bullying. Thus, they are able to establish a relationship with other group members and learn how to put their personal situations into perspective. They realise that they are not alone with their experiences, that others have faced similar situations. A "common language" is developed by defining the behaviour therapy framework, thereby also making the therapeutic procedures transparent. Transparency is particularly important because often, bullied patients have made their experience of being bullied into ambivalent and nontransparent processes, like being exposed to rumours, unclear instructions, and such (cf. Meschkutat et al., 2002). By illustrating the dysfunctional model that has to be developed for every patient, the therapist is attempting to initiate in the bullied patients a change of perspective, namely, from feeling trapped to having a sense that the bullying problem can be solved. This problem-solving model should help to establish distance and allow the patients to take their first steps to understanding the bullying situation.

For the success of therapy, it is crucial that patients accept the dysfunctional model as an explanation of their bullying situation. To make acceptance easier, it has proven useful to focus first on external aspects—that is, on the organisational or structural problems of the organisation—before dealing with such internal aspects as the victim's potential contribution to the bullying process. First, the therapist provides general information on bullying; second, the patient describes the organisation or company; and last, the focus is set on the patient's own contribution to the problem. This course of action is consistent with the patient's perception that such external factors as the malicious behaviour of a bully or certain company structures, including conflicting responsibilities, are the main causes of his or her complex problems (cf. Zapf, 1999b).

In the development of the dysfunctional model, the foundation is laid for the patient's change of perspective, but without directly postulating it. Within the *behavioural analysis* (Kanfer and Schefft, 1988), it is easier for the patients to identify the *contributions of the organisation* to the bullying—these are elaborated with the help of organisational charts—as well as the *contributions of the bullies*. In identifying these elements, it is important to keep an eye on the therapeutic process, as described earlier. In the beginning, only indirect references are made to the patient's own contributions. Understanding the various contributions to the bullying process is the patient's first step in "looking behind the scenes" of the bullying events. Throughout the therapy, the patient's own contributions are focused on little by little. This approach implies to patients that a change in how they experience the bullying events can take place only through their own change because changing themselves is the only thing that is under their control. To make the patient's hierarchical position in the organisation transparent, an organisational chart is used to point out the patient's relationship to the most important other persons on the chart. The visual presentation allows the participants of the bullying group to better comprehend the patient's role in the organisation, and usually inconsistencies in the structure of the working situation become apparent. Numerous questions are raised, among them these: "Who has a say to whom from an organisational point of view?" "How is the hierarchical structure designed?" "Who of the colleagues assigns work tasks to other colleagues at the same hierarchical level, thus exerting power that goes beyond the colleague's formal position?" "What kind of informal roles have developed over the course of time?"

In the following step, the bullies' supposed motives, intentions, and problems are brought up and analysed within the group. The analysis of the bullies from the patient's point of view is less concerned with being "objective" or developing a neutral position, which cannot be achieved anyway, than with establishing distance by a form of role reversal. If the patients successfully put themselves in the position of another person, they will learn to put their own behaviour into perspective more easily. The patients have to be treated cautiously and respectfully, as sometimes the stressful memories and still present experiences can make a role reversal impossible. However, other members of the bullied patients' group often provide suggestions about the bullies' possible motives, suggestions that can be relieving for the victims— for example, fear, envy, resentment, and competitiveness, but also "unofficial personnel work" (Zapf, 1999a). The latter refers to measures by leaders and the management aimed at the victim to reach organisational goals by means that are not allowed.

As a third component, the *patients' own contributions* are elaborated. In doing so, patients make apparent such things as injured feelings, disappointment, a lack of problem-solving strategies, or the tendency of being unable to say no. The individual behaviour is analysed while taking personality features into account. The patient's thoughts, emotions, bodily reactions, and actions are at the core of the dysfunctional model. It is most important that this model is acceptable, feasible, and constructive to the bullied patient.

Inpatient Bullying Therapy: The Dysfunctional Model

Behavioural analysis forms the basis for the elaboration of a dysfunctional model, which is the key element of the diagnostic and therapeutic process (Kanfer et al., 2006). The dysfunctional model describes the patient's problematic behaviour in the bullying process, thereby considering behavioural, physiological, cognitive, and emotional aspects (cf. Ellis and Dreyden, 2007). This microanalysis based on the patient's own contributions concentrates on the four levels of problematic behaviour linked to the bullying events, as shown in Figure 18.4. First, *observable behaviour* in the bullying process is outlined. For example, do the patients withdraw? Do they isolate themselves from other colleagues or rather irritate their colleagues? Will

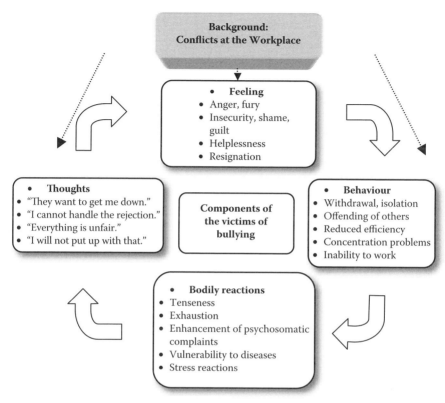

FIGURE 18.4
Behaviour analysis at the micro level (extract).

they "attack" and thus start a quarrel, thereby contributing to the escalation of the conflict? Often concentration and performance problems can be observed. Physical reactions are also recorded. For example, is the patient tense? Are there signs of physical symptoms like headache and/or backache? Second, *cognitions*, or rather *automatic thoughts*, that play a role in a situation of bullying are elaborated. Irrational beliefs (e.g., "I've got to solve that problem alone," "I have to be well behaved," or "I must not say no"), which are a form of automatic thoughts, function as internal driving forces. They are associated with the person's personality and attitude towards work. Third, *feelings* such as anger, rage, helplessness, or fear experienced by the patient in the course of the bullying events are both described and actualised within the group session and also recorded on a flip-chart. In many cases, the development of the dysfunctional model enables the patients for the first time to perceive their former—often confusing—experiences in a structured manner. This well-directed confrontation with the patient's own case often upsets the patient, but at the same time it offers the possibility to establish distance and to work on the therapy goals. In this early stage of therapy, neither a moral appraisal nor a clarification of the question of legal guilt is of interest. On the contrary, it is specifically pointed out that taking legal measures such as collecting evidence by taking notes and keeping a bullying diary or seeking advice from a lawyer—measures frequently recommended to bullying victims by victim support groups—often, from a psychological point of view, impedes the clarification of the bullying process. Whether legal measures will ultimately be taken is dealt with at a later point in the therapy process.

Moreover, to support the development of the dysfunctional model, explanatory models used in bullying research are introduced (cf. Einarsen et al., this volume), among them the scapegoat phenomenon, known in the context of social identity theories (see Tajfel and Turner, 1986) and used to assist the patients in understanding their experiences. Another useful element in the analysis of the bullying situation is a differentiated analysis of the patient's strengths and weaknesses at work. Thereby occupational knowledge, skills, and competences, among them method competence, social competence, and personal competence (Sonntag, 2004), are addressed and allow a juxtaposition of the patient's capabilities and shortcomings. Possible deficits in a patient's social competence can thus be accepted more easily if at the same time specific professional competences are acknowledged.

Based on this analysis, it also helps to resolve the patient's own expectations from work ("What is it that I want?"), the expectations of the employer or supervisor ("What is it that I should do?"), and the patient's own abilities ("What is it that I am capable to do?"). Patients can integrate themselves in this triangle of *can*, *should*, and *want* and thereby identify their own strengths and weaknesses. For instance, a patient can realise that a supervisor demanded him or her to work more than the patient was able to, which can become manifest in feelings of high stress, and be experienced as bullying.

In addition, the elaborated behavioural analysis is illustrated in various ways, for example, as a picture, a concise sentence, a myth or saying, a literary reference, or a narrative (cf. Angus and McLeod, 2004), to facilitate recall and to offer the patient a way of visualising the tangled situation at any time. Examples of themes of such illustrations include "Work as a family substitute," "Loss of work as a kind of divorce drama," "Bullying as a lifelong task," or "Prometheus," an exemplar from the ancient world.

Because of depressive processing of their problems, patients who are victims of bullying tend to generalise to their entire working life any recently experienced difficulties at work. This way of reacting often results in the patients' seeing their entire occupational career as an "experience of failure." The patients often forget that the bullying events in fact made up only a small part of their previous working life. To help the patients develop a more realistic perspective, their overall working life is visualised, for example, with the help of a cake model that has an inedible last piece, namely, the time the patient was bullied. Especially those patients who almost exclusively regard themselves to be the cause of the problem are asked to illustrate their previous work experiences in a picture containing, for instance, black for the bullying situations experienced and another colour for the rest of their past working lives. It is the goal to develop a realistic, reasonable assessment without playing down the depressive periods caused by the bullying events.

In the course of the bullying analysis, patients resolve for themselves whether or not they still fit into the organisation or workplace, given their individuality, their occupational qualifications and capabilities, and their new view of the bullying situation. It is essential, at this stage of the therapy, to elaborate therapeutic approaches for change and for finding ways out of the difficult situation. An indispensable element in cognitive behaviour therapy is reappraisal by cognitive restructuring (A. T. Beck, 1976; J. S. Beck, 1995; Beck et al., 1979; Ellis and MacLaren, 2005), which means identifying dysfunctional or irrational thoughts related to the difficult experience in the bullying situation. Driving thoughts, which influence the bullying experience as an inner force, are elaborated within the group. Examples include "I am responsible for everything," "I am not worth it," or "I have to be better than others." Thinking such irrational thoughts to the very end is damaging, and so the therapist must point out to the patients what it must mean for them to feel constantly responsible for everything. Normally, they will realise that life cannot go on as it has. When evaluating and classifying irrational thoughts, it has to be kept in mind that the expression of a particular cognitive pattern (e.g., "I have to be good") was developed to function as a survival strategy within a social setting. For instance, "being nice to everybody" could have been a protection against rejection, though it is an unreasonable strategy to stabilize one's self-worth. The therapy group is actively involved in the treatment of such irrational thoughts. The goal is to better understand individual reactions and to work on alternative and more functional cognitions, which allow for a "better" coping with life, but also especially with conflicts in the workplace. To achieve long-term coping with the bullying situation, the patient learns to answer the following questions: "What do I want for the rest of my life?" "What do I want for the close and distant future?" "What coping strategies do I want to learn?" This approach results in a relativisation of the significance of work compared to other areas of life. It initiates a willingness to accept previous, partly vain endeavours and opens new opportunities for the patient. In the end, a balance among job-related requirements, domestic duties, and reasonable leisure activities should be worked out with the patient. Helpful are such questions as "How did others cope with similar problems?" and "What are you going to tell your grandchildren about this incident in ten years' time?" For the patients, a chance to change things also means the opportunity to break away from old habits or attitudes and learn to reorientate themselves.

Inpatient Bullying Therapy: Perspective and Motto

After building a basis and precondition for coping in the initial phases of distancing and understanding, it is necessary for patients to develop a perspective, or motto, for their future career. Priority is given to the answering of two, often provocatively asked questions: "How much time do I still have left, considering the average life expectancy?" and "What do I consider to be important for my future?" Whereas the patients' perspectives were mainly oriented towards the past during their unsuccessful trials to cope with their bullying situations, the focus is now set on the future. This approach resembles a change of paradigm for many patients. This change of paradigm addresses the goals of the patients' further career and the significance that the patients ascribe to their occupational future. This clarification of perspective and the development of new goals and finding a purpose in life address the patient's need for meaning (Frankl, 1997). Thereby, questions of securing one's livelihood or of possibly waiving financial gratifications play an important role. The patients are to become proactive and to regain control of their lives again, instead of remaining patients and victims. They have to let go of old patterns of making sense of their lives in order to develop new ones. This future perspective, or so-called motto, is conceptualised in a suitable picture or story (therapeutic narrative) similar to the diagnostic narrative. Examples of perspective themes are "I will make a well-controlled withdrawal" and "I will acquire a thick skin." To resolve the patient's new perspective, it is helpful to answer together with the patient the *motivational questions* offered by Kanfer and Schefft (1988, p. 128): (1) What will it be like if I change? (2) How will I be better off if I change? (3) Can I change? (4) What will it cost to change? and (5) Can I trust this therapist and setting to help me get there? Examples for these clarifications of goals and values can be found in Kanfer et al. (2006). In summary, patients answer the question "What for?" and decide on their further occupational career.

It is important that after deciding on a future occupational direction (e.g., resigning from the company or going into retirement), patients reappraise their situations and translate new plans into action. While doing so, patients should avoid coping by wanting nothing more to do with their former workplace but should instead ignore the conflicts. Rather, therapists must explain to their patients the importance of actively coping with the unsettled crisis even if they resign from their jobs. Missing closure or having an unsatisfactory parting can have the consequence that the pending issues associated with the old workplace close in on the patients over and over again. Overcoming helplessness and finding sensible closure can happen through a previously settled ritual; for instance, a patient can organise a farewell celebration for former employees with a specific course of events, including a token gesture toward the bullies.

A series of further elements are integrated into the therapy concept: elements of forgiving (Kämmerer and Kapp, 2006); elements of acceptance therapy (Hayes et al., 1999)—focussing on helping the patients to accept personal characteristics or unchangeable circumstances—and components of wisdom therapy (Schippan et al., 2004). Furthermore, role plays are applied to strengthen assertive behaviour (see Fox and Boulton, 2003; Hollin and Trower, 1986). Moreover, if necessary, problem-solving strategies are developed and consolidated (D'Zurilla and Nezu, 2001). Elements of occupational therapy (developing a basic attitude towards work), sociotherapy (qualifications or occupational retraining), sports, movement therapy, and relaxation trainings (Bernstein and Borkovec, 1973) complete the therapy (for details, see Schwickerath, 2009).

Inpatient Bullying Therapy: An Evaluation Study

Here, we will report on an evaluation study of the therapy programme described earlier (for further details, see Schwickerath, 2009). This study was based on a sample of 102 patients. Measures were received immediately before (T1) and after (T2) the inpatient treatment. Moreover, data from 51 patients collected approximately one year later (T3) could be used for further analyses. Results showed a significant improvement of the patients' health as indicated in a significant reduction of complaints caused by different health symptoms, depressive moods, psychosomatic complaints, and a significantly lower rate of disability. Furthermore, a significant increase of having an optimistic point of view could be observed (Schwickerath, 2009). The descriptive analysis showed that patients who had been victims of bullying were very satisfied with the therapy. Moreover, the patients benefitted the most from being able to set themselves new goals and values, which was an important part of the therapy in connection with the elaboration of a stable perspective (motto). Because there was no control sample as a consequence of practical, legal, and ethical reasons, analyses were carried out in reference to subgroups of patients who were victims of bullying. There were no substantial differences between people who took part in the follow-up survey (responders) and people who did not answer (nonresponders).

A first question of interest was whether the health impairments caused by bullying could be improved if the patients left the bullying situation by being on sick leave or because of disability. In these cases, the victims would no longer be exposed to the bullying. No significant differences emerged between bullied patients who worked until they started the therapy programme and were thus continuously exposed to the bullying and those who were not exposed to the bullying situation during the last weeks. This result could mean that a strict time-out alone does not lead to a real improvement of the situation, at least not for those individuals waiting for therapy. By remaining in an unsettled situation while waiting for therapy, victims of bullying cannot solve the problem because they have little distance from the bullying events and lack active problem-solving strategies. However, those victims who benefit from stepping out of the nasty game of bullying at an early stage (Zapf and Gross, 2001), because they may not take the occupational situation as seriously as do those individuals who remain, tend not to go into therapy. These victims do not necessarily become ill, whereas those who remain and do become ill and seem to have little possibility of changing their situation are eligible for the therapy programme.

Patients who were examined for a second time after a year (N = 51) could be assigned to the following groups: (1) the "changers" (N = 20): these are patients who took a new job after rehabilitation. (2) the "stayers" (N – 10): those patients who returned to the unchanged situation in their old workplace; (3) The "bullies left" group (N = 7): patients who returned to their old workplace, but the bullies no longer worked in their close environment; (4) The "retirees" (N = 8): patients who went into retirement in the year following the therapy; and (5) the "unemployables" (N = 6): patients who were incapacitated at the time of the follow-up survey. As an example, we report the results regarding the development of depressive symptoms (see Figure 18.5).

First of all, the BDI depression scores (Beck's Depression Inventory; Hautzinger et al., 1995) at time 1 were marginally significant between groups. The changers and the unemployables showed higher scores than

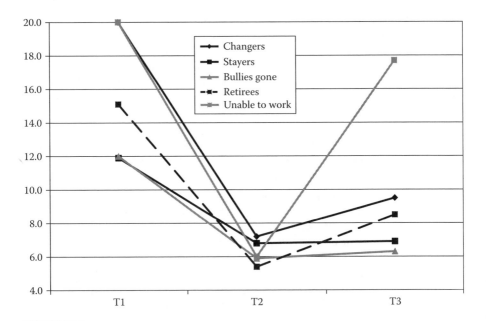

FIGURE 18.5
Evaluation study: BDI scores of different groups of bullying victims.
Note: T1: before treatment; T2: after treatment; T3: one year after treatment; BDI: Beck's Depression Inventory.

the other groups. No such significant differences occurred at time 2 and time 3. Moreover, all groups showed a significant improvement in BDI scores between T1 (before) and T2 (after the treatment). Significant improvement also occurred comparing T1 and T3 scores (10% level), with one exception: A significant (10%-level) change for the worse occurred for the unemployables between T2 and T3.

An interpretation of the data is that all patients except the unemployables profited from the treatment, even if they had different base levels. However, unemployable patients profited by neither depressive moods nor physical symptoms, nor with regard to developing an optimistic perspective of the therapy in the long run. Even though an improvement of BDI scores could be observed directly after therapy, after being interviewed for the second time they returned to their base level without an observable trend. A reason for these results could be the lack of active coping, which was described in various components of the therapy. The stayers compared to the other groups showed lower scores regarding physical symptoms and depressive moods at T1 before the therapy started, and they benefitted from the therapy programme as indicated by the significant reduction of BDI scores from T1 to T2. These patients probably returned to their old workplace because the psychological strain was not as distinctive, making it possible to return without great difficulties and with newly gained coping strategies. The comparatively lower BDI scores could be an indication that the bullying situation of these patients was less severe than the situation of the other groups. Therefore, they were able to manage the situation after the therapy.

Patients who changed their employment or were no longer confronted with the bullies at their former workplace benefitted the most during the course of treatment. Thus, the therapy goal to encourage patients to analyse their work situation adequately and consider leaving the organisation can be legitimised ex post by our data. This conclusion is important because victims of bullying often feel trapped at their workplace and see no possibilities to distance themselves or ways to change their problematic situation. Therefore, learning to make a decision in a difficult situation—namely, to stay or to leave the organisation—is an essential part of the therapy programme.

Because the research design of the study was limited because there was no control group, conclusions have to be drawn with care. On the other hand, we evaluated a comparable six-week cognitive behaviour therapy programme, run at the Vogelsberg Clinic in Germany (Düvel, 2008), with measures of before and after the treatment. As in the foregoing study, we found a significant reduction of BDI and Symptom Checklist-90-Revised (SCL-90-R) scores. We are, therefore, confident that cognitive behaviour therapy programmes that include findings of bullying research are capable means of helping the victims of bullying.

Within a health system, which is under considerable financial strain, monetary aspects of the newly established therapy programme are also of importance. A cost-benefit analysis shows a considerable "return on investment"

(Wittmann et al., 2002), and the investment into the therapy programme is also profitable, considering financial expenses. This finding is supported by investigations carried out by Zielke et al. (2004), who concluded that investing one euro in inpatient medical rehabilitation means a long-term gain of over three euros. For victims of bullying, the introduced programme offers effective and efficient therapy, which is tailored to the particular needs of this patient group.

The Treatment of Bullied Patients: Integrating Psychotherapy and Bullying Research

In the past, victims of bullying frequently reported that the psychotherapeutical treatment they received was of little help for them. A reason for this is that psychotherapists tend to have limited knowledge in work and organisational psychology and thus have a limited understanding of workplace conflicts. This reason also applies to the issue of bullying. In this chapter, we tried to overcome this problem by describing a therapy programme that integrates knowledge of work and organisational psychology, and in particular research on workplace bullying, and a cognitive behaviour therapy approach. In the final section of this chapter, we will summarize how work and organisational psychology issues and, in particular, findings of bullying research were integrated into the described inpatient psychotherapy programme.

1. Bullying plays a major role in the first phase of the programme where psycho-educative elements are in the foreground. The patients receive an overview on what bullying is, its definition, typical bullying behaviours, causes, and consequences.

2. This knowledge is also used for the analysis of the dysfunctional model. Here, knowledge about organisational structures and relationships with supervisors and colleagues, as well as knowledge about the bullying concept, is used to reconstruct and subjectively explain the victim's bullying case.

3. Specific findings such as the victims' problems distancing themselves from their work (Niedl, 1995; Schmiga and Rammsayer, 2004; Schwickerath, 2009) or the high importance of work to them (Schwickerath, 2009) are integrated into the programme.

4. It is acknowledged that bullying has a variety of causes that can reside in the organisation, in the social system at work, in the perpetrator, or in the victim (e.g., Zapf, 1999b). These various causes are used in the analysis of the bullying case. In addition, it is taken

into account that victims often have difficulties facing their case. To make it easier for patients, the therapy first starts with the organisation and then moves on to the patients' own contributions.

5. It is acknowledged that bullying often is an escalated conflict that is a no-control situation where a series of otherwise reasonable conflict management strategies have been used in vain and where leaving the organisation is often the only option (e.g., Zapf and Gross, 2001). The decision to leave the organisation is picked up in the later phases of the therapy and is thoroughly prepared in the previous phases.

6. The change of negative coping strategies (Rammsayer et al., 2006) and the improvement of lowered self-esteem (Vartia, 2003) are taken into account.

7. Finally, transparency is an issue in the therapy, as victims of bullying are very sensitive and easily regard things to be directed against themselves (Zapf and Einarsen, this volume).

In this chapter, we have described a therapy programme for bullying victims. Psychotherapy has to concentrate on the individual because there is no possibility by the means of the therapy to directly influence and change the organisation where the bullying took place. Focusing on the individual by no means implies that we consider the victim as the major problem in bullying scenarios. However, the victim may also have contributed to the bullying problem, and there are reasons to believe that the contribution of victims who take part in an inpatient therapy is larger than the contributions of other victims. But even if it becomes clear that bullying is purely the problem of the organisation and of some perpetrators, it is still the bullied patients who are treated in the therapy. Their understanding of the bullying, their capabilities to cope with the situation, and their physical and psychological health have to be addressed. Further, support may be given to help the patients consider legal actions against the bullies or their employer.

References

American Psychiatric Association (2000) *Diagnostic and statistical manual of mental disorders*, 4th ed. Washington, DC: American Psychiatric Association.

Angus, L. E., and McLeod, J. (eds.) (2004) *The handbook of narrative and psychotherapy: Practice, theory and research.* Thousand Oaks, CA: Sage.

Beck, A. T. (1976) *Cognitive therapy and the emotional disorders.* New York: International Universities Press.

Beck, A. T., Rush, A. J., Shaw, B. F., and Emery, G. (1979) *Cognitive therapy of depression.* New York: Guilford.

Beck, J. S. (1995) *Cognitive therapy: Basics and beyond.* New York: Guilford.

Bernstein, D. A., and Borkovec, T. D. (1973) _Progressive relaxation training._ Champaign, IL: Research Press.

Chambless, D. L., Sanderson, W. C., Shoham, V., Johnson, S. B., Pope, K. S., Crits-Christoph, P., Baker, M., Johnson, B., Woody, S. R., Sue, S., et al. (1997) _An update on empirically validated therapies._ Washington, DC: Division of Clinical Psychology, American Psychological Association.

Dilling, H., Mombour, W., and Schmidt, M. H. (eds.) (1993) _Internationale Klassifikation psychischer Störungen: ICD-10: Kapitel V (F); Klinisch-diagnostische Leitlinien/ Weltgesundheitsorganisation_ [The ICD-10 international classification of mental and behavioural disorders: Clinical descriptions and diagnostic guidelines/ WHO]. Bern, Switzerland: Verlag Hans Huber.

Düvel, H. (2008) Evaluation einer Mobbingtherapie [Evaluation of a therapy of bullied patients]. Unpublished diploma thesis, Goethe University, Frankfurt, Germany.

D'Zurilla, T. J., and Nezu, A. M. (2001) Problem solving therapies. In K. S. Dobson (ed.), _Handbook of cognitive-behavioural therapies,_ 2nd ed. (pp. 211–245). New York: Guilford.

Ellis, A., and Dryden, W. (2007) _The practice of rational emotive behavior therapy,_ 2nd ed. New York: Springer.

Ellis, A., and MacLaren, C. (2005) _Rational emotive behavior therapy: A therapist's guide._ Atascadero, CA: Impact.

Fox, C. L., and Boulton, M. J. (2003) Evaluating the effectiveness of a social skills training (SST) programme for victims of bullying. _Educational Research, 45,_ 231–247.

Frankl, V. E. (1997) _Man's search for meaning: An introduction to logotherapy._ New York: Pocket Books.

Grawe, K. (2002) _Psychological therapy._ Cambridge, UK: Hogrefe and Huber.

Gross, C. (2004) Analyse sozialer Konflikte und Mobbing am Arbeitsplatz—eine Tagebuchstudie (ASKA-Projekt) [Analysis of social conflicts and bullying at the workplace—a diary study]. Unpublished dissertation, Johann Wolfgang Goethe University, Frankfurt, Germany.

Hautzinger, M., Bailer, M., Worall, H., and Keller, F. (1995) _Beck-Depressions-Inventar (BDI) Testhandbuch,_ 2nd ed. Bern, Switzerland: Hans Huber.

Hayes, S. C., Strosahl, K. D., and Wilson, K. G. (1999) _Acceptance and commitment therapy: An experiential approach to behavior change._ New York: Guilford.

Hollin, C. R., and Trower, P. (eds.) (1986) _Handbook of social skills training,_ vols. 1 and 2. Oxford, UK: Pergamon.

Kämmerer, A., and Kapp, F. (2006) Vergebung als therapeutisches Ziel [Forgiveness as therapeutic target]. In S. Fliegel and A. Kämmerer (eds.), _Psychotherapeutische Schätze: 101 bewährte Übungen und Methoden für die Praxis_ (pp. 202–204). Tübingen, Germany: dgvt-Verlag.

Kanfer, F. H., Reinecker, H., and Schmelzer, D. (2006) _Selbstmanagement-therapie_ [Self-management therapy], 4th ed. Berlin: Springer.

Kanfer, F. H., and Schefft, B. K. (1988) _Guiding the process of therapeutic change._ Champaign, IL: Research Press.

Kivimäki, M., Virtanen, M., Vartia, M., Elovainio, M., Vahtera, J., and Keltikangas-Järvinen, L. (2003) Workplace bullying and the risk of cardiovascular disease and depression. _Occupational and Environmental Medicine, 60,_ 779–783.

Mahoney, M. (1989) _Human change processes: The scientific foundations of psychotherapy._ New York: Basic Books.

Margraf, J. (2000) Therapieindikation [Indication of therapy]. In J. Margraf (ed.), *Lehrbuch der Verhaltenstherapie*, 2nd ed., vol. 1 (pp. 145–154). Berlin: Springer Verlag.

Matthiesen, S. B., and Einarsen, S. (2001) MMPI-2 configurations among victims of bullying at work. *European Journal of Work and Organizational Psychology*, 10, 467–484.

Meschkutat, B., Stackelbeck, M., and Langenhoff, G. (2002) *Der mobbing-Report: Eine Repräsentativstudie für die Bundesrepublik Deutschland* [Report of bullying: A representative study for the Federal Republic of Germany]. Schriftenreihe der Bundesanstalt für Arbeitsschutz und Arbeitsmedizin. (Book series of the Bundesanstalt für Arbeitsschutz und Arbeitsmedizin) Berlin: Wirtschaftsverlag NW.

Niedl, K. (1995) *Mobbing/Bullying am Arbeitsplatz: Eine empirische Analyse zum Phänomen sowie zu personalwirtschaftlich relevanten Effekten von systematischen Feindseligkeiten* [Mobbing/bullying at work: An empirical analysis of the phenomenon and the effects of systematic harassment on human resource management]. Munich: Hampp.

Notelaers, G., Einarsen, S., De Witte, H., and Vermunt, J. (2006) Measuring exposure to bullying at work: The validity and advantages of the latent class cluster approach. *Work & Stress*, 20 (4), 289–302.

O'Donohue, W. T., Fisher, J. E., and Hayes, S. C. (eds.) (2003) *Cognitive behavior therapy: Applying empirically supported techniques in your practice*. New York: John Wiley.

Rammsayer, T., Stahl, J., and Schmiga, K. (2006) Grundlegende Persönlichkeitsmerkmale und individuelle Stressverarbeitungsstrategien als Determinanten der Mobbing-Betroffenheit [Basic personality dimensions and stress-related coping strategies in victims of workplace bullying]. *Zeitschrift für Personalpsychologie*, 5 (2), 41–52.

Schiller, A., Schwickerath, J., and Kneip, V. (2004) Stressoren der beruflichen Tätigkeit als Prädiktoren depressiver Verstimmung von Patienten infolge von Mobbing im Vergleich zu psychosomatischen Patienten. [Stressors of the occupational activity as predictors of depressive mood in patients as a result of bullying compared to psychosomatic patients]. In J. Schwickerath, W. Carls, M. Zielke, and W. Hackhausen (eds.), *Mobbing am Arbeitsplatz—Grundlagen, Beratungs- und Behandlungskonzepte* (pp. 86–106). Lengerich, Germany: Pabst Science.

Schippan, B., Baumann, K., and Linden, M. (2004) Weisheitstherapie—cognitive Therapie der posttraumatischen Verbitterungsstörung [Wisdom therapy: Cognitive strategies of posttraumatic embitterment disorder]. *Verhaltenstherapie*, 14, 284–293.

Schmiga, K., and Rammsayer, T. (2004) Mobbing und Persönlichkeit: Unterschiede in habituellen Stressverarbeitungsweisen zwischen Mobbing-Betroffenen und Nicht-Betroffenen [Bullying and personality: Differences in coping with stress between victims and nonvictims]. *Wirtschaftspsychologie*, 6 (1), 84–92.

Schwickerath, J. (2001) Mobbing am Arbeitsplatz: Aktuelle Konzepte zu Theorie, Diagnostik und Verhaltenstherapie [Bullying at work: Current concepts on theory, diagnostics and behaviour therapy]. *Psychotherapeut*, 46, 199–213.

——— (2009) *Mobbing am Arbeitsplatz: Stationäre Verhaltenstherapie von Patienten mit Mobbingerfahrungen*. [Bullying at the workplace: Inpatient treatment of victims of bullying]. Lengerich, Germany: Pabst Science.

———— (2010) Psychosomatische Reaktionsbildungen bei Mobbing am Arbeitsplatz: Hintergründe, therapeutisches Konzept und Indikation für stationäre Rehabilitation [Psychosomatic reactions due to bullying at the workplace: Background, conception and indication to inpatient treatment]. In M. Zielke (ed.), *Indikation zur stationären Behandlung und Rehabilitation bei psychischen und psychosomatischen Erkrankungen* (pp. 525–550). Lengerich, Germany: Pabst Science.

Sonntag, K. H. (2004) Personalentwicklung [Personnel development]. In H. Schuler (eds.), *Enzyklopädie der Psychologie: Organisationspsychologie—Grundlagen und Personalpsychologie* (pp. 827–890). Göttingen, Germany: Hogrefe.

Steffanowski, A., Löschmann, C., Schmidt, J., Wittmann, W. W., and Nübling, R. (2007) *Metaanalyse der Effekte stationärer psychosomatischer Rehabilitation* (MESTA Studie) [Meta-analysis of the outcome quality of inpatient psychosomatic therapy]. Bern, Switzerland: Hans Huber.

Tajfel, H., and Turner, J. (1986) The social identity theory of intergroup behaviour. In S. Worchel and W. G. Austin (eds.), *Psychology of intergroup relations* (pp. 7–24). Chicago: Nelson.

Tehrani, N. (2003) Counselling and rehabilitating employees involved with bullying. In S. Einarsen, H. Hoel, D. Zapf, and C. L. Cooper (eds.), *Bullying and emotional abuse in the workplace: International perspectives in research and practice* (pp. 270–284). London: Taylor & Francis.

Vartia, M. (2003) *Workplace bullying: A study on the work environment, well-being and health.* People and Work Research Reports 56. Helsinki: Finnish Institute of Occupational Health.

Vartia, M., Korppoo, L., Fallenius, S., and Mattila, M. L. (2003) Workplace bullying: The role of occupational health services. In S. Einarsen, H. Hoel, D. Zapf, and C. L. Cooper (eds.), *Bullying and emotional abuse in the workplace: International perspectives in research and practice* (pp. 285–298). London: Taylor & Francis.

WHO (2007) *International statistical classification of diseases and related health problems, 10th revision version for 2007.* http://apps.who.int/classifications/apps/icd/icd10online (accessed May 4, 2010).

Wittmann, W. W., Nübling, R., and Schmidt, J. (2002) Evaluationsforschung und programmevaluation im gesundheitswesen [Evaluation research and evaluation of programmes in public health]. *Zeitschrift für Evaluation*, 1, 39–60.

Yalom, I. D. (1970) *The theory and practice of group psychotherapy.* New York: Basic Books.

Zapf, D. (1999a) Mobbing in Organisationen: Ein Überblick zum Stand der Forschung [Mobbing in organisations: A state-of-the-art review]. *Zeitschrift für Arbeits- und Organisationspsychologie, 43*, 1–25.

———— (1999b) Organisational, work group related and personal causes of mobbing/bullying at work. *International Journal of Manpower, 20* (1/2), 70–85.

Zapf, D., and Einarsen, S. (2005) Mobbing at work: Escalated conflicts in organizations. In S. Fox and P. E. Spector (eds.), *Counterproductive work behavior: Investigations of actors and targets* (pp. 237–270). Washington, DC: American Psychological Association.

Zapf, D., and Gross, C. (2001) Conflict escalation and coping with workplace bullying: A replication and extension. *European Journal of Work and Organizational Psychology, 10*, 497–522.

Zapf, D., Knorz, C., and Kulla, M. (1996) On the relationship between mobbing factors and job content, the social work environment and health outcomes. *European Journal of Work and Organizational Psychology, 5,* 215–237.

Zapf, D., and Semmer, N. K. (2004) Stress und Gesundheit in Organisationen [Stress and health in organisations]. In H. Schuler (ed.), *Enzyklopädie der Psychologie, Themenbereich D, Serie III, Band 3, Organisationspsychologie* (pp. 1007–1112). Göttingen, Germany: Hogrefe.

Zielke, M. (1994) Indikation zur stationären Verhaltenstherapie [Indication to inpatient- behavior therapy]. In M. Zielke and J. Sturm (eds.), *Handbuch Stationäre Verhaltenstherapie* (pp. 193–249). Weinheim, Germany: Psychologie Verlags Union.

———— (ed.) (2010) *Indikation zur stationären Behandlung und Rehabilitation bei psychischen und psychosomatischen Erkrankungen* [Indication to inpatient treatment in mental disorders and psychosomatic illness]. Lengerich, Germany: Pabst Science.

Zielke, M., Borgart, E. J., Carls, W., Herder, F., Lebenhagen, J., Leidig, S., Limbacher, K., Meermann, R., Reschenberg, I., and Schwickerath, J. (2004) *Ergebnisqualität und Gesundheitsökonomie verhaltensmedizinischer Psychosomatik in der Klinik— Krankheitsverhalten und Ressourcenverbrauch von Patienten mit psychischen und psychosomatischen Erkrankungen: Ergebnisse verhaltensmedizinischer Behandlung und Rehabilitation im Langzeitverlauf* [Quality of outcome and the economy of mental health of inpatient behavioural treatment—illness behaviour and consumption of resources by patients with mental and psychosomatic symptoms: Results of inpatient behavioural treatment in long-time process]. Lengerich, Germany: Pabst Science.

Zielke, M., and Sturm, J. (1994) Chronisches Krankheitsverhalten: Entwicklung eines neuen Krankheitsparadigmas [Concept of abnormal illness behaviour: Development of a new paradigm of illness]. In M. Zielke and J. Sturm (eds.), *Handbuch Stationäre Verhaltenstherapie* (pp. 42–60). Weinheim, Germany: Psychologie Verlags Union.

19

Conflict, Conflict Resolution, and Bullying

Loraleigh Keashly and Branda L. Nowell

CONTENTS

Introduction

In their review of the workplace bullying literature, Hoel et al. (1999) argued for the importance of taking a conflict perspective on the problem of bullying. They suggested that the dyadic conflict literature is rich in insights on conflict development and escalation as well as the various procedures and processes for resolving conflicts. Their belief is premised on an implicit connection between conflict and bullying where severe bullying is likened to "destructive conflicts going beyond the point of no return" (p. 221). Zapf and Gross (2001) concur, describing bullying situations as "long-lasting and badly managed conflicts" (p. 499). From a workplace aggression perspective, Raver and Barling (2008) have argued that *conflict* is the umbrella term and that workplace aggression (of which bullying is a special case) should be considered as a particular form of workplace conflict.

Other scholars have viewed the association between conflict and bullying as less synonymous. For example, Einarsen and Skogstad (1996) make the connection between bullying and conflict, but as distinctive constructs hinging on the ability of the involved parties to respond to or defend against hostile actions. A key feature of bullying is the inability to defend oneself. If the parties involved are equally able to defend themselves, then the situation may well be a serious conflict, but it is not bullying. Einarsen (1999) further refines his earlier distinction by proposing that there are at least two types of bullying: predatory and dispute-related. Predatory bullying occurs when the victim has done nothing provocative that would reasonably invoke or justify the bully's behavior. Dispute-related bullying, however, develops out of grievances between two or more parties and involves retaliatory reactions to some perceived harm or wrongdoing. If one of the parties becomes "disadvantaged" during the dispute, he or she may become a victim of bullying. So a dispute may trigger bullying. In making this argument, Einarsen supports the idea that conflict and bullying are distinct yet related constructs.

Thus, while it is clear various authors consider conflict in their writings and research on workplace bullying and aggression, the connection between these two constructs remains unclear. In order to assess the value of a conflict perspective on bullying, we need to clarify what is the relationship between bullying and conflict. Thus, in the first section of this chapter, we will compare definitions of conflict and bullying in an effort to articulate their connection. We will then present and discuss several concepts from the conflict literature that may prove insightful for workplace bullying: individual conflict management strategies and the influence of power, escalation and deescalation, and coordinated intervention approaches. We will sum up with some cautions regarding the application of conflict concepts to the study and amelioration of workplace bullying.

Conflict and Bullying: Same or Different?

Like so many concepts in the social sciences, *conflict* has many definitions, and there is no consensus on a common definition (Thomas, 1992). However, Putnam and Poole (1987) note that there are three general properties reflected in general definitions of conflict: (1) interdependence between the parties; that is, each has the potential to interfere with the other; (2) perception by at least one party that there is opposition or incompatibility among the parties' concerns; (3) some form of interaction between two or more parties. There are also a number of definitions of *bullying*, with the term often encompassing a variety of situations. Like conflict definitions, these various definitions share certain key features (Einarsen, 1999; Keashly, 1998; Rayner et al., 1999): a pattern of repeated hostile behaviors over an extended period of time; actual or perceived intent to harm on the part of the actor; one party being unable to defend him- or herself; a power imbalance between the parties. A cursory glance at the general properties of these two constructs reveals some broad commonalties and some differences. In terms of what is shared, it is clear that both conflict and bullying are referring to some form of interaction between two or more parties who are interdependent in some way. Thus, it is not a conflict or bullying if the scene is played out only in someone's mind. From a stress perspective, conflict and bullying as social interactions are similar in that they are examples of negative social stressors stemming from relationships between people (Zapf, 1999). Second, there is negative tone associated with both constructs, albeit of different strengths. In bullying, the reference to hostile behavior and intent to harm is clearly and strongly negative. This global assessment has recently been challenged by Ferris, Zinko, Brouer, Buckley, and Harvey (2007), who suggest that in the hands of a politically skilled leader, bullying can be strategically utilized with low-performing employees and result in positive consequences for the organization, the leader, and the employees. Beyond this specific situation, however, the overwhelming research and writing highlights the negative and harmful aspects and effects of these behaviors. In conflict, the tone is subtler, captured to some extent in the perception of opposition or incompatibility. There is the potential for negativity, but it is not a central defining feature for conflict. Thomas (1992) has indicated that general definitions of conflict permit the opportunity for a variety of nonadversarial strategies to be used by parties to deal with the conflict. Thus, conflict can be a constructive and positive process rather than a destructive process, unlike bullying (De Dreu, 1997; Jehn, 1994, 1995; for a critique of this perspective, see De Dreu, 2007).

In terms of differences, time is clearly central to bullying, but this is not so for conflict. Conflict can be quickly overcome, as when a misunderstanding over a work task is clarified as soon as it arises (i.e., a single episode). Conflict can also be very long-standing (a series of episodes), as indicated by the number of intractable conflicts that exist interpersonally (e.g., a prolonged marital dispute), organizationally (e.g., negative union-management relations), and internationally (e.g., Northern Ireland, the Middle East, Rwanda, and the former Yugoslavia). Bullying is, by definition, long-standing; it is the outcome of a series of episodes. In that sense, bullying dynamics may hold important parallels to intractable conflicts that are noted for their long-standing and recurring nature (e.g., Coleman, 2006; Kriesberg et al., 1989). Another way to think of the element of time is that it provides a sense of a process of development from an initial episode to a series of events (Thomas, 1992). This is a place where the conflict literature can provide useful insight into the development of workplace bullying in terms of conflict stages, escalation, and intractability. We will discuss this more in a later section.

Intent plays an interesting role in both bullying and conflict. Einarsen (1999) specifically notes that bullying involves actual or perceived intent to harm. Intent figures prominently in workplace aggression and abuse research in North America (see Keashly and Jagatic, this volume; Raver and Barling, 2008). Intent is considered a necessary defining element, as it distinguishes these abusive and aggressive interactions from such other forms of harmful behavior as incivility or accidental harm (Andersson and Pearson, 1999; Neuman and Baron, 1997). With regards to conflict more generally, intent is not a defining element; however, perceived intent may figure in the attributions of behavior as the conflict proceeds, thus affecting response (e.g., Fisher, 1990). For example, attributions of intent to harm are associated with the enactment of increasingly provocative behavior that fuels an escalatory spiral (Fisher, 1990; Rubin et al., 1994). Thus, intent (actual or perceived) is a key defining feature of workplace bullying and, indeed, of workplace aggression, but its significance for conflict lies in its ability to explain conflict escalation rather than as a definitional element of conflict itself.

The role of power differences is key to bullying but not essential to defining a situation as conflictual. Certainly, there are conflicts between parties of unequal power such as between child and parent, boss and subordinate, or majority and minority groups. While being unequal or equal in power does not define whether a conflict can be said to exist, it does define whether bullying can be said to exist. Even though power imbalance may not be a key defining feature of conflict, it is very influential in the progress of and response to conflict (e.g., Musser, 1982). We will discuss the influence of power on managing conflict and its implications for bullying in a later section.

Power imbalance is relevant to the ability to defend (Keashly, 2001; Keashly and Jagatic, this volume; Zapf, 1999). The inability to defend is not an inherent feature of conflict. In fact, the idea of interdependence, of being able to interfere with the activities of the other, speaks to the ability of parties to defend themselves and to manifest some degree of influence and power. However, power imbalance does affect the types of responses one can or should use in the face of conflict with another. A feature of conflict escalation that is related to power and one's ability to defend oneself is reciprocity (Andersson and Pearson, 1999; Fisher, 1990). Mutuality or reciprocity is perhaps the key distinction between bullying and conflict as it is generally defined. In conflict, parties mutually engage in exchange of behaviors and are simultaneously actors and targets (Andersson and Pearson, 1999; Glomb, 2002). In the traditional conceptualization of bullying, there is a clear actor who is the instigator, and a target who either cannot respond or can respond only in a limited manner, which does not protect him or her from harm or stop the actor's behaviors. In dispute-related bullying, the bully-victim distinction is initially blurred as such interaction generates from grievances gone wrong. However, as the interaction progresses, a shift in ability to defend results in one party being at the mercy of the other (Keashly, 1998).

Bullying as defined is not a mutually engaged in, reciprocal process like conflict. However, some authors (e.g., Aquino, 2000; Aquino and Lamertz, 2004; Einarsen, 1999; Zapf, 1999; Zapf and Gross, 2001) have argued that the targets or victims can be contributory to the bullying experience. While this argument has to be pursued cautiously because it runs the risk of victim blaming, it is suggestive of the value of examining the extent to which reciprocity characterizes the bullying process from its early stages of "not-yet-bullied"

(Rayner, 1999) to later stages of stigmatization and traumatization (Einarsen, 1999). In the conflict literature, reciprocity also encompasses a notion of mutual impact of these actions. In other words, both parties are affected in conflicts, often negatively. This is particularly the case in the context of escalated conflicts (e.g., Fisher and Keashly, 1990). Glomb (2002) has provided evidence that actors of aggressive behaviors show similar negative effects as the targets of these aggressive actions. Thus, the conflict literature offers the study of workplace bullying a fuller consideration of the contributory activities of, and effects on, both actors and targets, particularly in the development of escalated conflicts (Raver and Barling, 2008). For example, from a conflict escalation perspective, the inability of one party to defend him- or herself portends the figurative and literal disappearance and perhaps destruction of the other party (Fisher and Keashly, 1990). This description certainly maps on to the effects of severe workplace bullying such as psychological withdrawal (reduced organizational commitment and job satisfaction) and actual withdrawal (leaving the job), as well as post-traumatic stress disorder (e.g., Barling, 1996; Hoel et al., 1999; Keashly and Jagatic, this volume). This "effects" connection between escalated and prolonged conflict and bullying highlights the value of discussing the characteristics of escalation processes and the role of various conflict management strategies, both individually and by third parties, that may alter these processes in more constructive ways.

While we have articulated several differences, they appear to be more matters of breadth rather than of qualitative differences. The variety of conflict situations that fall under the general definition of conflict suggests to the observer that bullying as defined is most like intractable, escalating violent conflicts between unequals. As noted, this then suggests that the concepts of individual conflict management strategies, escalation, and intervention approaches would be the most useful to discuss in terms of their implications for the study and amelioration of workplace bullying.

Individual Conflict Management Strategies

Research into conflict strategies has been conceptualized to differ across two dimensions: concern for self (satisfying one's own needs) and concern for others (satisfying the needs of others; Blake and Mouton, 1964; Rahim and Buntzman, 1990). When these two dimensions are crossed, five main strategies are identified:

> *Problem solving*: this strategy represents a high concern for self and the other. Through open exchange of information, common interests are identified to create integrative solutions meeting both parties' needs.

Obliging (accommodating, yielding): this strategy signifies a low concern for self but a high concern for the other by emphasizing commonalties and downplaying differences.

Dominating (competing): representing a high concern for self with low concern for the other. This style focuses on fulfilling one's own interests at the expense of the other.

Avoiding (withdrawing): represents a low concern for self and the other. The objective of this approach is not to acknowledge or engage in the conflict situation.

Compromising: involves an intermediate concern for self and the other. By developing solutions that meet somewhere in the middle, both parties get some, but not complete, satisfaction.

Although helpful in creating a framework to think about conflict behavior, these conflict dimensions fail to predict actual conflict behavior. This is because other salient elements of the conflict situation may limit strategy availability. Two elements of particular relevance to bullying are the type of conflict issue and relative status of the parties.

Type of Conflict Issue

It has been suggested that there are two broad types of issues over which conflicts arise: cognitive and affective (De Dreu, 1997). Cognitive conflicts are conflicts over ideas and tasks. For example, a work group might have a conflict deciding between what strategy to pursue or how to allocate responsibilities. These types of conflicts are not only unavoidable but also often highly fruitful and rejuvenating if managed correctly (see De Dreu and Weingart, 2003 for exceptions). Problem-solving approaches, which allow participants to vigorously debate ideas through strategies that communicate high respect and concern for the other party, are extremely productive in creating new solutions and enhancing relationships (Jehn, 1997).

Affective conflicts, on the other hand, involve issues that threaten one's identity and value system and are often characterized by intense negativity, friction, frustration, and personality clashes. These types of issues are often perceived as nonnegotiables and set the stage for a win-lose interaction. Research has found that affective conflicts are likely to reduce performance and satisfaction, as well as lead to aggressive behavior on the part of one or both parties (Berkowitz, 1989; Jehn, 1997). Thus, these socioemotional conflicts have a high risk for serious damage to both parties (De Dreu, 1997).

Because of the different nature of cognitive and affective conflicts, it is not surprising that they differ in their response to management strategies as well. Indeed, De Dreu (1997) found that although problem-solving strategies were effective in the productive management of cognitive conflicts, they were ineffective and even harmful in managing affective conflicts. Bullying can be construed as a form of relationship based on affective conflict because it involves an ongoing relationship characterized by negative emotions, hostile actions, and threats to identity; so much so that the target's identity is diminished as a result of prolonged exposure (Harlos, 2005). Extending the findings of managing affective conflict to bullying, one sees that problem solving by the target will not be successful in managing bullying. Indeed, Rayner (1999) found that open discussion and information sharing with the bully increased the likelihood of the bully taking retaliatory action against the target.

Relative Status

In addition to the type of conflict issue, the status of the parties in relation to one another will also influence the choice and effectiveness of a given management strategy. In the context of unequal status relationships within organizations, Musser (1982) proposes that bullied subordinates will base their choice of strategy on three criteria: (1) their desire to remain with the organization, (2) the degree of perceived congruence between their attitudes and beliefs and those of the supervisor, and (3) their perceived protection from arbitrary actions by the superior with whom the conflict exists.

Musser's model (1982) is useful when applied to bullying situations because it highlights some of the assumptions inherent in the strategic choice of conflict behavior under conditions of unequal power. In addition, this model further dictates what strategies are available depending on the conditions enumerated earlier. *Problem solving*, although generally perceived to be the most appropriate and effective strategy (Blake and Mouton, 1964), requires that both parties are able to participate openly in a nonhierarchical manner (Filley, 1975). If the subordinate values his or her position and perceives low congruence of beliefs as well as limited organizational protection, it is likely that problem solving will not be a viable option because of potential risks. In support of this prediction, research looking at bullying and harassing behaviors notes that few targets directly confront the actor, often for fear of retaliation (Cortina and Magley, 2003; Hoel et al., 1999; Keashly and Jagatic, this volume; Rayner and Cooper, 2006).

Bargaining or compromising as a strategy option exists only to the extent that the subordinate has leverage in the situation. Since it is quite possible that the superior may control any available power resources, the subordinate may have no leverage to effectively utilize such an approach. *Competing* as a strategic option is inherently risky in any conflict situation because it defines the conflict as a win-lose situation. Given that conflict behavior is often reciprocated in a like manner (Deutsch, 1973), the competitive strategy is more likely to intensify rather than diminish the conflict. To the extent the conflict is occurring in an unequal power relationship, the risks are multiplied for the party of lesser power unless those risks are offset by a low desire to remain with the organization or there is a significantly high degree of congruence between the two parties. Thus, *obliging or yielding* may be the only strategy available to a subordinate who has a strong desire to stay in his or her position but lacks power and/or protection. *Withdrawing* from the relationship either psychologically or physically may become an option as one's desire to remain with the organization is diminished through the conflict. Withdrawal may be characterized by increased apathy or actually terminating employment and results from the perception that there is no chance of winning and costs incurred in staying in the relationship have begun to outweigh any benefits gained from employment. This management strategy is consistent with findings from bullying research that many victims leave their organizations as a result of the bullying experience (e.g., Einarsen et al., this volume; Keashly and Jagatic, this volume; Lutgen-Sandvik, 2006; Zapf and Gross, 2001).

Thus, Musser's model (1982) offers some insight into explaining findings in the bullying literature. For example, Aquino (2000) found, contrary to expectation, that integrating was *positively* correlated with victimization when the individual held a lower power position. Similarly, Richman et al. (2001) found that active coping strategies, such as problem solving, were not only ineffective in stopping harassment but also increased negative personal outcomes for the targets. Zapf and Gross (2001) likewise found that bullied targets' initial attempts at active problem solving were ineffective and were eventually abandoned for other strategies. Cortina and Magley (2003) found that giving "voice" to one's mistreatment placed the individual at risk for both work and social retaliation. The implication of these findings is highly significant for both the field of bullying and that of conflict resolution. Regardless of the fact that problem solving has long been heralded as the right way to manage conflicts, research suggests that such an approach on an unequal playing field is not necessarily the most effective or the most appropriate strategy, and it may actually make things worse.

An Interesting Aside: The Paradox of the Bullying Coworker

In discussions of the centrality of power imbalance to bullying, it has been stated that this imbalance could be present a priori or develop during the course of the interaction (Einarsen, 1999; Zapf, 1999). Einarsen (1999) suggests that bullying may evolve out of a conflict between equals (e.g., coworkers) if one party becomes disadvantaged during the process. Supportive of this bullies-as-equals phenomenon, several studies have found that coworkers are the most frequent source of hostile workplace behaviors (e.g., Cortina et al., 2001; Keashly and Neuman, 2004; Neuman and Baron, 1997; Richman et al., 1999; Schat et al., 2006). To this point, we have examined conflict literature in which unequal power status between disputants existed prior to the conflict. So what does the conflict literature have to offer in understanding how equals become unequal and hence increase the risk of bullying? We will look first from the perspective of the soon-to-be target and then from the perspective of the soon-to-be actor.

Regarding the soon-to-be target, Deutsch (1985) suggests that when an individual experiences a frustration, he or she judges whether the actions involved are normatively or personally unjust. This assessment will influence the individual's sense of capacity to respond. A frustration is assessed as normatively unjust if it is interpreted as a threat against a larger group to which the individual belongs (e.g., negative comments about workers generally). If the frustration cannot be attributed to group threat, it is experienced as directed at the self and is judged as personally unjust. In the case of normatively unjust, the person feels empowered and emboldened by the group reference and, hence, experiences an enhanced sense of a personal ability to respond. Personally unjust actions make the individual more aware of the limitations of his or her own personal power to resolve the issue of frustration. This awareness makes the person feel weaker and less competent to deal directly with the source of the conflict. Behaviorally, when an individual is threatened at this level, his or her information processing is affected such that the person loses his or her repertoire of conflict management strategies and resorts to base levels of behavior even if such actions are less appropriate or effective (Mack et al., 1998). The resultant poor conflict performance puts the soon-to-be target at an increasing disadvantage in the conflict situation, widening the power disparity and leading to possible escalation and/or eventual withdrawal.

In terms of the soon-to-be actor, Papa and Pood (1988; Papa and Canary, 1995) found that in the context of a power imbalance, individuals are less likely to seek out others' views. Since actively seeking to understand the other's position is a prominent feature of productive resolution, lack of understanding of the other (i.e., empathy and perspective taking) on the part of the soon-to-be actor may lead to increasingly aggressive behaviors on his or her part and, hence, escalation. In short, coworkers who become cognizant that they hold the upper hand are less likely to pursue mutually constructive resolutions because of their greater leverage in the situation. Thus, by examining the attribution processes and subsequent choices of management behavior within a disputing dyad, it may be possible to track the development of inequality in resources to respond and, hence, make an assessment of the risk for bullying (Hoel et al., 1999; Raver and Barling, 2008).

This discussion on individual conflict management styles provides a framework for considering some of the influences on the target's and the actor's strategic and behavioral choices that are relevant to examinations of the dynamics of bullying. Unfortunately, this discussion has also revealed that recent evidence indicates none of the interpersonal management strategies available to targets are effective in stopping a bullying situation. For example, Zapf and Gross (2001), in examining the management strategies of bullying victims throughout the process of escalation, concluded that for some victims no strategy, whether active or passive, was successful and that leaving the organization was ultimately the only pragmatic option available. Richman et al. (2001) have argued that the problem may lie in expecting individual strategies to address a structural phenomenon. Given that the dynamic of bullying is largely hinged on someone of greater power acting on someone of lesser power, looking to the lower-power person to be empowered with the ability to remedy the situation is not likely to be fruitful. Thus, we turn our attention to ways in which individuals other than the disputants (bully and target) may influence the development of the interaction.

Conflict Processes: Escalation and Intervention

Bullying has been characterized as a process moving from subtle, low-level aggression (not-yet-bullied), to bullying (direct and intense aggression over a period of time), to stigmatization, and finally to traumatization (diminishment or destruction; e.g., Einarsen, 1999). The hope of many workplace bullying researchers is that there are ways to intervene early enough in this process either to prevent the interaction from reaching the stage of bullying or to stop the bullying, or at least to reduce its effects (Hoel et al., 1999; Rayner, 1999; Zapf, 1999). Earlier, we likened the bullying phase and beyond to an intractable, escalated conflict characterized by violence. To the extent this comparison is appropriate given our earlier caveats, a discussion of conflict escalation and the work linking type of conflict intervention to stage of conflict development will provide some insight into possible ways of dealing with workplace bullying.

Escalation

A fundamental tenet of many conflict theories is that conflict is prone to escalate—that is, to become more intense, hostile, and competitive; to include more issues; to undermine trust; and to involve more powerful attempts at control such as engaging other parties in alliances (Fisher, 1990; Thomas, 1976). A variety of social psychological processes can fuel a conflict's intensity. The processes include elements such as negative and simplified stereotypes, selective perception, self-fulfilling prophecies, negative attributions, communication problems, zero-sum thinking, and overcommitment and mistrust that build on each other—what Deutsch (1973) has characterized as a malignant social process.* In essence, conflicts can take on a life of their

---

* A detailed discussion of the dynamics of escalation is beyond the scope of this chapter. Interested readers are strongly encouraged to read the reviews of Fisher (1990), Rubin et al. (1994), and Thomas (1976) for further detail.

own, spiraling beyond even the parties' control to increasing levels of violence and destruction (Rubin et al., 1994). Even when conflicts can be deescalated to a more peaceful and constructive place, the experience of the conflict has produced fundamental structural changes in the parties, changes we may characterize as "sticky" conflict residues. Unless these residues are specifically recognized and addressed, they will encourage further contentious and hostile responses and inhibit efforts at resolution, generating the conflict anew. Conflict residues map on to the accumulative model of stress as articulated by Hoel et al. (1999) in which over time, the person is fundamentally and irrevocably changed. This is an important concept to consider in identifying what kinds of intervention efforts are needed to alter the conflict situation generally and workplace bullying specifically.

The perspective of conflict escalation has stimulated some interesting theorizing and research into hostile workplace behaviors. Pearson and her colleagues (Andersson and Pearson, 1999) draw on the concept of conflict escalation processes to illustrate how even minor acts of workplace incivility can spiral down into increasingly hostile and violent behavior on the part of coworkers. Glomb (2002), utilizing an escalation framework, has found evidence for the linkage of seemingly low-level verbally aggressive behaviors with increasingly severe physical behaviors. It is important to note that the focus of these researchers was on understanding the development of increasing hostility between equal-power coworkers. Thus, perceptions, orientations, attitudes, and behaviors of both parties were characterized similarly: as mirror images. As noted earlier, given that bullying involves a power imbalance, it cannot be assumed that both parties will behave and perceive in a similar fashion. However, changes in these elements marking increasing escalation can be very helpful in assessing the current state of the parties to the bullying. For example, to determine where bullying in its most extreme form might fall, Zapf and Gross (2001) utilized Glasl's description of behaviors, attitudes, and images at various stages of conflict escalation (1982). They suggest that severe bullying could be classified as a conflict at the boundary between the phase in which the relationship between the parties is severed and dominated by threats and the phase in which destruction of the other becomes paramount.

Given the extreme detrimental effects of escalated conflicts, it is no wonder that many conflict researchers have focused on ways to alter the conflict situation in order to deescalate to some more manageable and less damaging level. To facilitate their thinking regarding what to do and when, many theorists have chosen to package conflict into discrete yet related stages in the escalation process. There are numerous models available regarding stages of conflict escalation (e.g., Fisher and Keashly, 1990; Glasl, 1982; Rubin et al., 1994; Thomas, 1976). Although these models tend to differ in the number of stages utilized, they essentially define each stage by changes in overt behavior, patterns of interaction, perceptions, and attitudes (Fisher, 1990; Glasl, 1982). The move to each stage heralds a new and more pervasive level of intensity.

These stages of escalation form the basis of the contingency approach to conflict intervention (e.g., Fisher and Keashly, 1990; Glasl, 1982; Prein, 1984). The basic premise of the contingency approach is that different management or intervention strategies would be appropriate and effective at different points in time (Fisher, 1990). Indeed, one of the reasons for the failure of particular interventions in particular conflicts may be inappropriate application with respect to the stage of escalation. One of the earliest and most comprehensive models of this genre is Glasl's contingency model (1982), which consists of nine stages of escalation and six types of third-party intervention approaches. Zapf and Gross (2001) have drawn on this model in their discussions of how severe workplace bullying can be conceived as a particularly escalated form of conflict. Building on the formative work of Glasl (1982), Prein (1984), and Kriesberg (1989), Fisher and Keashly (1990) developed a four-stage model highlighting four main types of intervention strategies (see Table 19.1 and Figure 19.1). Although this model was developed in the particular context of international disputes, it is unique in that it goes beyond the idea of matching to suggest that intervention approaches can and should be sequenced and coordinated in order to deescalate and resolve the conflict. An additional reason for failure of some interventions may be the lack of coordinated follow-up interventions to deal with elements not addressed by the initial intervention (Bendersky, 2003; Olson-Buchanan and Boswell, 2008). Because of their focus on coordinated action, we have chosen to describe the Fisher and Keashly (1990) model in more detail.

Table 19.1 details the significant negative changes that occur in parties' communication, perceptions and images of each other and their relationship, overt issues in dispute, perceived possible outcomes, and hence, the

TABLE 19.1

Stages of Conflict Escalation

Dimensions of conflict				
Stage	**Communication / interaction**	**Perceptions / relationship**	**Issues**	**Outcome / Management**
I. Discussion	Discussion / debate	Accurate / trust, respect, commitment	Interests	Joint gain / mutual decision
II. Polarization	Less direct / deeds not words	Stereotypes / other still important	Relationship	Compromise / Negotiation
III. Segregation	Little / direct threats	Good vs. evil / distrust, lack of respect	Basic Needs	Win-lose / defensive competition
IV. Destruction	Nonexistent / direct attacks	Other nonhuman / hopeless	Survival	Lose-lose / destruction

Source: Reprinted from Fisher, R. J. (ed.), *The Social Psychology of Intergroup and International Conflict Resolution* (chap. 9, 211–238). New York: Springer-Verlag. With permission.

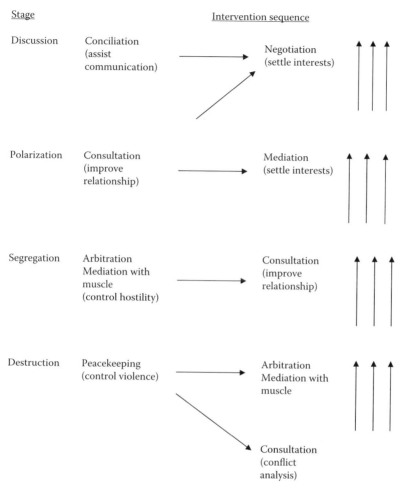

FIGURE 19.1
A contingency model of third-party intervention. (Reprinted from Fisher, R. J. (ed.), *The Social Psychology of Intergroup and International Conflict Resolution* (chap. 9, 211–238). New York. Springer-Verlag. With permission.)

appropriate approach to managing the conflict as the conflict escalates. The stunning part of this downward-spiraling process is how images of the other devolve to something bad, evil, and other than human. The exclusion of the other from the human condition opens up the possibility to treat the other in inhumane ways (Opotow, 2006). Descriptions of severe bullying could be similarly characterized where the actor expresses contempt and disgust for the other and wants to be rid of him or her, either through exit or destruction (e.g., Leymann, 1996; Zapf and Gross, 2001).

Keashly and Fisher (1990) propose that effective efforts at deescalation recognize the need to move the conflict stage by stage rather than attempting to move directly from violence to rational discussion. Our presentation of the model will follow this deescalatory sequence. At the *destructive* stage, the primary intent is to destroy or at least control the other by violence. Third parties function as peacekeepers by forcefully setting norms, defining unacceptable violence, and isolating parties when necessary to keep the violence under control. In the workplace, these types of activities include zero-tolerance policies, moving parties to separate departments, and enforcing behavioral contracts handled through personnel or human resources departments. If the relationship can be stabilized and a commitment to joint effort is made, the way is cleared for using other strategies depending on the parties' receptivity and sense of critical issues (substantive or relational). At the *segregation* stage, competition and hostility predominate; here, the conflict is perceived as threatening basic identity and security needs. Therefore, some immediate form of control is necessary to halt escalation and to demonstrate that agreement is still possible on substantive issues. Thus, either arbitration or power mediation can come into play. Many managers tend to use these latter styles quite readily. Sheppard (1984) describes the "providing impetus" intervention style, where a manager tells employees either to resolve the situation themselves or the manager will resolve it for them. Once hostility is controlled, consultation would be provided to assist parties in examining the dynamics of their conflict and ground rules for improving the relationship toward one of trust and mutual respect. This style of intervention is quite foreign in organizations because of its clear and direct focus on relational issues. However, the introduction of teams as a critical element of organizational functioning may be suggestive of the capacity among some organizational members to deal with these more sensitive yet critical issues. Employee assistance programs as well as process-minded management may be able to provide this expertise, assuming their focus is at the level of the relationship as opposed to the person.

At the *polarization* stage, relationship issues are central, for it is here that trust and respect are threatened and distorted perceptions and stereotypes emerge. At this point, consultation becomes most appropriate because it uniquely deals directly with relationship issues. Once a problem-solving orientation is established, work on the substantive issues can be handled by mediation and, one hopes, the parties can move to negotiating on their own. At the *discussion* stage, the key challenge is to ensure that communication is accurate and perceptions are grounded in reality. When needed, a third party can take a conciliation approach to facilitate clear and open communication on interests so that the parties can begin negotiating directly themselves. In organizations, this type of intervention is reflected in the myriad informal processes and skills that coworkers utilize with each other. For example, the use of summary statements and open-ended questions can often facilitate meetings becoming more productive.

Taking a Contingency Perspective on Addressing Bullying

So what does a contingency approach to conflict intervention have to offer the study and amelioration of workplace bullying? First, it highlights the critical role that other people outside the disputants can play in helping management of the conflict. We know from both the organizational conflict and bullying literatures that these interactions spill over and affect the people around them. So other people do have a stake in these situations being resolved constructively. The challenge is in identifying specific actions that can be taken and then mobilizing and coordinating the efforts of these critical others. In addition to the approaches suggested by Fisher and Keashly (1990), William Ury's book *The Third Side* (2000) identifies a number of roles others can take on to address conflict. His approach can be instructive in the developing capacities of others in the workplace to prevent, manage, and in some cases stop hostile behaviors.

Second, a contingency approach emphasizes the need to thoroughly assess the history and current status of the bullying situation. Only by having a sense of what has occurred and where things are at present will it be possible to select methods of intervening that increase the chances of at least minimizing the damage and, at most, rebuilding the parties, particularly the target, and possibly the working relationship as well as the working environment for others. Although this is a seemingly simple suggestion, it is very difficult to implement as evidenced by the apparent blind eye that organizations often turn toward many conflicts in general and to bullying, specifically (Keashly, 2001; Lutgen-Sandvik, 2006; Zapf and Gross, 2001).

Third, the contingency approach may explain why some intervention and management strategies may fail; that is, the approach may suggest that the strategies were inappropriate for the circumstances. At this point, we would like specifically to address suggestions that have been made that mediation may be the appropriate intervention for bullying (e.g., Hoel et al., 1999). Mediation has certainly gained privileged status among the dispute resolution processes, helping to facilitate solutions to seemingly unsolvable problems. Mediation involves a neutral third party working with the antagonistic parties to help them constructively discuss the issues between them so they can develop a solution that meets both parties' needs. Evidence of mediation effectiveness—in contrast to litigation in a variety of settings (divorce, the environment, the corporate sector, etc.)—has been strong (e.g., Kressel, 2006). However, mediation has been criticized for a number of failings. First, a basic assumption of mediation is that parties to the dispute are sufficiently capable of negotiating with one another as relative equals. This assumption is questionable in situations in which violence is involved, for instance, in domestic abuse cases, sexual and racial harassment, and violent crimes (e.g., Field, 2001; Ver Steegh, 2003). The victims of these behaviors are often diminished and disempowered as a result of the experience, undermining their abilities to be assertive in dealing directly with the actor. As noted earlier in the discussion on individual conflict management strategies, problem-solving approaches often fail to stop affective conflicts and in some cases serve to exacerbate them. Thus, the power imbalance, the inability to defend, and the extensive undermining of the target's personal resources that characterize severe bullying relationships suggest that mediation may not only be inappropriate but that it may also be harmful.

A second criticism concerns mediation's focus on present and future relationships. Specifically, mediation does not address or punish past behaviors. In situations of mutually engaged-in conflicts, this fact is likely agreeable to both parties. However, in the situation of workplace bullying where one person is clearly the victim, failure to redress past behavior may do little to address the bullying target's concerns for justice and recognition of the harm done. As damage to the targets is often cumulative in nature, the extent of the harm could easily be missed or delegitimized if the selected intervention focuses exclusively on a single episode without consideration of the history. In addition, the actor may well favor this approach because it does not require him or her to acknowledge responsibility for his or her actions.

The final criticism of mediation is that it keeps wrongdoings outside public scrutiny. One of the hallmarks of mediation is that it is a private, confidential process. Thus, information shared and decisions reached during the mediation are not available to anyone else. This confidentiality works against the identification of systematic patterns of conflict associated with a particular party, a particular unit within an organization, or across the organization (Rayner, 1999). Discussions of structural influences on conflict make clear that there are organizational ways of handling disputes that may not effectively and constructively result in necessary changes (e.g., Donnellon and Kolb, 1994). Thus, we disagree with Hoel et al.'s suggestion that mediation could help identify bullying patterns in an organization. In fact, it may work to obscure them. This is particularly true if such case-by-case processing is used to cover up systematic or repeated violations. A related implication of using a confidential process such as mediation once bullying has been established is that witnesses in the workplace receive little or no information regarding the dispensation of the situation. This may contribute to their sense that the organization tolerates these behaviors, a sense that would hence reduce their likelihood of reporting bullying situations. Thus, an important source of information to enable an organization to take corrective action may be stifled and employee trust in the organization reduced.

While we are critical of the potential of mediation in bullying situations, the contingency model does suggest its value in the early stages of its development, that is, in the not-yet-bullied stage (Rayner, 1999). At this point, the dynamic has not yet been established and the assumption of equal power is operational. There are challenges even in these early-on situations if the actor is more powerful than the other (e.g., the boss). The conflict literature suggests that the less powerful party is less willing to engage in direct processes such as mediation in these cases for fear of retaliation or the perception that change is unlikely. While mediation has been touted as being able to "level the playing field," the evidence is mixed as to how successful this is (see Mayer, 2004).

The fourth advantage of taking a contingency approach is that it highlights the need to view dealing with bullying as a comprehensive and coordinated effort of a number of different activities and a number of different parties (Olson-Buchanan and Boswell, 2008; Opotow, 2006). It requires the recognition of the need to coordinate short-term crisis management interventions such as separation with longer-term methods directed at fundamentally altering the parties' relationship specifically and the system generally (Coleman, 2006; Fiol et al., 2009). Such coordinated and comprehensive efforts require organizational awareness of bullying and a commitment to dealing with it directly (Hoel et al., 1999). An instructive literature for these efforts regarding bullying would be dispute resolution system design (e.g., Bendersky, 2003; Costantino and Merchant, 1996; Ury et al., 1988). This work speaks to how organizations can develop and implement complementary, multilevel and multifaceted systems for dealing with conflict in all its various forms and stages.

The notion of organizational involvement in dealing systematically with conflict and, by extension, bullying is worthy of further comment. A critical element of the development of such systems is that the organization must examine the way in which its structures and methods of operations may cause, or contribute to, the proliferation of conflict among its members. This critical and reflective process clearly has relevance for workplace bullying. For example, several researchers (e.g., Keashly and Neuman, 2005; Rayner, 1999; Richman et al., 2001) have argued that successfully dealing with bullying requires the recognition of structural culpability in permitting or protecting an actor (bully) through offering them broader resources of power. For example, it could be argued that flatter organizational structures where power is more widely shared would offer more protection against, and less support for, bullying than would rigid, hierarchical structures where power is narrowly channeled. From a conflict perspective, flatter power structures do not reduce the amount of conflict and may in fact increase it (Kabanoff, 1991). The equal power relationships encourage individuals to engage in open communication, resulting in a greater exchange of information and hence the greater possibility for opposing ideas to be expressed. However, such conflicts are cognitive in nature and are likely to facilitate cohesion and solidarity among workers. Hierarchical structures with large power disparities tend to result in more hidden forms of conflict that manifest themselves in more covert and potentially counterproductive ways (Kolb et al., 1992). Since bullying is initially characterized as consisting of highly covert and indirect behaviors, these findings from the conflict literature are supportive of the value of pursuing the connection between organizational structure and bullying and considering the implications for managing bullying situations (Aquino and Thau, 2009).

Finally, the contingency approach and our awareness of the sticky residues of conflict further highlight the limited means of handling severe bullying. This recognition underscores the importance of preventive measures and of addressing the harmful interaction early (the not-yet-bullied stage) before more damage occurs and when there is a chance of (re)building a productive working relationship (Rayner, 1999; Ury, 2000; Zapf and Gross, 2001). Individual skill development, whether communication, anger management, perspective taking, or negotiation, is considered one of the primary preventive dispute resolution efforts that may be relevant for the not-yet-bullied situations (for further information, see Deutsch, 1993). Sheehan (1999) cogently argues for such skill development for managers who use bullying as a way of managing.

A Final Cautionary Note

In an effort to identify places of connection between the conflict literature and workplace bullying, we feel that we have underplayed some critical differences that have less to do with defining or influential elements and more to do with how conflict and bullying are generally perceived. Researchers and practitioners alike have expended considerable effort to promote and support the idea that conflict is inherent in social interaction and thus is common to organizational life. The corollary to this is that conflict in organizations can be a very good thing. When effectively stimulated and managed, conflict can result in improved relationships, greater creativity and innovation, and enhanced task performance. Thus, conflicts are normal and can lead to positive outcomes for all. To suggest that bullying is a conflict without attaching the numerous qualifiers we have noted in this chapter runs the risk of normalizing this form of relationship and potentially providing justification that it makes people and organizations function better (for this perspective, see Ferris et al., 2007). While researchers and practitioners who engage people who have been bullied may not fall victim to this thinking, such thinking is functional for others who may wish to deny the extent and seriousness of bullying, as it minimizes their need to take action or, indeed, their need to desist from such treatment.

Applying the label of conflict wholesale without qualification also creates the sense of shared responsibility for the bullying, and the victim may be expected to manage the situation on his or her own or, in some cases, be held accountable for the hostility exhibited by the other person. As discussed, this can only result in further damage to the target. The notion of shared responsibility is particularly troublesome if certain tenets of principled negotiation theory (e.g., Fisher et al., 1991) are applied to "bullying conflicts." For example, one tenet is to "separate the person from the problem." As an approach to conflict, particularly between equals, this reflects an awareness that frustration over tasks (cognitive conflict) can easily generalize to more personalized and antagonistic frustrations with the other (affective conflict), frustrations that are difficult to resolve. Therefore, disputants are encouraged to focus on the task or substantive issues and not the ways in which the person is handling issues or dealing with them. But in the case of bullying, the person is the problem, and hence the two cannot be separated. Trying to frame the situation otherwise can only be disastrous for the target, who will experience the inability to do so as failure on his or her part, contributing further to the diminishment of the target's sense of competence as a worker and as a person (Keashly, 2001).

For these reasons we caution researchers, practitioners, and organizational members not to describe bullying as a conflict but, instead, to consider what a conflict perspective might offer in understanding and ameliorating this extremely hostile and devastating phenomenon.

References

Andersson, L. M., and Pearson, C. M. (1999). Tit for tat? The spiraling effect of incivility in the workplace. *Academy of Management Review, 24,* 452–471.

Aquino, K. (2000). Structural and individual determinants of workplace victimization: The effects of hierarchical status and conflict management style. *Journal of Management, 26,* 171–193.

Aquino, K., and Lamertz, K. (2004). A relational model of workplace victimization: Social roles and patterns of victimization in dyadic relationships. *Journal of Applied Psychology, 89,* 1023–1034.

Aquino, K., and Thau, S. (2009). Workplace victimization: Aggression from the target's perspective. *Annual Review of Psychology, 60,* 717–741.

Barling, J. (1996). The prediction, psychological experience, and consequences of workplace violence. In G. VandenBos and E. Q. Bulatao (eds.), *Violence on the job: Identifying risks and developing solutions* (pp. 29–50). Washington, DC: American Psychological Association.

Bendersky, C. (2003). Organizational dispute resolution systems: A complementarities model. *Academy of Management Review, 28,* 643–656.

Berkowitz, L. (1989). Frustration, appraisals, and aversively stimulated aggression. *Aggressive Behavior, 14,* 3–11.

Blake, R. R., and Mouton, J. S. (1964). *Managerial grid.* Houston, TX: Gulf.

Coleman, P. T. (2006). Intractable conflict. In M. Deutsch, P. T. Coleman, and E. C. Marcus (eds.), *The handbook of conflict resolution: Theory and practice* (pp. 533–559). San Francisco: Jossey-Bass.

Cortina, L. M., and Magley, V. J. (2003). Raising voice, risking retaliation: Events following interpersonal mistreatment in the workplace. *Journal of Occupational Health Psychology, 8,* 247–265.

Cortina, L. M., Magley, V. J., Williams, J. H., and Langhout, R. D. (2001). Incivility in the workplace: Incidence and impact. *Journal of Occupational Health Psychology, 6*, 64–80.

Costantino, C. A., and Merchant, C. S. (1996). *Designing conflict management systems.* San Francisco: Jossey-Bass.

De Dreu, C. K. W. (1997). Productive conflict: The importance of conflict management and conflict issue. In C. K. W. De Dreu and E. Van de Vliert (eds.), *Using conflict in organizations* (pp. 9–22). Thousand Oaks, CA: Sage.

———— (2007). The virtue and vice of workplace conflict: Food for (pessimistic) thought. *Journal of Organizational Behavior, 29*, 5–18.

De Dreu, C. K. W., and Weingart, L. R. (2003). Task versus relationship conflict, team performance and team member satisfaction: A meta-analysis. *Journal of Applied Psychology, 88*, 741–749.

Deutsch, M. (1973). *The resolution of conflict.* New Haven, CT: Yale University Press.

———— (1985). *Distributive justice: A social-psychological perspective.* New Haven, CT: Yale University Press.

———— (1993). Educating for a peaceful world. *American Psychologist, 48*, 510–517.

Donnellon, A., and Kolb, D. M. (1994). Constructive for whom? The fate of diversity disputes in organizations. *Journal of Social Issues, 50*, 139–155.

Einarsen, S. (1999). The nature and causes of bullying at work. *International Journal of Manpower, 20*, 16–27.

Einarsen, S., and Skogstad, A. (1996). Bullying at work: Epidemiological findings in public and private organizations. *European Journal of Work and Organizational Psychology, 5*, 185–201.

Ferris, G. R., Zinko, R., Brouer, R. L., Buckley, M. R., and Harvey, M. G. (2007). Strategic bullying as a supplementary, balanced perspective on destructive leadership. *Leadership Quarterly, 18*, 195–206.

Field, R. (2001). Convincing the policy makers that mediation is often an inappropriate dispute resolution process for women: A case of being seen but not heard. *National Law Review* (Australia), *1*, 1–19. http://pandora.nla.gov.au/parchive/2001/Z2001-Mar-13/web.nlr.com.au/nlr/HTML/Articles/field/field.htm (retrieved June 29, 2009).

Filley, A. C. (1975). *Interpersonal conflict resolution.* Glenview, IL: Scott Foresman.

Fiol, C. M., Pratt, M. G., and O'Connor, E. J. (2009). Managing intractable identity conflicts. *Academy of Management Review, 34*, 32–55.

Fisher, R., Ury, W., and Patton, B. (1991). *Getting to yes: Negotiating agreement without giving in.* New York: Penguin.

Fisher, R. J. (ed.) (1990). *The social psychology of intergroup and international conflict resolution.* New York: Springer-Verlag.

Fisher, R. J., and Keashly, L. (1990). Third party consultation as a method of intergroup and international conflict resolution. In R. J. Fisher (ed.), *The social psychology of intergroup and international conflict resolution* (pp. 211–238). New York: Springer-Verlag.

Glasl, F. (1982). The process of conflict escalation and roles of third parties. In G. B. J. Bomers and R. B. Peterson (eds.), *Conflict management and industrial relations* (pp. 119–140). Boston: Kluwer-Nijhof.

Glomb, T. M. (2002). Workplace anger and aggression: Informing conceptual models with data from specific encounters. *Journal of Occupational Health Psychology, 7* (1), 20–36.

Harlos, K. (2005). How disputants manage interpersonal conflict at work. Unpublished manuscript, McGill University, Montreal, Canada.

Hoel, H., Rayner, C., and Cooper, C. L. (1999). Workplace bullying. *International Review of Industrial Organizational Psychology, 14*, 195–229.

Jehn, K. (1994). Enhancing effectiveness: An investigation of advantages and dis-advantages of value-based intragroup conflict. *International Journal of Conflict Management, 5*, 223–238.

———— (1995). A multi-method examination of the benefits and detriments of intra-group conflict. *Administrative Science Quarterly, 40*, 256–282.

———— (1997). Affective and cognitive conflict in work groups: Increasing perfor-mance through value-based intragroup conflict. In C. K. W. De Dreu and E. Van de Vliert (eds.), *Using conflict in organizations* (pp. 87–100). Thousand Oaks, CA: Sage.

Kabanoff, B. (1991). Equity, equality, power, and conflict. *Academy of Management Review, 16* (2), 416–441.

Keashly, L. (1998). Emotional abuse in the workplace: Conceptual and empirical issues. *Journal of Emotional Abuse, 1* (1), 85–117.

———— (2001). Interpersonal and systemic aspects of emotional abuse at work: The target's perspective. *Violence and Victims, 16* (3), 233–268.

Keashly, L., and Neuman, J. H. (2004). Bullying in the workplace: Its impact and man-agement. *Employee Rights and Employment Policy Journal, 8* (2), 335–373.

Kolb, D. M., Bartunek, J. M., and Putnam, L. L. (1992). *Hidden conflict in organizations.* Thousand Oaks, CA: Sage.

Kressel, K. (2006). Mediation revisited. In M. Deutsch, P. T. Coleman, and E. C. Marcus (eds.), *The handbook of conflict resolution: Theory and practice* (pp. 726–756). San Francisco: Jossey-Bass.

Kriesberg, L. (1989). Varieties of mediation activities. Paper presented at the annual meeting of the International Society of Political Psychology, June, Tel Aviv, Israel.

Kriesberg, L., Northrup, T. A., and Thorson, S. J. (1989). *Intractable conflicts and their transformation.* Syracuse, NY: Syracuse University Press.

Leymann, H. (1996). The content and development of mobbing. *European Journal of Work and Organizational Psychology, 5* (2), 165–184.

Lutgen-Sandvik, P. (2006). Take this job and . . .: Quitting and other forms of resistance to workplace bullying. *Communication Monographs, 73*, 406–433.

Mack, D., Shannon, C., Quick, J. D., and Quick, J. C. (1998). Stress and the preventive management of workplace violence. In R. W. Griffin, A. O'Leary-Kelly, and J. M. Collins (eds.), *Dysfunctional behaviors in organizations* (pp. 119–141). Greenwich, CT: JAI Press.

Mayer, B. (2004). *Beyond neutrality: Confronting the crisis in conflict resolution.* San Francisco: Jossey-Bass.

Musser, S. J. (1982). A model for predicting the choice of conflict management strate-gies by subordinates in high-stakes conflicts. *Organizational Behavior and Human Performance, 29*, 257–269.

Neuman, J. H., and Baron, R. A. (1997). Aggression in the workplace. In R. A. Giacalone and J. Greenberg (eds.), *Antisocial behavior in organizations* (pp. 37–67). Thousand Oaks, CA: Sage.

Olson-Buchanan, J. B., and Boswell, W. R. (2008). An integrative model of experienc-ing and responding to mistreatment at work. *Academy of Management Review, 33*, 76–96.

Opotow, S. (2006). Aggression and violence. In M. Deutsch, P. T. Coleman, and E. C. Marcus (eds.), *The handbook of conflict resolution: Theory and practice* (pp. 509–532). San Francisco: Jossey-Bass.

Papa, M., and Canary, D. J. (1995). Conflict in organizations: A competence-based approach. In A. M. Nicotera (ed.), *Conflict and organizations: Communication processes* (pp. 153–179). New York: State University of New York Press.

Papa, M., and Pood, E. A. (1988). Co-orientation accuracy and differentiation in the management of conflict. *Communication Research, 15,* 400–425.

Prein, H. (1984). A contingency approach for conflict intervention. *Group and Organization Studies, 9* (1), 81–102.

Putnam, L. L., and Poole, M. S. (1987). Conflict and negotiation. In F. M. Jablin, L. L. Putnam, K. H. Roberts, and L. W. Porter (eds.), *Handbook of organizational communication* (pp. 549–599). Beverly Hills, CA: Sage.

Rahim, M. A., and Buntzman, G. F. (1990) Supervisory power bases, styles of handling conflict with subordinates, and subordinate compliance and satisfaction. *Journal of Psychology, 123,* 195–210.

Raver, J. L., and Barling, J. (2008). Workplace aggression and conflict: Constructs, commonalities, and challenges for future inquiry. In C. K. W. De Dreu and M. J. Gelfand (eds.), *The psychology of conflict and conflict management in organizations* (pp. 211–244). Mahwah, NJ: Lawrence Erlbaum.

Rayner, C. (1999). From research to implementation: Finding leverage for prevention. *International Journal of Manpower, 20,* 28–38.

Rayner, C., and Cooper, C. (2006). Workplace bullying. In E. K. Kelloway, J. Barling, and J. J. Hurrell Jr. (eds.), *Handbook of workplace violence* (pp. 121–146). Thousand Oaks, CA: Sage.

Rayner, C., Sheehan, M., and Barker, M. (1999). Theoretical approaches to the study of bullying at work. *International Journal of Manpower, 20,* 11–15.

Richman, J., Rospenda, K. M., Flaherty, J. A., and Freels, S. (2001). Workplace harassment, active coping and alcohol-related outcomes. *Journal of Substance Abuse, 13* (3), 347–366.

Richman, J., Rospenda, K. M., Nawyn, S. J., Flaherty, J., Fendrich, M., Drum, M. L., and Johnson, T. P. (1999). Sexual harassment and generalized workplace abuse among university employees: Prevalence and mental health correlates. *American Journal of Public Health, 89* (3), 358–363.

Rubin, J. Z., Pruitt, D. G., and Kim, S. H. (1994). *Social conflict: Escalation, stalemate, and settlement,* 2nd ed. New York: McGraw-Hill.

Schat, A. C. H., Frone, M. R., and Kelloway, E. K. (2006). Prevalence of workplace aggression in the U.S. workforce: Findings from a national study. In E. K. Kelloway, J. Barling, and J. J. Hurrell Jr. (eds.), *Handbook of workplace violence* (pp. 47–90). Thousand Oaks, CA: Sage.

Sheehan, M. (1999). Workplace bullying: Responding with some emotional intelligence. *International Journal of Manpower, 20,* 57–69.

Sheppard, B. (1984). Third party conflict intervention: A procedural framework. *Research in Organizational Behavior, 6,* 141–190.

Thomas, K. W. (1976). Conflict and conflict management. In M. D. Dunnette (ed.), *Handbook of industrial psychology* (pp. 889–935). Chicago: Rand-McNally.

——— (1992). Conflict and negotiation processes in organizations. In M. D. Dunnette (ed.), *Handbook of industrial organizational psychology,* 2nd ed., vol. 3 (pp. 651–718). Palo Alto, CA: Consulting Psychologists Press.

Ury, W. L. (2000). *The third side: Why we fight and how we can stop.* New York: Penguin.

Ury, W. L., Brett, J. M., and Goldberg, S. B. (1988). *Getting disputes resolved: Designing systems to cut the costs of conflict.* San Francisco: Jossey-Bass.

Ver Steegh, N. (2003). Yes, no, and maybe: Informed decision making about divorce mediation in the presence of domestic violence. *William & Mary Journal of Women and the Law, 9* (2), 145.

Zapf, D. (1999). Organizational, work group related and personal causes of mobbing/bullying at work. *International Journal of Manpower, 20,* 70–85.

Zapf, D., and Gross, C. (2001). Conflict escalation and coping with workplace bullying: A replication and extension. *European Journal of Work and Organizational Psychology, 10,* 497–522.

20

Challenging Workplace Bullying in the United States: An Activist and Public Communication Approach

Gary Namie, Ruth Namie, and Pamela Lutgen-Sandvik

CONTENTS

Introduction

In this chapter, we discuss the Workplace Bullying Institute's (WBI) multifaceted and 12-year-old campaign to raise awareness and to reverse acceptance of workplace bullying in the United States (see http://workplacebullying. org). The organization has a long history of assistance for bullied workers, legislative advocacy, and collaboration with academics (e.g., Lutgen-Sandvik, Namie, and Namie, 2009; Neuman, 2000; Yamada, 2002, 2008). The chapter describes our efforts directed at three principal constituent groups; persons suffering because of bullying, lawmakers who have the power to mandate worker protections against psychological violence at work, and organizational decision makers responsible for work environments. We report the current state of progress with each group as well as the barriers we continue to face in meeting our main goals. We close with work yet to be done and future directions to continue these U.S. endeavors.

An Applicable Persuasion Theory

One theory of persuasion readily applicable to the U.S. campaign against workplace bullying is the elaboration likelihood model (ELM) (Petty and Cacioppo, 1986). This cognitive process model derives its name from the likelihood that a person thinks deeply (elaborates) about a message when exposed to it. The premise of ELM is that the route by which a message persuades recipients depends on their involvement with the message. Two routes are posited: the central and the peripheral. With central route processing, people have both the motivation (the strength of desire to process the message, the love of cognitive engagement) and the ability to critically evaluate the message.

When people lack the motivation or ability to evaluate the message, they are more likely to respond to cues associated with the message (peripheral route processing), cues such as entertainment value or association with a celebrity spokesperson, rather than to the actual content of the arguments.

Petty and Cacioppo (1986) considered attitudes that are the product of central route processing to be more accessible, persistent, resistant to change, and a better predictor of behavior than when the peripheral route is taken. Conditions that promote high elaboration can also affect the extent to which a person has confidence in, and thus trusts, her or his own thoughts in response to a message (Petty, Briñol, and Tormala, 2002). After cognitive effort is expended, positions and attitudes held serve a self-validating role.

However, high elaboration is difficult to achieve for individuals bullied at work. Targets strained by the stresses of bullying are capable of little more than minimal involvement. Most targets learn initially about bullying on the Internet, on television, or from a newspaper article and thus may be more responsive to peripheral cue complexity. Contemporary website design takes this response into account—moving images, multiple columns, colors, embedded videos, graphics—to pique the attention of minimally involved web browsers. For example, the website designer for the Workplace Bullying Institute (WBI), an organization discussed in detail later in this chapter, changed the site design from its original voluminous, barely navigable format to a newer one with keen attention to peripheral details to facilitate target searches for answers to fundamental queries.

The ELM information complexity variable is also related to the phase of the bullying episode when the help-seeking visitor discovers the website. In the beginning of bullying episodes, targeted workers are consumed with stabilizing and sense-making tasks to cope with the uninvited assault that disrupted their psychological comfort (Lutgen-Sandvik, 2008). Bombardment with information (central route processing in the ELM model) during acute phases is ineffective. Next, targets begin to respond to the trauma and stigma attached to bullying by neutralizing and countering accusations purported by the bully. Repairing one's reputation comes next, as shame is gradually reversed. In the postbullying phase, when targets are no longer vulnerable to bullying, grieving over the losses (e.g., belief in justice) and major life and career restructuring take precedence. At this point, targets may be able to incorporate information necessary for recovery.

The lesson for communicating effectively to bullied targets is that when they are able to be involved—that is, when they are calm enough to digest more than a few paragraphs—and sufficiently motivated to understand the complexity of their bullying problem, comprehensive, substantive resources should be available for them. Designers of websites on bullying have to consider the different phases through which bullied targets pass in order to optimize the utility of their sites for emotional visitors who demand immediacy as well as for visitors capable of contemplative, in-depth information processing.

For several reasons, a majority of U.S. lawmakers also have difficulty incorporating the message that a law against bullying at work should be enacted. Applying ELM theory to their receptivity, we conclude that few legislators are sufficiently motivated. A lawmaker's likely motivation to advance workers' rights is blocked by a countercampaign to protect and enlarge employers' rights by business groups, who outspend labor activists by a 40:1 ratio in election campaign contributions. Further, the ability of lawmakers to attend to the details of the persuasive arguments in favor of antibullying legislation is undermined by their hectic schedules during short legislative seasons (varying from 60 to 180 days per year) in most states. Few legislators have time to study any issue in depth.

Lawmakers are swayed more by vivid, televised tales of egregious crimes for which laws are hastily crafted. Bullying stories do not make news in the United States. Therefore, when lobbying for legislation, we are careful to devote most face-to-face meeting time to emotionally charged individual tales that enhance attention through peripheral cue complexity. Prevalence statistics and reports are left behind with law-makers for their subsequent perusal (and, one hopes, for elaboration and incorporation).

For employers, both motivation and ability to address workplace bullying in the United States are lacking. When bullying is reported, 44% of employers do nothing and 18% worsen the situation for the targeted worker (Namie, 2007).

A Bullying-Tolerant Society

A commonly accepted societal explanation for employer indifference is the preference for individualistic, aggressive, and abusive responses to interpersonal conflict. It is normative when all types of interpersonal mistreatment are rationalized as necessary because "it's just business," as if there were no personal consequences for the actions taken. For instance, Levitt (2009) wrote for a financial sector publication, "In a competitive environment, an assertive and 'take charge' style is usually rewarded. If a manager exhorts and pushes subordinates to perform . . . those people who are laconic by nature, may view the exhortations as bullying." From this perspective, bullied workers are evidently the rude, discourteous, and unsuccessful ones.

A Tennessee appellate court decision stated in a 2007 case that without proof of discrimination, "the fact that a supervisor is mean, hard to get along with, overbearing, belligerent or otherwise hostile and abusive does not violate civil rights statutes" (*Frye v. St. Thomas Health Services*, 227 S.W.3d 595, as cited in Davis, 2008). The decision implies that anything goes if the conduct is not explicitly illegal. Hence, an explicit law against bullying is clearly needed.

Corporate employment law attorneys frequently defend bullying perpetrators in cases and are their best apologists. Mathiason and Savage (2008) told a revealing story about a bully in their own law office. "Clearly there is a type of abusive treatment that exceeds the standards of our firm . . . conduct that destroys teamwork and office morale . . . [but] we do accept and value an individual teaching style that is very demanding of new associates." In other words, abuse is an allowable "style" difference.

In short, American employers exert unchallenged control over most work conditions with only 7.5% of the nongovernmental workforce represented by a union. American workers are "at will" employees facing immediate termination without a just-cause requirement. Challenging bullying in the United States defies societal norms. Bullying is not yet taboo. It is an acceptable operational tactic in the corporate world. Workers dare not complain. This is the context of unbalanced employer–employee power facing the U.S. campaign against workplace bullying.

Despite the hurdles, we have enjoyed modest success with goal attainment. We next report progress in the U.S. campaign with respect to each of the three involved constituent groups: the general public, lawmakers, and employers.

Group 1: The General Public

The benefits of an informed public are twofold. First, familiarity with the topic of bullying helps remove its stigma. Second, people will feel empowered to challenge the current acceptance of bullying.

Starting the Movement

We begin with a traumatic bullying experience that affected the authors of this chapter. Dr. Ruth Namie's tale was the inaugural story for the movement. Her mistreatment came at the hands of a fellow woman professional in a psychiatry clinic. Approximately one year after resolution of the case, we discovered the European term *workplace bullying*. Then, in 1997, we started WBI, the Workplace Bullying Institute, which had its origins in the Campaign Against Workplace Bullying, to help individuals.

Originally, WBI provided three paths for bullied individuals to find support: (1) a toll-free telephone crisis line; (2) a dedicated website including a collection of articles about the phenomenon, online surveys, and research findings; and (3) a self-help book published one year after our start (now Namie and Namie, 2009a).

In January 2000, we staged the first U.S. workplace bullying conference in Oakland, California. It was an unfunded two-day event. Many of the international speakers and presenters who graciously attended at their own expense are authors of several chapters in this book: Michael Sheehan, Charlotte Rayner, David Yamada, and Loraleigh Keashly. In September 2000, Suffolk University Law School hosted a second conference in Boston that focused on the legal challenges facing the workplace bullying movement.

The crisis line was publicized first in two national newspapers. We coached over 5,500 emotionally wounded people one hour at a time in six years. We learned that it is important to establish limits for telephone counselors because the risk of vicarious trauma is high. We had to stop the inordinately expensive service. Subsequently, we charged a modest fee for coaching that, over time, reduced the number of callers.

The founders of WBI brought to the movement prior academic preparation in social and clinical psychology; experience in behavioral research methodology, survey design, and statistical analyses; experience in family systems therapy, chemical dependency, and domestic violence treatment; years of university teaching in management and psychology; and business consulting and corporate management experience. David Yamada's legal expertise (see his chapter, this volume) complemented the work soon after the organization began. Future advocacy groups should not rely solely on veterans of the bullying wars, however. Expertise is needed from individuals who have not personally experienced bullying. These experts can learn about all aspects of bullying and are less likely than bullying victims to be adversely affected from working with, and on behalf of, traumatized individuals.

Website visitors expect information to be *free*. Bullied workers often lose their jobs (Namie, 2007) and cannot afford to pay for necessary legal or mental health services. Groups desiring to emulate our nonprofit organization's commitment to helping bullied workers are therefore advised to secure funding to sustain their efforts. Whereas, encouragingly, consulting and training services for employers and fees for professional speeches support WBI's work, in 2009 the United Kingdom's pioneering organization Andrea Adams Trust closed its charitable operation after 15 years because of a lack of funding.

The Media as Communication Partners

Thanks to more than 800 media interviews and appearances, workplace bullying in the United States is now publicly recognized. Through a wide distribution of national and regional television, radio, and newspapers and magazines, WBI is able to reach Americans at no cost. The Internet also helps carry the message that workplace bullying is a common, unconscionable, but legal form of mistreatment.

Workplace bullying has begun to take its rightful place among such better-known topics as domestic violence, post-traumatic stress disorder (PTSD), and other forms of trauma and abuse in the United States. A typical media story begins with the "human interest" angle. For instance, a targeted worker (prescreened by us to ensure psychological stability and referred to the reporter) describes her or his bullying experience. The story is then traditionally shortened to three to five on-air minutes or short paragraphs in print. In the early years, stories had a mostly anecdotal focus.

Media necessarily provide a narrow focus. In recent years, for example, the media love a woman-on-woman bullying story (Meece, 2009) to the exclusion of covering other forms of bullying. However, in the United States, only 29% of all bullying is between a woman perpetrator and a woman target; men, in fact, represent 60% of the bullies (Namie, 2007). Nonetheless, these newspaper articles prompt 300–500 reader comments per article, and televised segments on woman-on-woman bullying garner high ratings.

The Bully Boss

The American public, if not the business media, seems ready for candor about destructive people who make work life a living hell for others. An example is the bestseller *The No Asshole Rule*, a book related to bullying written by Stanford Business School professor Robert Sutton (2007). The public has embraced its frankness and simplicity. It provides a cathartic venting of pent-up frustrations with bullies.

Business media frequently cite the statistic that 72% of bullies outrank their targets (Namie, 2007). Thus, the alliterative stereotype of "bully boss" is an accurate headline. Of course, bullying originates with, and affects, individuals at most organizational levels. Executives experience the least amount of bullying (5%). The portrayal of exploitation by bullying is more vivid when it is managerial rather than internecine to the work team. Thus, the media spotlight is on the quirky or aberrant boss as an individual (without interviewing actual perpetrators) and is absent of reportage on the work environment that sustains that boss. Questions to WBI about what individuals can do when faced with a bully boss outnumber questions about why and how employers should deal with systemic bullying. The burden for finding a solution tends to fall on the victimized target. When media experts are management consultants or executive coaches, they give poor advice to workers to subordinate themselves, to attempt no change to the toxic work environment that fosters bullying.

Some business reporters doubt the targeted workers' accounts of their bullying. A few television interviews of bullied individuals did not air because producers were reluctant to believe the target's account or a lawsuit was threatened. The bullying victim's account was said to be "only one side" of the story. Seeking balance, the producers wanted to hear and see the bully's side of the story. Understandably, the bullies were reluctant to come forward with a public admission of wrongdoing.

Research Bolsters the Message

Since 2000, we were able to supplement anecdotal tales with empirical study data. The institute conducted descriptive large-sample surveys of website visitors (n = 1,335, Namie, 2000; and n = 1,000, Namie, 2003). The self-selected sample studies were not extrapolated to describe national trends or national prevalence. However, several metrics did approximate estimates from the large representative study WBI conducted later (Namie, 2007).

The first credible estimate for U.S. bullying prevalence was one in six Michigan workers (Keashly, 2001). The study's sampling techniques afforded external validity. But there were only approximately 100 individuals who reported "very bothersome" mistreatment. This estimate was the best one available until 2007.

In 2007, WBI, with support from the Waitt Institute for Violence Prevention, commissioned the polling firm Zogby International to conduct the first U.S. survey of workplace bullying. The stratified sample was large enough (n = 7,740) to represent the experiences of *all* adult Americans. The 20-item survey (Namie, 2007) used the WBI definition of bullying without explicit inclusion of the term *bullying*. Instead, the term was replaced with a descriptor: "repeated mistreatment: sabotage by others that prevented work from getting done, verbal abuse, threatening conduct, intimidation, or humiliation."

The WBI-Zogby survey found 12.6% of U.S. workers were either being bullied currently or had been within the year, 24.2% were previously but not currently bullied, 12.3% witnessed bullying but never experienced it, and 44.9% of respondents reported never witnessing and never experiencing it. Of 7,740 survey respondents, only 22 people admitted being a perpetrator, despite the anonymity granted by the survey (Namie, 2007).

Thereafter, media quoted the finding that 37% of the population, representing 54 million Americans, has been bullied. The media took a keen interest in the finding that women bullies choose women as targets in 71% of cases. Men bullies choose women targets (46%) less frequently than they target men. Women are the slight majority of targeted individuals (57%).

The WBI-Zogby Survey also demonstrated that the experiences of bullied workers was similarly reported by witnesses. There was a great deal of congruence, confirming that the experiences are not imaginary (Namie and Lutgen-Sandvik, 2010).

It is common in the United States to blame victims for their fate. This denigration is an example of the fundamental attribution error committed by observers (Ross, 1977). However, targets themselves *underestimate* the negativity of their situation and often blame themselves for their fate. The mischaracterization of targets as whiners or complainers is unwarranted. We know from our own WBI anecdotal data that targets are reluctant to characterize themselves as victims. Lutgen-Sandvik, Tracy, and Alberts (2007) discovered a disparity between the researcher-defined prevalence of bullying based on an operational definition (28%) and the survey respondents' self-identification as a bullied person (9.4%). This result was true for a group of Americans as well for a Danish sample group in the same study (see also Nielsen et al., this volume).

Framing the Message

Commercial media reflect the values of American business culture as seen from the top rather than as lived by subordinate workers. It will be interesting to see if the credibility of chief executive officers, or CEOs, diminishes in light of the global economic crisis that is partly blamed on CEO failures. Any anti-CEO sentiment during tough times presents the opening for populist stories about the plight of trapped workers who face a nearly certain escalation of cruelty because few employment alternatives exist. Bullying cannot exist without tacit approval from executives and owners.

In 2009, WBI surveyed 400 respondents asking whether bullying escalated after the recognized start date of the worldwide economic recession in September 2008. For 27.5% of the respondents, the bullying became "more abusive/severe/frequent"; 67% reported no change; and 3.4% reported a decrease in bullying since the onset of recessionary times (Namie, 2009).

Workplace bullying activists often characterize the workplace bullying movement as "antiabuse," whereas defenders of individual bullies and the practice of systemic bullying describe the movement as "anticorporate." The pejorative mischaracterization makes the activists' public education goals harder to accomplish. Activists need to emphasize that bullying hurts business in addition to hurting people.

ELM and the Media

Commercial television is the ultimate forum for persuasive appeals employing peripheral cues, according to ELM (Petty and Cacioppo, 1986). Soap is not sold by listing ingredients, which would require central route processing by viewers. Instead, it is sold as an indispensable route to a desirable lifestyle with distracting emotionally evocative images. News stories are treated likewise, becoming "infotainment" as newscasters and newsmakers respond to pressure to make news, including bullying stories, entertaining.

A low-involvement TV viewer is unlikely to remember anything about stories and their associated content. There is competition among the several visual frames on a TV screen—station logos, borders with colorful motion backgrounds, and text crawls with additional headlines. Just as problematic, print media have limited space, favoring short 500–700-word accounts rather than lengthier in-depth stories.

Bullying is a complex phenomenon with multiple aspects. To accommodate the media trends, as noted earlier, and to make stories memorable, we simplify advice for bullied targets and reasons for bullying and we adopt slogans. We use Bullies Are Too Expensive to Keep; Work Shouldn't Hurt; and Good Employers Purge Bullies, Bad Ones Promote 'Em. Academic activists would benefit from media training, for it is through the media one can reach the people who stand to benefit most from knowledge about bullying.

Group 2: Educating Lawmakers

Rationale for a Law

All social movements that have sought to stop psychological violence—from child abuse to domestic violence, discriminatory harassment (gender, race, etc.), and schoolyard bullying—have been able to eventually convince legislators to pass state or federal legislation to negatively sanction misconduct. Although these types of mistreatment continue, laws compel negative consequences for offenders. The workplace bullying phenomenon most closely resembles domestic violence (Janoff-Bulman, 2002) with respect to the interaction between abuser and the abused, witnesses' nonintervention, and societal-institutional denial and rationalizations to excuse it. For legal purposes, however, bullying falls under the rubric of employment law, akin to antidiscrimination laws for the workplace.

Regarding employment law, existing civil rights laws compel employers to create policies to prevent future occurrences. In addition, they must have procedures in place to correct discrimination once reported, investigated, and confirmed. If there were no laws in effect, would employers voluntarily stop the mistreatment of women workers with internal procedures? Evidence suggests that they did not do so before the Civil Rights Act of the 1960s. After enactment of laws, employers took steps to comply. The sequence is clear. Laws drive internal policies. Enforcement of those policies is most likely when there exists a threat of punishment for negligent employers. Credible policy enforcement results in prevention and correction. The power of a law derives from employers' internal preventive actions that protect workers.

Perusal of Suffolk University law professor David Yamada's chapter in this volume reveals that as of 2009, there are no state or federal laws to satisfactorily address workplace bullying in the United States. Therefore, bullying is nearly always legal.

The Antibullying Healthy Workplace Bill

In 2000, David Yamada wrote the text for the original Healthy Workplace Bill (HWB). The bill addresses workplace bullying by prohibiting an "abusive work environment." The proposed legislation does not mandate employer actions. It gives employers multiple opportunities to escape liability for a bully's abusive conduct. The requirements to file a lawsuit using this bill are strict. Evidence of malice is required in addition to documented physical or psychological health harm. There is no government intervention or enforcement. Individual plaintiffs must find and pay for private legal counsel. Though the HWB provides redress for people where current laws do not, its ultimate purpose is to convince employers to stop bullying proactively.

The Legislative Campaign

In 2001, WBI expanded its efforts by adding a separate division. The Workplace Bullying Institute–Legislative Campaign (WBI-LC) goal is to enact state laws. It was decided from the outset to focus on the 50 states rather than to seek a federal law with significantly different features. Congress and presidential administrations in the last 30 years have not expanded labor rights. So, with the help of a network of volunteer state coordinators, the WBI-LC aims to mobilize citizen lobbyists in the states. To date, 28 of the 50 states are represented by at least one coordinator.

In 2003, after two years of lobbying by volunteers, California became the first state to introduce the HWB. To date, 17 states through 200 state legislators have introduced 55 bills representing some variation of the HWB. No state has yet passed any bill into law. The HWB website (www.healthywork placebill.org) is the repository of the bill's history and current activity.

Unpaid WBI-LC State Coordinators compete with professional advocates for employers' attention. To date, coordinators include attorneys, a physician, mental health professionals, professors, nurses, teachers, social workers, community organizers, and advocates who have worked for other social causes. The WBI-LC provides coordinators with all necessary materials to customize a lobbying campaign, an information kit for their state legislators, a private listserv, a private website, copies of the HWB, training tapes, and periodic teleconferences for the group to stay current. Whenever possible WBI leaders give expert testimony at public hearings for HWB. It is a collaborative, creative group that grows in size and effectiveness every year.

On the public HWB website, citizen lobbyists from all states willing to support the bill can volunteer. Coordinators then work with those volunteers to mount writing, telephoning, and e-mailing lobbying campaigns. Coordinators orchestrate one or two in-person lobbying days at their respective state capitals. Some coordinators have formed in-person groups and maintain state-specific websites in addition to ongoing virtual communication with volunteers in their state.

When organizing a group of activists such as the Legislative Campaign Coordinators, it is important to screen members for emotional and behavioral disturbances which may or may not be linked to a personal bullying experience. Experience is valuable, but lobbyists must represent the thousands or millions of bullied workers in their state or province. They cannot use the lobbying platform to tell their personal stories or to vent to lawmakers. We incorporate a rule that state coordinators must be at least two years postbullying to participate. Also, with a group of veterans of bullying, some of whom suffer periodic retraumatization, there is a risk of group dysfunction from emotional flare-ups. It is helpful to establish an intragroup code of conduct to prevent bullying from within.

HWB Supporters

Bullying at work ignores political party affiliation. Targeted workers have not reported personal politics as a reason for being targeted. The HWB is nonpartisan. Sponsors of the HWB include members of both major political parties—Democrats and Republicans. However, in the U.S. survey, Democrats were more likely than Republicans to report direct and witnessed bullying (Namie, 2007).

Coordinators solicit support and endorsements for the HWB from local and state groups. Unions for state government workers, teachers, and nurses have backed the bill. Endorsements have also come from women's groups. In addition, the Illinois Association of Minorities in Government is the sponsor for the first Illinois bill.

The HWB enjoys the support of one national group—the National Association for the Advancement of Colored People (NAACP), the largest U.S. advocacy organization for the rights of African Americans. According to the WBI-Zogby survey, 91% of African Americans want additional workplace protections to supplement existing antidiscrimination laws. Data show that Hispanics suffer the highest rate of ever being bullied, with African Americans second highest (Namie, 2007).

HWB Opponents

Membership in industry trade associations gives employers access to professional lobbyists who oppose the HWB. Opposition is based on one or more of these grounds: (1) in times of economic crises, businesses should not be regulated, for government's only role is to help business operate freely and profitably; (2) employers can control bullying voluntarily, so let them alone and they will do what is best for their business; (3) whining employees will file frivolous, baseless, expensive-to-defend lawsuits that will only clog the courts; (4) current laws provide sufficient protections; and (5) bullying or abusive conduct cannot be precisely defined, for it is too subjective.

The WBI-LC counters with the following reasonable propositions. (1) Business leaders' decisions led to the recent global financial calamity; it is reasonable to question their judgment. (2) Employers have the chance to voluntarily stop bullying whenever they become aware of it. They historically respond inappropriately. (3) Financial and emotional hurdles to file private lawsuits overwhelm aggrieved workers. The reality is that only 3% of mistreated employees file a lawsuit in the United States (Namie, 2007). On the other hand, employers routinely carry employment practices liability insurance to provide legal defense in the event of a harassment or misconduct lawsuit. The HWB provides sufficient affirmative defenses for good employers who take steps to prevent bullying. (4) Law professor David Yamada concludes that current U.S. laws are inadequate. We trust his legal expertise. (5) Prior to the 2007 WBI-Zogby survey, lobbyists for employers argued that bullying did not exist in the workplace. Since the survey is indisputable, they now complain that bullying cannot be precisely defined. The Healthy Workplace Bill requires that the plaintiff's health harm from malicious conduct be proven. The high standard rebuts the subjectivity objection.

The fundamental question about legal reform for bullying is whether it will take a law to compel compliance or employers will voluntarily choose to abandon abuse as routine practice. The nascent intolerance of the assault on an employee's dignity at work in the United States may force an answer.

Persuasion Theory and Lawmakers

For HWB sponsors, bullying is not an abstraction. Those lawmakers who support the HWB agree to champion it because family members, legislative aides, or they themselves have been bullied. For the sake of others, they want bullying to stop. For early adopting lawmakers, the introduction of their bill is personal. Facilitating the personal connection to bullying spells the difference between successful and failed lobbying efforts.

Given the elaboration likelihood model, ELM (Petty and Cacioppo, 1986), one might expect that the lawmaking process is deliberate, based on facts and reasoning, and message content dependent. That is, lawmaking should tap central route processing with reduced susceptibility to peripheral cues. Marshaling facts to support your position is the underpinning of amateur citizen lobbying. Indeed, WBI-LC coordinators refer constantly to the scientific U.S. survey showing that 13% of workers are currently bullied, with an additional 24% having been bullied at some time in their careers (Namie, 2007). Our use of the survey marked a sea change in lawmakers' reactions to workplace bullying. They stopped denying that bullying happens. Credible survey results are thus an essential tool for communicating with public policy makers. So, we have facts on our side and also use the power of compelling anecdotal tales told by bullied individuals (peripheral cues).

Opponents to HWB maintain year-round contact with lawmakers. Facing multiple lobbyists, lawmakers hear the rationale for employer opposition to our bill repeatedly from different sources. Because of the ongoing presence of full-time paid lobbyists (who donate money to election campaigns), opposing arguments are made more memorable. It is not surprising that no state has yet passed our bill into law.

In contrast, WBI-LC Coordinators act primarily during the legislative season and work their regular jobs the remainder of the year. The WBI-LC does not give money to lawmakers.

To augment coordinators' efforts, the WBI-LC has begun to form coalitions of supporting and endorsing groups that do have full-time lobbyists advocating for labor and human rights. This is a relatively new strategy and has yet to yield clear results.

In 2010, the campaign added a significant element. We proposed a version of the HWB for consideration by federal lawmakers. The bill, if enacted, cannot supersede the autonomy of states, but would provide protection for employees of the federal government who would not be eligible for state law protections.

In addition, we have asked that federal lawmakers commission a scientific survey of the prevalence of workplace bullying among federal employees. Results are expected in 2011.

Group 3: Convincing Employers

Employers are the third constituent group in the campaign to stop workplace bullying. They, more than bullied individuals or lawmakers, have the leverage to provide safety for millions of employees under their control.

Employers determine the size and composition of the workforce, the workplace culture, and every aspect of the work environment. The responsibility for the correction and prevention of bullying lies with the top management because senior-level administrators and managers shape the culture of the organization through decisions made (Liefooghe and Davey, 2001). European empirical studies have established an association between leadership, or its absence, and workplace bullying. For example, Leymann (1996) and Einarsen, Raknes, and Matthiesen (1994) found that bullying among colleagues was often associated with "weak" or "inadequate" leadership by the most senior managers. Similarly, Hoel and Cooper (2000) showed that bullying was associated with high scores on a laissez-faire style of leadership. A lack of organizational coherence (integrated, functioning production procedures), only token accountability (few consequences for wrongdoing), and low security (apprehension about layoffs) all combine to foster a chaotic workplace climate that gives opportunistic abusers of authority the chance to harm others (Hodson, Roscigno, and Lopez, 2006). Conversely, Cortina, Magley, Williams, and Langhout (2001) found that in a workplace climate in which fair, respectful treatment prevailed, bullying was rare.

The "Business Case" for Bullying

It is natural to assume that if a rational "business case" argument is made detailing the financial impact of bullying, then rational employers will pursue self-interest and stop costly bullying. Because of employers' costs associated with bullying—productivity loss, costs regarding interventions by third parties, high turnover, increased sick leave, workers compensation and disability insurance claims, and legal liability—employers should logically be motivated to stop bullying (Hoel and Einarsen, 2009). One health care industry intervention that improved employee perceptions of trust and fair treatment was estimated to potentially save $1.2 million annually for a single organization (Keashly and Neuman, 2004).

Logic suggests termination of costly offenders. In spite of ascertainable loss patterns, offenders are retained while targeted workers who report the mistreatment are the ones who lose their jobs. Alleged offenders were punished in less than 2% of cases (Namie, 2009). But because of bullying, 40% of targets quit, 24% are terminated, and 13% transfer to safer positions with the same employer (Namie, 2007).

To whom should the business case be made? Bullying is typically perceived as a human resources (HR) department problem because HR receives the majority of complaints. In the United States, 80% of those complaints do not require employers to respond; they are legal actions (Namie, 2007). Often, HR representatives fail to act because they are not legally compelled to do so. In addition, HR often lacks the credibility with executives to autonomously effect organizational changes.

Bullying is the responsibility of executive leadership (Einarsen, Raknes, and Matthiesen, 1994). To the contrary, many executives support bullies within their organizations. According to Namie (2007), sources of a bully's support are executive sponsors (43%), management peers (33%), and HR (14%). So, rather than prevent financial losses attributable to bullies, executives frequently support the offenders.

Another persuasion theory, social judgment theory (Sherif and Sherif, 1968), posits that opinions (e.g., about a bully) linked to a person's (e.g., an executive's) self-identity are unlikely to change. The executive's allegiance to the bully is derived from a high degree of personal ego involvement because the Machiavellian bully massaged that ego over a protracted period of time (Paulhus and Williams, 2002). The executive can easily reject evidence about the bully's negative conduct that disconfirms a strongly held positive attitude about the bully. Thus, the fact-based, rational business case pales in comparison.

U.S. Employers' Reactions to Bullying

When bullying incidents are reported to employers, the most frequent response (in 44% of cases) is to do nothing (Namie, 2007). A more complete description of employer responses comes from a small (n = 400) online survey of bullied individuals (Namie, 2008). Respondents reported that employers did nothing to stop the reported mistreatment (53%), and they predominantly retaliated against the person who dared to report it (71%). In 40% of cases, targets considered the employer's investigation to be inadequate or unfair, with less than 2% of investigations described as fair and safe for the bullied person. Filing complaints led to retaliation resulting in lost jobs (24%). Alleged offenders were punished in only 6.2% of cases.

A National Institute for Occupational Safety and Health (NIOSH) research team (Grubb, Roberts, Grosch, and Brightwell, 2004) assessed employers' perceptions about the prevalence of bullying within their own organizations. Researchers used a pair of nationally representative government surveys—one of U.S. residents and another sample of organizations. Researchers identified one contact person for each of 516 organizations. The majority (75.5%) of employer representatives—typically HR or owners—said bullying never happened at their site. Only 1.6% said it happened frequently. The second most frequent response was that it was rare (17.4%), with 5.5% acknowledging that bullying happened sometimes. Employees were seen as the most frequent aggressor (in 39.2% of cases), as well as being the most frequent victim (55.2%). Two assessed measures of workplace climate were associated with increased levels of bullying—lack of job security and lack of trust in management (Grubb et al., 2004).

Remarkably, in Sweden, where the regulatory ordinance has been in effect 15 years, only one of out of nine businesses had voluntarily implemented policies and procedures against bullying (Hoel and Einarsen, 2009). The lack of employer initiative in the Scandinavian antibullying pioneering nations suggests modest expectations about American employers' attitudes toward bullying, even if laws are passed.

Not only do employers do very little to stop bullying, but also coworkers who witness bullying are similarly ineffective. From an online study, we know that self-identified bullied individuals reported in 46% of bullying cases that coworkers abandoned them and 15% aggressed against them along with the bully (Namie, 2008). Coworkers did nothing in 16% of cases. In less than 1% of cases, coworkers rallied to the defense of an attacked target and confronted the bully as a group. There are several potential explanations for these ostensibly heartless coworker reactions that are explored elsewhere in detail (Namie and Namie, 2009a). Suffice it to say that most of the time fear, real or imagined, prevents coworkers from getting involved.

Employers are reluctant to allow practitioners to assess preintervention levels of bullying in their workplaces, for they assume that the findings will become public and perceived as pejorative. Estimates of employer perceptions are limited to the perceptions of targeted individuals (Namie, 2007, 2008) and lone employer representatives (Grubb et al., 2004). The one American exception is the large-scale, multiyear project within the Veterans Administration (Keashley and Neuman, 1994). However, that intervention was broader in scope than workplace bullying. Therefore, there is a paucity of empirical, organization-based studies in the United States. Instead, we rely most on our experience as consultant-practitioners.

The Motivation to Act

Because there is no law to compel U.S. employers to act, when an American employer requests help with bullying, it is a rare event. The Workplace Bullying Institute's principals were consultants to employers years before starting the nonprofit organization discussed in this chapter. Since 1998, the consulting focus has exclusively been the refinement of a comprehensive, proprietary approach to prevent and correct workplace bullying (Namie and Namie, 2009b; see also http://workdoctor.com).

Here is a sampling of positive, proactive reasons employers voluntarily address bullying based on our American clientele. Some are early adopters wanting to be first in their industry. Some clients seek congruence with espoused organization values of respect and dignity for all, to "do the right thing." Whereas mission statements do not hold organizations accountable, policies can. Some clients seek media coverage and notoriety for their willingness to address bullying. Some CEOs want to leave a positive legacy at the end of their careers.

In 2009, the Sioux City, Iowa, public school district implemented a comprehensive antibullying system for teachers and staff in the schools, thereby becoming the first in the nation to do so. It is a logical step to see that the quality of interpersonal relationships among the adults is the context for student bullying. This WBI "Workplace Bullying in Schools Project" combined a policy, enforcement procedures, impact assessment, staff education, training a group of peer experts, and community education.

Dispositional versus Systemic Solutions

Solutions are driven by the definition of problems. Which is the stronger causal explanatory factor, actors' personalities or the work environment? If greater weight is given to an offender's personality, solutions may include skills-based training, anger management, mental health counseling, or executive coaching. However, recidivism is predictable when bullying-prone work conditions are not addressed.

For long-term success, the organization needs a new behavioral standard, that is, a new policy or code of conduct, to which alleged misconduct can be compared to determine whether or not a violation occurred. Procedures to enforce the standard widely and fairly must be created. Executives must defer to the process and not meddle to protect friends.

Our preferred approach is a dispositional-systemic hybrid. We first facilitate the collaborative creation of a policy and operational procedures. Then, when a high-profile person's offense is confirmed as a policy violation, there is a personalized program, complete with behavioral monitoring, for that individual.

German workplace bullying consultants described three approaches adopted in their practices: moderation or mediation; coaching; and organization development, or OD (Saam, 2010). Moderation is a clarification process to allow the parties to move beyond misunderstandings or misperceptions. Mediation is the traditional conflict resolution process. Moderation or mediation works only when conflict does not escalate to a level for which only a power intervention is appropriate (Ferris, 2004). Coaching necessarily develops solutions on a case-by-case basis. Coaching is support—tactical, emotional, career development, personalized skills education, and rehearsal.

Organization development approaches make change of culture the primary goal. This goal is accomplished by redesigning work processes and hierarchy within the organization. The preferred tool of the OD bullying consultant is the proscription of bullying behavior by policy and enforcement procedures (Namie and Namie, 2009b). Ferris (2004) similarly contends that helpful, responsive organizations provide coaching for the bully, counseling for targets, and policies that clearly define unacceptable conduct.

Finally, we implement our signature process with organizational clients—the creation of an internal group of peers trained to be experts in workplace bullying. The team members are chosen for their empathic tendencies and trust as gauged by peers. Then, the team is given immersion training in the phenomenon. Skills in problem triage, clarification, coaching, informal resolution strategies, and incident interventions are also taught. It is the peer team that sustains the antibullying initiative, better than policy and formal enforcement alone. True culture change is effected by a credible team.

Conclusion

The inherent barriers through which the antibullying campaign must still break are linked to U.S. norms, beliefs, and values. The most important of these are historical legislative patterns, market ideologies, values of individualism, reverence for hierarchy, and dehumanizing notions of workers.

The history of U.S. labor and employment law, built upon the inequitable master-servant relationship, is the foundation for an employer-employee power imbalance. Advocates for civil rights law organized around gender and racial identities. The civil rights laws of the 1960s were the result. Bullied workers have a less organized constituent base. The legislative campaign has found weaker purchase with lawmakers reluctant to expand protections beyond prohibitions against discrimination.

Market ideologies (i.e., veneration for capitalism as sustained by ubiquitous business lobbying groups) also impede both legislative efforts and organizational interventions because these ideologies esteem profit and productivity over all other, equally valuable ends. In this cultural environment, worker treatment is often a secondary consideration and given attention only when it affects the productivity that contributes to the organization's economic advantage. Sadly, this situation is true even in not-for-profit organizations where "numbers of clients" served operates as a productivity marker.

The U.S. value of individualism, fueled by the mythical notion of rugged pioneers, often serves primarily to blame the victim of any unfortunate event, including workplace bullying. Simultaneously, conditions that define the work environment are difficult to identify and thus underestimated. The bully is a toxic component of the environment. As such, rather than organizations retraining or removing aggressors, targeted workers often lose their jobs when bullied because they are blamed for their fate (Namie, 2007).

The reverence for hierarchy is also a barrier to the antibullying campaign—and it is so for all stakeholder groups. In markets led by economic-capitalistic values and in work groups socialized into the classic management chain of command, persons in top organizational positions have taken on a nearly godlike social position, one that is rarely questioned, even by targets, and one that gives primary voice to upper echelons in organizations. This belief system denigrates worker perspectives, accounts, and experiences.

Finally, justifying worker subordination (e.g., because workers are indolent) is alive and well in many U.S. workplaces. The idea that workers are untrustworthy, lazy, and lack initiative shores up "organizations' use [of] terror tactics to drive human resources" (Lutgen-Sandvik and McDermott, 2008, p. 320). Clearly, there is much work still to do.

References

Cortina, L. M., Magley, V. J., Williams, J. H., and Langhout, R. D. (2001) Incivility in the workplace: Incidence and impact. *Journal of Occupational Health Psychology*, *6*, 64–80.

Davis, W. H. (2008) No putting up with putdowns. *American Bar Association Journal*, February. http://www.abajournal.com/magazine/no_putting_up_with_putdowns (accessed February 3, 2008).

Einarsen, S., Raknes, B. I., and Matthiesen, S. B. (1994) Bullying and harassment at work and their relationships to work environment quality: An exploratory study. *European Journal of Work and Organizational Psychology, 4* (4), 381–401.

Ferris, P. (2004) A preliminary typology of organizational response to allegations of workplace bullying: See no evil, hear no evil, speak no evil. *British Journal of Guidance and Counselling, 32* (3), 389–395.

Grubb, P. L., Roberts, R. K., Grosch, J. W., and Brightwell, W. S. (2004) Workplace bullying: What organizations are saying. *Employee Rights and Employment Policy Journal, 8* (2), 407–422.

Hodson, R., Roscigno, V. J., and Lopez, S. H. (2006) Chaos and the abuse of power: Workplace bullying in organizational and interactional context. *Work and Occupations, 33* (4), 382–416.

Hoel, H., and Cooper, C. L. (2000) Working with victims of workplace bullying. In H. Kemshall and J. Pritchard (eds.), *Good practice in working with victims of violence* (pp. 101–118). London: Jessica Kingsley.

Hoel, H., and Einarsen, S. (2009) Shortcomings of anti-bullying regulations: The case of Sweden. *European Journal of Work and Organizational Psychology*. doi: 10.1080/13594320802643665.

Janoff-Bulman, R. (2002) *Shattered assumptions: Towards a new psychology of trauma*. New York: Free Press.

Keashly, L. (2001) Interpersonal and systemic aspects of emotional abuse at work: The target's perspective. *Violence and Victims, 16* (2), 211–245.

Keashly, L., and Neuman, J. H. (2004) Bullying in the workplace: Its impact and management. *Employee Rights and Employment Policy Journal, 8* (2), 335–373.

Levitt, H. (2009) When does a robust leader become a bully? *Financial Post*, May 13. http://www.financialpost.com/story.html?id=1590364 (accessed May 15, 2009).

Leymann, H. (1996) The content and development of mobbing at work. *European Journal of Work and Organizational Psychology, 5* (2), 165–184.

Liefooghe, A. P. D., and Davey, K. M. (2001) Accounts of workplace bullying: The role of the organization. *European Journal of Work and Organizational Psychology, 10* (4), 375–392.

Lutgen-Sandvik, P. (2008) Intensive remedial identity work: Responses to workplace bullying trauma and stigmatization. *Organization*, 15 (1), 97–119.

Lutgen-Sandvik, P., and McDermott, V. (2008) The constitution of employee-abusive organizations: A communication flows theory. *Communication Theory*, 18 (2), 304–333.

Lutgen-Sandvik, P. Namie, G., and Namie, R. (2009) Workplace bullying: Causes, consequences, and corrections. In P. Lutgen-Sandvik and B. D. Sypher (eds.), *Destructive organizational communication: Processes, consequences, and constructive ways of organizing*. New York: Routledge/Taylor & Francis.

Lutgen-Sandvik, P., Tracy, S. J., and Alberts, J. K. (2007) Burned by bullying in the American workplace: Prevalence, perception, degree, and impact. *Journal of Management Studies*, 44 (6), 837–862.

Mathiason, G., and Savage, O. (2008) Defining and legislating bullying. *The Complete Lawyer*, 4 (1). http://workplacebullying.org/press/2008tcl4.html (accessed January 4, 2008).

Meece, M. (2009) Backlash: Women bullying women at work. *New York Times*, May 9. http://www.nytimes.com/2009/05/10/business/10women.html (accessed May 10, 2009).

Namie, G. (2000) U.S. Hostile Workplace Survey. *Workplace Bullying Institute*, http://workplacebullying.org/research.html (accessed May 15, 2010).

——— (2003) The WBI report on abusive workplaces. Paper presented at the Gauteng International Conference on the Management of Psychosocial Problems in the Workplace, Johannesburg, South Africa, November 18–20.

——— (2007) *The Workplace Bullying Institute 2007 U.S. Workplace Bullying Survey*. http://workplacebullying.org/research/WBI-Zogby2007Survey.html (retrieved September 16, 2007).

——— (2008) Employers' response study. *Workplace Bullying Institute*. http://workplacebullying.org/res/N-N-2008A.pdf (accessed May 9, 2010).

——— (2009) Bullying with impunity. *Workplace Bullying Institute*. http://workplacebullying.org/res/N-N-2009D.pdf (accessed May 9, 2010).

Namie, G., and Lutgen-Sandvik, P. E. (2010) Active and passive accomplices: The communal character of workplace bullying. *International Journal of Communication*, 4, 343–373.

Namie, G., and Namie, R. (2009a) *The bully at work*, 3rd ed. Naperville, IL: Sourcebooks.

——— (2009b) U.S. workplace bullying: Some basic considerations and consultation interventions. *Consulting Psychology Journal*, 61 (3), 202–219.

Neuman, J. H. (2000) Injustice, stress, and bullying can be expensive! Paper presented at the Workplace Bullying Conference, Oakland, California, January 28.

Paulhus, D. L., and Williams, K. M. (2002) The dark triad of personality: Narcissism, Machiavellianism, and psychopathy. *Journal of Research in Personality*, 36, 556–563.

Petty, R. E., Briñol, P., and Tormala, Z. L. (2002) Thought confidence as a determinant of persuasion: The self-validation hypothesis. *Journal of Personality and Social Psychology*, 82, 722–741.

Petty, R. E., and Cacioppo, J. T. (1986) *Communication and persuasion: Central and peripheral routes to attitude change*. New York: Springer-Verlag.

Ross, L. (1977) The intuitive psychologist and his shortcomings: Distortions in the attribution process. In L. Berkowitz (ed.), *Advances in experimental social psychology*. New York: Academic Press.

Saam, N. J. (2010) Interventions in workplace bullying: A multilevel approach. ✓ *European Journal of Work and Organizational Psychology, 19* (1), 51–75.

Sherif, M., and Sherif, C. W. (1968) Attitude as the individuals' own categories: The social judgment–involvement approach to attitude and attitude change. In M. Sherif and C. W. Sherif (eds.), *Attitude, ego-involvement, and change* (pp. 105–139). New York: John Wiley.

Sutton, R. I. (2007) *The no asshole rule: Building a civilized workplace and surviving one that isn't.* New York: Warner Business.

Yamada, D. (2002) A policy analysis perspective on the role of the law in responding to workplace bullying. Paper presented at the International Conference on Bullying and Harassment at Work, September 23–24, London.

———— (2008) Multidisciplinary responses to workplace bullying: Systems, synergy and sweat. Paper presented at the International Conference on Bullying and Harassment in the Workplace, June 6–8, Montreal, Canada.

21

Workplace Bullying and the Law: Emerging Global Responses

David C. Yamada

CONTENTS

Introduction

As the chapters in this volume have demonstrated, we are moving toward a global understanding that workplace bullying poses a serious threat to workers and employers alike. The work being done to comprehend and address workplace bullying is multidisciplinary in nature, and one of the emerging focal points is the law. In nations around the world, there is a growing belief that the law should respond to the harm caused by workplace bullying. This chapter will identify some of the central themes concerning bullying and the law and examine legal and policy responses in several nations over the past decade.

Any analysis of the law's role in addressing bullying at work should start by identifying the public policy objectives that should be advanced. At least four figure prominently: First, prevention is the most important goal. The law should encourage employers to use preventive measures to reduce the likelihood of workplace bullying. These include developing and enforcing policies, educating employees, and supporting a workplace culture that values employee dignity. If bullying is prevented, then workers and employers alike will benefit, and all stakeholders are spared burdensome litigation. Second, the legal system should provide a means of relief to targets who are subjected to severely abusive treatment. This should include monetary damages, mental health counseling, and where applicable, reinstatement to the target's original position.

Third, the law should encourage prompt, internal resolution of bullying disputes, with procedures designed to be fair to all parties, and with strong protections against retaliation for workers who report instances of bullying. Finally, bullies and employers who enable them should be held responsible for their actions. This will have a deterrent effect and further encourage the use of preventive measures to discourage bullying behavior.

It must be said at the outset that when weighed against these policy goals and viewed from a global perspective, the law does not adequately respond to workplace bullying. However, the situation is changing. During the past decade, we have seen a tangible increase in legislative, administrative, and judicial responses to workplace bullying. These legal and policy initiatives fit into five general (albeit sometimes overlapping) categories:

1. Provisions that amend or supplement existing occupational safety and health laws
2. Judicial and administrative recognition of bullying grounded in existing common, statutory and administrative law
3. Voluntarily adopted employee policies and collective bargaining agreements that may create legal obligations and contractual rights
4. Enactment of statutory tort protections incorporating elements of tort law and harassment law

5. Recognition of workplace bullying by transnational bodies that influence or engage in policy making

Although it is beyond the scope of this chapter to provide a complete summary of all legal responses to workplace bullying, the following considers relevant developments in several nations, as well as the emerging policy-shaping role of the European Union.

Australia

Australia has made genuine progress since organizational behavior specialists Robyn Kieseker and Teresa Marchant concluded in 1999 that "workplace bullying is not effectively addressed under current Australian legislation" (Kieseker and Marchant, 1999). After surveying existing discrimination, occupational health and safety, and unfair dismissal claims, they opined that "[u]ntil legislation is changed to make workplace bullying illegal, the costs associated with it will continue to rise, governments will be pressured to implement changes, and unions will be called on to assist members at work and before industrial tribunals." Although Australia does not yet have a comprehensive national law that provides targets with a clear path to compensation and relief, Australian states and territories all have addressed workplace bullying in one form or another, typically through their workplace safety agencies.

State Initiatives

In 2005, South Australia took a leading role in enacting protections against workplace bullying when it amended its Occupational Healthy, Safety and Welfare Act to include bullying among the inappropriate workplace behaviors covered by an employer's duty of care to its employees (South Australia, 2005). An employer's failure to "adequately manage" bullying behaviors can lead to prosecution and fine. The South Australian government has engaged in extensive public education efforts to explain the new law and hired additional industrial inspectors to handle complaints about bullying at work.

Other states have adopted a more educative approach, incorporating existing regulatory standards. For example, in 2004, the Queensland government adopted a "Prevention of Workplace Harassment Code of Practice" that "provides practical advice about ways to prevent or control exposure to the risk of death, injury or illness created by workplace harassment" (Queensland, 2004). The code defines workplace harassment as repeated behavior, other than sexual harassment, that "is unwelcome and solicited" and considered to be "offensive, intimidating, humiliating or threatening" to both the target and the "reasonable person." It goes on to explain responsibilities under the state's occupational safety and health laws and to identify best practices for preventing and handling bullying complaints.

Victoria's occupational safety and health agency has produced a guide to workplace violence and bullying prevention and response (Victoria, 2005). It summarizes the relevant provisions of the state's workplace safety laws and provides advice to employers and employees, including legal options. It also explains relevant criminal law provisions and sexual harassment liability.

Common Law and Statutory Obligations Concerning Occupational Health and Safety

Even in the absence of stand-alone national legislation concerning bullying, existing law offers some protections to targets. Australian common law imposes upon an employer a "contractual duty to take reasonable care for the safety of employees while in the course of their employment" (Brooks, 1996). Employers also owe a duty of care for the safety of their employees under the tort law, which governs negligence and other claims for personal injuries. In addition, federal and state occupational safety and health statutes establish general standards of care owed to employees and provide specific measures for workplace safety.

Typically, legal disputes invoking an employer's obligations to safeguard its workers have involved physical injuries. However, there are encouraging signs that this combination of an employer's common law and statutory obligations may provide grounds of legal relief for psychological injuries wrought by bullying as well. A significant example is a 1998 Queensland Supreme Court decision, *Arnold v. Midwest Radio Limited*. The targeted employee had preexisting psychiatric problems that were exacerbated by being subjected to repeated verbal abuse by her supervisor. In holding the employer liable for the harm under duties of care prescribed by both common law tort theories and the applicable workplace health and safety statute, the court recognized the viability of actions "brought by employees claiming to have suffered psychological or psychiatric damage as a result of being unnecessarily exposed to stressful situations in the course of their employment."

Canada

During the new millennium, Canada has joined the forefront in enacting responses to workplace bullying. Since 2002, the provinces of Quebec and

Saskatchewan and the Canadian national government have added bullying provisions to their respective occupational safety and health laws.

Quebec Legislation

In 2002, the province of Quebec became the first North American governmental entity to enact antibullying legislation. Quebec's Psychological Harassment at Work Act provides that "[e]very employee has a right to a work environment free from psychological harassment" and that "[e]mployers must take reasonable action to prevent psychological harassment and, whenever they become aware of such behaviour, to put a stop to it" (Quebec, 2002). The act defines psychological harassment as

> any vexatious behaviour in the form of repeated and hostile or unwanted conduct, verbal comments, actions or gestures, that affects an employee's dignity or psychological or physical integrity and that results in a harmful work environment for the employee.... A single serious incidence of such behaviour that has a lasting harmful effect on an employee may also constitute psychological harassment.

The Quebec legislation became effective in 2004. Research into its impact and effectiveness is still in its infancy, but early indications suggest that complainants are not enjoying an easy path to recovery under the provision (Soares, 2006). While this may not be the best news for bullying targets, it also may serve to rebut concerns that antibullying laws open the floodgates to frivolous claims and unjust damage awards.

After Quebec

Following the Quebec legislation, both the province of Saskatchewan and the Canadian national government entered the fray. In 2007, Saskatchewan amended its Occupational Safety and Health Act to protect against personal harassment "that constitutes a threat to the health or safety of the worker" (Saskatchewan, 2007). The next year, the Canadian government enacted regulatory amendments to the national occupational health and safety regulations requiring employers in federally regulated workplaces "to dedicate sufficient attention, resources and time to address factors that contribute to workplace violence including but not limited to, bullying, teasing, and abusive and other aggressive behaviour and to prevent and protect against it" (Canada, 2008).

It may require several years before the Saskatchewan and national initiatives can be adequately evaluated. However, with two provinces and the national government enacting responses to bullying through their respective workplace safety laws, Canada has joined with Australia and Sweden in becoming an important jurisdiction for studying the efficacy of the occupational safety and health regulatory route for addressing workplace bullying.

France

France's Social Modernisation Law, enacted in 2002, prohibits "moral harassment" in the workplace, making such behavior a violation of the nation's labor and criminal codes (Guerrero, 2004). The Labor Code "now provides that no employee shall suffer repeated acts of moral harassment, which have the purpose of causing a deterioration in working conditions by impairing the employee's rights and dignity, affecting the employee's physical or mental health, or compromising the employee's professional future." The legislation provides harassed employees with job security and antiretaliation protections. The Penal Code and the Labor Code allow imprisonment and fines to be imposed on offending parties.

At this juncture, France is the only nation to criminalize bullying behavior at work. The legal history of this prong of the French statute is slim. Guerrero (2004) reported that the first criminal case to be brought for bullying resulted in the Criminal Court of Paris holding for the employer and narrowing the scope of the French statute. It does not appear that the criminalization approach is gaining adherents in other nations.

Sweden

The home of pioneering researcher Heinz Leymann, Sweden was the first nation to fashion a direct legal response to bullying in the form of its Victimisation at Work ordinance, promulgated in 1993 by the National Board of Occupational Safety and Health. The ordinance characterizes victimization as forms of behavior such as "adult bullying, mental violence, social rejection and harassment—including sexual harassment." Under its provisions, employers are obligated to institute measures to prevent victimization and to act responsively if "signs of victimisation become apparent," including providing prompt assistance to targets of abusive behavior.

Although much hope has been invested in the ordinance as an effective check on bullying at work, Hoel and Einarsen (2009) reported, "There is evidence to suggest that [it] has not been as effective as anticipated." The authors conducted in-depth interviews with 18 individuals closely familiar with the ordinance, representing a span of stakeholder interests. They concluded that "despite contributing to raising awareness about a difficult and complex social problem, the Swedish regulation against bullying has important shortcomings, frequently failing those it was meant to protect." Among the possible reasons are "the ordinance itself, the problems victims faced when seeking redress...the responses of the employers, the trade-unions and the Labour Inspectorate...and...cultural and socio-economic factors." The last reason may include stakeholder attitudes toward litigation, liability exposure, and proactive prevention.

United Kingdom

The United Kingdom is where the term *workplace bullying* first became popularized, thanks to the work of Andrea Adams in the early 1990s. Although the United Kingdom has not enacted comprehensive bullying legislation, existing statutory schemes have been invoked to encourage employers to act preventively and to provide relief to targets.

Protection from Harassment Act

The Protection from Harassment Act (PHA) of 1997 was enacted largely as a response to personal stalking, but it has recently been cited as grounds for relief in bullying cases. In addition to providing for criminal sanctions, the PHA imposes civil liability where a defendant engages in a "course of conduct (1) which amounts to harassment of another, and (2) which he knows or ought to know amounts to harassment of another." It took a well-publicized 2006 court decision, *Green v. DB Group Services (UK) Ltd.*, to highlight the PHA's applicability to workplace bullying scenarios.

A recent analysis of this decision summarized the experience of London Deutsche-Bank secretary Helen Green (Harthill, 2008). Green's colleagues "subjected her to a sustained campaign of emotional abuse. They constantly made it difficult for Green to perform her work by moving her papers, hiding her mail, removing her from document circulation lists, and ignoring and excluding her in meetings and social functions." They would "burst out laughing when she walked past them and made crude and lewd remarks." Her supervisor, one of the tormenters, "increased her workload to unreasonable and arbitrary levels." After her complaints to management went unheeded, "Green eventually developed a major depressive disorder and at one point was taken to hospital and put on suicide watch." On these facts, a British court found Deutsche-Bank vicariously liable under the PHA and awarded Green approximately £800,000 in total damages.

The judicially approved application of the PHA to employment situations may have created a de facto statutory tort remedy for workplace bullying targets, thereby undercutting momentum for passage of a "Dignity at Work" bill, the main legislative proposal to address bullying specifically. In 1997, a British labor union drafted the bill, which has been introduced in the House of Commons but has not yet become law. The bill imposes civil liability on an employer for bullying and similar acts, including "behaviour on more than one occasion which is offensive, abusive, malicious, insulting or intimidating." The proposed law is designed to provide protections to bullied employees on par with protections extended to targets of sexual or racial harassment.

Occupational Safety and Health

Under the Health and Safety at Work Act (1974), an employer must "ensure, so far as is reasonably practicable, the health, safety and welfare at work of all his employees." The broad wording of this mandate appears to open a wide door for regulatory involvement concerning bullying, but the applicability of the act in this regard is largely untested. Britain's Health and Safety Executive, the governmental agency responsible for enforcing the act, recognizes bullying as a cause of work-related stress, but its role at present is largely an educative one through publications and website information.

Unfair Dismissal

The British law governing unfair dismissal also offers a promising source of legal protection for bullied employees. The Employment Rights Act (1996) provides that employees may not be unfairly dismissed. Poor performance, improper conduct, and redundancy are among the chief reasons that may justify dismissal.

Of particular relevance to bullying is the concept of constructive dismissal. An employee is constructively dismissed when she voluntarily leaves her employment because the employer has fundamentally breached an express or implied term of the employment contract. Subjecting an employee to severe mistreatment such as bullying can be a form of breach of an implied contractual term. For example, in *Abbey National PLC v. Janet Elizabeth Robinson* (2000), an Employment Appeal Tribunal upheld a finding of constructive dismissal where the worker's manager "had been bullying and harassing her in the workplace to a degree she found insufferable."

In another case, *Roger Storer v. British Gas PLC* (2000), an appeals court reinstated a claim of constructive dismissal that had been dismissed because the complainant failed to file within the statutory time limit. The complainant had been "victimised and bullied" by his manager, resulting in stress and depression that culminated in symptoms of post-traumatic stress disorder. The court ordered an employment tribunal to consider whether the complainant's mental condition rendered him unable to file in a timely fashion. Equally encouraging is the decision of an Employment Appeal Tribunal in

Ezekiel v. The Court Service (2000), which held that an employee was properly dismissed under the Employment Rights Act because he engaged in severe bullying and mistreatment of several coworkers.

United States

The term *workplace bullying* did not begin to enter the vocabulary of American employment relations until the late 1990s. Awareness of the phenomenon has grown rapidly since then, but at present, there are limited remedies for bullying targets under American law. However, advocacy groups and unions are beginning to organize around legal reform.

Tort Claims for Emotional Distress

In the United States, the favored (albeit seldom successful) tort claim for emotionally abusive treatment at work has been intentional infliction of emotional distress (IIED). Typically, plaintiffs have sought to impose liability for IIED on both their employers and the specific workers, often supervisors, who engaged in the alleged conduct. The tort of IIED can be defined this way (*Kroger Co. v. Willgruber*, 1996):

1. The wrongdoer's conduct must be intentional or reckless.
2. The conduct must be outrageous and intolerable in that it offends against the generally accepted standards of decency and morality.
3. There must be a causal connection between the wrongdoer's conduct and the emotional distress.
4. The emotional distress must be severe.

Although on the surface the tort of IIED appears to be an ideal legal protection against workplace bullying, this author's extensive analysis of judicial decisions deciding IIED claims showed that typical workplace bullying, especially conduct unrelated to sexual harassment or other forms of status-based discrimination, seldom resulted in liability for IIED (Yamada, 2000). The most frequent reason given by courts for rejecting workplace-related IIED claims was that the complained-of behavior was not sufficiently extreme and outrageous to meet the requirements of the tort.

Perhaps the most stunning example of this reasoning came in *Hollomon v. Keadle*, a 1996 Arkansas Supreme Court case that involved a female employee, Hollomon, who worked for a male physician, Keadle, for two years before she voluntarily left the job. Hollomon claimed that during this period of employment, "Keadle repeatedly cursed her and referred to her with offensive terms, such as 'white nigger,' 'slut,' 'whore,' and 'the igno-rance of Glenwood, Arkansas.'" Keadle repeatedly used profanity in front of his employees and patients, and he frequently remarked that women working outside the home were "whores and prostitutes." According to Hollomon, Keadle "told her that he had connections with the mob" and mentioned that "he carried a gun," allegedly to "intimidate her and to sug-gest that he would have her killed if she quit or caused trouble." Hollomon claimed that as a result of this conduct, she suffered from "stomach prob-lems, loss of sleep, loss of self-esteem, anxiety attacks, and embarrass-ment." On these allegations, the Arkansas Supreme Court ruled for the defendant Keadle, holding that Hollomon's failure to establish that Keadle was made aware of her peculiar vulnerability to emotional distress was fatal to her claim.

Collective Bargaining and Labor Advocacy

Labor unions and the legal frameworks that protect collective employee action constitute potentially important avenues for addressing workplace bullying. Organized labor remains one of the strongest and most vocal sources of advocacy on behalf of working people, and there are encouraging signs that some labor unions are responding to workplace bullying faced by their members. These unions are raising concerns about workplace bullying at the bargaining table and in grievances, and they are supporting efforts toward law reform.

An excellent example of labor advocacy came about in the United States in 2009, when a coalition of unions representing over 21,000 public employees proposed, negotiated, and approved a new collective bargaining provision with the Commonwealth of Massachusetts that covers workplace bully-ing and abusive supervision (Yamada, 2009). Dubbed the "mutual respect" provision in the new contract, it is believed to be one of the first major American collective bargaining agreements to include express protections against bullying at work. Under the provision, behaviors "that contribute to a hostile, humiliating or intimidating work environment, including abusive language or behavior," may be the subject of a valid grievance (Yamada, 2009).

There are, of course, impediments to utilizing unions and collective bargaining protections to combat bullying. Overall, union density in industrialized nations showed a steady decline during the last third of the twentieth century; with notable exceptions such as Sweden, union-ized employees constitute less than half the wage and salary earners in these countries. In addition, collective bargaining processes implicitly assume that major workplace conflicts are between employers and rank-and-file workers. Bullying scenarios between union members are not easily addressed within this structure; proactive union leadership is required to take on these situations.

Workplace Bullying Legislation: The Healthy Workplace Bill

In response to the shortcomings of existing employment law for targets of severe workplace bullying in the United States, this author drafted model antibullying legislation, now dubbed the "Healthy Workplace Bill," that has been introduced (but not yet enacted) in some 12 state legislatures since 2003. The legislation declares that it "shall be an unlawful employment practice … to subject an employee to an abusive work environment," which is found to exist "when the defendant, acting with malice, subjects the employee to abusive conduct so severe that it causes tangible harm to the employee" (Yamada, 2004). Successful plaintiffs will be able to recover lost wages and medical expenses, and even emotional distress and punitive damages where merited.

Grassroots lobbying and advocacy efforts on behalf of the Healthy Workplace Bill have both generated strong opposition from business trade associations and demonstrated the potential preventive impacts of such legislation. Organizations such as the chambers of commerce are quick to oppose the bill in any state in which it is introduced. However, the growing possibility of enacting legislation has galvanized many lawyers who represent employers to advise their clients to develop workplace bullying policies as part of their human resources operations.

European Union

The European Union, which adopts regulatory structures that affect both commercial exchange and social protections within and among member nations, is the leading example of a multinational body that has recognized the significance of workplace bullying. In 2002, a committee report issued by the European Parliament endorsed the development of antibullying legislation (European Parliament, 2002). In addition, the European Foundation for the Improvement of Living and Working Conditions, an EU research unit, gave significant attention to bullying in a study of workplace violence and harassment (Di Martino, Hoel, and Cooper, 2003). The report acknowledged that emerging research "may reflect the fact that the problem of bullying and harassment is still not fully appreciated by employers and policy-makers."

Many of these efforts came to a head in 2007, when the European social partners, comprised of coalitions of employers and unions operating under the aegis of the EU, signed a "Framework Agreement on Harassment and Violence at Work" designed to prevent and manage problems of workplace bullying, sexual harassment, and workplace violence (Framework Agreement, 2007). Under the terms of the agreement,

European companies are obliged to adopt zero-tolerance policies toward the offending behaviors and to develop in-house procedures for handling situations that arise. The agreement further identifies features that must be incorporated into these policies, such as confidentiality and due process provisions.

Additional Legal and Policy Considerations

This discussion has concentrated on statutory and judicial developments specifically related to workplace bullying. However, two other areas of employment law and policy provisions should be acknowledged in this chapter as being relevant to workplace bullying.

Safety Net of No-Fault Employee Benefits

Income replacement, counseling, and transitional assistance are often necessary for severely bullied workers who leave or are terminated from their employment. For the many workers who do not wish to invoke legal process to obtain compensation for personal and economic injuries related to workplace bullying, a standard menu of no-fault employee and public benefits should be available to them. These include workers' compensation, unemployment assistance, health care provision (including mental health counseling), and disability payments. The provision of such benefits for bullying situations varies widely among nations and merits a much deeper examination than is possible in this chapter.

Employment Discrimination Laws

Employment discrimination laws may provide bullying targets with legal protections when harassment behaviors are motivated by race, sex, or membership in other commonly protected classes. In addition, disability discrimination laws may come into play when bullying causes or aggravates mental illness. Retaliation in the form of bullying against complainants in discrimination cases also may form the basis of a legal claim.

The potential coexistence of bullying and discrimination laws raises interesting issues for legal system in countries such as the United States, where protected-class status remains the dominant paradigm of how legal issues of worker harassment and mistreatment are framed. Given the history and continuing experience of discrimination and difference in the workplace, it is important that we remain alert to the importance of these critical issues. However, there is plenty of room in our various legal regimes to include bullying along with status-based mistreatment.

Toward an Effective Legal Response

At this juncture, efforts toward law reform appear to be moving in two distinct directions. First, among nations that have adopted some tangible legal or policy responses to bullying at work (and there are many in addition to those examined in this chapter), occupational safety and health laws have been the predominant regulatory approach, with recent initiatives in Australia, Canada, and Sweden. Second, in the United Kingdom, through application of the Prevention of Harassment Act to bullying cases, and in the United States, through calls for the Healthy Workplace Bill, statutory tort remedies are emerging as the next most-favored approach. France is the outlier here, being the only nation so far to criminalize bullying conduct.

Amid these advocacy efforts and scholarly calls for the adoption of new legislation, others are examining the nascent body of existing statutory law and issuing less than enthusiastic reviews. Those of us who favor workplace bullying laws must be sharply attentive to potential shortcomings and limitations in existing legislation. Hoel and Einarsen (2009), whose critical assessment of the Swedish legislation is discussed earlier, recommend that in order to be effective, bullying laws must impose sanctions on employers that negligently or intentionally subject their employees to mistreatment. They also urge that effective training and prevention must complement legal intervention. This author suggests that mixed assessments of existing bullying legislation may be the result of inherent inadequacies in the predominant occupational safety and health regulatory model. The statutory tort approach, in contrast to the workplace safety regulatory approach, provides positive and negative incentives for employers to engage in preventive measures and to respond fairly and promptly to reports of bullying.

In any event, it is encouraging to see how these initiatives reflect changing attitudes toward workplace bullying. For too long, bullying at work has been seen as part of the cost of being employed, much like sexual harassment was regarded some 30 or 40 years ago. But now, legislative measures are being debated and enacted in countries that, 10 years ago, were only starting to consider the legal and policy implications of bullying, if at all.

Furthermore, as bullying has entered the mainstream of our examination of employment relations, transnational bodies such as the International Labour Organization (ILO) and the World Health Organisation (WHO) are paying attention. The ILO has examined workplace bullying in the broader context of violence at work, observing that bullying represents behavior that "by itself may be relatively minor but which cumulatively can become a very serious form of violence" (Chappell and Di Martino, 2000). The WHO has recognized bullying legislation as a useful method of tertiary prevention that can help targets "recover their health and dignity" (Cassitto et al., 2003).

Legal intervention comprises but one component in an overall societal response to workplace bullying, but it can and should play a meaningful role in combating this destructive phenomenon. The exact nature of this role, however, remains a new topic of research and analysis. There is a significant need for in-depth study of the effectiveness of existing regulatory measures, conducted by individuals with expertise in the fields of employment law and organizational behavior. Comparative examinations such as this one, while useful to a point, must be grounded in a deeper contextual discussion of different legal systems and diverse cultures of employment relations. In sum, as in so many other areas of inquiry concerning workplace bullying, we are only at the beginning of understanding the promise and limitations of this particular form of intervention.

Legislative, Administrative, and Judicial Materials

Australia

Arnold v. Midwest Radio Limited. Aust Torts Reports, para 81–472 (1998).
Occupational Health, Safety and Welfare Amendment Act, 2005 (South Australia).
Prevention of Workplace Harassment Code of Practice, 2004 (Queensland).
Sex Discrimination Act, 1984 (Commonwealth).
Worksafe Victoria (June 2005). *Workplace violence and bullying* (Victoria).

Canada

Commission des normes du travail, *An Act Respecting Labour Standards, Sec. 81–19, 2002* (Quebec).
Occupational Health and Safety (Harassment Prevention) Amendment Act, 2007 (Saskatchewan).
Regulations Amending the Canada Occupational Health and Safety Regulations, May 8, 2008 (Canada).

European Union

Framework Agreement on Harassment and Violence at Work, April 26, 2007.

Sweden

Victimization at Work, adopted September 21, 1993.

United Kingdom

Abbey National PLC v. Janet Elizabeth Robinson. WL 1741415 (EAT 2000).
Dignity at Work Bill, 1997.

Employment Rights Act, 1996.
Ezekiel v. The Court Service. WL 1274032 (EAT 2000).
Green v. DB Group Services (UK) Ltd. EWHC 1898 (QB 2006).
Health and Safety at Work Act, 1974.
Protection from Harassment Act, 1997.
Roger Storer v. British Gas PLC. WL 191091 (CA 2000).

United States

Hollomon v. Keadle, 931 S.W.2d 413 (Ark. 1996).
Kroger Co. v. Willgruber, 920 S.W.2d 61 (Ky. 1996).

References

Brooks, A. (1996). Occupational health and safety. In J. Golden and D. Grozier (eds.), *Labour law*, vol. 26 of *The Laws of Australia* (p. 13). Sydney: Law Book Company.

Cassitto, M., Fattorini, E., Gilioli, R., and Rengo, C. (2003). *Raising awareness of psychological harassment at work*. Geneva: World Health Organisation.

Chappell, D., and Di Martino, V. (2000). *Violence at work*, 2nd ed. Geneva: International Labour Office.

Di Martino, V., Hoel, H., and Cooper, C. (2003). *Preventing violence and harassment in the workplace*. Luxembourg: Office for Official Publications of the European Communities.

European Parliament (2002). *Communication from the Commission: Adapting to change in work and society: A new community strategy on health and safety at work, 2002–2006*.

Guerrero, M. (2004). The development of moral harassment (or mobbing) law in Sweden and France as a step towards EU legislation. *Boston College International and Comparative Law Review, 27*, 477–500.

Harthill, S. (2008). Bullying in the workplace: Lessons from the United Kingdom. *Minnesota Journal of International Law, 17*, 247–301.

Hoel, H., and Einarsen, S. (2009). Shortcomings of anti-bullying regulations: The case of Sweden. *European Journal of Work and Organisational Psychology*. doi: 10.1080/13594320802643665

Kieseker, R., and Marchant, T. (1999). Workplace bullying in Australia: A review of current conceptualisations and existing research. *Australian Journal of Management and Organisational Behaviour, 2* (5) 61–75.

Soares, A. (2006). *The anti-bullying law: The Quebec experience*. Miami, FL: Work, Stress, and Health Conference.

Yamada, D. (2000). The phenomenon of "workplace bullying" and the need for status-blind hostile work environment protection. *Georgetown Law Journal, 88*, 475–536.

——— (2004). Crafting a legislative response to workplace bullying. *Employee Rights and Employment Policy Journal, 8,* 475–519.

——— (2009). Massachusetts public employee unions successfully negotiate workplace bullying provision. http://newworkplace.wordpress.com/2009/03/03/massachusetts-public-employee-unions-successfully-negotiate-workplace-bullying-provision

Index